Emerging Role of Lipids in
Metabolism and Disease

Emerging Role of Lipids in Metabolism and Disease

Editors

Marco Segatto
Valentina Pallottini

MDPI • Basel • Beijing • Wuhan • Barcelona • Belgrade • Manchester • Tokyo • Cluj • Tianjin

Editors
Marco Segatto
University of Molise
Italy

Valentina Pallottini
University Roma Tre
Italy

Editorial Office
MDPI
St. Alban-Anlage 66
4052 Basel, Switzerland

This is a reprint of articles from the Special Issue published online in the open access journal *International Journal of Molecular Sciences* (ISSN 1422-0067) (available at: https://www.mdpi.com/journal/ijms/special_issues/Lipids_Disease).

For citation purposes, cite each article independently as indicated on the article page online and as indicated below:

LastName, A.A.; LastName, B.B.; LastName, C.C. Article Title. *Journal Name* **Year**, *Article Number, Page Range*.

ISBN 978-3-03943-701-6 (Pbk)
ISBN 978-3-03943-702-3 (PDF)

Cover image courtesy of Marco Segatto.

Contents

About the Editors

Marco Segatto (Assistant Professor) has developed a great interest in cholesterol homeostasis since completing his master's degree, focusing his attention on the physiopathological role played by lipid homeostasis in different organs and tissues such as liver, muscle, and brain. He continued to acquire new skills and knowledge about cholesterol metabolism in the brain as a Ph.D. student, and as a Postdoctoral Fellow, broadened his experience in the field of metabolism. He is currently Assistant Professor at University of Molise, and his main research interests remain focused on the involvement of lipid metabolism in brain and muscle physiopathology.

Valentina Pallottini (Professor), The research activity of Professor Pallottini has been focused on changes in cholesterol metabolism under different physiological and pathological conditions. Her research has been principally carried on HMGCR (3 hydroxy 3-methylglutaryl coenzyme A reductase), the rate-limiting enzyme of cholesterol biosynthetic pathway, and on cholesterol metabolism in different physiological states. Her research on cholesterol metabolism has been principally focused on aged rat livers, models with high concentration of free radicals, and gender differences. During the last 10 years, her research has been focused on the role of cholesterol metabolism in the central nervous system and its involvement in neurodevelopmental and neurodegenerative disorders.

International Journal of
Molecular Sciences

Editorial

Facts about Fats: New Insights into the Role of Lipids in Metabolism, Disease and Therapy

Marco Segatto [1] and Valentina Pallottini [2],*

[1] Department of Biosciences and Territory, University of Molise, Contrada Fonte Lappone, 86090 Pesche (Is), Italy; marco.segatto@unimol.it
[2] Department of Science, University Roma Tre, Viale Marconi 446, 00146 Rome, Italy
* Correspondence: valentina.pallottinini@uniroma3.it

Received: 31 August 2020; Accepted: 7 September 2020; Published: 11 September 2020

Keywords: Cholesterol; Fatty acids; Lipid mediators; Lipids; Lipophagy; Sphingolipids

---

Although initially regarded as a passive system to store energy, lipids are now considered to play crucial, structural and functional roles in almost all the biological processes involved in the regulation of physiological and pathological conditions. For instance, they are pivotal constituents of cell membranes, where they essentially contribute to the assembly of the bilayer configuration. Lipid species are not uniformly distributed in cell membranes, as they are mainly concentrated in specialized sphingolipid- and cholesterol-rich domains called lipid rafts and caveolae, which serve as a matrix for the attachment and the interaction of several proteins implicated in membrane-initiating signal transduction pathways [1]. Besides the involvement in the organization of membrane domains, lipids may influence numerous cellular processes by directly participating as both primary and secondary messengers. In recent decades, lipids derived from both dietary sources and endogenous biosynthesis gained considerable clinical relevance for their implications in a plethora of human diseases. Hyperlipidemia often results in a premature and increased risk of cardiovascular diseases [2]. Increasing evidence highlights that the metabolic reprogramming of cancer cells also involves many lipid compounds, such as fatty acids and cholesterol, which contribute to key oncogenic functions [3,4]. Furthermore, a variety of experimental and clinical studies demonstrated that lipids, particularly cholesterol, play essential roles in brain physiology [5]. Thus, it is not surprising that numerous neurological and neurodegenerative disorders are often accompanied by misbalances in lipid homeostasis [6].

Despite a satisfying level of knowledge being reached in the field of lipid research, several questions still remain unanswered; a deeper comprehension will be important to fully elucidate the molecular mechanisms regulating lipid homeostasis in health and loss of homeostasis in disease, and to design innovative therapeutic strategies. This Special Issue, entitled "Emerging Role of Lipids in Metabolism and Disease", comprises a collection of seven research articles, seven review articles and one concept paper reporting new insights into the biological role of lipids.

Lipid homeostasis is guaranteed by a delicate equilibrium among biosynthesis, uptake and catabolism. Kloska and colleagues [7] systematically reviewed the current knowledge on lipophagy and cytosolic lipolysis, two selective lipid catabolic processes whose activity is essential for the proper regulation of energy homeostasis in cells. The authors provide an in-depth discussion of the transcriptional modulation of these pathways by the mammalian target of rapamycin complex 1 (mTORC1) and by the transcription factor EB (TFEB). In addition, alterations of both lipophagy and lipolysis in pathological conditions are also documented. Even though basic molecular pathways underlying the maintenance of lipid homeostasis are now well established, several regulatory mechanisms still remain unknown. In this context, the experimental work provided by Tonini and colleagues report a novel epigenetic pathway by which cells physiologically control lipid homeostasis [8]. Notably, the authors unravel that bromodomain and extraterminal domain (BET) proteins, which are epigenetic readers particularly committed to the regulation of cell cycle progression, govern lipid metabolism by modulating the expression of proteins and enzymes implicated in fatty acid and cholesterol biosynthesis, trafficking and uptake. Coherently, BET protein blockade by the small inhibitor JQ1 strongly reduces the amount of intracellular lipids. In addition, Tonini and colleagues also showed that the decreased proliferation of HepG2 cancer cells induced by JQ1 is dependent on the modulation of cholesterol metabolism. The connection between lipid metabolism and the biology of cancer cells is also investigated by Avril et al., who studied the impact of epinephrine-infiltrated adipose tissue (AT) on MCF-7 breast cancer cells [9]. They observed that epinephrine-infiltrated AT secretes different factors, including lipids, which may act as signaling molecules. Indeed, both epinephrine-infiltrated AT and its corresponding conditioned medium (CM) enhance the proliferation of MCF-7 breast cancer cells in vitro. The secreted factors contained in CM also appear to increment the in vivo growth of MCF7 cells in mice. However, injection of whole epinephrine-infiltrated AT did not induce any change in the progression of MCF-7- tumor in mice, suggesting that the employment of CM to mimic the secretome of cells or tissues may explain some divergences observed among in vitro, pre-clinical and clinical data using AT samples.

Besides cancer biology, alterations of lipid metabolism are a major public health concern because of their association to cardiovascular diseases. Atherosclerosis, a complex process characterized by progressive inflammation and build-up of cholesterol and other lipids in the artery walls, is the leading cause of cardiovascular diseases. The exact etiopathogenesis of atherosclerosis is not fully elucidated, however, several risk factors have been identified, such as high blood pressure, hypercholesterolemia, obesity and diabetes. In their review article, Poznyak and colleagues provided up-to-date information about the use of knockout mice as experimental models to investigate genes involved in atherosclerosis and dyslipidemia [10]. In addition, Poznyak et al. discuss the current knowledge concerning the mechanisms by which diabetes mellitus promotes the atherogenic process, and summarize the physiopathological hallmarks linking atherosclerosis and diabetes mellitus, such as protein kinase signaling, oxidative stress, miRNA alterations and epigenetic changes [11]. The breakdown of homeostatic regulation of lipids occurring during diabetes has also been examined by Primer and collaborators [12]. In particular, they summarize the current understanding of endothelial cell metabolism and its dysregulation during diabetes and discuss the different mechanisms by which high-density lipoproteins (HDL) modulate endothelial cell metabolic reprogramming and counteract diabetes-impaired angiogenesis.

Lipid metabolism is also finely regulated within the brain: it maintains neuronal functionality and signals the nutrient status to regulate the whole-body metabolism by modulating key peripheral organs and tissues, such as the liver and adipose tissue. Alterations of lipid metabolism have been frequently associated with disturbances in brain functioning, such as neurodegeneration, neuroinflammation, cognitive alterations and neurodevelopmental problems. It is becoming increasingly clear that high levels of dietary fats and sugars, which are typically comprised in a western diet, may be detrimental to brain health. Mazzoli and colleagues provided an experimental work aimed at analyzing the putative effects of a western diet on the metabolic response of nervous and adipose tissue [13]. For instance, they reported that the expression of specific cyto/adipokines, such as TNFα and adiponectin, are significantly affected in both brain and adipose tissue of rats fed with a diet high in saturated fats and fructose (HFF). The observed changes are accompanied by a reduction in brain-derived neurotrophic factor (BDNF) and synaptotagmin I levels, and by an increase in the expression of the post-synaptic density protein, PSD95, in HFF-fed animals. When evaluated as a whole, these results underline that a western diet may induce similar metabolic alterations in adipose tissue and brain. Chun and Chung further confirmed the involvement of dietary lipids in brain physiology [14]. Specifically, they show experimental evidence that a high-cholesterol diet significantly decreases the expression levels of phospholipase C β1 (PLCβ1) and of phosphatidylinositol 4,5-bisphosphate (PIP$_2$). Interestingly, there is no direct correlation between the amount of cholesterol and of PIP$_2$, suggesting that PIP$_2$ levels are modulated by cholesterol through changes in the expression of PLCβ1. Since the reduction in PIP$_2$ levels has been associated with β-amyloid production, these results indicate that a high cholesterol diet may influence brain cholesterol, which reflects in PIP$_2$ changes that could contribute to the pathogenesis of neurodegenerative conditions. Considering that the blood brain barrier prevents the uptake of lipoprotein-bound cholesterol from the bloodstream, it will be a stimulating challenge to comprehend how diet-derived cholesterol can influence neuronal processes.

Among lipids, sphingolipids represent a major subcategory. These molecules are not only structural components, but also act as bioactive compounds in the mediation of physiological processes involved in cell proliferation, survival, inflammation, senescence and death. A number of experimental evidence sustains that the metabolic pathway, which governs sphingolipid metabolism, exerts a pivotal role in the regulation of the response to DNA damage. Francis and colleagues analytically reviewed how sphingolipid signaling influences the DNA damage response (DDR) induced by metabolic stress, ionizing radiation or other genotoxic stimuli [15]. In particular, they illustrate how different sphingolipid metabolites interact with the mediators of DDR to define cell fate. In the field of sphingolipid research, Cataldi and colleagues provided new interesting insights about the involvement of sphingomyelinases in the effects of ionizing radiations. Notably, experimental data highlighted that ionizing radiations cause altered hepatic cell structure and increased caspase-1 expression in mice. These effects are attenuated by the administration of recombinant manganese superoxide dismutase (rMnSOD). Importantly, rMnSOD counteracts the radiation-induced liver damage exerting a protective role via acid sphingomyelinase (aSMase), and a preventive role via neutral sphingomyelinase (nSMase) [16]. Cataldi et al. also demonstrated that nSMase is responsible for the preventive and protective effect elicited by rMnSOD against radiation-induced damage in the brain [17].

As already mentioned above, lipids not only serve as structural components, but also exert crucial biological roles as signaling molecules. This aspect is extensively discussed in the review article proposed by Kloska and collaborators, who focused their attention on the role of lipid mediators in the risk and pathology of ischemic stroke [18]. Notably, a lively metabolism of polyunsaturated fatty acid has been documented in ischemic brain, and different lipid mediators are implicated in the neuroprotective or neurodegenerative effects occurring in the post-stroke brain tissue. Among signaling molecules, eicosanoids seem to play crucial roles in the disease pathology, and a variety of reports suggest them as useful molecular targets for innovative therapeutic interventions. The involvement of lipid molecules as signaling mediators is further examined in the comprehensive review by Suryadevara et al., which collects up-to-date knowledge about the pathways mediated by lipid mediators in pulmonary fibrosis [19]. Indeed, the metabolism of phospholipids, sphingolipids, and polyunsaturated fatty acids may generate key molecules capable of signaling properties, which exhibit pro- and anti-fibrotic effects in patients and preclinical models of idiopathic pulmonary fibrosis (IPF). In light of this evidence, it is not surprising that prostanoids, lysophospholipids, sphingolipids and their metabolizing enzymes are currently under active investigation as potential pharmacological targets to treat IPF.

Finally, Carbone and collaborators contributed to this Special Issue with a research article aimed at assessing the association of circulating C-reactive protein (CRP) levels with epicardial and visceral fat depots in women with one or more defining criteria for metabolic syndrome [20]. The main findings highlight that men and women have a different epicardial fat deposition and systemic inflammation. Intriguingly, a correlation between visceral/epicardial fat depots and chronic low-grade inflammation was also noted, suggesting that sex may play an essential role in the stratification of obese individuals and dysmetabolic patients.

In conclusion, the experimental data summarized and presented in this Special Issue further strengthen the centrality of lipids in a plethora of biological processes and underline the importance of lipid research in physiopathology. Indeed, lipids may represent useful biomarkers for a number of diseases, and alterations in their metabolism may concur to the development of different disorders. Lipids can also be combined or conjugated to drug compounds, determining several benefits in terms of treatment effect. In this context, further basic, translational, and clinical research are imperative to discover novel mechanisms controlling lipid metabolism in health and disease, and to set up optimal drug design. All these concepts are clearly debated in the concept paper by Markovic et al. [21].

Author Contributions: Writing—original draft preparation, M.S. and V.P.; writing—review and editing, M.S. and V.P. All authors have read and agreed to the published version of the manuscript.

Funding: This research received no external funding.

Conflicts of Interest: The authors declare no conflict of interest.

References

1. Escribá, P.V.; González-Ros, J.M.; Goñi, F.M.; Kinnunen, P.K.; Vigh, L.; Sánchez-Magraner, L.; Fernández, A.M.; Busquets, X.; Horváth, I.; Barceló-Coblijn, G. Membranes: A meeting point for lipids, proteins and therapies. *J. Cell Mol. Med.* **2008**, *12*, 829–875. [CrossRef] [PubMed]
2. Kuller, L.H. Nutrition, lipids, and cardiovascular disease. *Nutr. Rev.* **2006**, *64*, S15–S26. [CrossRef]
3. Butler, L.M.; Perone, Y.; Dehairs, J.; Lupien, L.E.; de Laat, V.; Talebi, A.; Loda, M.; Kinlaw, W.B.; Swinnen, J.V. Lipids and cancer: Emerging roles in pathogenesis, diagnosis and therapeutic intervention. *Adv. Drug Deliv. Rev.* **2020**. [CrossRef] [PubMed]
4. Pesiri, V.; Totta, P.; Segatto, M.; Bianchi, F.; Pallottini, V.; Marino, M.; Acconcia, F. Estrogen receptor α L429 and A430 regulate 17β-estradiol-induced cell proliferation via CREB1. *Cell. Signal.* **2015**, *27*, 2380–2388. [CrossRef]
5. Cartocci, V.; Segatto, M.; Di Tunno, I.; Leone, S.; Pfrieger, F.W.; Pallottini, V. Modulation of the Isoprenoid/Cholesterol Biosynthetic Pathway During Neuronal Differentiation In Vitro. *J. Cell Biochem.* **2016**, *117*, 2036–2044. [CrossRef]
6. Segatto, M.; Tonini, C.; Pfrieger, F.W.; Trezza, V.; Pallottini, V. Loss of Mevalonate/Cholesterol Homeostasis in the Brain: A Focus on Autism Spectrum Disorder and Rett Syndrome. *Int. J. Mol. Sci.* **2019**, *20*, 3317. [CrossRef]
7. Kloska, A.; Węsierska, M.; Malinowska, M.; Gabig-Cimińska, M.; Jakóbkiewicz-Banecka, J. Lipophagy and Lipolysis Status in Lipid Storage and Lipid Metabolism Diseases. *Int. J. Mol. Sci.* **2020**, *21*, 6113. [CrossRef]
8. Tonini, C.; Colardo, M.; Colella, B.; Di Bartolomeo, S.; Berardinelli, F.; Caretti, G.; Pallottini, V.; Segatto, M. Inhibition of Bromodomain and Extraterminal Domain (BET) Proteins by JQ1 Unravels a Novel Epigenetic Modulation to Control Lipid Homeostasis. *Int. J. Mol. Sci.* **2020**, *21*, 1297. [CrossRef]
9. Avril, P.; Vidal, L.; Barille-Nion, S.; Le Nail, L.R.; Redini, F.; Layrolle, P.; Pinault, M.; Chevalier, S.; Perrot, P.; Trichet, V. Epinephrine Infiltration of Adipose Tissue Impacts MCF7 Breast Cancer Cells and Total Lipid Content. *Int. J. Mol. Sci.* **2019**, *20*, 5626. [CrossRef]
10. Poznyak, A.V.; Grechko, A.V.; Wetzker, R.; Orekhov, A.N. In Search for Genes Related to Atherosclerosis and Dyslipidemia Using Animal Models. *Int. J. Mol. Sci.* **2020**, *21*, 2097. [CrossRef]
11. Poznyak, A.; Grechko, A.V.; Poggio, P.; Myasoedova, V.A.; Alfieri, V.; Orekhov, A.N. The Diabetes Mellitus-Atherosclerosis Connection: The Role of Lipid and Glucose Metabolism and Chronic Inflammation. *Int. J. Mol. Sci.* **2020**, *21*, 1835. [CrossRef] [PubMed]
12. Primer, K.R.; Psaltis, P.J.; Tan, J.T.M.; Bursill, C.A. The Role of High-Density Lipoproteins in Endothelial Cell Metabolism and Diabetes-Impaired Angiogenesis. *Int. J. Mol. Sci.* **2020**, *21*, 3633. [CrossRef]
13. Mazzoli, A.; Spagnuolo, M.S.; Gatto, C.; Nazzaro, M.; Cancelliere, R.; Crescenzo, R.; Iossa, S.; Cigliano, L. Adipose Tissue and Brain Metabolic Responses to Western Diet-Is There a Similarity between the Two? *Int. J. Mol. Sci.* **2020**, *21*, 786. [CrossRef] [PubMed]
14. Chun, Y.S.; Chung, S. High-Cholesterol Diet Decreases the Level of Phosphatidylinositol 4,5-Bisphosphate by Enhancing the Expression of Phospholipase C (PLCβ1) in Rat Brain. *Int. J. Mol. Sci.* **2020**, *21*, 1161. [CrossRef]
15. Francis, M.; Abou Daher, A.; Azzam, P.; Mroueh, M.; Zeidan, Y.H. Modulation of DNA Damage Response by Sphingolipid Signaling: An Interplay that Shapes Cell Fate. *Int. J. Mol. Sci.* **2020**, *21*, 4481. [CrossRef]
16. Cataldi, S.; Borrelli, A.; Ceccarini, M.R.; Nakashidze, I.; Codini, M.; Belov, O.; Ivanov, A.; Krasavin, E.; Ferri, I.; Conte, C.; et al. Acid and Neutral Sphingomyelinase Behavior in Radiation-Induced Liver Pyroptosis and in the Protective/Preventive Role of rMnSOD. *Int. J. Mol. Sci.* **2020**, *21*, 3281. [CrossRef] [PubMed]

17. Cataldi, S.; Borrelli, A.; Ceccarini, M.R.; Nakashidze, I.; Codini, M.; Belov, O.; Ivanov, A.; Krasavin, E.; Ferri, I.; Conte, C.; et al. Neutral Sphingomyelinase Modulation in the Protective/Preventive Role of rMnSOD from Radiation-Induced Damage in the Brain. *Int. J. Mol. Sci.* **2019**, *20*, 5431. [CrossRef]
18. Kloska, A.; Malinowska, M.; Gabig-Cimińska, M.; Jakóbkiewicz-Banecka, J. Lipids and Lipid Mediators Associated with the Risk and Pathology of Ischemic Stroke. *Int. J. Mol. Sci.* **2020**, *21*, 3618. [CrossRef]
19. Suryadevara, V.; Ramchandran, R.; Kamp, D.W.; Natarajan, V. Lipid Mediators Regulate Pulmonary Fibrosis: Potential Mechanisms and Signaling Pathways. *Int. J. Mol. Sci.* **2020**, *21*, 4257. [CrossRef]
20. Carbone, F.; Lattanzio, M.S.; Minetti, S.; Ansaldo, A.M.; Ferrara, D.; Molina-Molina, E.; Belfiore, A.; Elia, E.; Pugliese, S.; Palmieri, V.O.; et al. Circulating CRP Levels Are Associated with Epicardial and Visceral Fat Depots in Women with Metabolic Syndrome Criteria. *Int. J. Mol. Sci.* **2019**, *20*, 5981. [CrossRef]
21. Markovic, M.; Ben-Shabat, S.; Aponick, A.; Zimmermann, E.M.; Dahan, A. Lipids and Lipid-Processing Pathways in Drug Delivery and Therapeutics. *Int. J. Mol. Sci.* **2020**, *21*, 3248. [CrossRef] [PubMed]

International Journal of
Molecular Sciences

Review

Lipophagy and Lipolysis Status in Lipid Storage and Lipid Metabolism Diseases

Anna Kloska [1], Magdalena Węsierska [1], Marcelina Malinowska [1], Magdalena Gabig-Cimińska [1,2,*] and Joanna Jakóbkiewicz-Banecka [1,*]

[1]   Department of Medical Biology and Genetics, Faculty of Biology, University of Gdańsk, Wita Stwosza 59, 80-308 Gdańsk, Poland; anna.kloska@ug.edu.pl (A.K.); magdalena.wesierska@phdstud.ug.edu.pl (M.W.); marcelina.malinowska@ug.edu.pl (M.M.)
[2]   Laboratory of Molecular Biology, Institute of Biochemistry and Biophysics, Polish Academy of Sciences, Kładki 24, 80-822 Gdańsk, Poland
*   Correspondence: magdalena.gabig-ciminska@ug.edu.pl (M.G.-C.); joanna.jakobkiewicz-banecka@ug.edu.pl (J.J.-B.); Tel.: +48-585-236-046 (M.G.-C.); +48-585-236-043 (J.J.-B.)

Received: 6 July 2020; Accepted: 21 August 2020; Published: 25 August 2020

Abstract: This review discusses how lipophagy and cytosolic lipolysis degrade cellular lipids, as well as how these pathway ys communicate, how they affect lipid metabolism and energy homeostasis in cells and how their dysfunction affects the pathogenesis of lipid storage and lipid metabolism diseases. Answers to these questions will likely uncover novel strategies for the treatment of aforementioned human diseases, but, above all, will avoid destructive effects of high concentrations of lipids—referred to as lipotoxicity—resulting in cellular dysfunction and cell death.

Keywords: lipophagy; lipolysis; lipid metabolism; lipid droplets; lipid storage diseases; lipid metabolism diseases; mTORC1; TFEB

1. Introduction

Lipids are water-insoluble biological macromolecules that are essential for maintaining cellular structure, function, signaling and energy storage. They are basic components of all cellular membranes which separate cell compartments in eukaryotic cells and provide a permeability barrier. These membrane boundaries are necessary for maintaining cellular homeostasis [1,2]. Moreover, lipids affect the function of membrane proteins. Lipid rafts play a specific role in protein segregation; membrane proteins can interact with lipids, which serve as cofactors [3,4]. Finally, changes in lipid organization influence signal transduction and membrane trafficking [2]. Cholesterol serves as a precursor for steroid hormones and bile acid biosynthesis [5]. Lipids are also molecules that serve as a source of energy when tissue energy is depleted [6]. Despite their role in essential cellular functions, incorrect lipid distribution or metabolism can result in abnormal concentrations of lipids being toxic because of their limited solubility and amphipathic nature, their adverse impact on cellular homeostasis and their ready transformation into highly bioactive, cytotoxic lipid species. These effects have serious consequences for cellular function and homeostasis and may even lead to cell death [2].

In this review, we provide information about lipid metabolism in health and disease, focusing on lipid storage diseases and lipid metabolism diseases. We summarize the current knowledge about the role of two cytosolic pathways designed for lipid selective catabolism—lipophagy and lipolysis—and discuss the transcriptional regulation of these processes by the mechanistic target of rapamycin kinase complex 1 (mTORC1)—transcription factor EB (TFEB) signaling. We also characterize lipid storage and lipid metabolism diseases, highlighting the latest research on the contribution of mTORC1-TFEB signaling in the regulation of lipophagy, a subtype of macroautophagy, and lipolysis, an enzymatic hydrolysis process, in the selected human dysfunctions.

2. Lipids in Eukaryotic Cells

Based on the chemical origin (i.e., whether ketoacyl groups or isoprene groups serve as fundamental "building blocks"), lipids are divided into eight categories: fatty acids, glycerolipids, glycerophospholipids, sphingolipids, saccharolipids, polyketides, sterols and prenols [7].

2.1. Fatty Acids and Cholesterol—Essential and Toxic

Fatty acids (FAs) are hydrophobic molecules consisting of an aliphatic hydrocarbon chain terminating in a carboxylic acid moiety. FAs usually contain 16–18 carbons, and the chain can be fully saturated (saturated FA) or may contain one or more double bonds (unsaturated FAs). The main source of FAs for humans and other animals are dietary fats and oils, but they can also be synthesized de novo from metabolites of sugar and protein catabolic pathways [8]. Fatty acids can be harmful to cells because of lipotoxicity; thus, cells convert FAs into neutral lipids for storage in organelles called lipid droplets (LDs). Biogenesis of LDs is stimulated upon the increase in cellular free FA levels. Different cell types have LDs of various sizes and numbers, potentially reflecting the capacity of the cell for managing lipid storage. Moreover, these organelles are often heterogeneous within a population of a single cell type. It is believed that LDs not only serve as lipid storage organelles, but also interact with most, if not all, cellular organelles, mediating lipid transfer via direct contact [9].

Cholesterol is an elementary component of mammalian cell membranes. It interacts with phospholipids and sphingolipid fatty acyl chains in order to maintain appropriate membrane fluidity. Interactions between these lipids also regulate water and ion membrane permeability [10]. Cholesterol is required for normal prenatal development, as embryonic and fetal cells demonstrate high membrane formation rates [11,12]. At both fetal and adult stages of development, cholesterol is the precursor for biosynthesis of five major classes of steroid hormones (i.e., androgens, estrogens, glucocorticoids, mineralocorticoids and gestagens), vitamin D and bile acids [10,11,13]. Mammalian cells require cholesterol for proliferation. Moreover, cholesterol is specifically required for the transition from G1 to S during cell cycle progression [14,15]. Additionally, cholesterol is essential for mitosis progression and its deficiency leading to aberrant mitosis and polyploid cell formation [15,16].

The cell synthesizes cholesterol de novo or internalizes it from exogenous sources. Interestingly, cells do not have any enzymes to break down the sterol core to acetyl-CoA units; thus, cholesterol cannot be used as an energy source [10]. Regardless of its source, free cholesterol must be esterified; otherwise, it has a toxic effect on cellular membranes and induces cell death. Esterified cholesterol is stored in cells in cytoplasmic lipid droplets [2,5].

2.2. Lipid Droplets—Storage of Neutral Lipids

LDs are highly dynamic cellular organelles responsible for the storage of neutral lipids. They are found in most eukaryotic cell types. The size of LDs varies within the range 0.4–100 µm in different cell types or within the same cells, depending on physiological conditions. Lipid droplets originate from the endoplasmic reticulum (ER) and have a unique architecture consisting of a hydrophobic core of neutral lipids which is enclosed by a phospholipid monolayer with hundreds of resident and transient proteins that influence LD metabolism and signaling, known generically as perilipins (PLINs) (Figure 1A). Organization of these organelles is quite different than any other because the core of a LD is hydrophobic, the hydrophobic acyl chains of the monolayer's phospholipids are in contact with neutral lipids and the polar head groups face the aqueous cytosol [17]. Furthermore, LDs can also be found in the nuclei, where they are thought to regulate nuclear lipid homeostasis and modulate signaling through lipid molecules [18]. Cells preserve lipids by converting them into neutral lipids such as triacylglycerols (TAGs) and sterol esters (ESs), which are in various ratio deposit in LDs. Depending on the cell type, many other endogenous neutral lipids, such as retinyl esters, ether lipids and free cholesterol, can be stored in the LD core. Defects in LD biogenesis lead to insufficient or excess storage. Beyond the main function in energy metabolism, LDs play an important role in various cellular events, such as protein degradation, sequestration of transcription factors and chromatin components, generation of lipid ligands for certain nuclear receptors and serving as fatty acid trafficking nodes [19].

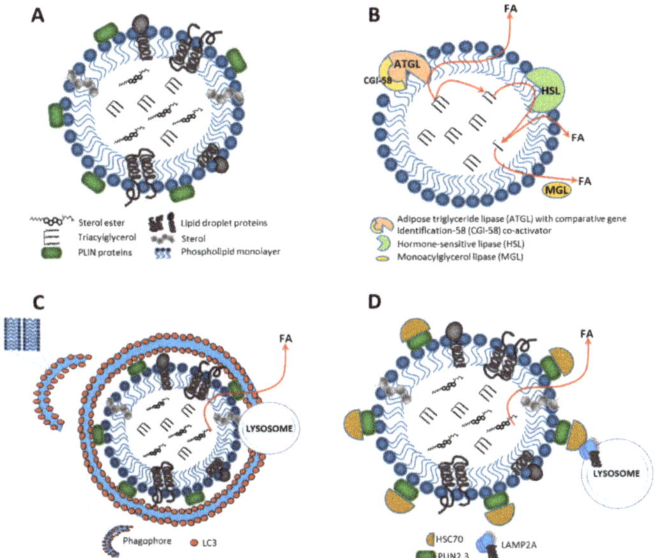

Figure 1. Structure and catabolism of a lipid droplet (LD). (**A**) LD is surrounded by the phospholipid monolayer enclosing a core filled with neutral lipids, e.g., triacylglycerol (TAG) and sterol esters. Polar heads of phospholipids are oriented toward the cytosol, whereas their acyl chains contact the hydrophobic lipid core. The LD surface is associated with various proteins, e.g., members of the perilipin (PLIN) family. There are two major types of LD catabolism: lipolysis—an enzymatic hydrolysis of lipids in cytosol, and lipophagy—an autophagic/lysosomal pathway in the form of macroautophagy or chaperone-mediated autophagy (CMA). (**B**) In lipolysis, protein kinase A (PKA) phosphorylates PLIN1 proteins, leading to their proteasomal degradation and activating adipose triglyceride lipase (ATGL), which then initiates TAG hydrolysis to generate diacylglycerols (DAGs) and free fatty acids (FAs). Further degradation of DAGs occurs through activation of the hormone sensitive lipase (HSL), leading to monoacylglycerol (MAG) and FAs production. MAGs are released to the cytosol and cleaved by monoacylglycerol lipase (MGL) to generate glycerol and FAs. (**C**) In macroautophagy, the phagophore is formed and LC3 positive membranes engulf small LD or sequester portions of a large LD to form the autophagosome, which later fuses with lysosome where LD degradation and neutral lipid catabolism occur. (**D**) In chaperone-mediated autophagy, lipid droplet-coat proteins—PLIN2 and PLIN3—are degraded through a coordinated action of Hsc70 protein and lysosome-associated membrane protein 2A (LAMP2A) receptor; this makes the LD surface accessible to cytosolic lipases, which hydrolyze LD cargo to generate FAs, which next are released to the cytosol and undergo subsequent mitochondrial β-oxidation.

In mammalian cells, the phospholipid composition of LD membranes differs from that of the ER and other organelles. The main constituent is phosphatidylcholine (PC), followed by phosphatidylethanolamine (PE), phosphatidylinositol (PI), phosphatidylserine (PS) and sphingomyelin (SM), as well as free cholesterol and phosphatidic acid in minor amounts. The unique phospholipid membrane composition affects LD synthesis, size and catabolism. The homeostasis of these organelles under physiological conditions is maintained through changes in membrane phospholipid ratios in various cell types [20].

In addition to the composition of phospholipids, LD membrane surface proteins are another key factor that is important for their homeostasis and intracellular interactions. Each LD has many different structural and functional proteins on its surface. In mammalian LDs, predominant proteins are PLINs, adipophilin (ADRP) and a tail-interacting protein of 47 kDa (TIP47)—all belonging to the PAT (PLIN/ADRP/TIP47) protein family, which was named after its members [21]. Of these proteins, the structure and function of PLINs that regulate lipase access to the LD core is best known, and increased lipolysis in adipocytes is observed in their absence [22]. Furthermore, there are many other proteins involved in the maintenance of lipid homeostasis, including signaling and membrane trafficking proteins, chaperones and proteins associated with cellular organelles. Mitogen-activated protein kinase (MAPK) and phosphatidylinositol 3-kinase (PI3K) play the major role among signaling proteins associated with the LD surface [23]. Caveolin 1 (CAV1) and 2 (CAV2) are other proteins present on the LD surface; they generate membrane domains that serve as regulators of signaling proteins. In general, caveolins form a coat by making invaginations in surrounding cellular membranes. The coats are called caveolae and they function in endocytosis, signal transduction, cholesterol transport and growth control [24]. Amongst membrane trafficking-related proteins, five subgroups are distinguished: small GTPases governing vesicle formation and motility; proteins that carry LDs on the cytoskeleton, such as kinesin and myosin; proteins that mediate membrane docking and fusion, such as soluble N-ethylmaleimide-sensitive factor attachment receptor (SNARE); proteins that regulate cargo sorting and vesicle budding, such as ADP-ribosylation factor (ARF)-related proteins and coat proteins (COPs); and other proteins of miscellaneous function [19].

Generally, once LDs are synthesized they keep growing because of excessive amounts of intracellular FAs until they reach a final size. It has been shown that many proteins, e.g., PLIN1 and lipids, such as PC, are involved in LD growth mechanisms [25].

3. Lipophagy and Lipolysis—Two Pathways that Play a Crucial Role in Lipid Metabolism

Mobilization of fat stores from LDs is regulated by the metabolic and energy demands of the cell. This process usually appears in the form of lipophagy or lipolysis. They are the catabolic pathways that have a fundamental role in breaking down lipids during nutrient deprivation. Both have an impact on cellular energetic balance: directly through their important role in the early steps of lipid breakdown and indirectly by regulating food intake. Defects in lipophagy and lipolysis have been linked to many metabolic disorders; among them are lipid storage and lipid metabolism diseases.

3.1. Catabolism of Lipid Droplets

Catabolism of LDs into free FAs is a crucial cellular pathway that is required to generate energy in the form of ATP. Their catabolism is strictly under the control of hormone and enzyme activation. Moreover, it is required to provide building blocks for biological membranes and precursors in hormone synthesis. Degradation of LDs is strictly regulated by the protein composition on the surface of the vesicle and generally occurs by two mechanisms: lipolysis or lipophagy.

Lipolysis is a biochemical catabolic pathway that relies on direct activation of LD-associated lipases, such as adipose triglyceride lipase (ATGL), hormone-sensitive lipase (HSL) and monoglyceride lipase (MGL), which, together with regulatory protein factors (ATGL activators and inhibitors), constitute the basis for this process [26]. Under fed conditions, LDs mainly store TAGs in adipose tissue, and lipolytic hydrolysis is based on the hydrolysis of ester bonds between long-chain FAs and the glycerol backbone. During the first step of this process, protein kinase A (PKA) phosphorylates PLIN1, leading to its proteasomal degradation. This results in the release of an ATGL activator protein—comparative gene identification-58 (CGI-58)—which selectively activates ATGL, which then catalyzes TAG hydrolysis to diacylglycerols (DAGs) and free FAs. The next step of the process depends on the activation of a multifunctional enzyme, hormone sensitive lipase (HSL), that hydrolyzes DAGs and produces monoacylglycerol (MAG) and FAs (Figure 1B). HSL functions as a rate-limiting enzyme for DAG catabolism. HSL also retains specificity to other lipid ester bonds, such as cholesteryl esters, retinyl esters and short-chain carbonic acid esters [27]. It is responsible for mediating the hydrolysis of diacylglycerol and triacylglycerol. In testis, HSL is the only esterase that can hydrolyze cholesteryl ester, and the loss of this activity results in cholesteryl ester and diacylglycerol accumulation [28], as well as altered lipid homeostasis [29]. In the last step of the lipolysis cascade, MAGs are released into the cytosol and cleaved by MGL, generating glycerol and FAs [30,31]. Products of lipolysis secreted from adipose tissue are transported to other tissues and used for β-oxidation and ATP production. In non-adipose tissues, mitochondria or peroxisomes can directly oxidize products of lipolytic hydrolysis through β-oxidation and release acetyl CoA [32].

In turn, the lysosomal–autophagic pathway that plays an important role in the early steps of lipid degradation has been termed lipophagy. In general, autophagy is one of the major degradation pathways that enables the cell to survive under stress conditions by recycling metabolic components. This process is initiated by sequestering cytosolic organelles or macromolecules in a double membrane vesicle, which is then delivered to lysosomes for degradation by lytic enzymes. Degradation products that can be reused by the cell in synthesis processes are then released into the cytosol. Proper functioning of autophagy allows the cell to maintain homeostasis [33]. Due to the mechanism of the process, we can distinguish three types of autophagy: macroautophagy, which targets large substrates in a selective or nonselective manner to form autophagosomes prior to fusion with lysosomes [34]; microautophagy, which degrades molecules through direct engulfment by membranes of lytic compartments (lysosomes or late endosomes); and chaperone-mediated autophagy (CMA), which is a selective form of autophagy, targeting specific proteins through the recognition activity of chaperone protein heat shock cognate 70 (Hsc70) [35].

Uptake of LDs by macroautophagy is an alternative route for the mobilization of lipid storage and degradation of intracellular LDs (Figure 1C). Such a process is called lipophagy, in which LDs are selectively delivered to a lytic compartment for degradation via actions of autophagic (Atg) proteins. This process was first described in mouse hepatocytes under starvation, when LDs were mobilized in order to generate free FAs [36]. LC3, a classical marker of the autophagosome, was able to directly interact with ATGL and HSL at the surface of LDs. LC3 binds ATGL via an LC3 interaction region (LIR) and, under fed deprivation and LIR deficiency conditions, reduced basal ATGL localization to LDs, preventing the ATGL translocation to the LD surface, was observed. When we consider the above, it seems that LC3 is required for translocation of ATGL to the surface of LDs, to facilitate TAG hydrolysis [37].

Numerous Rab proteins have been identified on LDs. In general, Rab proteins are a family of small GTPases, acting as important mediators of endosomal trafficking events. They are molecular switches, cycling between active GTP-bound and inactive GDP-bound states, regulating the vesicular trafficking network within the cell. Perturbation to some members of the Rab family proteins has deleterious effects on LD turnover in response to classical lipophagy-inducing causes [38]. The most predominant Rab protein on the LD surface is Rab7, which is a well-characterized marker of the late endocytic pathway and a participant in the process of autophagosomal maturation. This protein assists in the regulation of lysosome–autophagosome interaction. Rab7 GTPase on the surface of LDs becomes active upon nutrient deprivation, resulting in its increased activity for GTP over GDP. Such an activated state promotes the requirement of lysosomes near LDs and their target degradation via lipophagy [39]. Another LD-localized protein from the Rab family that potentially participates in lipophagy is Rab10, which in its active state is significantly redistributed on the LD surface under nutrient deprivation conditions. This GTPase co-localizes with autophagic membrane markers such as LC3 and Atg16. However, it seems that Rab10 acts downstream of Rab7 as a part of a complex that promotes the envelopment of LD during lipophagy progression [40]. There are several other Rab proteins that have been studied to determine their role in LD catabolism via conventional lipolysis and selective lipophagy, such as Rab32, Rab18 and Rab25.

In LD catabolism, a link between PLIN proteins and CMA has been identified (Figure 1D). For the CMA process, LAMP2A is required and lack of LAMP2A leads to LD accumulation. Moreover, Hsc70 binds to CMA recognition motif (KFERQ) within PLIN proteins, acting as a signal for CMA-mediated degradation in the lysosome. It appears that degradation of PLIN proteins is required to promote LD catabolism by allowing ATGL and autophagic proteins to access the LD surface. Blocking the CMA process reduces lipase-mediated lipolysis and lipophagy. Therefore, CMA-mediated degradation of PLIN proteins seems to be a crucial event for initiating lipophagy [41,42].

3.2. Energy Release from Fatty Acids

Triacylglycerols are highly concentrated forms of metabolic energy because they are reduced and anhydrous. They are made up of three FAs that are ester-linked to a single glycerol. Complete oxidation of FAs provides more than twice as much energy than is obtained from carbohydrates or proteins. TAGs have much lower toxicity compared to FAs; thus, they can reach much higher concentrations (e.g., in plasma). For this reason, TAGs are the major form of FA storage and transport [6,43].

Before FAs can be used as a source of energy, they must be released from TAGs by lipolysis or lipophagy. Products of lipid stores that are broken down by lipolysis or lipophagy are subsequently utilized in β-oxidation for ATP production (Figure 2). At the outer mitochondrial membrane, FAs are activated by thioesterification to acyl-CoA esters. Next, carnitine, together with acylcarnitine translocase, transports FAs across the inner mitochondrial membrane to the matrix, where β-oxidation takes place. FA β-oxidation consists of four cyclic biochemical reactions. First, acyl-CoA is oxidized with the participation of flavin-adenine dinucleotide (FAD). At this stage, electrons from $FADH_2$ reduce ubiquinone to ubiquinol, which transfers them on the respiratory chain and leads to the generation of 1.5 ATP molecules. Acyl-CoA oxidation introduces a double bond in the FA chain, which is hydrated in the second step of β-oxidation. The third reaction involves oxidation with creating ketone group at C-3 and reduction of NAD^+ to NADH (to generate 2.5 ATP molecules). In the fourth and final step, the thiol group of the next CoA molecule resolves 3-ketoacyl-CoA into acyl-CoA (two carbon atoms shortened) and acetyl-CoA by thiolytic cleavage. B-oxidation is repeated until the initial FA chain is converted into acetyl-CoA, which enters the Krebs cycle (to generate 10 ATP molecules) [44,45].

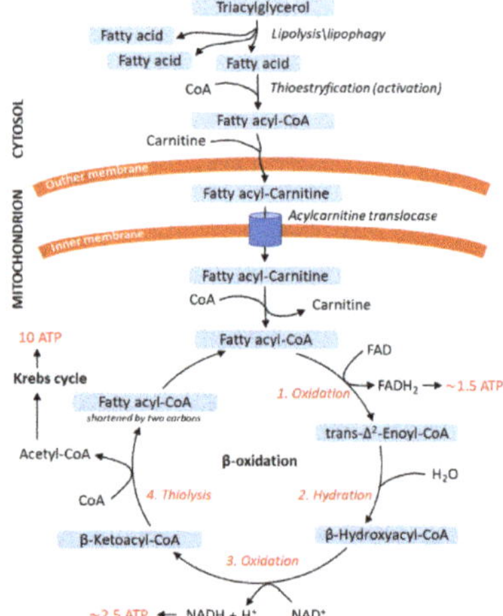

Figure 2. Energy release from saturated fatty acids in mitochondrial β-oxidation. Fatty acids are released from triacylglycerol by lipolysis or lipophagy and translocated into the mitochondrion. Fatty acid is shortened by two carbons in one β-oxidation cycle; the β-oxidation steps are repeated until only two carbon units remain. The $FADH_2$ and NADH are utilized to generate ATP in the electron transport chain and acetyl-CoA enters the Krebs cycle. The β-oxidation steps are shown in red italics, numbered 1–4. The number of ATP molecules obtained from β-oxidation and Krebs cycle is shown in red. ATP, adenosine triphosphate; CoA, coenzyme A; FAD and $FADH_2$, flavin-adenine dinucleotide, oxidized and reduced forms, respectively; NAD^+ and NADH, nicotinamide adenine dinucleotide, oxidized and reduced forms, respectively.

Mitochondrial β-oxidation is more complex with unsaturated FAs or odd-chain FAs being the source of energy. Degradation of unsaturated FAs involves the participation of additional enzymes. Isomerase converts configuration of the double bond in mono- and poly-unsaturated FA. Next, mono-unsaturated FA is hydrated, and β-oxidation progresses as for saturated FA. In turn, poly-unsaturated FA is oxidized with the participation of FAD and then reduced by mitochondrial NADPH-dependent reductase. Another double bond is formed and FA is again converted by isomerase until a regular intermediate of β-oxidation pathway is obtained, which is entered into the cycle at the stage of hydration [46]. At the end of odd-chain FA β-oxidation, acetyl-CoA and propionyl-CoA are produced in place of two molecules of acetyl-CoA. Propionyl-CoA enters into the Krebs cycle after it has been converted into succinyl-CoA [47].

3.3. Transcriptional Regulation of Lipophagy, Lipolysis and Lipid Metabolism

Lipophagy control depends on several transcription factors, activators and nuclear receptors, which, in response to nutrient status, enhance or decrease the process of lipid breakdown, in order to support current energy demands of the cell. Expression of many autophagy- and lipophagy-related genes is controlled by transcription factors belonging to the microphthalmia (MiT/TFE) family. One of these factors is TFEB, which regulates not only the general autophagic process, but lipophagy and lipid metabolism, as well [48]. Phosphorylation and dephosphorylation of TFEB determines its cellular localization and activity. Both events are mainly controlled by the nutrient or lysosomal storage status of the cell [48]. Under nutrition-rich conditions, the phosphorylated form of TFEB is retained in the cytoplasm (Figure 3A). However, nutrient depletion or aberrant lysosomal storage results in dephosphorylation of TFEB, causing its translocation from the cytoplasm to the nucleus, where it induces the transcription of target genes (Figure 3B). Promoters of many lysosome-related genes share a common 10-base E-box-like palindromic sequence; they compose the coordinated lysosomal expression and regulation (CLEAR) gene network that TFEB directly targets and controls the transcription process of [49]. Chromatin immunoprecipitation assays identified over 600 endocytic genes regulated by TFEB; among them are genes related to lysosomal biogenesis and autophagy, as well as lipid catabolism. TFEB phosphorylation is mainly exerted by two kinases: mTORC1 (one of two complexes having mTOR kinase as a core component) or extracellular signal-regulated kinase 2 (ERK2, also known as MAPK1). MTORC1 is the main negative regulator of autophagy, acting in response to amino acids, growth factors or cellular energy status [50]. Under sufficient intracellular availability of nutrients, mTORC1 is activated and inhibits autophagy, but as nutrients are deprived, this kinase is switched off, leading to autophagy activation and inhibition of anabolic processes. Active mTORC1 phosphorylates TFEB, preventing its translocation to the nucleus and thus indirectly inhibiting autophagy as the TFEB-dependent transcription of genes related to autophagy and lysosomal biogenesis stays suppressed [51].

The role of TFEB in lipophagy was firmly revealed by Settembre et al. [52], as they demonstrated that TFEB regulates lipid degradation in the mouse liver. They showed that, upon nutrient depletion, TFEB deficiency lead to accumulation of LDs and impairment of FA oxidation in the liver [52]. In animals receiving a high-fat diet, overexpression of TFEB prevented the development of obesity and improved the metabolic syndrome phenotype by reducing abnormal levels of circulating triglycerides, cholesterol, glucose and insulin [52]. High TFEB activity was also able to revert the metabolic syndrome when it was already present [52]. A functional autophagic pathway was required to observe TFEB-mediated lipid degradation, as overexpression of TFEB was not able to decrease the lipid droplet number, liver weight gain or lipid content in livers of mice with blocked autophagy [52]. Upon starvation, TFEB enhanced the expression of genes related to lipid metabolism and lipophagy [52]. In mice, TFEB-dependent transcriptional upregulation of monocarboxylic acid, FA, ketone catabolism and FA oxidation pathways was observed [52]. Several genes involved in lysosome organization and autophagy were also upregulated by TFEB, upon reduced nutrient availability; these include ATPase subunits, proteases, membrane proteins and fusion proteins [52]. Additionally, TFEB downregulated gene expression of lipid biosynthetic pathways; these included pathways of steroid, lipid, isoprenoid and FA biosynthesis [52]. TFEB was shown to regulate expression of genes involved in several steps of lipid catabolism, including genes related to FA transport across the plasma membrane (e.g., *Cd36* and *Fabps*), β-oxidation of free FAs in mitochondria (e.g., *Cpt1*, *Crat*, *Acadl*, *Acads* and *Hdad*) and peroxisomes (*Cyp4a* genes) [52]. Moreover, TFEB controls its own expression in an autoregulatory feedback loop [52]. CLEAR elements are present in the *TFEB* gene promoter; thus, as TFEB translocates to the nucleus, it is able to bind to its own promoter, enhancing transcription of itself in response to cellular status [52]. Interestingly, the role of TFEB in lipophagy appears to be evolutionarily conserved. In *Caenorhabditis elegans*, a gene encoding the worm's TFEB orthologue, *HLH-30*, regulates the expression of fat catabolism enzymes and autophagy genes in response to nutrient availability. Moreover, loss-of-function mutations of this gene result in impairment of lipophagy [53]. Transcriptional control of genes involved in lipid metabolism by TFEB is exerted through the peroxisome proliferator-activated receptor gamma coactivator 1α (PPARGC1α; also known as PGC1α) and the downstream nuclear receptor peroxisome proliferator activated receptor 1α (PPAR1α) [52]. The promoter of the PGC1α-encoding gene was shown to have three CLEAR sites and upon starvation TFEB bound to two of them [52]. Thus, TFEB induces PGC1α expression upon starvation and regulates the expression of genes related to lipid metabolism by controlling the downstream PPARα activity [52,54]. Independently from TFEB, lipophagy control is also mediated by nuclear farnesoid X receptor (FXR) and transcriptional activator cAMP response element-binding protein (CREB) [55]. Under nutrient-deprived conditions, CREB promotes lipophagy by the upregulation of autophagy-related genes, including *Atg7*, *Ulk1* and *Tfeb*, but FXR, a fed-state sensing gene regulator, inhibits this response after feeding [55].

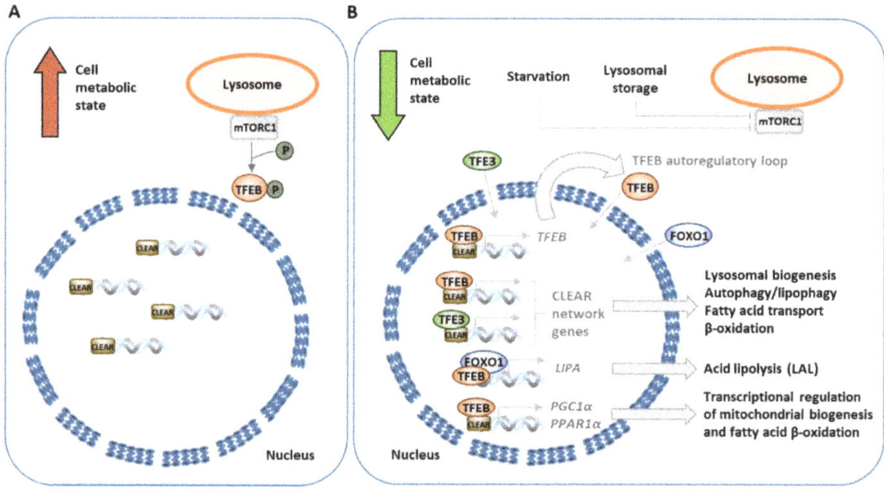

Figure 3. Transcriptional regulation of autophagy/lipophagy, lipolysis and lipid metabolism by transcription factor EB (TFEB) under nutrition-rich conditions (**A**) and nutrient depletion or aberrant lysosomal storage (**B**). Bold red arrow indicates the fed metabolic state during nutrient sufficiency, while bold green arrow shows low metabolic state due to nutrient limitations or abnormal lysosomal storage.

Lipolysis, another process involved in intracellular lipid breakdown, was shown to be transcriptionally modulated by forkhead homeobox type O (FOXO) and TFEB transcription factors. Nutrient restriction upregulates FOXO1, which then activates lipid catabolism by inducing lysosomal acid lipase (LAL) expression [56]. In this case, colocalization of LDs with lysosomes was observed. In response to nutrient restriction, FOXO1 and TFEB were shown to induce the expression of *LIPA*, a gene encoding LAL. In a mouse atherosclerosis model, lysosomal stress conditions induced by atherogenic lipids were shown to promote TFEB translocation to the nucleus and upregulation of *LIPA* gene expression and lysosomal biogenesis [57]. TFEB overexpression in macrophages loaded with acetylated low-density lipoprotein (LDL) enhanced cholesterol efflux [57].

Lipophagy and lipid metabolism are also regulated by transcription factor E3 (TFE3), which is another transcription factor belonging to the MiT family. In the liver, TFE3 was shown to induce lipophagy as its overexpression alleviated steatosis of this organ in mice [58]. Similarly to TFEB, TFE3 induces expression of genes that modulate mitochondrial fatty acid β-oxidation; an increased mRNA level of PGC1α and PPARα was found upon TFE3 overexpression [58]. Another study also showed that TFE3 deficiency resulted in altered mitochondrial morphology and function [59]. *Tfe3*-knockout mice show abnormalities in energy balance and alterations in systemic glucose and lipid metabolism, resulting in high-fat-diet-induced obesity and diabetes [59]. However, overexpression of TFE3, as well as TFEB, was shown to improve this metabolic outcome [59]. Because TFE3 and TFEB were able to compensate for each other's deficiency, the authors suggested that both play a cooperative role in controlling metabolism.

Generally, autophagy was thought to contribute to lipid oxidation by increasing the supply of free FAs by lipophagy. However, a recent study demonstrated that autophagy regulates lipid metabolism by participating in regulation of PPARα activity through the degradation of nuclear receptor co-repressor 1 (NCoR1) [60]. Autophagic degradation of NCoR1 was shown to contribute to PPARα activation to effectively promote β-oxidation in response to physiological fasting [60]. Defects of liver autophagy were accompanied by accumulation of NCoR1 and suppression of PPARα activity leading to defective β-oxidation and ketogenesis [60].

4. Lipid Metabolism and Diseases

Lipids have been found necessary in tissues such as adipose tissue, intestine and liver for energy storage or lipid turnover, but they are accreted in skeletal muscles, macrophages, mammary glands and the adrenal cortex. Under energy-poor conditions, lipid accumulation allows organisms to survive, and stored lipids are then used to produce energy [61]. Abnormal lipid metabolism is associated with many diseases, including type 2 diabetes, obstructive sleep apnea, non-alcoholic fatty liver disease, coronary artery disease and cancer. A number of studies have been published to reveal the important role of lipid metabolism in energy homeostasis and metabolic diseases [62]. Inherited metabolic disorders are genetic conditions that cause metabolism problems. There are hundreds of different genetic metabolic disorders, and their symptoms, treatment and prognosis vary significantly. Genetic disorders associated with abnormal lipid turnover belong to two groups of inherited metabolic disorders: lipid storage diseases and lipid metabolism diseases [63].

The pathological accumulation of undigested biomaterials in the lysosome, including lipids, leads to the development of metabolic disorders collectively called lysosomal storage diseases (LSDs). LSDs are traditionally classified due to the nature of undigested materials in the lysosome; in this group, mucopolysaccharidosis, cystinosis, mannosidosis and lipid storage diseases can be distinguished.

4.1. Characterization of Lipid Storage Diseases and Lipid Metabolism Diseases

Lipid storage diseases are the most common LSDs and constitute the largest group of these diseases [64]. However, undigested lipids can also accumulate due to secondary mechanisms, e.g., mechanisms secondary to carbohydrate or protein accumulation or membrane trafficking. Most LSDs are caused by lysosomal hydrolase mutations. Lipid storage diseases are a genetically determined group of disorders characterized by excessive lipid (fatty acids, cholesterol or complex lipids) accumulation due to inherited abnormalities in lipid metabolism. Excessive lipid deposition ultimately causes damage to cells and tissues, resulting in neurodegeneration and also often heart, liver, spleen and kidney problems [65]. In most lysosomal lipid storage diseases, the accumulation of one or more lipids leads to the co-precipitation of other hydrophobic substances in the endolysosomal system, such as lipids and proteins [66]. The progressive accumulation of undigested lipids in lysosomes leads to the accumulation of enlarged (>500 nm) but dysfunctional lysosomes [67]. These swollen lysosomes are mainly endolysosomes and autolysosomes; therefore, LSD is a state of endocytic and autophagic "block" or "arrest". Although the total number of lysosomes is not reduced in LSDs, the overall lysosomal function in the cell is blocked, which can lead to serious cell consequences [68]. The accumulation of undigested materials in lysosomes can cause a deficiency of building block precursors for biosynthetic pathways, while the accumulation of various membrane-associated lipids can affect the properties and integrity of the cell membrane. In addition, lipid storage may alter the functionality of lysosomal membrane proteins, such as lysosomal ion channels or catabolite exporters, affecting the physiology and ionic composition of lysosomes. In turn, altered heavy metal ion homeostasis can increase oxidative stress, causing lipid peroxidation and affecting membrane integrity. Over time, excessive lipid storage can cause permanent damage to cells, tissues in the brain and peripheral nervous system and other organs [64]. The brain is particularly sensitive to lipid accumulation, as any increase in fluid or deposits can lead to changes in pressure and disruption of normal neurological function [69]. Symptoms may appear early in life or develop in teenage years or even adulthood. Neurological complications of lipid storage diseases depend on the type of storage material and may include: lack of muscle coordination, brain degeneration, seizures, loss of muscle tone, spasticity, difficulty feeding and swallowing, pain in arms and legs and corneal opacity [70].

Congenital lipid metabolism errors are a heterogeneous group of disorders characterized by problems with the breakdown synthesis of lipids. Diseases that affect lipid metabolism can be caused by defects in the structural proteins of lipoprotein particles, in the cell receptors that recognize different types of lipoproteins or in fat-breaking enzymes [71]. As a result of such defects, excessively accumulating lipids can deposit in the walls of blood vessels, which can lead to atherosclerosis and, as a consequence, strokes or coronary heart diseases. Lipid metabolism diseases are associated with an increase in plasma lipoprotein levels, such as LDL cholesterol, very low-density lipoprotein (VLDL) and triglycerides, or combinations thereof. The historical framework for the classification of lipoprotein disorders is dominated by the Fredrickson classification, which is based on the pattern of lipoproteins on electrophoresis or ultracentrifugation [72]. The phenotypic classification of lipid metabolism diseases, which is widely used and has been accepted internationally, is based on the affected lipoprotein; however, the clinical approach is to classify dyslipidemia according to the high lipid content fraction: cholesterol (hypercholesterolemia, Fredrickson class IIa), triglycerides (hypertriglyceridemia, Frederickson classes I, IV and V) or a combination of the two (hypercholesterolemia and hypertriglyceridemia, Fredrickson classes III or IIb) [73]. In addition, it is crucial to consider the etiological aspects of dyslipidemias, which may help in the diagnosis or initial treatment.

Table 1 contains a list of examples of diseases classified as lipid storage diseases and lipid metabolism diseases. The classification was based on data contained in Mammalian Phenotype Ontology [63], with the exception of sitosterolemia. This disease is sometimes classified as rarer inherited metabolic disorder [72].

Table 1. Characterization of lipid storage diseases and lipid metabolism diseases.

Disease	Gene Deficient Enzyme/Protein	Accumulated Products	Symptoms	Perturbations in Autophagy/Lipophagy/Lipolysis	Reference
Lysosomal storage diseases					
Lipid storage diseases					
Sphingolipidoses					
Niemann–Pick disease types A and B	*SMPD1* sphingomyelinase	Sphingomyelin in brain and red blood cells (RBCs)	Hepatosplenomegaly, psychomotor regression, clumsiness and difficulty walking, dystonia, sleep disturbances, difficulty swallowing and eating, recurrent pneumonia, thrombocytopenia, a cherry-red spot inside the eye, frequent respiratory infections, slow mineralization of bone	Impaired autolysosomal clearance; formation of late endosome/lysosome (LE/LY)-like storage organelles (LSOs) and the misdirection of lipids to the LSOs; defect in autophagosome maturation; accumulation of autophagosomes	[74–77]
Niemann–Pick disease type C	*NPC1* or *NPC2* intracellular cholesterol transporters located within lysosomal and endosomal membranes (NPC1) or inside lysosomes (NPC2)	Free cholesterol, sphingomyelin and glycosphingolipid storage in lysosomes or late endosomes	Hepatosplenomegaly, problems with speech and swallowing, dementia, seizures, ataxia, vertical supranuclear gaze palsy, dystonia, severe liver disease, interstitial lung disease	Defective amphisome formation; impaired maturation of autophagosomes; accumulation of autophagosomes and autolysosomes	[78–82]
Fabry disease	*GLA* α-galactosidase A	Glycolipids, particularly ceramide trihexoside, in brain, heart and kidney	Episodes of pain (particularly acroparesthesias), angiokeratomas, hypohidrosis, corneal opacity or corneal verticillate, problems with the gastrointestinal system, tinnitus, hearing loss, kidney damage, heart attack, stroke	Impairment of the autophagic pathway	[83–86]
Krabbe disease (globoid cell leukodystrophy)	*GALC* galactocerebrosidase	Glycolipids, particularly galactocerebroside, in oligodendrocytes	Irritability, muscle weakness, feeding difficulties, stiff posture, delayed mental and physical development, spasticity, hypertonia, blindness, hyperreflexia, deafness, neurodegeneration (leading to death)	Impairment of autophagy; lysosomal dysfunction; partial blocking and saturation of the autophagy flux	[87–90]
Gaucher disease	*GBA* glucocerebrosidase	Glucocerebrosides in RBCs, liver and spleen	Hepatosplenomegaly, pancytopenia, Erlenmeyer flask deformity, anemia, lung disease, bone abnormalities such as bone pain, fractures, arthritis	Impaired autophagosome maturation; accumulation of autophagosomes; autophagy block	[91–93]
Tay–Sachs disease	*HEXA* β-hexosaminidase A	GM2 gangliosides in neurons	Neurodegeneration, seizures, vision and hearing loss, cherry-red spot, muscle weakness, ataxia, intellectual disability, paralysis, early death	Altered lipid trafficking; impaired autophagy	[94–97]
Tay–Sachs Disease, AB Variant (AB-variant GM2)	*GM2A* GM2 ganglioside activator	GM2 ganglioside in neurons in the brain and spinal cord	Psychomotor deterioration, seizures, vision and hearing loss, intellectual disability, paralysis, cherry-red spot, early death	Impaired autophagy	[95,98,99]
Metachromatic leukodystrophy (MLD)	*ASA* or *PSAP* arylsulfatase A or prosaposin	Sulfatide compounds in neural tissue	Demyelination in central and peripheral nervous systems (peripheral neuropathy, mental retardation, motor dysfunction, ataxia, hyporeflexia), seizures, incontinence, paralysis, inability to speak, blindness, hearing loss	Affected trafficking due to altered chain length of the lipids; defective autophagosome–lysosome fusion, impaired autophagy	[100–103]
Sandhoff disease	*HEXB* β-hexosaminidase A and β-hexosaminidase B	GM2 ganglioside in neurons of the brain and spinal cord	Progressive nervous system deterioration, muscle weakness, ataxia, speech problems, mental retardation, blindness, seizures, spasticity, macrocephaly, cherry-red spots in the eyes, frequent respiratory infections, doll-like facial appearance, hepatosplenomegaly	Disruption of autophagy, aberrant lysosomal–autophagic turnover	[104–107]
Multiple sulfatase deficiency	*SUMF1* formylglycine-generating enzyme (FGE)	Sulfatides, sulfated glycosaminoglycans, sphingolipids and steroid sulfates in tissues	Leukodystrophy, movement problems, seizures, developmental delay, slow growth, ichthyosis, hypertrichosis, skeletal abnormalities (scoliosis, joint stiffness, dysostosis multiplex), hypotonia, coarse facial features, mild deafness, hepatomegaly, progressive neurologic deterioration, hydrocephalus	Accumulation of autophagosomes, defective autophagosome–lysosome fusion	[108–110]

Table 1. *Cont.*

Disease	Gene Deficient Enzyme/Protein	Accumulated Products	Symptoms	Perturbations in Autophagy/Lipophagy/Lipolysis	Reference
GM1 gangliosidosis	*GLB1* β-galactosidase	GM1 ganglioside in tissues and organs, particularly in the brain	Hepatosplenomegaly, skeletal abnormalities, seizures, profound intellectual disability, cherry-red spot, gingival hypertrophy, cardiomyopathy, dysostosis multiplex, coarsened facial features	Accumulation of autophagosomes, impaired lysosomal flux	[101,111,112]
Schindler disease	*NAGA* α-N-acetylgalactosaminidase	Glycosphingolipids, glycoproteins and oligosaccharides with terminal or preterminal N-acetylgalactosaminyl residues in the lysosomes of most tissues	Developmental regression, blindness, seizures, loss of awareness of surroundings, unresponsive, cognitive impairment, sensorineural hearing loss, weakness and loss of sensation, angiokeratomas	No data	[113]
Sea-blue histiocytosis (inherited lipemic splenomegaly)	*APOE* apolipoprotein E	Cholesterol, triglycerides and beta-very-low-density lipoproteins (beta-VLDLs) in the blood; glycosphingolipids, particularly sphingomyelins in the histocytes	Hypertriglyceridemia, splenomegaly, liver function abnormalities, heart disease, sea-blue histiocytes in many organs (bone marrow, liver and spleen)	No data	[114]
			Neuronal ceroid lipofuscinosis		
Batten disease (juvenile neuronal ceroid lipofuscinosis, CLN3 disease)	*CLN3* battenin, hydrophobic transmembrane protein involved in lysosomal function	Lysosomal autofluorescent storage material (AFSM) in the cells of the brain, central nervous system, and retina in the eye	Progressive blindness, seizures, mental and cognitive decline, dementia, speech and motor skills problems, premature death	Disruption of autophagy, vacuole maturation and impaired mitophagy; impaired autophagic clearance, defective autophagosome maturation	[115–118]
Jansky–Bielschowsky disease (late infantile neuronal ceroid lipofuscinosis, LINCL, CLN2 disease)	*TPP1* tripeptidyl-peptidase 1	Lipopigments in neurons, primarily in the cerebral and cerebellar cortices	Epilepsy, ataxia, myoclonus, vision loss, speech and motor skills problems (e.g., sitting and walking), developmental regression, intellectual disability, behavioral problems	Reduction in autophagic flux, inhibition of autophagosome formation, reduction in autophagosomes and autophagic degradation	[119,120]
			Lysosomal and lipase deficiency		
Lysosomal acid lipase deficiency (Wolman disease, cholesteryl ester storage disease)	*LIPA* lysosomal acid lipase	Cholesteryl esters, triglycerides, and other lipids within lysosomes of most tissues	Hepatosplenomegaly, ascites, calcified adrenal glands, vomiting, diarrhea with steatorrhea, progressive psychomotor degradation, anemia, cachexia, low muscle tone, jaundice, vomiting, developmental delay, anemia, poor absorption of nutrients from food	Impairment of the lipophagic pathway	[121–124]
			Mucolipidosis		
Mucolipidosis IV	*MCOLN1 (TRPML1)* mucolipin-1	Sphingolipids, phospholipids, mucopolysaccharides and glycoproteins in cells of almost all tissues, including liver, spleen and in fibroblasts	Intellectual disability, psychomotor retardation, hypotonia, retinal degeneration, strabismus, photophobia, myopia, amblyopia or blindness, iron-deficiency anemia, achlorhydria with elevated blood gastrin levels	Impairment of autophagy and lipolysis; accumulation of lysosomes, autophagosomes and autophagy substrates	[125–128]
Sialidosis (mucolipidosis I)	*NEU1* neuraminidase 1	Sialic acid–containing compounds (sialyloligosaccharides and sialolipids) in lysosomes in bodily tissues	Type I: progressive neurological impairment without bone or joint abnormalities; type II: mental retardation, severe hepatosplenomegaly, coarse facial features, dysostosis multiplex, seizures, myoclonus, ataxia, aminoaciduria, corneal opacity, macular cherry-red spot, skeletal abnormalities	Impairment of lipolysis and autophagy	[129–131]
			Neutral lipid storage disease		
Neutral lipid storage disease with myopathy	*PNPLA2* adipose triglyceride lipase (ATGL)	Triglycerides in muscle and other tissues	Myopathy, fatty liver, cardiomyopathy, pancreatitis, hypothyroidism, type 2 diabetes	Impairment of lipolysis	[132–134]

Table 1. *Cont.*

Disease	Gene Deficient Enzyme/Protein	Accumulated Products	Symptoms	Perturbations in Autophagy/Lipophagy/Lipolysis	Reference
Chanarin–Dorfman syndrome (neutral lipid storage disease type I, neutral lipid storage disease with ichthyosis)	*ABHD5* abhydrolase domain containing 5 (activator of ATGL)	Triglycerides in organs and tissues, including skin, liver, muscles, intestine, eyes and ears	Ichthyosis, hepatomegaly, cataracts, ataxia, hearing loss, short stature, myopathy, nystagmus, mild intellectual disability	Impaired long-chain fatty acid oxidation; impaired BECN1-induced autophagic flux	[135–137]
			Xanthomatosis		
Cerebrotendinous xanthomatosis (CTX)	*CYP27A1* sterol 27-hydroxylase	Cholestanol and bile alcohols in the blood	Neonatal cholestasis, childhood-onset cataract, tendon and brain xanthomata, neurologic dysfunction (dementia, psychiatric disturbances, pyramidal and/or cerebellar signs, seizures and neuropathy), liver dysfunction, intellectual impairment, neuropsychiatric symptoms (hallucinations, aggression and depression)	Induced autophagy	[138–140]
			Plant sterol storage disease		
Sitosterolemia	*ABCG5* or *ABCG8* sterolin	Plant sterols, such as sitosterol, and LDL in the blood	Atherosclerosis, increased chance of a heart attack, stroke or sudden death, xanthomas, joint stiffness and pain, hemolytic anemia, macrothrombocytopenia	Accumulation of autophagic vacuoles	[141,142]
			Farber lipogranulomatosis		
Farber disease (Farber lipogranulomatosis)	*ASAH1* acid ceramidase	Lipids in cells and tissues throughout the body, particularly around the joints.	Lipogranulomas, swollen and painful joint deformity, subcutaneous nodules, hoarseness, difficulty breathing, hepatosplenomegaly, developmental delay, vomiting	Impairment of autophagic flux	[143]
			Fucosidosis		
Fucosidosis	*FUCA1* alpha-L-fucosidase	Fucose containing glyco-lipids and polysaccharides in the brain, liver, spleen, skin, heart, pancreas and kidneys	Intellectual disability, dementia, delayed development of motor skills, impaired growth, dysostosis multiplex, seizures, spasticity, angiokeratomas, coarse facial features, recurrent respiratory infections, visceromegaly	Induction of the autophagic cell death	[144]
Lipid metabolism diseases					
Familial hyperlipidemia					
Hyperlipoproteinemia					
Familial dysbetalipoproteinemia (hyperlipoproteinemia type III)	*APOE* apolipoprotein E	Chylomicrons and VLDL remnants in plasma	Palmar and tuberoeruptive xanthomas, coronary heart disease, peripheral vascular disease	Decreased lipolysis	[145–148]
Familial hypercholesterolemia (hyperlipoproteinemia type IIa)	*LDLR* LDL receptor	LDL in plasma	Tendon xanthomas, coronary heart disease, increased chance of a heart attack, stroke or sudden death	Impairment of autophagic flux; altered autophagy flux by persistent mitophagy	[149–151]
Familial defective apoB-100 (hyperlipoproteinemia type IIa)	*APOB* apolipoprotein B-100	LDL in plasma	Tendon xanthomas, coronary heart disease, increased chance of a heart attack, stroke or sudden death	Impairment of autophagic flux; altered autophagy flux by persistent mitophagy	[151,152]
Familial chylomicronemia syndrome					
ApoA-V deficiency	*APOA5* apolipoprotein A-V	Chylomicrons and VLDL in blood	Eruptive xanthomas, hepatosplenomegaly, pancreatitis	Impairment of lipolysis	[153,154]

Table 1. *Cont.*

Disease	Gene Deficient Enzyme/Protein	Accumulated Products	Symptoms	Perturbations in Autophagy/Lipophagy/Lipolysis	Reference
GPIHBP1 deficiency	*GPIHBP1* glycosylphosphatidylinositol-anchored high-density lipoprotein binding protein 1	Chylomicrons in plasma	Eruptive xanthomas, pancreatitis	Impairment of lipolysis	[155,156]
Lipoprotein lipase deficiency (hyperlipoproteinemia type I)	*LPL* lipoprotein lipase	Chylomicrons in plasma	Eruptive xanthomas, abdominal pain, lipemia retinalis, hepatosplenomegaly, pancreatitis	Impairment of lipolysis	[157,158]
Familial apolipoprotein C-II deficiency (hyperlipoproteinemia type I)	*APOC2* apolipoprotein C-II (LPL cofactor)	Chylomicrons in plasma	Eruptive xanthomas, abdominal pain, lipemia retinalis, hepatosplenomegaly, pancreatitis	Impairment of lipolysis	[159–161]
Familial hepatic lipase deficiency	*LIPC* hepatic lipase	VLDL remnants and IDLs in plasma	Pancreatitis, coronary heart disease, increased chance of a heart attack, stroke or sudden death	Impairment of lipolysis	[162]
Familial hypercholesterolemia					
Autosomal recessive hypercholesterolemia	*LDLRAP1 (ARH)* low-density lipoprotein receptor adaptor protein 1	LDL in plasma	Tendon xanthomas, coronary heart disease, increased chance of a heart attack, stroke or sudden death	Induced autophagy	[73,163,164]
Autosomal dominant hypercholesterolemia	*PCSK9* proprotein convertase subtilisin/kexin type 9	LDL in plasma	Tendon xanthomas, coronary heart disease, increased chance of a heart attack, stroke or sudden death	Increased autophagic flux	[73,165,166]

Groups (bold regular font), subgroups (regular font) or classes (italic font) of disorders related to abnormal lipid storage or lipid metabolism are indicated in the lines with a gray background. VLDL, Very Low Density Lipoprotein; LDL, Low Density Lipoprotei.

4.2. Dysregulation of Autophagy or Lipolysis in Diseases

Due to the important role of the lysosome in autophagy, this pathway is an obvious candidate in the pathogenesis of LSDs [167]. Autophagy has been identified as the primary pathway of lipid metabolism in cells [36]; therefore, it is believed that perturbances in autophagy, particularly lipophagy, are responsible for cellular lipid accumulation in patients with lipid storage diseases [168]. Studies have documented autophagic dysregulation in patient samples and disease models of various LSDs (i.e., Tay–Sachs disease, Fabry disease and Krabbe disease) [77,169]. Although impaired autophagy has been seen in many storage diseases, the defects observed relate to different stages of the autophagic pathway (Table 1 and Figure 4). While in GM1 gangliosidosis and Niemann–Pick disease, the impairment is due to overactivation of autophagy, in other LSDs, e.g., MSD and MLD, the autophagosome–lysosome fusion is defective [96]. Similar abnormalities of autophagy can be observed in diseases with secondary lipid accumulation (Figure 4). Although there is an increasing evidence of dysregulation of autophagy in lipid storage disorders, the role of these abnormalities in the pathogenesis of the diseases is still not well understood and requires further research. Impairment of autophagy has also been observed in lipid metabolism diseases; for example, autophagy induction occurs in the familial hypercholesterolemia and familial defective apoB-100. Impaired lipolysis has been reported in two groups of diseases—the neutral lipid storage diseases and the familial chylomicronemia syndromes (lipid metabolism diseases).

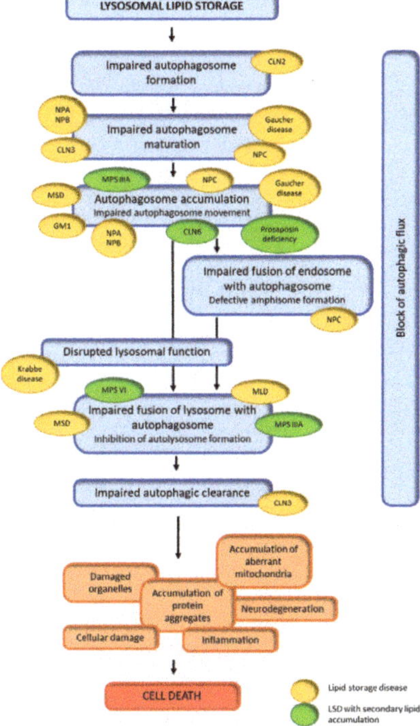

Figure 4. Alterations in different stages of autophagy in the pathogenesis of lipid storage diseases. Lysosomal lipid storage leads to a reduced ability to autophagosome formation, maturation, or fusion of lysosomes with autophagosomes. This results in a block of the autophagic flux. The steps of these abnormalities are presented in blue boxes. Consequently autophagy substrates (orange boxes) such as protein aggregates and dysfunctional mitochondria accumulate and promote cell death. The inflammatory response, cellular damage or neurodegeneration (orange boxes) further contribute to cell death (red box).

4.3. Secondary Lipid Accumulation in Lysosomal Storage Diseases

Storage processes in LSDs are much more complex than one would expect from the deficiency of a single enzyme altering a single substrate degradation in a particular catabolic pathway. Multiple substrates at variable ratios are detected as the storage material; this may result from metabolic links, e.g., when one enzyme is committed in the catabolism of multiple compounds. Secondary storage compounds can be actively involved in the pathogenesis of LSDs and the most common group of compounds that are subject to secondary storage are lipids. Phospholipids, glycosphingolipids and cholesterol are mainly identified as secondary storage materials. In numerous LSDs, one or more of these compounds, in various proportions, may accumulate inside cells (Table 2).

Table 2. Secondary lipid storage in lysosomal storage diseases. The individual classes of lipids are indicated in the lines with a gray background.

Secondary Storage Lipid	Disease	Compartment	Cellular Disturbance	Reference
Phospholipids				
Sphingomyelin	*Sphingolipidoses*: Niemann–Pick type C	Lysosomes	Altered membrane lipids trafficking	[170,171]
Bis(monoacylglycero)phosphate (BMP)	*Sphingolipidoses*: Niemann–Pick type C, Fabry disease, Gaucher disease, GM1 gangliosidosis, GM2 gangliosidosis *Mucopolysaccharidoses*: Hurler syndrome, Hunter syndrome *Neuronal ceroid lipofuscinoses*: NCL 10	Endosomes, lysosomes	Altered membrane lipids trafficking, lamellar bodies formation	[171,172]
Glycosphingolipids				
Gangliosides—GM1, GM2, GM3, GD1a, GD2, GD3	*Sphingolipidoses*: Niemann–Pick type A, B and C, Gaucher disease, prosaposin deficiency *Mucopolysaccharidoses*: Hurler syndrome, Hunter syndrome, Sanfilippo syndrome, Maroteaux–Lamy syndrome, Sly syndrome *Glycoproteinoses*: Galactosialidosis, α-mannosidosis, sialidosis *Mucolipidoses*: mucolipidosis II/III, mucolipidosis IV *Neuronal ceroid lipofuscinoses*: NCL 3, NCL 6, NCL 10	Late endosomes, lysosomes, cytoplasmic vesicles	Alteration of lysosomal pH, autophagy dysregulation, rupture of H^+/Ca^{2+} homeostasis, altered vesicle trafficking, dysregulation of signaling pathways, accumulation of polyubiquitinated proteins, reduced capacity of immune cells to produce cytokines and antibodies, neurodegeneration (gliosis, demyelination of white matter, astrocyte and microglial activation)	[173–181]
Cholesterol				
Cholesterol	*Sphingolipidoses*: Niemann–Pick type A and B *Mucopolysaccharidoses*: Hurler syndrome, Hunter syndrome, Sanfilippo syndrome, Maroteaux–Lamy syndrome *Glycoproteinoses*: α-mannosidosis	Late endosomes, lysosomes, cytoplasmic vesicles	Impaired vesicle trafficking, abnormal sequestration of materials, foam cells in cerebral blood vessels and liver	[171,174,176–178]

Subgroups (regular font) or classes (italic font) of disorders related to abnormal lipid storage or lipid metabolism are indicated in the lines with a gray background.

Two phospholipids, i.e., sphingomyelin and bis(monoacylglycero)phosphate (BMP), are identified as the secondary storage lipids in LSDs. Sphingomyelin is a primary storage material in Niemann–Pick type A and B, but cholesterol is the primary material in Niemann–Pick type C, while sphingomyelin is the secondary storage material. A moderate sphingomyelin increase already occurs in livers from 20-week-old fetuses with Niemann–Pick type C and remains at this level; sphingomyelin levels in the spleen are much more elevated, as compared to those in the liver [171]. Interestingly, the main organ of sphingomyelin accumulation in mice is the liver [182]. Currently, no secondary sphingomyelin storage has been identified in the brain from LSDs. Accumulation of BMP occurs in the liver and spleen of humans with all three types of Niemann–Pick disease, but it has not been identified in the brain [171,182]. BMP storage in the brain was described for humans with infantile neuronal-ceroid lipofuscinosis (CLN1 disease) [183] and for mouse models of other types of neuronal-ceroid lipofuscinosis, CLN6 and CLN10 diseases [184].

Secondary accumulation of GM2 and GM3 gangliosides is very often observed in diseases with progressive neurodegeneration. In the brains of healthy humans or wild-type mice, GM2 and GM3 constitute only 1–2% of total gangliosides in humans and even less in mice. In immunostaining, these gangliosides appear as punctate, granular structures in the cytoplasm, suggesting that they are sequestered in vesicles [171]. More precise analysis has shown that GM2 and GM3 are found in separate vesicle populations in the same cell; this may suggest that they are metabolized in separate cell compartments or are generated by independent processes [173]. GM2 and GM3 gangliosides accumulate in various neurons and glial cells in various LSDs, e.g., in Niemann–Pick type C [182]; mucopolysaccharidosis (MPS) types I, II, IIIA, IIIB, IIID, VI and VII [173]; and mucolipidosis type IV [185]. However, the presence of these gangliosides in non-nervous organs has been studied in only a few disorders. Elevated GM3 levels have been reported in the liver and spleen of patients with Niemann–Pick and Gaucher diseases [171].

Unesterified cholesterol is primary storage material in Niemann–Pick type C; this is in contrast to Niemann–Pick types A and B, where cholesterol is accumulated secondarily to sphingolipids. Sphingomyelin effectively inhibits the secretion of cholesterol from late endosomes and lysosomes, which results in secondary cholesterol storage. In turn, unesterified cholesterol affects sphingomyelin metabolism and regulates the trafficking of sphingolipids to other sites in the cell. Finally, perturbation in cholesterol homeostasis correlates with sphingolipid accumulation. Accumulation of unesterified cholesterol, which appears as storage-like granules inside cells, was also observed in MPS types I, II, IIIA and VI, as well as mucolipidosis types II and IV [171,185,186].

4.4. Consequences of Secondary Lipid Storage

Excessive accumulation of compounds in cellular compartments is often a pathological process. Storage of useless materials may be a result of a deficiency of catabolic enzymes, but disturbed catabolism may also occur without any apparent genetic defects leading to underlying enzyme deficiencies. When the primary storage in a particular disorder leads to secondary lipid accumulation, it may result in incorrect vesicular or protein trafficking, signal transduction and membrane disability.

Lysosomal storage of undegraded compounds leads to disturbed secretion of breakdown products from autolysosomes, consequently resulting in deficiency of precursors for cellular biosynthetic pathways. Lipid turnover has fundamental importance in maintaining membrane permeability to secure cell homeostasis. Cholesterol and glycosphingolipids are the main components of lipid rafts, which play an important role in determining membrane plasticity. Keeping the membrane plasticity is fundamental for fusion between autophagosomes and lysosomes in lipophagy. This process has an influence on lipid turnover that, as a consequence, may affect the properties and functionality of the membranes of other organelles. An example is a robust loss of mitochondrial membrane potential in multiple sulfatase deficiency (MSD) cells after starvation [110]. Abnormal lipid metabolism leads to disturbances in trafficking of synthesized proteins and lipids to their target destinations in the cell. Induction of cholesterol accumulation in Niemann–Pick type C cells perturbs the intra-endosomal trafficking [187]. Cholesterol stores correlate with primary or secondary storage of glycosphingolipids, suggesting that a molecular linkage between the sequestration of these two lipid classes may exist. In turn, blockade of GM2 and GM3 ganglioside synthesis results in an absence or dramatic reduction in free cholesterol in NPC1-deficient neurons [188]. Further investigations are needed to determine whether cholesterol sequestration depends on gangliosides or whether the storage of GM2 and GM3 gangliosides disturbs lipophagy.

Correct membrane function is also important for maintaining the physiology of the lysosome, the organelle crucial for lipophagy. Appropriate lysosomal pH and its regulation are essential for the activity of lysosomal acid hydrolases. Lipid dyshomeostasis may alter the functioning of lysosomal membrane proteins (e.g., V-ATPase), ion channels and catabolite exporters, affecting lysosomal physiology. Accumulation of primary storage molecules may inhibit catabolic pathways that are genetically unaffected, and, as a consequence, accumulation of that pathway's substrates as secondary storage materials begins. For example, primary storage of chondroitin sulfate (in mucopolysaccharidoses—Hurler disease, Hunter disease, Sanfilippo disease and Sly syndrome) or sphingomyelin (in Niemann–Pick types A and B) and cholesterol (in Niemann–Pick type C) inhibits several catabolic pathways of gangliosides and glycosphingolipids and causes secondary neuronal GM2 accumulation that triggers neurodegeneration [189,190]. Furthermore, primary sphingomyelin storage in Niemann–Pick type A and B results in strong inhibition of lysosomal cholesterol secretion [191], which may lead to its deficiency in internal circulation. Excessive material storage inside the lysosome affects lysosomal pH and impairs activities of various acid hydrolases, consequently leading to the accumulation of other components. Dysfunctional lysosomes also undergo defective fusion with autophagosomes, which has been observed in cell culture models or macrophages of MPS IIIA and MSD [110].

4.5. mTOR–TFEB Signaling Pathway and Dysregulation of Autophagy in Lipid Storage Diseases

Functional lysosomes are particularly important for autophagy; lysosomes, by fusing with autophagosomes, deliver digestion enzymes that are necessary for breaking down the stores. Disruption of the autophagy–lysosomal pathway affects the normal autophagic flux and leads to impaired cellular capacity to remove the stored materials. Deregulation of autophagy has been reported in many lysosomal storage diseases, including those characterized with lipids as the main storage material—lipid storage diseases.

Dysfunction of the autophagy–lysosomal pathway is indicated as the main pathogenic event associated with neurodegeneration in Gaucher disease. Impaired autophagosome maturation accompanied with downregulation of TFEB and reduction of lysosomal gene expression were found in neurons differentiated from induced pluripotent stem cells of Gaucher disease patients [93]. MTORC1 was shown to be hyperactivated by the accumulation of glycosphingolipids in Gaucher cells [192]. As a consequence, increased TFEB phosphorylation by mTORC1 lead to decreased TFEB stability in Gaucher cells. The authors proposed that glycosphingolipid accumulation in Gaucher disease leads to increased mTORC1 activity, which in turn results in increased TFEB phosphorylation. Phosphorylated TFEB is targeted for proteasomal degradation and downregulation of lysosomal functions is observed as a consequence. It was already shown that mTORC1 regulates lipid metabolism [193] by controlling lipophagy in response to the nutrition status of the cell [123]; this process is mediated by mTORC1 and TFEB. It is therefore possible that the lipid storage disrupts the proper signaling of autophagy/lipophagy pathways.

In Fabry disease, a disturbance of the autophagic pathway is observed in kidney cells, fibroblasts and lymphoblasts [86,194]. Interestingly, studies on female Fabry disease cases showed that mild symptoms correlate with normal autophagic flux, whilst severe symptoms correlate with abnormal autophagic flux with enlarged lysosomes [195]. Neuropathology and axonal neurodegeneration in a Fabry disease mouse model was shown to be associated with disruption of the autophagy–lysosome pathway [196]. Accumulation of intracellular globotriaosylsphingosine was found to cause increased autophagosome formation, loss of mTOR kinase activity and downregulation of Akt kinase activity in Fabry podocytes, suggesting that dysregulated autophagy in Fabry disease may result from deficient mTOR signaling, which possibly leads to podocyte damage [194]. Other studies on both Fabry and Gaucher disease blood mononuclear cells revealed that dysfunction of the mTOR pathway accompanies sphingolipid accumulation, but was shown to be partially improved by enzyme replacement therapy [197].

In mice, Niemann–Pick type C1 maturation of autophagosomes appears to be impaired due to defective amphisome formation caused by the failure in soluble N-ethylmaleimide-sensitive factor attachment receptor (SNARE) machinery [81]. Decreased cell viability, cholesterol accumulation and dysfunctional autophagic flux was characteristic for Niemann–Pick-type-C1-deficient human hepatic and neural cells [198]. Genetic correction of a disease-causing mutation rescued these defects and directly linked NPC1 protein function to impaired cholesterol metabolism and autophagy [198]. Recently, cholesterol was identified as an essential activator for the master growth regulator, mTORC1 kinase. Cholesterol promotes mTORC1 recruitment and activation at the lysosomal membrane, and a lysosomal transmembrane protein called SLC38A9 is required for this process [199].

Lysosomal cholesterol content was shown to regulate mTORC1 signaling in Niemann–Pick type C. ER–lysosome contacts enable cholesterol sensing by mTORC1, as was shown in a Niemann–Pick type C1 model [200]. Cholesterol trafficking mediated by oxysterol binding protein (OSBP), a protein responsible for cholesterol delivery across ER–lysosome contacts, results in constitutive mTORC1 activation in a Niemann–Pick type C model, while cells lacking OSBP show inhibition of mTORC1 recruitment by Rag GTPases as a result of impaired transport of cholesterol to lysosomes [200].

Increased autophagy was demonstrated in GM1-gangliosidosis mouse brains, and it was accompanied with enhanced Akt-mTOR and ERK signaling [201]. In this case, activation of autophagy was pointed to lead to mitochondrial dysfunction in the mouse brain as the mitochondria isolated from animals were morphologically abnormal and had a decreased membrane potential.

In mucolipidosis type IV, mTOR kinase directly targets and inactivates the transient receptor potential mucolipin 1 (TRPML1) channel, a lysosomal calcium channel, mutations of which cause this disease, thereby affecting functional autophagy [202]. Lysosomal calcium release through TRPML1 channel was shown to regulate autophagy by promoting TFEB dephosphorylation by calcineurin [203]. TRPML1 channel was also shown to regulate autophagosome biogenesis by a mechanism independent of TFEB. TRPML2 can act through activation of a signaling pathway of calcium/calmodulin-dependent protein kinase β (CaMKKβ) and AMP-activated protein kinase (AMPK), the induction of the Beclin1/VPS34 autophagic complex and the generation of phosphatidylinositol 3-phosphate (PI3P) [204].

Neuronal ceroid lipofuscinoses are also characterized with inhibition of autophagosome formation, reduction in autophagosomes and autophagic degradation, defects in autophagosome maturation, accumulation of autophagosomes and autophagic cargo [117,118,205]. Mechanisms involved in the autophagy deregulation include upregulation of mTOR signaling [120], intracellular calcium homeostasis and CLN3 protein (also named battenin) function [206].

Other lysosomal lipid storage diseases are also associated with observations of impaired autophagy, but only a limited number of studies have been performed to elucidate the mechanism. For example, autophagy was shown to be defective in Tay–Sachs disease due to either a reduction in the number of autophagosomes produced or the amount of autophagic flux; studies on pyrimethamine, a known pharmacological chaperone of β-hexosaminidase A, showed that the mechanism of action of pyrimethamine in reversing the defective lysosomal phenotype was by improving autophagy [207]. Autophagy dysregulation is also observed in Krabbe disease; expression of some fundamental autophagy markers (LC3, p62 and Beclin-1) was elevated in the brain and sciatic nerve of a murine model of the disease [90]. Treatment with rapamycin, an autophagy inducer, was shown to restore autophagy in vitro [90].

5. Conclusions

To sum up, despite the major progress in our understanding of how the different pathways—lipophagy and lipolysis—communicate with each other, how they contribute—separately and collectively—to cytosolic degradation of lipids, how they affect the human pathophysiology and pathogenesis of lipid storage and lipid metabolism diseases, many questions remain unanswered. Over the past few years, an increasing body of research in this subject—as we reported in this review—has radically refilled our knowledge. However, the structural and functional depiction of the lipophagic and lipolytic machinery is still incomplete. Thus, the functional link between lipophagy and lipolysis and their cross-talk in the regulation of lipid metabolism to prevent and treat lipid accumulation and lipotoxicity requires further interrogation.

Author Contributions: Conceptualization, J.J.-B.; writing—original draft preparation, A.K., J.J.-B., M.G.-C., M.M. and M.W.; writing—review and editing, A.K., J.J.-B., M.G.-C., M.M. and M.W.; visualization, M.M. All authors have read and agreed to the published version of the manuscript.

Funding: This work was supported by the Faculty of Biology of the University of Gdańsk (task grant no. 531-D130-D693/20) and the Institute of Biochemistry and Biophysics of the Polish Academy of Sciences (task grant no. T-32.1).

Conflicts of Interest: The authors declare no conflict of interest.

Abbreviations

ADRP	adipophilin
AFSM	autofluorescent storage material
AMPK	AMP-activated protein kinase
ARF	ADP-ribosylation factor
ATGL	adipose triglyceride lipase
BMP	bis(monoacylglycero)phosphate
CaMKKβ	calcium/calmodulin-dependent protein kinase β
CAV	caveolin
CGI-58	comparative gene identification-58
CLEAR	coordinated lysosomal expression and regulation
CLN	ceroid lipofuscinosis, neuronal
CMA	chaperon-mediated autophagy
CoA	coenzyme A
COP	coat protein
CREB	cAMP response element-binding
CTX	cerebrotendinous xanthomatosis
DAG	diacylglycerols
ER	endoplasmic reticulum
ERK2	extracellular signal-regulated kinase 2
ES	sterol ester
FA	fatty acid
FAD	flavin-adenine dinucleotide
FGE	formylglycine-generating enzyme
FOXO	forkhead homeobox type O
FXR	farnesoid X receptor
Hsc70	heat shock cognate 70
HSL	hormone-sensitive lipase
LAL	lysosomal acid lipase
LAMP2A	lysosome-associated membrane protein 2A
LC3	light chain 3
LD	lipid droplet
LDL	low-density lipoprotein
LINCL	late infantile neuronal ceroid lipofuscinosis
LIR	LC3 interaction region
LSD	lysosomal storage disease
MAG	monoacylglycerol
MAPK	mitogen-activated protein kinase
MGL	monoglyceride lipase
MiT	microphthalmia
MLD	metachromatic leukodystrophy
MPS	mucopolysaccharidosis
MSD	multiple sulfatase deficiency
mTOR	mechanistic target of rapamycin
mTORC1	mechanistic target of rapamycin complex 1
NCoR1	nuclear receptor co-repressor 1
OSBP	oxysterol binding protein
PAT	PLIN/ADRP/TIP47

PC	phosphatidylcholine
PE	phosphatidylethanolamine
PGC1α	peroxisome proliferator-activated receptor gamma coactivator 1α
PI	phosphatidylinositol
PI3K	phosphatidylinositol 3-kinase
PI3P	phosphatidylinositol 3-phosphate
PKA	protein kinase A
PLIN	perilipin
PPAR1α	peroxisome proliferator activated receptor 1α
PPARGC1α	peroxisome proliferator-activated receptor gamma coactivator 1α
PS	phosphatidylserine
RBC	red blood cell
SM	sphingomyelin
SNARE	soluble N-ethylmaleimide-sensitive factor attachment receptor
TAG	triacylglycerol
TFE3	transcription factor E3
TFEB	transcription factor EB
TIP47	tail-interacting protein of 47 kDa
TRPML1	mucolipin transient receptor potential 1
VLDL	very low-density lipoprotein

References

1. De Weer, P. A Century of Thinking about Cell Membranes. *Annu. Rev. Physiol.* **2000**, *62*, 919–926. [CrossRef]
2. Maxfield, F.R.; Tabas, I. Role of cholesterol and lipid organization in disease. *Nature* **2005**, *438*, 612–621. [CrossRef]
3. Lee, A.G. How lipids affect the activities of integral membrane proteins. *Biochim. Biophys. Acta BBA Biomembr.* **2004**, *1666*, 62–87. [CrossRef]
4. Holowka, D.; Gosse, J.A.; Hammond, A.T.; Han, X.; Sengupta, P.; Smith, N.L.; Wagenknecht-Wiesner, A.; Wu, M.; Young, R.M.; Baird, B. Lipid segregation and IgE receptor signaling: A decade of progress. *Biochim. Biophys. Acta BBA Bioenerg.* **2005**, *1746*, 252–259. [CrossRef]
5. Tabas, I. Consequences of cellular cholesterol accumulation: Basic concepts and physiological implications. *J. Clin. Investig.* **2002**, *110*, 905–911. [CrossRef]
6. Gibbons, G.F.; Islam, K.; Pease, R.J. Mobilisation of triacylglycerol stores. *Biochim. Biophys. Acta BBA Mol. Cell Biol. Lipids* **2000**, *1483*, 37–57. [CrossRef]
7. Fahy, E.; Cotter, D.; Sud, M.; Subramaniam, S. Lipid classification, structures and tools. *Biochim. Biophys. Acta BBA Mol. Cell Biol. Lipids* **2011**, *1811*, 637–647. [CrossRef]
8. Watkins, P.A. Fatty acids: Metabolism. In *Encyclopedia of Human Nutrition*; Caballero, B., Third, E., Eds.; Elsevier: Waltham, MA, USA, 2013; pp. 220–230. ISBN 978-0-12-384885-7.
9. Olzmann, J.A.; Carvalho, P. Dynamics and functions of lipid droplets. *Nat. Rev. Mol. Cell Biol.* **2019**, *20*, 137–155. [CrossRef]
10. Cortés, V.A. Physiological and pathological implications of cholesterol. *Front. Biosci.* **2014**, *19*, 416. [CrossRef]
11. Schmid, K.E.; Woollett, L.A. Differential effects of polyunsaturated fatty acids on sterol synthesis rates in adult and fetal tissues of the hamster: Consequence of altered sterol balance. *Am. J. Physiol. Liver Physiol.* **2003**, *285*, G796–G803. [CrossRef]
12. Yao, L.; Jenkins, K.; Horn, P.S.; Lichtenberg, M.H.; Woollett, L.A. Inability to fully suppress sterol synthesis rates with exogenous sterol in embryonic and extraembyronic fetal tissues. *Biochim. Biophys. Acta BBA Mol. Cell Biol. Lipids* **2007**, *1771*, 1372–1379. [CrossRef] [PubMed]
13. Shen, W.-J.; Azhar, S.; Kraemer, F.B. Lipid droplets and steroidogenic cells. *Exp. Cell Res.* **2016**, *340*, 209–214. [CrossRef] [PubMed]
14. Singh, P.; Saxena, R.; Srinivas, G.; Pande, G.; Chattopadhyay, A. Cholesterol Biosynthesis and Homeostasis in Regulation of the Cell Cycle. *PLoS ONE* **2013**, *8*, e58833. [CrossRef]
15. Lasunción, M.A.; Martín-Sánchez, C.; Canfrán-Duque, A.; Busto, R. Post-lanosterol biosynthesis of cholesterol and cancer. *Curr. Opin. Pharmacol.* **2012**, *12*, 717–723. [CrossRef]

16. Fernandez, C.; Lobo, M.D.V.T.; Gómez-Coronado, D.; Lasunción, M.A. Cholesterol is essential for mitosis progression and its deficiency induces polyploid cell formation. *Exp. Cell Res.* **2004**, *300*, 109–120. [CrossRef]
17. Jackson, C.L. Lipid droplet biogenesis. *Curr. Opin. Cell Biol.* **2019**, *59*, 88–96. [CrossRef]
18. Welte, M.A. Expanding roles for lipid droplets. *Curr. Biol.* **2015**, *25*, R470–R481. [CrossRef]
19. Zehmer, J.K.; Huang, Y.; Peng, G.; Pu, J.; Anderson, R.G.W.; Liu, P. A role for lipid droplets in inter-membrane lipid traffic. *Proteomics* **2009**, *9*, 914–921. [CrossRef]
20. Penno, A.; Hackenbroich, G.; Thiele, C. Phospholipids and lipid droplets. *Biochim. Biophys. Acta BBA Mol. Cell Biol. Lipids* **2013**, *1831*, 589–594. [CrossRef]
21. Bickel, P.E.; Tansey, J.T.; Welte, M.A. PAT proteins, an ancient family of lipid droplet proteins that regulate cellular lipid stores. *Biochim. Biophys. Acta BBA Mol. Cell Biol. Lipids* **2009**, *1791*, 419–440. [CrossRef]
22. Tansey, J.; Sztalryd, C.; Hlavin, E.; Kimmel, A.; Londos, C. The Central Role of Perilipin A in Lipid Metabolism and Adipocyte Lipolysis. *IUBMB Life Int. Union Biochem. Mol. Biol. Life* **2004**, *56*, 379–385. [CrossRef] [PubMed]
23. Welte, M.A. Proteins under new management: Lipid droplets deliver. *Trends Cell Biol.* **2007**, *17*, 363–369. [CrossRef]
24. Martin, S. Caveolae, lipid droplets, and adipose tissue biology: Pathophysiological aspects. *Horm. Mol. Biol. Clin. Investig.* **2013**, *15*, 11–18. [CrossRef]
25. Yang, H.; Galea, A.; Sytnyk, V.; Crossley, M. Controlling the size of lipid droplets: Lipid and protein factors. *Curr. Opin. Cell Biol.* **2012**, *24*, 509–516. [CrossRef]
26. Ducharme, N.A.; Bickel, P.E. Minireview: Lipid Droplets in Lipogenesis and Lipolysis. *Endocrinology* **2008**, *149*, 942–949. [CrossRef]
27. Zechner, R.; Zimmermann, R.; Eichmann, T.O.; Kohlwein, S.D.; Haemmerle, G.; Lass, A.; Madeo, F. FAT SIGNALS—Lipases and Lipolysis in Lipid Metabolism and Signaling. *Cell Metab.* **2012**, *15*, 279–291. [CrossRef]
28. Osuga, J.-I.; Ishibashi, S.; Oka, T.; Yagyu, H.; Tozawa, R.; Fujimoto, A.; Shionoiri, F.; Yahagi, N.; Kraemer, F.B.; Tsutsumi, O.; et al. Targeted disruption of hormone-sensitive lipase results in male sterility and adipocyte hypertrophy, but not in obesity. *Proc. Natl. Acad. Sci. USA* **2000**, *97*, 787–792. [CrossRef]
29. Casado, M.E.; Pastor, Ó.; García-Seisdedos, D.; Huerta, L.; Kraemer, F.B.; Lasunción, M.A.; Martin, A.; Busto, R. Hormone-sensitive lipase deficiency disturbs lipid composition of plasma membrane microdomains from mouse testis. *Biochim. Biophys. Acta BBA Mol. Cell Biol. Lipids* **2016**, *1861*, 1142–1150. [CrossRef]
30. Haemmerle, G.; Zimmermann, R.; Hayn, M.; Theussl, C.; Waeg, G.; Wagner, E.; Sattler, W.; Magin, T.M.; Wagner, E.F.; Zechner, R. Hormone-sensitive Lipase Deficiency in Mice Causes Diglyceride Accumulation in Adipose Tissue, Muscle, and Testis. *J. Biol. Chem.* **2001**, *277*, 4806–4815. [CrossRef]
31. Taschler, U.; Radner, F.P.; Heier, C.; Schreiber, R.; Schweiger, M.; Schoiswohl, G.; Preiss-Landl, K.; Jaeger, D.; Reiter, B.; Köfeler, H.; et al. Monoglyceride Lipase Deficiency in Mice Impairs Lipolysis and Attenuates Diet-induced Insulin Resistance. *J. Biol. Chem.* **2011**, *286*, 17467–17477. [CrossRef]
32. D'Andrea, S. Lipid droplet mobilization: The different ways to loosen the purse strings. *Biochimie* **2016**, *120*, 17–27. [CrossRef] [PubMed]
33. Oral, O.; Akkoc, Y.; Bayraktar, O.; Gozuacik, D. Physiological and pathological significance of the molecular cross-talk between autophagy and apoptosis. *Histol. Histopathol.* **2015**, *31*, 479–498. [PubMed]
34. Rabinowitz, J.D.; White, E. Autophagy and Metabolism. *Science* **2010**, *330*, 1344–1348. [CrossRef] [PubMed]
35. Kaushik, S.; Cuervo, A.M. The coming of age of chaperone-mediated autophagy. *Nat. Rev. Mol. Cell Biol.* **2018**, *19*, 365–381. [CrossRef] [PubMed]
36. Singh, R.; Kaushik, S.; Wang, Y.; Xiang, Y.; Novak, I.; Komatsu, M.; Tanaka, K.; Cuervo, A.M.; Czaja, M.J. Autophagy regulates lipid metabolism. *Nature* **2009**, *458*, 1131–1135. [CrossRef] [PubMed]
37. Martinez-Lopez, N.; García-Macia, M.; Sahu, S.; Athonvarangkul, D.; Liebling, E.; Merlo, P.; Cecconi, F.; Schwartz, G.J.; Singh, R. Autophagy in the CNS and Periphery Coordinate Lipophagy and Lipolysis in the Brown Adipose Tissue and Liver. *Cell Metab.* **2016**, *23*, 113–127. [CrossRef] [PubMed]
38. Kiss, R.S.; Nilsson, T. Rab proteins implicated in lipid storage and mobilization. *J. Biomed. Res.* **2014**, *28*, 169–177. [CrossRef]
39. Carmona-Gutierrez, D.; Zimmermann, A.; Madeo, F. A molecular mechanism for lipophagy regulation in the liver. *Hepatology* **2015**, *61*, 1781–1783. [CrossRef]

40. Li, Z.; Schulze, R.J.; Weller, S.G.; Krueger, E.W.; Schott, M.B.; Zhang, X.; Casey, C.A.; Liu, J.; Stöckli, J.; James, D.E.; et al. A novel Rab10-EHBP1-EHD2 complex essential for the autophagic engulfment of lipid droplets. *Sci. Adv.* **2016**, *2*, e1601470. [CrossRef]

41. Kaushik, S.; Cuervo, A.M. Degradation of lipid droplet-associated proteins by chaperone-mediated autophagy facilitates lipolysis. *Nat. Cell Biol.* **2015**, *17*, 759–770. [CrossRef]

42. Kaushik, S.; Cuervo, A.M. AMPK-dependent phosphorylation of lipid droplet protein PLIN2 triggers its degradation by CMA. *Autophagy* **2016**, *12*, 432–438. [CrossRef] [PubMed]

43. Berg, J.M.; Tymoczko, J.L.; Stryer, L. Triacylglycerols Are Highly Concentrated Energy Stores. In *Biochemistry*; W.H. Freeman: New York, NY, USA, 2002; ISBN 0-7167-3051-0.

44. Adeva, M.M.; Carneiro-Freire, N.; Seco-Filgucira, M.; Fernández-Fernández, C.; Mouriño-Bayolo, D. Mitochondrial β-oxidation of saturated fatty acids in humans. *Mitochondrion* **2019**, *46*, 73–90. [CrossRef] [PubMed]

45. Rinaldo, P.; Matern, D.; Bennett, M.J. Fatty Acid Oxidation Disorders. *Annu. Rev. Physiol.* **2002**, *64*, 477–502. [CrossRef] [PubMed]

46. Janssen, U.; Stoffel, W. Disruption of Mitochondrial β-Oxidation of Unsaturated Fatty Acids in the 3,2- trans -Enoyl-CoA Isomerase-deficient Mouse. *J. Biol. Chem.* **2002**, *277*, 19579–19584. [CrossRef]

47. Berg, J.M.; Tymoczko, J.L.; Stryer, L. Certain Fatty Acids Require Additional Steps for Degradation. In *Biochemistry*; W.H. Freeman: New York, NY, USA, 2002; ISBN 0-7167-3051-0.

48. Napolitano, G.; Ballabio, A. TFEB at a glance. *J. Cell Sci.* **2016**, *129*, 2475–2481. [CrossRef]

49. Palmieri, M.; Impey, S.; Pelz, C.; Kang, H.; Di Ronza, A.; Sardiello, M.; Ballabio, A. Characterization of the CLEAR network reveals an integrated control of cellular clearance pathways. *Hum. Mol. Genet.* **2011**, *20*, 3852–3866. [CrossRef]

50. Kim, Y.C.; Guan, K.-L. mTOR: A pharmacologic target for autophagy regulation. *J. Clin. Investig.* **2015**, *125*, 25–32. [CrossRef]

51. Settembre, C.; Fraldi, A.; Medina, D.L.; Ballabio, A. Signals from the lysosome: A control centre for cellular clearance and energy metabolism. *Nat. Rev. Mol. Cell Biol.* **2013**, *14*, 283–296. [CrossRef]

52. Settembre, C.; De Cegli, R.; Mansueto, G.; Saha, P.K.; Vetrini, F.; Visvikis, O.; Huynh, T.; Carissimo, A.; Palmer, N.; Klisch, T.J.; et al. TFEB controls cellular lipid metabolism through a starvation-induced autoregulatory loop. *Nat. Cell Biol.* **2013**, *15*, 647–658. [CrossRef]

53. O'Rourke, E.J.; Ruvkun, G. MXL-3 and HLH-30 transcriptionally link lipolysis and autophagy to nutrient availability. *Nat. Cell Biol.* **2013**, *15*, 668–676. [CrossRef]

54. Finck, B.N.; Kelly, D.P. PGC-1 coactivators: Inducible regulators of energy metabolism in health and disease. *J. Clin. Investig.* **2006**, *116*, 615–622. [CrossRef] [PubMed]

55. Seok, S.; Fu, T.; Choi, S.-E.; Li, Y.; Zhu, R.; Kumar, S.; Sun, X.; Yoon, G.; Kang, Y.; Zhong, W.; et al. Transcriptional regulation of autophagy by an FXR–CREB axis. *Nature* **2014**, *516*, 108–111. [CrossRef] [PubMed]

56. Lettieri-Barbato, D.; Tatulli, G.; Aquilano, K.; Ciriolo, M.R. FoxO1 controls lysosomal acid lipase in adipocytes: Implication of lipophagy during nutrient restriction and metformin treatment. *Cell Death Dis.* **2013**, *4*, e861. [CrossRef] [PubMed]

57. Emanuel, R.; Sergin, I.; Bhattacharya, S.; Turner, J.N.; Epelman, S.; Settembre, C.; Diwan, A.; Ballabio, A.; Razani, B. Induction of Lysosomal Biogenesis in Atherosclerotic Macrophages Can Rescue Lipid-Induced Lysosomal Dysfunction and Downstream Sequelae. *Arter. Thromb. Vasc. Biol.* **2014**, *34*, 1942–1952. [CrossRef] [PubMed]

58. Xiong, J.; Wang, K.; He, J.; Zhang, G.; Zhang, D.; Chen, F. TFE3 Alleviates Hepatic Steatosis through Autophagy-Induced Lipophagy and PGC1α-Mediated Fatty Acid β-Oxidation. *Int. J. Mol. Sci.* **2016**, *17*, 387. [CrossRef]

59. Pastore, N.; Vainshtein, A.; Klisch, T.J.; Armani, A.; Huynh, T.; Herz, N.J.; Polishchuk, E.V.; Sandri, M.; Ballabio, A. TFE 3 regulates whole-body energy metabolism in cooperation with TFEB. *EMBO Mol. Med.* **2017**, *9*, 605–621. [CrossRef]

60. Saito, T.; Kuma, A.; Sugiura, Y.; Ichimura, Y.; Obata, M.; Kitamura, H.; Okuda, S.; Lee, H.-C.; Ikeda, K.; Kanegae, Y.; et al. Autophagy regulates lipid metabolism through selective turnover of NCoR1. *Nat. Commun.* **2019**, *10*, 1567. [CrossRef]

61. Gross, D.A.; Silver, D.L. Cytosolic lipid droplets: From mechanisms of fat storage to disease. *Crit. Rev. Biochem. Mol. Biol.* **2014**, *49*, 304–326. [CrossRef]

62. Tang, Q.-Q. Lipid metabolism and diseases. *Sci. Bull.* **2016**, *61*, 1471–1472. [CrossRef]

63. Smith, C.L.; Eppig, J.T. The mammalian phenotype ontology: Enabling robust annotation and comparative analysis. *Wiley Interdiscip. Rev. Syst. Biol. Med.* **2009**, *1*, 390–399. [CrossRef]

64. Samie, M.A.; Xu, H. Lysosomal exocytosis and lipid storage disorders. *J. Lipid Res.* **2014**, *55*, 995–1009. [CrossRef]

65. Schulze, H.; Sandhoff, K. Lysosomal Lipid Storage Diseases. *Cold Spring Harb. Perspect. Biol.* **2011**, *3*, a004804. [CrossRef] [PubMed]

66. Kolter, T.; Sandhoff, K. Lysosomal degradation of membrane lipids. *FEBS Lett.* **2010**, *584*, 1700–1712. [CrossRef] [PubMed]

67. Yu, L.; McPhee, C.K.; Zheng, L.; Mardones, G.A.; Rong, Y.; Peng, J.; Mi, N.; Zhao, Y.; Liu, Z.; Wan, F.; et al. Termination of autophagy and reformation of lysosomes regulated by mTOR. *Nature* **2010**, *465*, 942–946. [CrossRef] [PubMed]

68. Xu, H.; Ren, D. Lysosomal Physiology. *Annu. Rev. Physiol.* **2015**, *77*, 57–80. [CrossRef] [PubMed]

69. Rieger, D.; Auerbach, S.; Robinson, P.; Gropman, A.L. Neuroimaging of lipid storage disorders. *Dev. Disabil. Res. Rev.* **2013**, *17*, 269–282. [CrossRef]

70. Lipid Storage Diseases Information Page. National Institute of Neurological Disorders and Stroke. Available online: https://www.ninds.nih.gov/Disorders/All-Disorders/Lipid-Storage-Diseases-Information-Page (accessed on 22 June 2020).

71. Semenkovich, C.F.; Goldberg, A.C.; Goldberg, I.J. Disorders of Lipid Metabolism. In *Williams Textbook of Endocrinology*; Melmed, S., Polonsky, K.S., Larsen, P.R., Kronenberg, H.M., Eds.; Elsevier: Philadelphia, PA, USA, 2016; pp. 1660–1700.

72. Sullivan, D.R.; Lewis, B. A classification of lipoprotein disorders: Implications for clinical management. *Clin. Lipidol.* **2011**, *6*, 327–338. [CrossRef]

73. Patni, N.; Ahmad, Z.; Wilson, D.P. *Genetics and Dyslipidemia*, 2000th ed.; Feingold, K.R., Anawalt, B., Boyce, A., Chrousos, G., Dungan, K., Grossman, A., Hershman, J.M., Kaltsas, G., Koch, C., Kopp, P., et al., Eds.; MDText.com, Inc.: South Dartmouth, MA, USA, 2000.

74. Gabandé-Rodríguez, E.; Boya, P.; Labrador, V.; Dotti, C.G.; Ledesma, M.D. High sphingomyelin levels induce lysosomal damage and autophagy dysfunction in Niemann Pick disease type A. *Cell Death Differ.* **2014**, *21*, 864–875. [CrossRef]

75. Schuchman, E.H.; Desnick, R.J. Types A and B Niemann-Pick disease. *Mol. Genet. Metab.* **2017**, *120*, 27–33. [CrossRef]

76. Pipalia, N.; Hao, M.; Mukherjee, S.; Maxfield, F.R. Sterol, Protein and Lipid Trafficking in Chinese Hamster Ovary Cells with Niemann-Pick Type C1 Defect. *Traffic* **2007**, *8*, 130–141. [CrossRef]

77. Seranova, E.; Connolly, K.J.; Zatyka, M.; Rosenstock, T.R.; Barrett, T.; Tuxworth, R.I.; Sarkar, S. Dysregulation of autophagy as a common mechanism in lysosomal storage diseases. *Essays Biochem.* **2017**, *61*, 733–749. [CrossRef] [PubMed]

78. Newton, J.; Milstien, S.; Spiegel, S. Niemann-Pick type C disease: The atypical sphingolipidosis. *Adv. Biol. Regul.* **2018**, *70*, 82–88. [CrossRef] [PubMed]

79. Hammond, N.; Munkacsi, A.B.; Sturley, S.L. The complexity of a monogenic neurodegenerative disease: More than two decades of therapeutic driven research into Niemann-Pick type C disease. *Biochim. Biophys. Acta BBA Mol. Cell Biol. Lipids* **2019**, *1864*, 1109–1123. [CrossRef] [PubMed]

80. Bräuer, A.U.; Kuhla, A.; Holzmann, C.; Wree, A.; Witt, M. Current Challenges in Understanding the Cellular and Molecular Mechanisms in Niemann-Pick Disease Type C1. *Int. J. Mol. Sci.* **2019**, *20*, 4392. [CrossRef] [PubMed]

81. Sarkar, S.; Carroll, B.; Buganim, Y.; Maetzel, R.; Ng, A.H.; Cassady, J.P.; Cohen, M.A.; Chakraborty, S.; Wang, H.; Spooner, E.; et al. Impaired autophagy in the lipid-storage disorder Niemann-Pick type C1 disease. *Cell Rep.* **2013**, *5*, 1302–1315. [CrossRef] [PubMed]

82. Elrick, M.J.; Yu, T.; Chung, C.; Lieberman, A.P. Impaired proteolysis underlies autophagic dysfunction in Niemann–Pick type C disease. *Hum. Mol. Genet.* **2012**, *21*, 4876–4887. [CrossRef] [PubMed]

83. Chan, B.; Adam, D.N. A Review of Fabry Disease. *Ski. Ther. Lett.* **2018**, *23*, 4–6.

84. Juchniewicz, P.; Kloska, A.; Tylki-Szymańska, A.; Jakóbkiewicz-Banecka, J.; Węgrzyn, G.; Moskot, M.; Gabig-Cimińska, M.; Piotrowska, E. Female Fabry disease patients and X-chromosome inactivation. *Gene* **2018**, *641*, 259–264. [CrossRef]

85. Cairns, T.; Müntze, J.; Gernert, J.; Spingler, L.; Nordbeck, P.; Wanner, C. Hot topics in Fabry disease. *Postgrad. Med. J.* **2018**, *94*, 709–713. [CrossRef]

86. Chévrier, M.; Brakch, N.; Lesueur, C.; Genty, D.; Ramdani, Y.; Moll, S.; Djavaheri-Mergny, M.; Brasse-Lagnel, C.; Barbey, F.; Bekri, S.; et al. Autophagosome maturation is impaired in Fabry disease. *Autophagy* **2010**, *6*, 589–599. [CrossRef]

87. Spassieva, S.; Bieberich, E. Lysosphingolipids and sphingolipidoses: Psychosine in Krabbe's disease. *J. Neurosci. Res.* **2016**, *94*, 974–981. [CrossRef] [PubMed]

88. Won, J.-S.; Singh, A.K.; Singh, I. Biochemical, cell biological, pathological, and therapeutic aspects of Krabbe's disease. *J. Neurosci. Res.* **2016**, *94*, 990–1006. [CrossRef]

89. Lin, D.-S.; Ho, C.-S.; Huang, Y.-W.; Wu, T.-Y.; Lee, T.-H.; Huang, Z.-D.; Wang, T.-J.; Yang, S.-J.; Chiang, M.-F. Impairment of Proteasome and Autophagy Underlying the Pathogenesis of Leukodystrophy. *Cells* **2020**, *9*, 1124. [CrossRef] [PubMed]

90. Del Grosso, A.; Angella, L.; Tonazzini, I.; Moscardini, A.; Giordano, N.; Caleo, M.; Rocchiccioli, S.; Cecchini, M. Dysregulated autophagy as a new aspect of the molecular pathogenesis of Krabbe disease. *Neurobiol. Dis.* **2019**, *129*, 195–207. [CrossRef] [PubMed]

91. Stirnemann, J.; Belmatoug, N.; Camou, F.; Serratrice, C.; Froissart, R.; Caillaud, C.; Levade, T.; Astudillo, L.; Serratrice, J.; Brassier, A.; et al. A Review of Gaucher Disease Pathophysiology, Clinical Presentation and Treatments. *Int. J. Mol. Sci.* **2017**, *18*, 441. [CrossRef]

92. Nguyen, Y.; Stirnemann, J.; Belmatoug, N. Gaucher disease: A review. *Rev. Med. Interne* **2019**, *40*, 313–322. [CrossRef] [PubMed]

93. Awad, O.; Sarkar, C.; Panicker, L.M.; Sgambato, J.A.; Lipinski, M.M.; Miller, D.; Zeng, X.; Feldman, R.A. Altered TFEB-mediated lysosomal biogenesis in Gaucher disease iPSC-derived neuronal cells. *Hum. Mol. Genet.* **2015**, *24*, 5775–5788. [CrossRef] [PubMed]

94. Tsuji, D. Molecular Pathogenesis and Therapeutic Approach of GM2 Gangliosidosis. *Yakugaku Zasshi J. Pharm. Soc. Jpn.* **2013**, *133*, 269–274. [CrossRef] [PubMed]

95. Cachón-González, M.B.; Zaccariotto, E.; Cox, T.M. Genetics and Therapies for GM2 Gangliosidosis. *Curr. Gene Ther.* **2018**, *18*, 68–89. [CrossRef]

96. Vitner, E.B.; Platt, F.M.; Futerman, A.H. Common and Uncommon Pathogenic Cascades in Lysosomal Storage Diseases. *J. Biol. Chem.* **2010**, *285*, 20423–20427. [CrossRef]

97. Xu, Y.-H.; Barnes, S.; Sun, Y.; Grabowski, G. Multi-system disorders of glycosphingolipid and ganglioside metabolism. *J. Lipid Res.* **2010**, *51*, 1643–1675. [CrossRef] [PubMed]

98. Li, S.-C.; Hama, Y.; Li, Y.-T. Ineraction of GM2 Activator Protein with Glycosphingolipids. *Adv. Exp. Med. Biol.* **2001**, *491*, 351–367.

99. Sandhoff, K. Neuronal sphingolipidoses: Membrane lipids and sphingolipid activator proteins regulate lysosomal sphingolipid catabolism. *Biochimie* **2016**, *130*, 146–151. [CrossRef] [PubMed]

100. Gieselmann, V.; Krägeloh-Mann, I. Metachromatic Leukodystrophy—An Update. *Neuropediatrics* **2010**, *41*, 1–6. [CrossRef] [PubMed]

101. Breiden, B.; Sandhoff, K. Lysosomal Glycosphingolipid Storage Diseases. *Annu. Rev. Biochem.* **2019**, *88*, 461–485. [CrossRef] [PubMed]

102. Mahmood, A.; Berry, J.; Wenger, D.A.; Escolar, M.; Sobeih, M.; Raymond, G.; Eichler, F. Metachromatic Leukodystrophy: A Case of Triplets with the Late Infantile Variant and a Systematic Review of the Literature. *J. Child Neurol.* **2010**, *25*, 572–580. [CrossRef]

103. Maegawa, G. Patil Developing therapeutic approaches for metachromatic leukodystrophy. *Drug Des. Devel. Ther.* **2013**, *7*, 729. [CrossRef]

104. Hendriksz, C.J.; Corry, P.C.; Wraith, J.E.; Besley, G.T.N.; Cooper, A.; Ferrie, C.D. Juvenile Sandhoff disease—Nine new cases and a review of the literature. *J. Inherit. Metab. Dis.* **2004**, *27*, 241–249. [CrossRef]

105. Kolodny, E.H. *Tay–Sachs Disease. Encyclopedia of Neuroscience*; Squire, L.R., Ed.; Academic Press: Cambridge, MA, USA, 2009; pp. 895–902. ISBN 9780080450469.

106. Tamboli, I.Y.; Hampel, H.; Tien, N.T.; Tolksdorf, K.; Breiden, B.; Mathews, P.M.; Saftig, P.; Sandhoff, K.; Walter, J. Sphingolipid Storage Affects Autophagic Metabolism of the Amyloid Precursor Protein and Promotes Aβ Generation. *J. Neurosci.* **2011**, *31*, 1837–1849. [CrossRef]

107. Keilani, S.; Lun, Y.; Stevens, A.C.; Williams, H.N.; Sjoberg, E.R.; Khanna, R.; Valenzano, K.J.; Checler, F.; Buxbaum, J.D.; Yanagisawa, K.; et al. Lysosomal Dysfunction in a Mouse Model of Sandhoff Disease Leads to Accumulation of Ganglioside-Bound Amyloid-Peptide. *J. Neurosci.* **2012**, *32*, 5223–5236. [CrossRef]

108. Annunziata, I.; Bouché, V.; Lombardi, A.; Settembre, C.; Ballabio, A. Multiple sulfatase deficiency is due to hypomorphic mutations of theSUMF1 gene. *Hum. Mutat.* **2007**, *28*, 928. [CrossRef] [PubMed]

109. Schlotawa, L.; Adang, L.; Radhakrishnan, K.; Ahrens-Nicklas, R.C. Multiple Sulfatase Deficiency: A Disease Comprising Mucopolysaccharidosis, Sphingolipidosis, and More Caused by a Defect in Posttranslational Modification. *Int. J. Mol. Sci.* **2020**, *21*, 3448. [CrossRef] [PubMed]

110. Settembre, C.; Fraldi, A.; Jahreiss, L.; Spampanato, C.; Venturi, C.; Medina, D.L.; De Pablo, R.; Tacchetti, C.; Rubinsztein, D.C.; Ballabio, A. A block of autophagy in lysosomal storage disorders. *Hum. Mol. Genet.* **2007**, *17*, 119–129. [CrossRef] [PubMed]

111. Brunetti-Pierri, N.; Scaglia, F. GM1 gangliosidosis: Review of clinical, molecular, and therapeutic aspects. *Mol. Genet. Metab.* **2008**, *94*, 391–396. [CrossRef] [PubMed]

112. Boland, B.; Smith, D.A.; Mooney, D.; Jung, S.S.; Walsh, D.M.; Platt, F.M. Macroautophagy Is Not Directly Involved in the Metabolism of Amyloid Precursor Protein. *J. Biol. Chem.* **2010**, *285*, 37415–37426. [CrossRef]

113. Schindler, D.; Desnick, R.J. Schindler Disease: Deficient α-N-acetylgalactosaminidase Activity. In *Rosenberg's Molecular and Genetic Basis of Neurological and Psychiatric Disease*; Rosenberg, R.N., Pascual, J.M., Eds.; Elsevier: Boston, MA, USA, 2015; pp. 431–439. ISBN 9780124105294.

114. Wu, T.T.; Hoff, D.S. Fish Oil Lipid Emulsion-Associated Sea-Blue Histiocyte Syndrome in a Pediatric Patient. *J. Pediatr. Pharmacol. Ther.* **2015**, *20*, 217–221.

115. Mirza, M.; Vainshtein, A.; DiRonza, A.; Chandrachud, U.; Haslett, L.J.; Palmieri, M.; Storch, S.; Groh, J.; Dobzinski, N.; Napolitano, G.; et al. The CLN3 gene and protein: What we know. *Mol. Genet. Genom. Med.* **2019**, *7*, e859. [CrossRef]

116. Mukherjee, A.B.; Appu, A.P.; Sadhukhan, T.; Casey, S.; Mondal, A.; Zhang, Z.; Bagh, M.B. Emerging new roles of the lysosome and neuronal ceroid lipofuscinoses. *Mol. Neurodegener.* **2019**, *14*, 4. [CrossRef]

117. Cao, Y.; Espinola, J.A.; Fossale, E.; Massey, A.C.; Cuervo, A.M.; Macdonald, M.E.; Cotman, S.L. Autophagy Is Disrupted in a Knock-in Mouse Model of Juvenile Neuronal Ceroid Lipofuscinosis. *J. Biol. Chem.* **2006**, *281*, 20483–20493. [CrossRef]

118. Lojewski, X.; Staropoli, J.F.; Biswas-Legrand, S.; Simas, A.M.; Haliw, L.; Selig, M.K.; Coppel, S.H.; Goss, K.A.; Petcherski, A.; Chandrachud, U.; et al. Human iPSC models of neuronal ceroid lipofuscinosis capture distinct effects of TPP1 and CLN3 mutations on the endocytic pathway. *Hum. Mol. Genet.* **2013**, *23*, 2005–2022. [CrossRef]

119. Nita, A.A.; Mole, S.E.; Minassian, B.A. Neuronal ceroid lipofuscinoses. *Epileptic Disord* **2016**, *18*, 73–88. [CrossRef] [PubMed]

120. Vidal-Donet, J.M.; Carcel-Trullols, J.; Casanova, B.; Aguado, C.; Knecht, E. Alterations in ROS Activity and Lysosomal pH Account for Distinct Patterns of Macroautophagy in LINCL and JNCL Fibroblasts. *PLoS ONE* **2013**, *8*, e55526. [CrossRef] [PubMed]

121. Aguisanda, F.; Thorne, N.; Zheng, W. Targeting Wolman Disease and Cholesteryl Ester Storage Disease: Disease Pathogenesis and Therapeutic Development. *Curr. Chem. Genom. Transl. Med.* **2017**, *11*, 1–18. [CrossRef] [PubMed]

122. Pericleous, M.; Kelly, C.; Wang, T.; Livingstone, C.; Ala, A. Wolman's disease and cholesteryl ester storage disorder: The phenotypic spectrum of lysosomal acid lipase deficiency. *Lancet Gastroenterol. Hepatol.* **2017**, *2*, 670–679. [CrossRef]

123. Settembre, C.; Ballabio, A. Lysosome: Regulator of lipid degradation pathways. *Trends Cell Biol.* **2014**, *24*, 743–750. [CrossRef]

124. Schulze, R.J.; Sathyanarayan, A.; Mashek, D.G. Breaking fat: The regulation and mechanisms of lipophagy. *Biochim. Biophys. Acta BBA Mol. Cell Biol. Lipids* **2017**, *1862*, 1178–1187. [CrossRef]

125. Ruivo, R.; Anne, C.; Sagné, C.; Gasnier, B. Molecular and cellular basis of lysosomal transmembrane protein dysfunction. *Biochim. Biophys. Acta BBA Bioenerg.* **2009**, *1793*, 636–649. [CrossRef]

126. Jezela-Stanek, A.; Ciara, E.; Stepien, K.M. Neuropathophysiology, Genetic Profile, and Clinical Manifestation of Mucolipidosis IV—A Review and Case Series. *Int. J. Mol. Sci.* **2020**, *21*, 4564. [CrossRef]

127. Venkatachalam, K.; Long, A.A.; Elsaesser, R.; Nikolaeva, D.; Broadie, K.; Montell, C. Motor Deficit in a Drosophila Model of Mucolipidosis Type IV due to Defective Clearance of Apoptotic Cells. *Cell* **2008**, *135*, 838–851. [CrossRef]

128. Venugopal, B.; Mesires, N.T.; Kennedy, J.C.; Laplante, J.M.; Dice, J.F.; Slaugenhaupt, S.A.; Curcio-Morelli, C. Chaperone-mediated autophagy is defective in mucolipidosis type IV. *J. Cell. Physiol.* **2009**, *219*, 344–353. [CrossRef]

129. Khan, A.; Sergi, C. Sialidosis: A Review of Morphology and Molecular Biology of a Rare Pediatric Disorder. *Diagnostics* **2018**, *8*, 29. [CrossRef] [PubMed]

130. Natori, Y.; Nasui, M.; Kihara-Negishi, F. Neu1 sialidase interacts with perilipin 1 on lipid droplets and inhibits lipolysis in 3T3-L1 adipocytes. *Genes Cells* **2017**, *22*, 485–492. [CrossRef] [PubMed]

131. Davaadorj, O.; Akatsuka, H.; Yamaguchi, Y.; Okada, C.; Ito, M.; Fukunishi, N.; Sekijima, Y.; Ohnota, H.; Kawai, K.; Suzuki, T.; et al. Impaired Autophagy in Retinal Pigment Epithelial Cells Induced from iPS Cells obtained from a Patient with Sialidosis. *Cell Dev. Biol.* **2017**, *6*, 1–7. [CrossRef]

132. Missaglia, S.; Coleman, R.A.; Mordente, A.; Tavian, D. Neutral Lipid Storage Diseases as Cellular Model to Study Lipid Droplet Function. *Cells* **2019**, *8*, 187. [CrossRef]

133. Massa, R.; Pozzessere, S.; Rastelli, E.; Serra, L.; Terracciano, C.; Gibellini, M.; Bozzali, M.; Arca, M. Neutral lipid-storage disease with myopathy and extended phenotype with novelPNPLA2mutation. *Muscle Nerve* **2016**, *53*, 644–648. [CrossRef]

134. Angelini, C.; Nascimbeni, A.C.; Cenacchi, G.; Tasca, E. Lipolysis and lipophagy in lipid storage myopathies. *Biochim. Biophys. Acta BBA Bioenerg.* **2016**, *1862*, 1367–1373. [CrossRef]

135. Yoneda, K. Inherited ichthyosis: Syndromic forms. *J. Dermatol.* **2016**, *43*, 252–263. [CrossRef]

136. Mogahed, E.A.; El-Hennawy, A.; El-Sayed, R.; El-Karaksy, H. Chanarin–Dorfman syndrome: A case report and review of the literature. *Arab. J. Gastroenterol.* **2015**, *16*, 142–144. [CrossRef]

137. Peng, Y.; Miao, H.; Wu, S.; Yang, W.; Zhang, Y.; Xie, G.; Xie, X.; Li, J.; Shi, C.; Ye, L.; et al. ABHD5 interacts with BECN1 to regulate autophagy and tumorigenesis of colon cancer independent of PNPLA2. *Autophagy* **2016**, *12*, 2167–2182. [CrossRef]

138. Nie, S.; Chen, G.; Cao, X.; Zhang, Y. Cerebrotendinous xanthomatosis: A comprehensive review of pathogenesis, clinical manifestations, diagnosis, and management. *Orphanet J. Rare Dis.* **2014**, *9*, 179. [CrossRef]

139. Salen, G.; Steiner, R. Epidemiology, diagnosis, and treatment of cerebrotendinous xanthomatosis (CTX). *J. Inherit. Metab. Dis.* **2017**, *40*, 771–781. [CrossRef] [PubMed]

140. Li, J.; Xu, E.; Mao, W.; Qiao, H.; Zhou, Y.; Yang, Q.; Liu, S.; Chan, P. Parkinsonism with Normal Dopaminergic Presynaptic Terminals in Cerebrotendinous Xanthomatosis. *Mov. Disord. Clin. Pr.* **2019**, *7*, 115–116. [CrossRef] [PubMed]

141. Liebeskind, A.; Wilson, D.P. Sitosterolemia in the Pediatric Population. In *Endotext [Internet]*; Feingold, K.R., Anawalt, B., Boyce, A., Chrousos, G., Dungan, K., Grossman, A., Hershman, J.M., Kaltsas, G., Koch, C., Kopp, P., et al., Eds.; MDText.com, Inc.: South Dartmouth, MA, USA, 2000.

142. Bao, L.; Li, Y.; Deng, S.-X.; Landry, D.; Tabas, I. Sitosterol-containing Lipoproteins Trigger Free Sterol-induced Caspase-independent Death in ACAT-competent Macrophages. *J. Biol. Chem.* **2006**, *281*, 33635–33649. [CrossRef] [PubMed]

143. Alves, M.Q.; Le Trionnaire, E.; Ribeiro, I.; Carpentier, S.; Harzer, K.; Levade, T.; Ribeiro, M.G. Molecular basis of acid ceramidase deficiency in a neonatal form of Farber disease: Identification of the first large deletion in ASAH1 gene. *Mol. Genet. Metab.* **2013**, *109*, 276–281. [CrossRef]

144. Wali, G.; Wali, G.M.; Sue, C.M.; Kumar, K. A Novel Homozygous Mutation in the FUCA1 Gene Highlighting Fucosidosis as a Cause of Dystonia: Case Report and Literature Review. *Neuropediatrics* **2019**, *50*, 248–252. [CrossRef]

145. Koopal, C.; Marais, A.D.; Westerink, J.; Visseren, F.L. Autosomal dominant familial dysbetalipoproteinemia: A pathophysiological framework and practical approach to diagnosis and therapy. *J. Clin. Lipidol.* **2017**, *11*, 12–23. [CrossRef]

146. Koopal, C.; Marais, A.D.; Visseren, F.L.J. Familial dysbetalipoproteinemia. *Curr. Opin. Endocrinol. Diabetes Obes.* **2017**, *24*, 133–139. [CrossRef]

147. Evans, D.; Beil, F.U. Genetic factors that modify the expression of type III hyperlipidemia in probands with apolipoprotein E ε2/2 genotype. *Future Lipidol.* **2009**, *4*, 137–140. [CrossRef]

148. Henneman, P.; Beer, F.V.D.S.-D.; Moghaddam, P.H.; Huijts, P.; Stalenhoef, A.F.; Kastelein, J.J.; Van Duijn, C.M.; Havekes, L.M.; Frants, R.R.; Van Dijk, K.W.; et al. The expression of type III hyperlipoproteinemia: Involvement of lipolysis genes. *Eur. J. Hum. Genet.* **2009**, *17*, 620–628. [CrossRef]

149. Hendricks-Sturrup, R.; Clark-LoCascio, J.; Lu, C.Y. A Global Review on the Utility of Genetic Testing for Familial Hypercholesterolemia. *J. Pers. Med.* **2020**, *10*, 23. [CrossRef]

150. Pang, J.; Sullivan, D.R.; Brett, T.; Kostner, K.M.; Hare, D.L.; Watts, G.F. Familial Hypercholesterolaemia in 2020: A Leading Tier 1 Genomic Application. *Hear. Lung Circ.* **2020**, *29*, 619–633. [CrossRef] [PubMed]

151. Suárez-Rivero, J.M.; De La Mata, M.; Pavón, A.D.; Villanueva-Paz, M.; Povea-Cabello, S.; Cotán, D.; Álvarez-Córdoba, M.; Villalón-García, I.; Ybot-Gonzalez, P.; Salas, J.J.; et al. Intracellular cholesterol accumulation and coenzyme Q10 deficiency in Familial Hypercholesterolemia. *Biochim. Biophys. Acta BBA Mol. Basis Dis.* **2018**, *1864*, 3697–3713. [CrossRef] [PubMed]

152. Andersen, L.H.; Miserez, A.R.; Ahmad, Z.; Andersen, R.L. Familial defective apolipoprotein B-100: A review. *J. Clin. Lipidol.* **2016**, *10*, 1297–1302. [CrossRef] [PubMed]

153. Sharma, V.; Forte, T.M.; Ryan, R.O. Influence of apolipoprotein A-V on the metabolic fate of triacylglycerol. *Curr. Opin. Lipidol.* **2013**, *24*, 153–159. [CrossRef] [PubMed]

154. Kwiterovich, P.O. Diagnosis and Management of Familial Dyslipoproteinemias. *Curr. Cardiol. Rep.* **2013**, *15*, 371. [CrossRef] [PubMed]

155. Chait, A.; Eckel, R.H. The Chylomicronemia Syndrome Is Most Often Multifactorial: A Narrative Review of Causes and Treatment. *Ann. Intern. Med.* **2019**, *170*, 626–634. [CrossRef] [PubMed]

156. Young, S.G.; Davies, B.S.J.; Voss, C.V.; Gin, P.; Weinstein, M.M.; Tontonoz, P.; Reue, K.; Bensadoun, A.; Fong, L.G.; Beigneux, A.P. GPIHBP1, an endothelial cell transporter for lipoprotein lipase. *J. Lipid Res.* **2011**, *52*, 1869–1884. [CrossRef]

157. Feingold, K.R. Triglyceride Lowering Drugs. In *Endotext [Internet]*; Feingold, K.R., Anawalt, B., Boyce, A., Chrousos, G., Dungan, K., Grossman, A., Hershman, J.M., Kaltsas, G., Koch, C., Kopp, P., et al., Eds.; MDText.com, Inc.: South Dartmouth, MA, USA, 2000.

158. Burnett, J.R.; Hooper, A.J.; Hegele, R.A. *Familial Lipoprotein Lipase Deficiency*; Adam, M., Ardinger, H., Pagon, R., Wallace, S., Bean, L., Stephens, K., Amemiya, A., Eds.; University of Washington: Seattle, WA, USA, 1993; ISBN 0444810781.

159. Wolska, A.; Dunbar, R.L.; Freeman, L.A.; Ueda, M.; Amar, M.J.; Sviridov, D.O.; Remaley, A.T. Apolipoprotein C-II: New findings related to genetics, biochemistry, and role in triglyceride metabolism. *Atherosclerosis* **2017**, *267*, 49–60. [CrossRef]

160. Wilson, C.; Oliva, C.P.; Maggi, F.; Catapano, A.L.; Calandra, S. Apolipoprotein C-II deficiency presenting as a lipid encephalopathy in infancy. *Ann. Neurol.* **2003**, *53*, 807–810. [CrossRef]

161. Desnick, R.J.; Guntinas-Lichius, O.; Padberg, G.W.; Schonfeld, G.; Lin, X.; Averna, M.; Yue, P.; Schnog, J.-J.B.; Gerdes, V.E.A.; Cutillas, P.R.; et al. Familial Lipoprotein Lipase Deficiency. In *Encyclopedia of Molecular Mechanisms of Disease*; Springer: Berlin/Heidelberg, Germany, 2009; p. 635. ISBN 0444810781.

162. Kobayashi, J.; Miyashita, K.; Nakajima, K.; Mabuchi, H. Hepatic Lipase: A Comprehensive View of its Role on Plasma Lipid and Lipoprotein Metabolism. *J. Atheroscler. Thromb.* **2015**, *22*, 1001–1011. [CrossRef]

163. Warden, B.A.; Fazio, S.; Shapiro, M.D. Familial Hypercholesterolemia: Genes and Beyond. In *Endotext [Internet]*; Feingold, K.R., Anawalt, B., Boyce, A., Chrousos, G., Dungan, K., Grossman, A., Hershman, J.M., Kaltsas, G., Koch, C., Kopp, P., et al., Eds.; MDText.com, Inc.: South Dartmouth, MA, USA, 2000.

164. Fellin, R.; Arca, M.; Zuliani, G.; Calandra, S.; Bertolini, S. The history of Autosomal Recessive Hypercholesterolemia (ARH). From clinical observations to gene identification. *Gene* **2015**, *555*, 23–32. [CrossRef] [PubMed]

165. Foody, J.M.; Vishwanath, R. Familial hypercholesterolemia/autosomal dominant hypercholesterolemia: Molecular defects, the LDL-C continuum, and gradients of phenotypic severity. *J. Clin. Lipidol.* **2016**, *10*, 970–986. [CrossRef]

166. Sun, H.; Krauss, R.M.; Chang, J.T.; Teng, B.-B. PCSK9 deficiency reduces atherosclerosis, apolipoprotein B secretion, and endothelial dysfunction. *J. Lipid Res.* **2018**, *59*, 207–223. [CrossRef] [PubMed]

167. Ballabio, A.; Gieselmann, V. Lysosomal disorders: From storage to cellular damage. *Biochim. Biophys. Acta BBA Bioenerg.* **2009**, *1793*, 684–696. [CrossRef] [PubMed]

168. Ward, C.; Martinez-Lopez, N.; Otten, E.G.; Carroll, B.; Maetzel, R.; Singh, R.; Sarkar, S.; Korolchuk, V.I. Autophagy, lipophagy and lysosomal lipid storage disorders. *Biochim. Biophys. Acta BBA Mol. Cell Biol. Lipids* **2016**, *1861*, 269–284. [CrossRef] [PubMed]

169. Lieberman, A.P.; Puertollano, R.; Raben, N.; Slaugenhaupt, S.; Walkley, S.U.; Ballabio, A. Autophagy in lysosomal storage disorders. *Autophagy* **2012**, *8*, 719–730. [CrossRef]

170. Patterson, M.C.; Vanier, M.T.; Suzuki, K.K.K.; Morris, J.A.; Carstea, E.; Neufeld, E.B.; Blanchette-Mackie, E.J.; Pentchev, P.G.; Blanchette-Mackie, J.E.; Pentchev, P.G. Niemann-Pick Disease Type C: A Lipid Trafficking Disorder. In *The Online Metabolic & Molecular Bases of Inherited Disease*; Valle, D.L., Antonarakis, S., Ballabio, A., Beaudet, A.L., Mitchell, G.A., Eds.; McGraw-Hill: New York, NY, USA, 2004; pp. 1–44. ISBN 9780071459969.

171. Walkley, S.U.; Vanier, M.T. Secondary lipid accumulation in lysosomal disease. *Biochim. Biophys. Acta BBA Bioenerg.* **2009**, *1793*, 726–736. [CrossRef]

172. Akgoc, Z.; Sena-Esteves, M.; Martin, U.R.; Han, X.; D'Azzo, A.; Seyfried, T.N. Bis (monoacylglycero) phosphate: A secondary storage lipid in the gangliosidoses. *J. Lipid Res.* **2015**, *56*, 1006–1013. [CrossRef]

173. Walkley, S.U. Secondary accumulation of gangliosides in lysosomal storage disorders. *Semin. Cell Dev. Biol.* **2004**, *15*, 433–444. [CrossRef]

174. Schuchman, E.H.; Wasserstein, M.P. Types A and B Niemann-Pick disease. *Best Pract. Res. Clin. Endocrinol. Metab.* **2015**, *29*, 237–247. [CrossRef]

175. Hulkova, H.; Cervenková, M.; Ledvinova, J.; Tocháčková, M.; Hrebícek, M.; Poupětová, H.; Befekadu, A.; Berná, L.; Paton, B.; Harzer, K.; et al. A novel mutation in the coding region of the prosaposin gene leads to a complete deficiency of prosaposin and saposins, and is associated with a complex sphingolipidosis dominated by lactosylceramide accumulation. *Hum. Mol. Genet.* **2001**, *10*, 927–940. [CrossRef] [PubMed]

176. Campos, D.; Monaga, M. Mucopolysaccharidosis type I: Current knowledge on its pathophysiological mechanisms. *Metab. Brain Dis.* **2012**, *27*, 121–129. [CrossRef] [PubMed]

177. Fecarotta, S.; Tarallo, A.; Damiano, C.; Minopoli, N.; Parenti, G. Pathogenesis of Mucopolysaccharidoses, an Update. *Int. J. Mol. Sci.* **2020**, *21*, 2515. [CrossRef] [PubMed]

178. Tessitore, A.; Pirozzi, M.; Auricchio, A. Abnormal autophagy, ubiquitination, inflammation and apoptosis are dependent upon lysosomal storage and are useful biomarkers of mucopolysaccharidosis VI. *Pathogenetics* **2009**, *2*, 4. [CrossRef] [PubMed]

179. Annunziata, I.; D'Azzo, A. Galactosialidosis: Historic aspects and overview of investigated and emerging treatment options. *Expert Opin. Orphan Drugs* **2016**, *5*, 131–141. [CrossRef] [PubMed]

180. Seyrantepe, V.; Poupětová, H.; Froissart, R.; Pshezhetsky, A.V. Molecular pathology of NEU1 gene in sialidosis. *Hum. Mutat.* **2003**, *22*, 343–352. [CrossRef]

181. Kyttälä, A.; Lahtinen, U.; Braulke, T.; Hofmann, S.L. Functional biology of the neuronal ceroid lipofuscinoses (NCL) proteins. *Biochim. Biophys. Acta BBA Mol. Basis Dis.* **2006**, *1762*, 920–933. [CrossRef]

182. Sleat, D.E.; Wiseman, J.A.; El-Banna, M.; Price, S.M.; Verot, L.; Shen, M.M.; Tint, G.S.; Vanier, M.T.; Walkley, S.U.; Lobel, P. Genetic evidence for nonredundant functional cooperativity between NPC1 and NPC2 in lipid transport. *Proc. Natl. Acad. Sci. USA* **2004**, *101*, 5886–5891. [CrossRef]

183. Käkelä, R.; Somerharju, P.; Tyynelä, J. Analysis of phospholipid molecular species in brains from patients with infantile and juvenile neuronal-ceroid lipofuscinosis using liquid chromatography-electrospray ionization mass spectrometry. *J. Neurochem.* **2003**, *84*, 1051–1065. [CrossRef]

184. Jabs, S.; Quitsch, A.; Kkel, R.; Koch, B.; Tyynel, J.; Brade, H.; Glatzel, M.; Walkley, S.; Saftig, P.; Vanier, M.T.; et al. Accumulation of bis(monoacylglycero)phosphate and gangliosides in mouse models of neuronal ceroid lipofuscinosis. *J. Neurochem.* **2008**, *106*, 1415–1425. [CrossRef]

185. Micsenyi, M.C.; Dobrenis, K.; Stephney, G.; Pickel, J.; Vanier, M.T.; Slaugenhaupt, S.A.; Walkley, S.U. Neuropathology of the Mcoln1−/− Knockout Mouse Model of Mucolipidosis Type IV. *J. Neuropathol. Exp. Neurol.* **2009**, *68*, 125–135. [CrossRef] [PubMed]

186. Otomo, T.; Higaki, K.; Nanba, E.; Ozono, K.; Sakai, N. Lysosomal Storage Causes Cellular Dysfunction in Mucolipidosis II Skin Fibroblasts. *J. Biol. Chem.* **2011**, *286*, 35283–35290. [CrossRef] [PubMed]

187. Sobo, K.; Le Blanc, I.; Luyet, P.-P.; Fivaz, M.; Ferguson, C.; Parton, R.G.; Gruenberg, J.; Van Der Goot, F.G. Late Endosomal Cholesterol Accumulation Leads to Impaired Intra-Endosomal Trafficking. *PLoS ONE* **2007**, *2*, e851. [CrossRef]

188. Gondré-Lewis, M.C.; McGlynn, R.; Walkley, S.U. Cholesterol accumulation in NPC1-deficient neurons is ganglioside dependent. *Curr. Biol.* **2003**, *13*, 1324–1329. [CrossRef]

189. Anheuser, S.; Breiden, B.; Sandhoff, K. Ganglioside GM2 catabolism is inhibited by storage compounds of mucopolysaccharidoses and by cationic amphiphilic drugs. *Mol. Genet. Metab.* **2019**, *128*, 75–83. [CrossRef] [PubMed]

190. Anheuser, S.; Breiden, B.; Sandhoff, K. Membrane lipids and their degradation compounds control GM2 catabolism at intralysosomal luminal vesicles. *J. Lipid Res.* **2019**, *60*, 1099–1111. [CrossRef]

191. Oninla, V.O.; Breiden, B.; Babalola, J.O.; Sandhoff, K. Acid sphingomyelinase activity is regulated by membrane lipids and facilitates cholesterol transfer by NPC2. *J. Lipid Res.* **2014**, *55*, 2606–2619. [CrossRef]

192. Brown, R.A.; Voit, A.; Srikanth, M.P.; Thayer, J.A.; Kingsbury, T.J.; Jacobson, M.A.; Lipinski, M.M.; Feldman, R.A.; Awad, O. mTOR hyperactivity mediates lysosomal dysfunction in Gaucher's disease iPSC-neuronal cells. *Dis. Model. Mech.* **2019**, *12*, dmm038596. [CrossRef]

193. Ricoult, S.J.H.; Manning, B.D. The multifaceted role of mTORC1 in the control of lipid metabolism. *EMBO Rep.* **2013**, *14*, 242–251. [CrossRef]

194. Liebau, M.C.; Braun, F.; Höpker, K.; Weitbrecht, C.; Bartels, V.; Müller, R.-U.; Brodesser, S.; Saleem, M.A.; Benzing, T.; Schermer, B.; et al. Dysregulated Autophagy Contributes to Podocyte Damage in Fabry's Disease. *PLoS ONE* **2013**, *8*, e63506. [CrossRef]

195. Yanagisawa, H.; Hossain, M.A.; Miyajima, T.; Nagao, K.; Miyashita, T.; Eto, Y. Dysregulated DNA methylation of GLA gene was associated with dysfunction of autophagy. *Mol. Genet. Metab.* **2019**, *126*, 460–465. [CrossRef] [PubMed]

196. Nelson, M.P.; Tse, T.E.; O'Quinn, D.B.; Percival, S.M.; Jaimes, E.A.; Warnock, D.G.; Shacka, J.J. Autophagy-lysosome pathway associated neuropathology and axonal degeneration in the brains of alpha-galactosidase A-deficient mice. *Acta Neuropathol. Commun.* **2014**, *2*, 20. [CrossRef] [PubMed]

197. Ivanova, M.M.; Changsila, E.; Iaonou, C.; Goker-Alpan, O. Impaired autophagic and mitochondrial functions are partially restored by ERT in Gaucher and Fabry diseases. *PLoS ONE* **2019**, *14*, e0210617. [CrossRef] [PubMed]

198. Maetzel, D.; Sarkar, S.; Wang, H.; Abi-Mosleh, L.; Xu, P.; Cheng, A.W.; Gao, Q.; Mitalipova, M.; Jaenisch, R. Genetic and Chemical Correction of Cholesterol Accumulation and Impaired Autophagy in Hepatic and Neural Cells Derived from Niemann-Pick Type C Patient-Specific iPS Cells. *Stem Cell Rep.* **2014**, *2*, 866–880. [CrossRef] [PubMed]

199. Castellano, B.M.; Thelen, A.M.; Moldavski, O.; Feltes, M.; Van Der Welle, R.E.N.; Mydock-McGrane, L.; Jiang, X.; Van Eijkeren, R.J.; Davis, O.B.; Louie, S.M.; et al. Lysosomal cholesterol activates mTORC1 via an SLC38A9–Niemann-Pick C1 signaling complex. *Science* **2017**, *355*, 1306–1311. [CrossRef]

200. Lim, C.-Y.; Davis, O.B.; Shin, H.R.; Zhang, J.; Berdan, C.A.; Jiang, X.; Counihan, J.L.; Ory, D.S.; Nomura, D.K.; Zoncu, R. ER-lysosome contacts enable cholesterol sensing by mTORC1 and drive aberrant growth signalling in Niemann-Pick type C. *Nat. Cell Biol.* **2019**, *21*, 1206–1218. [CrossRef]

201. Takamura, A.; Higaki, K.; Kajimaki, K.; Otsuka, S.; Ninomiya, H.; Matsuda, J.; Ohno, K.; Suzuki, Y.; Nanba, E. Enhanced autophagy and mitochondrial aberrations in murine G(M1)-gangliosidosis. *Biochem. Biophys. Res. Commun.* **2008**, *367*, 616–622. [CrossRef]

202. Onyenwoke, R.U.; Sexton, J.Z.; Yan, F.; Díaz, M.C.H.; Forsberg, L.J.; Major, M.B.; Brenman, J.E. The mucolipidosis IV Ca^{2+} channel TRPML1 (MCOLN1) is regulated by the TOR kinase. *Biochem. J.* **2015**, *470*, 331–342. [CrossRef]

203. Medina, D.L.; Di Paola, S.; Peluso, I.; Armani, A.; De Stefani, D.; Venditti, R.; Montefusco, S.; Rosato, A.S.; Prezioso, C.; Forrester, A.; et al. Lysosomal calcium signalling regulates autophagy through calcineurin and TFEB. *Nat. Cell Biol.* **2015**, *17*, 288–299. [CrossRef]

204. Rosato, A.S.; Montefusco, S.; Soldati, C.; Di Paola, S.; Capuozzo, A.; Monfregola, J.; Polishchuk, E.; Amabile, A.; Grimm, C.; Lombardo, A.L.; et al. TRPML1 links lysosomal calcium to autophagosome biogenesis through the activation of the CaMKKβ/VPS34 pathway. *Nat. Commun.* **2019**, *10*, 1–16. [CrossRef]

205. Leinonen, H.; Keksa-Goldsteine, V.; Ragauskas, S.; Kohlmann, P.; Singh, Y.; Savchenko, E.; Puranen, J.; Malm, T.; Kalesnykas, G.; Koistinaho, J.; et al. Retinal Degeneration in a Mouse Model of CLN5 Disease Is Associated with Compromised Autophagy. *Sci. Rep.* **2017**, *7*, 1597. [CrossRef] [PubMed]

206. Chandrachud, U.; Walker, M.W.; Simas, A.M.; Heetveld, S.; Petcherski, A.; Klein, M.; Oh, H.; Wolf, P.; Zhao, W.-N.; Norton, S.; et al. Unbiased Cell-based Screening in a Neuronal Cell Model of Batten Disease Highlights an Interaction between Ca^{2+} Homeostasis, Autophagy, and CLN3 Protein Function. *J. Biol. Chem.* **2015**, *290*, 14361–14380. [CrossRef] [PubMed]
207. Colussi, D.J.; Jacobson, M.A. Patient-Derived Phenotypic High-Throughput Assay to Identify Small Molecules Restoring Lysosomal Function in Tay–Sachs Disease. *SLAS Discov. Adv. Sci. Drug Discov.* **2019**, *24*, 295–303. [CrossRef] [PubMed]

International Journal of
Molecular Sciences

Article

Inhibition of Bromodomain and Extraterminal Domain (BET) Proteins by JQ1 Unravels a Novel Epigenetic Modulation to Control Lipid Homeostasis

Claudia Tonini [1], Mayra Colardo [2], Barbara Colella [2], Sabrina Di Bartolomeo [2], Francesco Berardinelli [1], Giuseppina Caretti [3], Valentina Pallottini [1] and Marco Segatto [2,*]

[1] Department of Science, University of Rome "Roma Tre", Viale Marconi 446, 00146 Rome, Italy; claudia.tonini@uniroma3.it (C.T.); francesco.berardinelli@uniroma3.it (F.B.); valentina.pallottini@uniroma3.it (V.P.)

[2] Department of Bioscience and Territory, University of Molise, Contrada Fonte Lappone, 86090 Pesche (Is), Italy; m.colardo@studenti.unimol.it (M.C.); b.colella@studenti.unimol.it (B.C.); sabrina.dibartolomeo@unimol.it (S.D.B.)

[3] Department of Biosciences, University of Milan, Via Celoria 26, 20133 Milan, Italy; giuseppina.caretti@unimi.it

* Correspondence: marco.segatto@unimol.it

Received: 6 January 2020; Accepted: 13 February 2020; Published: 14 February 2020

Abstract: The homeostatic control of lipid metabolism is essential for many fundamental physiological processes. A deep understanding of its regulatory mechanisms is pivotal to unravel prospective physiopathological factors and to identify novel molecular targets that could be employed to design promising therapies in the management of lipid disorders. Here, we investigated the role of bromodomain and extraterminal domain (BET) proteins in the regulation of lipid metabolism. To reach this aim, we used a loss-of-function approach by treating HepG2 cells with JQ1, a powerful and selective BET inhibitor. The main results demonstrated that BET inhibition by JQ1 efficiently decreases intracellular lipid content, determining a significant modulation of proteins involved in lipid biosynthesis, uptake and intracellular trafficking. Importantly, the capability of BET inhibition to slow down cell proliferation is dependent on the modulation of cholesterol metabolism. Taken together, these data highlight a novel epigenetic mechanism involved in the regulation of lipid homeostasis.

Keywords: BET proteins; cell proliferation; cholesterol; epigenetics; HMGCR; JQ1; LDLr; lipid metabolism; SREBP; TMEM97

---

1. Introduction

The homeostatic regulation of lipid metabolism is essential for the maintenance of key cellular processes involved in a plethora of biological functions. Fatty acids constitute the major energy source and are fundamental constituents of cell membranes [1]. In addition, they serve as substrates for cellular phospholipases, which convert these compounds into pivotal signalling factors implicated in anti- and pro-inflammatory actions [2]. Similarly, cholesterol exerts both structural and functional roles, being the precursor of steroid hormones, vitamin D and bile acids, and regulating the assembly of specialized membrane microdomains called lipid rafts and caveolae [3].

During last decades, it has become increasingly clear that lipid homeostasis is crucial for cell growth and proliferation [4,5], as membrane biosynthesis requires the coordinated assembly of different lipid species during cell division, involving the concerted regulation of lipid biosynthesis, uptake, subcellular localization and turnover [6]. For these reasons, the body employs an intricate homeostatic network to regulate the availability of lipids for cells and tissues. This network mainly operates in the liver, where the major part of lipid metabolism takes place [3]. Hepatic cells are the principal site for lipid biosynthesis, determined by the establishment of a complex series of enzymatic reactions; in particular, 3-hydroxy-3-methylglutaryl Coenzyme A reductase (HMGCR) and Acyl Coenzyme A carboxylase (ACC) represent the key and rate-limiting enzymes for cholesterol and fatty acid synthesis, respectively [7,8]. Considering their central role in lipid biosynthesis, both HMGCR and ACC are tightly regulated at both short- and long-term levels. Short-term regulation is controlled by phosphorylative events, which negatively affect the activation state of the enzymes [9]. Conversely, the long-term transcriptional regulation of the protein machinery involved in lipid metabolism is mediated by sterol regulatory element binding proteins (SREBPs). When intracellular sterol content is reduced, SREBPs precursors are proteolytically processed to form the NH(2)-terminal fragment that enters the nucleus (nuclear, nSREBP) and induces the transcription of genes coding for lipogenic enzymes, such as HMGCR and ACC [9,10]. Notably, SREBPs also promote the extracellular uptake of lipoprotein-derived lipids by eliciting the transcription of lipoprotein receptors, such as low density lipoprotein receptor (LDLr) and scavenger receptor class B type 1 (SR-B1) [11,12].

Lipid abnormalities play a crucial role in a plethora of pathological conditions, such as cardiovascular diseases, neurodevelopmental alterations, neurodegenerative and lipid storage disorders [13–15]. Despite the widespread use of lipid-lowering medications, effective pharmacological approaches are still lacking for several lipid metabolism-related disorders. Thus, there is still a need to better dissect the regulatory mechanisms of lipid homeostasis in order to identify novel molecular targets that could be useful for designing promising therapeutic treatments.

Bromodomain and extra-terminal domain (BET) proteins, comprising BRD2, BRD3 and BRD4, are epigenetic readers recruited to the chromatin by the presence of acetylated histones, thereby regulating gene expression [16]. The involvement of these epigenetic sensors has been extensively characterized in cancer and inflammation, because of their incontrovertible role in the transcriptional modulation of oncogenes and mediators of the immune response [17,18]. BET proteins attracted considerable interest in biomedical research, as they are extremely druggable by potent and specific inhibitors. In this context, BET inhibition exerts anti-proliferative activity in different cancer cells and shows outstanding anti-inflammatory properties in a number of physiopathological conditions [17,19,20]. Recently, experimental evidence also demonstrated that the activity of BET proteins may be extended to the regulation of metabolic processes. For instance, BET inhibition strongly affects protein homeostasis, autophagy induction, reactive oxygen species (ROS) and glucose metabolism [17,21–24]. Conversely, the prospective role exerted by BET proteins on lipid metabolism is still poorly characterized. Microarray analysis suggested that the BET inhibitor RVX-208 induces changes in ApoA1 and high density lipoprotein (HDL) levels [25]. Coherently, overexpression of BET proteins increased cholesterol biosynthesis [26]. Furthermore, it has been observed that BET inhibition efficiently counteracted plasma low density lipoprotein (LDL) alterations in a mouse model of cancer cachexia [17]. Despite this evidence, no studies systematically addressed the involvement of BET proteins in the modulation of the main proteins and enzymes controlling the regulatory machinery of lipid metabolism. A better comprehension of these mechanisms is essential to identify novel physiological regulatory pathways and to select innovative therapeutic targets. Here, we provide a proof of concept study, aimed at evaluating the putative role of BET modulation on lipid homeostasis. To reach this objective, we took advantage of a loss-of-function approach by using JQ1 as a potent and specific inhibitor of BET protein activity [27]. The effects induced by BET inhibition were mainly evaluated in HepG2 cell line, a human liver-derived cell culture model widely used to assess lipid homeostasis [28–30].

2. Results

2.1. BET Inhibition Decreases Lipid Content in HepG2 Cells

In order to evaluate the putative involvement of BET inhibition on lipid homeostasis, the effect of JQ1 on cellular lipid content was firstly assessed. HepG2 cells were treated with JQ1 for 48 hours, and Oil Red O staining was employed as a widely used and accurate method to measure neutral lipids [31]. Descriptive evaluation highlighted that the number of lipid droplets, as well as their size, appeared reduced in JQ1-treated cells (Figure 1A). This result corroborated the quantitative assessment of lipid content estimated by Oil Red O absorbance, which showed a significant five-fold decrease upon BET inhibition (Figure 1B). Coherently, immunofluorescence intensity of the lipid droplet marker perilipin-2 (Plin2) was found to be lower 48 hours after pharmacological BET blockade (Figure 1C). Notably, JQ1 was also effective in reducing the amount of intracellular cholesterol, as observable by filipin staining and quantification (Figure 1D). Thus, BET inhibition significantly lowers lipid content in HepG2 cells.

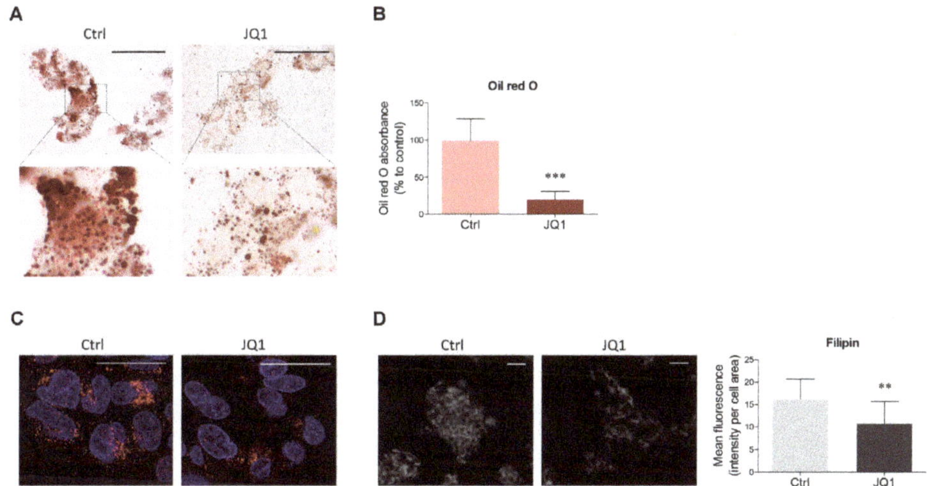

Figure 1. Effects of bromodomain and extraterminal domain (BET) inhibition by JQ1 on intracellular lipids in HepG2 cells. (**A**) HepG2 cells were treated with vehicle (Ctrl) or JQ1 (0.4 μM), and after 48 hours they were stained with Oil Red O as described in the Materials and Methods Section to visualize the intracellular content of neutral lipids. $n = 6$ different experiments. Scale bar: 50 μm (**B**) HepG2 cells were treated as in (A), and Oil Red O was extracted with isopropanol. The eluted dye was then quantified by spectrophotometry to evaluate the amount of neutral lipids. $n = 6$ different experiments. (**C**) Vehicle- and JQ1-treated HepG2 cells were fixed and stained with antibody against Plin2 (red). DAPI was used as a nuclear counterstain. Scale bar: 25 μm (**D**) Representative image (left panel) and quantification of the mean fluorescence intensity (right panel) of filipin staining performed on HepG2 cells treated with vehicle and JQ1 for 48 hours. $n = 5$ different experiments. Scale bar: 50 μm. Data represent means ± SD. Statistical analysis was performed by using unpaired Student's t test. ** $p < 0.01$; *** $p < 0.001$.

2.2. BET Inhibition by JQ1 Modulates the Expression of Proteins and Enzymes Involved in Lipid Metabolism

To understand the cellular mechanisms underlying the reduction of lipid content induced by BET inhibition, the prospective modulation of proteins belonging to the lipid metabolism machinery were assessed. The analysis was initially focused on ACC and HMGCR, the rate-limiting enzymes involved in fatty acid and cholesterol biosynthesis, respectively. Western blot analysis revealed that JQ1 treatment significantly decreased ACC protein expression if compared to vehicle-treated HepG2 cells (Figure 2A). However, no changes were observed in the ratio between the phosphorylated fraction of ACC and its total levels, suggesting that BET inhibition modulated the protein amount of the enzyme without influencing its activation state by inhibitory phosphorylation. Similar results were obtained by analyzing HMGCR; in fact, JQ1 administration strongly reduced the protein levels of the enzyme without affecting its phosphorylation state (Figure 2B). The effect of BET inhibition on

HMGCR expression was further confirmed by confocal analysis, showing an overall decrease of immunofluorescence intensity in JQ1-treated HepG2 with respect to control cells (Figure 2C).

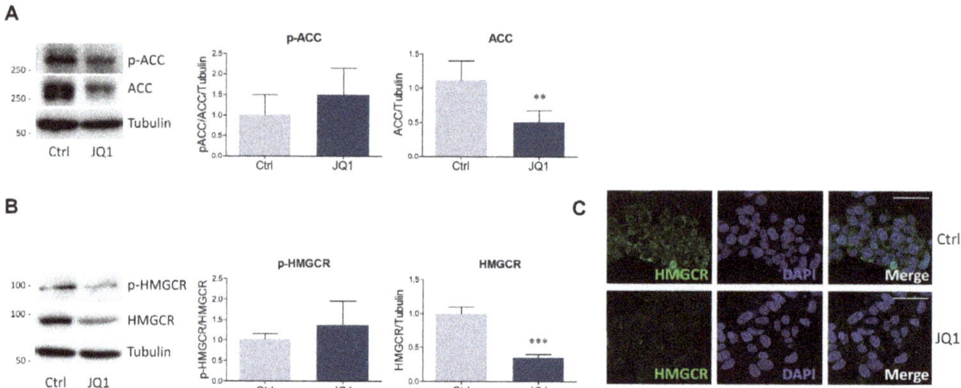

Figure 2. Evaluation of BET inhibition on lipid biosynthesis enzymes. (**A**) Representative Western blot (left panel) and densitometric analysis of phosphorylated Acyl Coenzyme A carboxylase (ACC) (P-ACC, ser79) and total ACC in HepG2 cells treated with vehicle (Ctrl) or JQ1 (0.4 µM) for 48 hours. n = 6 independent experiments. Tubulin was employed as a housekeeping protein to normalize protein loading. (**B**) Representative Western blot (left panel) and densitometric analysis (right panel) of phosphorylated 3-hydroxy-3-methylglutaryl Coenzyme A reductase (HMGCR) (p-HMGCR, ser872) and total HMGCR in HepG2 cells treated with vehicle (Ctrl) or JQ1 (0.4 µM) for 48 hours. n = 7 independent experiments. Tubulin served as a housekeeping protein to normalize protein loading. (**C**) Immunofluorescence staining of HMGCR (green) of HepG2 cells treated as in (B). Nuclei were counterstained with DAPI. n = 3 different experiments. Scale bar: 50 µm. Data are expressed as means ± SD. Statistical analysis was carried out by using unpaired Student's t test. ** $p < 0.01$; *** $p < 0.001$.

Lipid homeostasis is guaranteed by a delicate equilibrium between biosynthesis and extracellular uptake. The latter process is mainly operated by LDLr, which internalizes LDL through receptor-mediated endocytosis [32]. In addition to LDLr, hepatic cells also express SR-B1, a multiligand receptor that binds several lipoproteins, including HDL and LDL [33]. Considering their pivotal role in the physiological regulation of lipid metabolism, the prospective effects mediated by BET inhibition were also assessed for these two lipoprotein receptors. SR-B1 expression was significantly repressed by JQ1 treatment, as observed by Western blot and immunofluorescence data (Figure 3A,B). Similarly, BET inhibition determined a three-fold reduction in LDLr expression levels (Figure 3C). Immunofluorescence microscopy confirmed this result, being LDLr barely detectable in JQ1-treated HepG2 cells when compared to vehicle-treated cells (Figure 3D).

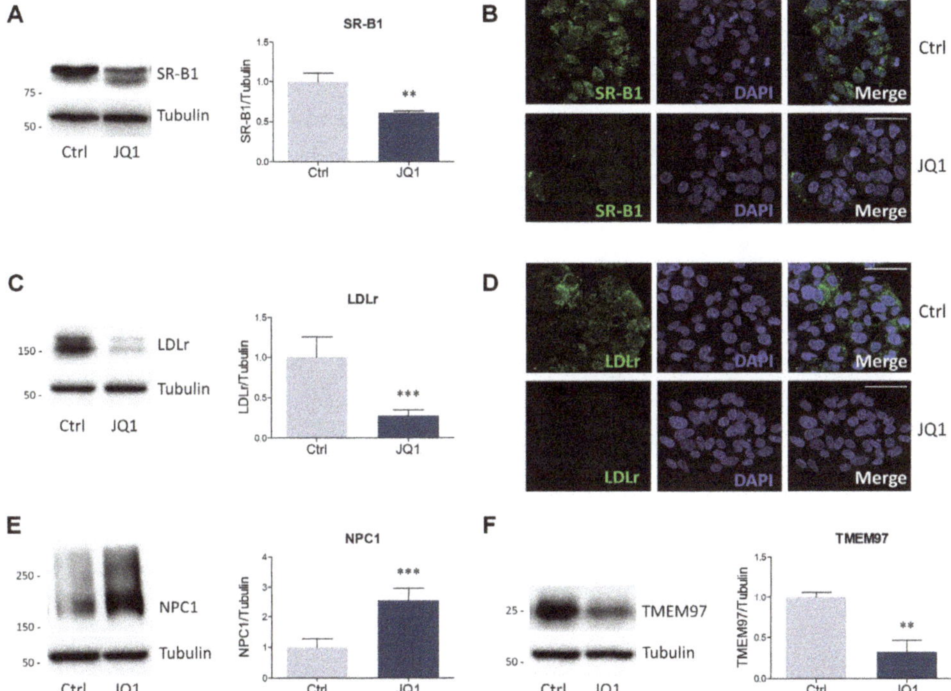

Figure 3. Expression of proteins involved in extracellular lipid uptake and intracellular cholesterol trafficking following JQ1 (0.4 μM) administration to HepG2 cells for 48 hours. (**A**) Representative Western blot (left panel) and densitometric analysis (right panel) of SR-B1. $n = 6$ independent experiments. (**B**) Immunofluorescence analysis of SR-B1 (green). Nuclei were counterstained with DAPI. $n = 3$ different experiments. Scale bar: 50 μm. (**C**) Representative Western blot (left panel) and densitometric analysis (right panel) of LDLr. $n = 5$ independent experiments. (**D**) LDLr immunofluorescence (green). Nuclei were counterstained with DAPI. $n = 3$ different experiments. (**E–F**) Representative Western blots and densitometric analysis of NPC1 and TMEM97. Tubulin was chosen as loading control. $n = 6$ independent experiments. Data represent means ± SD. Statistical analysis was performed by using unpaired Student's t test. ** $p < 0.01$; *** $p < 0.001$.

Upon binding to lipoprotein receptors, LDL are internalized and transported throughout the endocytic pathway to lysosomes, where cholesteryl esters can be hydrolyzed by acid lipases [34]. Subsequently, unesterified cholesterol exits the lysosomal compartment, through the activity of NPC1, and is delivered to the plasma membrane and the endoplasmic reticulum (ER) [35]. Because of its essential role in intracellular cholesterol trafficking, Niemann-Pick type C1 (NPC1) protein levels were assessed in this study. BET inhibition by JQ1 strongly enhances NPC1 protein expression in HepG2 cells (Figure 3E). Interestingly, the rise in NPC1 levels was accompanied by a significant reduction of transmembrane protein 97 (TMEM97) protein content, also known as the sigma-2 receptor (Figure 3F), which has already been involved in NPC1 regulation [36].

Most proteins involved in lipid metabolism are under the transcriptional control of SREBPs. Considering the effects of BET inhibition evaluated in this work, the expression of SREBP-1 and SREBP-2 was estimated. JQ1 treatment induced a significant increase of SREBP-1 precursor (full-length, FL SREBP-1). However, this effect was not paralleled by a concurrent modification in the nuclear and transcriptionally active fragment of SREBP-1 (nSREBP-1), as its expression is similar between the two experimental groups (Figure 4A). Morphological analysis corroborated this evidence, revealing a slightly higher fluorescence intensity in the cytosolic compartment in JQ1-treated cells, consistent with the increase of FL SREBP-1 observed by the Western blot. On the contrary, no differences were detected in the number and intensity of the stained nuclei, reflecting the lack of significant alterations in the amount of nSREBP-1 (Figure 4B). Differently from SREBP-1, BET inhibition reduced the expression of both precursor (FL SREBP-2) and nuclear SREBP-2 (nSREBP-2) (Figure 4C). SREBP-2 immunofluorescence revealed a predominant nuclear staining, and a weak signal in the cytoplasmic compartment. Notably, JQ1 administration markedly decreased the intensity and the number of nuclei stained for SREBP-2 (Figure 4D).

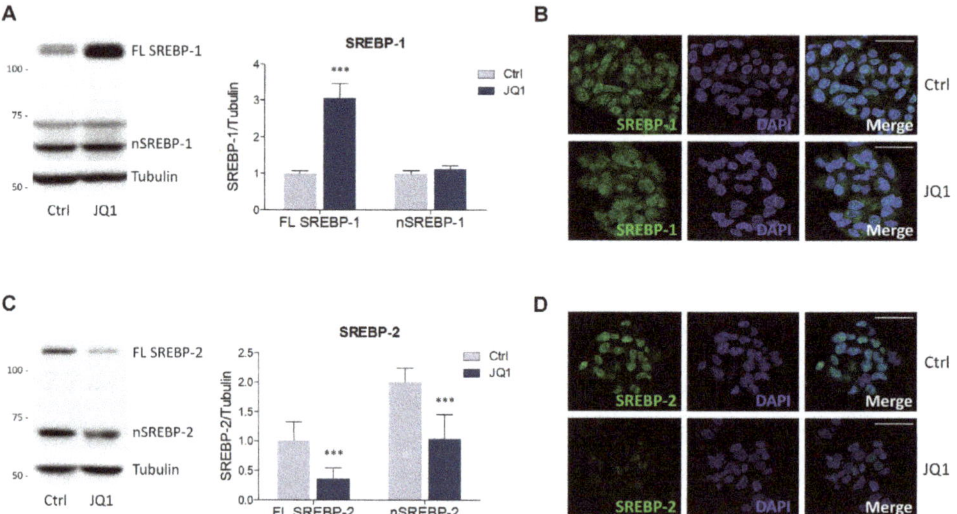

Figure 4. BET inhibition alters sterol regulatory element binding proteins (SREBPs) expression in HepG2 cells treated with JQ1 (0.4 µM) for 48 hours. (**A**) Representative Western blot (left panel) and densitometric analysis (right panel) of SREBP-1. FL SREBP-1 (Full-length SREBP-1); nSREBP-1 (nuclear SREBP-1). $n = 6$ independent experiments. Tubulin was employed for control loading. (**B**) SREBP-1 immunofluorescence staining (green) in HepG2 cells. Nuclei were counterstained with DAPI. $n = 3$ different experiments. Scale bar: 50 µm. (**C**) Representative Western blot (left panel) and densitometric analysis (right panel) of SREBP-2. FL SREBP-2 (Full-length SREBP-2); nSREBP-2 (nuclear SREBP-2). $n = 6$ independent experiments. Tubulin was used as a housekeeping protein. (**D**) Immunofluorescence analysis of SREBP-2 (green). Nuclei were counterstained with DAPI. $n = 3$ different experiments. Scale bar: 50 µm. Data represent means ± SD. Statistical analysis was assessed by using unpaired Student's t test. *** $p < 0.001$.

To further strengthen the effects induced by JQ1 on lipid homeostasis in HepG2 cells, we recapitulated the most relevant findings in different cell lines. As expected, BET blockade significantly suppressed the expression of SREBP-2 and of its targets LDLr and HMGCR also in the neuroblastoma cell line N1E-115 (Figure 5A,B) and in primary culture of human fibroblasts (Figure 5C,D).

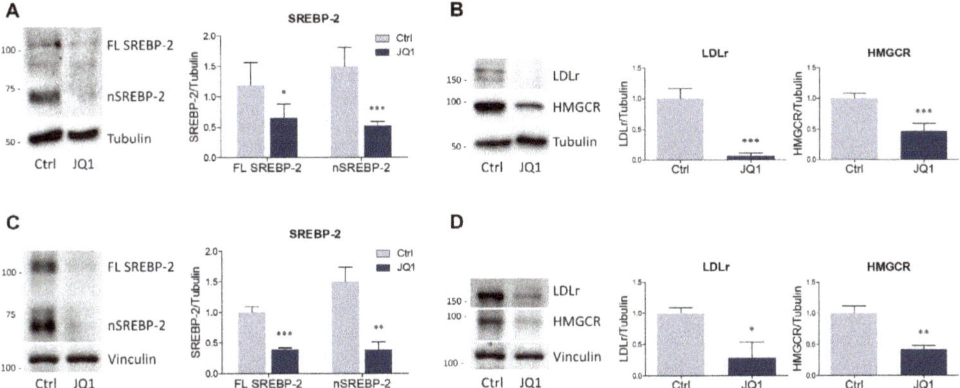

Figure 5. BET inhibition modulates the expression of SREBP-2, HMGCR and LDLr in different cell lines. (**A–B**) Representative Western blot and densitometric analysis of SREBP-2, HMGCR and LDLr in differentiated N1E-115 treated with JQ1 (0.1 µM) for 48 hours. (**C–D**) Representative Western blot and densitometric analysis of SREBP-2, HMGCR and LDLr in primary human fibroblasts treated with JQ1 (0.4 µM) for 48 hours. *n* = 3 different experiments. Tubulin and vinculin were used as loading control. Data represent means ± SD. Statistical analysis was assessed by using unpaired Student's t test. * $p < 0.05$, ** $p < 0.01$, *** $p < 0.001$.

Overall, these data indicate that BET inhibition deeply affects the main proteins involved in lipid biosynthesis, uptake and intracellular transport.

2.3. BET Inhibition Affects Cell Proliferation in a Cholesterol-Dependent Manner

The maintenance of a proper amount of lipids, and in particular of cholesterol, is crucial for several biological processes, such as cell growth and cell proliferation [4,37]. In addition, it has been extensively demonstrated that BET inhibition promotes the suppression of cell proliferation in a number of normal and cancer cell types [19,38,39]. Results collected in this work are in agreement with previous reports, demonstrating that JQ1 treatment significantly slowed down the proliferation rate of HepG2 cells starting at 48 hours from the pharmacological treatment (Figure 6A). Remarkably, the cell growth rate was rescued when JQ1 was co-administered with mevalonate (MVA), the product of the reaction catalyzed by HMGCR. The involvement of cholesterol biosynthesis in the anti-proliferative effects mediated by BET inhibition was further supported by the co-administration of cholesterol to JQ1-treated cells that, similarly to MVA addition, was able to restore cell proliferation. In order to delve deeper into the effects exerted by BET inhibition on cell proliferation and cholesterol modulation, we generated a HepG2 lineage with acquired resistance to JQ1 (HepG2-R) by continuously treating cells with increasing doses of the drug. While JQ1 decreased the number of JQ1-sensitive cells as previously shown, no effects were induced in HepG2-R upon drug administration (Figure 6B). Importantly, the restoration of cell proliferation observed in HepG2-R cells was completely abolished by 25-hydroxycholesterol (25OHC) treatment, a well-known methodological approach to mediate HMGCR degradation through the activation of a feedback mechanism [40]. Similar results were also obtained by blocking HMGCR activity with simvastatin, a potent inhibitor of cholesterol biosynthesis (Figure 6C). Because statins suppress the production of isoprenoid intermediates in the cholesterol biosynthetic pathway, they can exert a plethora of pleiotropic actions independently from cholesterol decrease [8]. To confirm a direct involvement, cholesterol was then administered to HepG2-R cells co-treated with JQ1 and simvastatin. Cholesterol supplementation efficiently prevented the reduction of cell proliferation induced by simvastatin in HepG2-R cells, thus restoring the acquired resistance to JQ1 as a function of cell growth (Figure 6D).

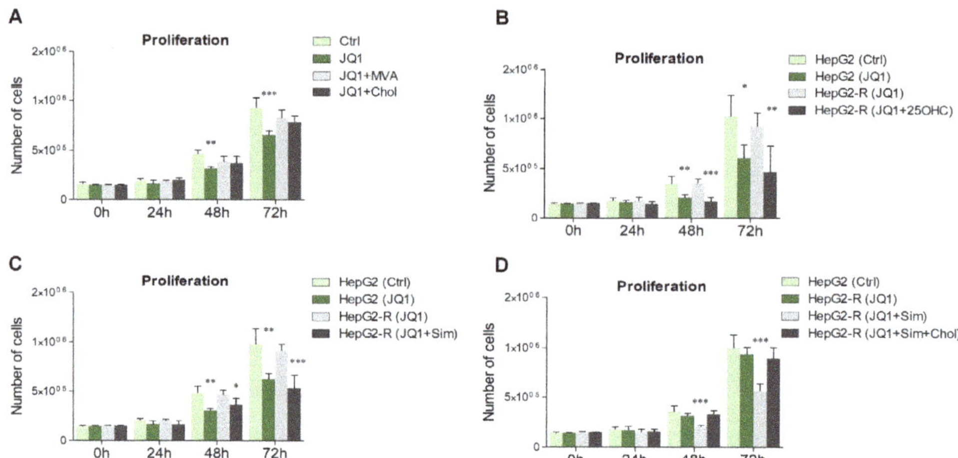

Figure 6. BET inhibition influences cell proliferation through cholesterol metabolism. (**A**) HepG2 cells were seeded in 6-well plates (150,000 cells for each well) and were treated with JQ1 (0.4 μM), mevalonate (MVA, 100 μM) and cholesterol (chol, 50 μM) for 72 hours. Cell counts were conducted over time with a hemocytometer. (**B–C**) Cell proliferation was evaluated in JQ1-sensitive (HepG2) and JQ1-resistant (HepG2-R) cells. JQ1-resistant and -sensitive HepG2 cells were treated with the BET inhibitor for 72 hours, and additional groups of HepG2-R cells were co-stimulated with JQ1+25-hydroxycholesterol (25OHC, 20 μM) or JQ1+simvastatin (Sim, 1 μM). Other cells were treated with vehicle and served as control (Ctrl). (**D**) Cell proliferation was evaluated in JQ1-sensitive (HepG2) and JQ1-resistant (HepG2-R) cells. JQ1-resistant cells were constantly stimulated with the BET inhibitor. Additional groups of HepG2-R cells were co-stimulated with JQ1+simvastatin (Sim, 1 μM) or JQ1+Sim+cholesterol (Chol, 50 μM). $n = 4$ independent experiments. Data represent means ± SD. Statistical analysis was assessed by using one-way ANOVA, followed by Dunnett's post hoc. * $p < 0.05$; ** $p < 0.01$; *** $p < 0.001$.

Overall, these results suggested that the anti-proliferative effects elicited by JQ1 can be mediated by the suppression of cholesterol metabolism and that the acquisition of resistance to BET inhibition may be accompanied by adaptive changes in the main proteins controlling cholesterol homeostasis. Western blot data confirmed this hypothesis, highlighting that both FL SREBP-2 and nSREBP-2 levels were increased in JQ1-resistant HepG2 cells (Figure 7A). Coherently, the expression of SREBP-2 target genes HMGCR, SR-B1 and LDLr was properly restored at the level of control cells in HepG2-R cells (Figure 7B–D). Taken together, these results suggest that the impact of HepG2 proliferation mediated by BET inhibition is dependent on cholesterol metabolism.

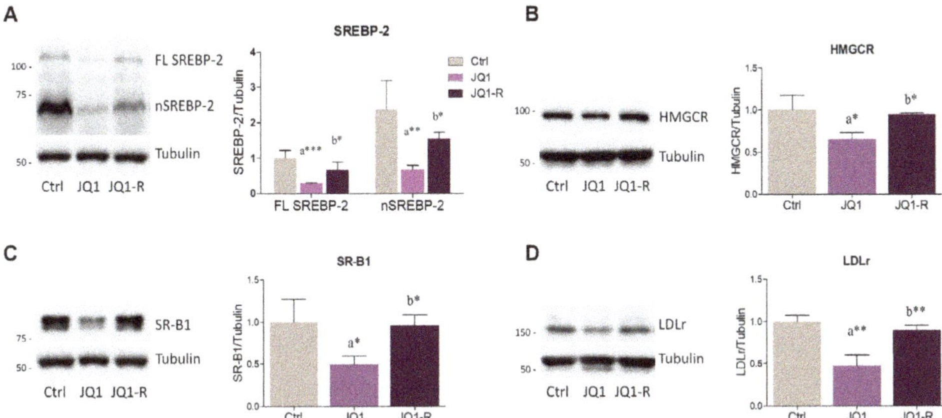

Figure 7. JQ1 resistance is accompanied by the upregulation of proteins controlling cholesterol metabolism. (**A–D**) Representative Western blots and densitometric analysis of SREBP-2, HMGCR, SR-B1and LDLr in JQ1-sensitive cells treated with vehicle (Ctrl) or JQ1 (0.4 µM), and in HepG2-resistant (JQ1-R) cells constantly stimulated with JQ1 (0.4 µM). Experiments were performed after 48 hours. $n = 4$ independent experiments. Tubulin served as loading control. Data represent means ± SD. Statistical analysis was assessed by using one-way ANOVA followed by Tukey's post-hoc. $* p < 0.05$; $** p < 0.01$; $*** p < 0.001$. "a" indicates statistical significance versus control group (Ctrl); "b" indicates statistical significance compared to JQ1 group.

3. Discussion

The homeostatic maintenance of lipid metabolism plays a fundamental role in assuring the correct development of cellular functions. As a consequence, alterations in the mechanisms controlling lipid balance may be associated to several pathological conditions, ranging from cardiovascular diseases and lipidosis to neurodegenerative and neurodevelopmental diseases [13–15,41]. The discovery of effective lipid-lowering medications, such as statins and fibrates, have revolutionized the treatment and the clinical outcomes of different lipid disorders [42]. Unfortunately, numerous diseases associated to lipid disturbances do not yet have a resolutive therapy. The obstacles in designing efficient pharmacological approaches relies on the fact that cellular and molecular mechanisms regulating lipid metabolism are not completely elucidated.

In this context, it is important to delve deeper into the regulatory mechanisms of lipid homeostasis, with the attempt to unravel potential etiopathological factors and novel molecular targets that could be useful to set up promising therapies. In the last few years, several reports highlighted that epigenetic factors may play crucial roles in the control of lipid homeostasis. For instance, it is becoming increasingly clear that histone deacetylases (HDAC) and microRNAs exert a crucial modulatory activity in lipid and energy metabolism [43–47]. On the contrary, the involvement of BET proteins in the regulation of lipid homeostasis is still elusive. Thus, in this work, we evaluated the effects of BET inhibition in the expression of the main proteins controlling lipid metabolism.

Collectively, our results demonstrated that BET inhibition induced an overall suppression of lipid metabolism. In particular, JQ1 administration decreased the intracellular lipid content and reduced the expression of biosynthetic enzymes (ACC and HMGCR) as well as of receptors involved in lipoprotein uptake (LDLr and SR-B1). BET inhibition also induced a strong downregulation in the expression of TMEM97. This conserved integral membrane protein has been identified as an important modulator of cholesterol levels and, similarly to other proteins coordinating lipid metabolism, is a SREBP-2 target gene [36,48,49]. It is well described that SREBPs isoforms possess different roles in lipid biosynthesis. SREBP-1 isoforms are mostly devoted to the regulation of fatty acid and triglyceride metabolism, whereas SREBP-2 is relatively selective in activating cholesterol-related genes [50,51]. Coherently with these notions, the JQ1-mediated suppression of HMGCR, LDLr, SR-B1 and TMEM97 was paralleled by a concurrent decrease of both the full-length and the nuclear active fraction of SREBP-2. Interestingly, BET inhibition was also responsible for a significant decrease of ACC expression, which, however, was not adequately accompanied by changes in nSREBP-1 levels. Indeed, the levels of SREBP-1 precursor were significantly augmented, probably representing a transcriptional attempt to counteract the marked decrease of the SREBP-2 isoform. Conversely, the expression of the nuclear and transcriptionally active fragment of SREBP-1 was unaltered upon JQ1 treatment, excluding its involvement in the modulation of ACC levels. Even though ACC transcription is preferentially controlled by SREBP-1 isoforms [50], it has been observed that SREBP-2 can equally influence ACC transcription [52], supporting the notion that the reduction of ACC levels observed in this work may be ascribable to the suppression of SREBP-2 mediated by JQ1. In addition, BET blockade led to a rise in NPC-1 protein levels. Literature data illustrated that a reduction of TMEM97 increases NPC-1 protein expression by a post-translational mechanism [36], and suggest that the suppression of TMEM97 could explain the build-up of NPC-1 observed in this study following JQ1 administration. Consistent with JQ1-mediated decrease of LDLr expression, NPC1 induction may also represent a refined compensatory response to contrast the prospective decrease of LDL-cholesterol uptake. Furthermore, it cannot be excluded that BET inhibition directly affects the transcription of *NPC1*, as well as of other genes involved in the maintenance of lipid homeostasis.

Subsequently, the prospective contribution of lipid metabolism in the anti-proliferative effects induced by JQ1 administration was evaluated. It has been extensively reported that BET inhibition exerts a remarkable reduction in cell proliferation by hindering the transcription of oncogenes such as c-Myc [53], and by suppressing the activation of pro-survival and growth signalling kinases like Akt [26]. In this study, we highlight for the first time a novel mechanism by which BET inhibition modulates cell proliferation in a cholesterol-dependent manner. Indeed, the administration of both MVA and cholesterol to culture medium efficiently abolished the delay in cell proliferation induced by JQ1. Consistently, the generation of HepG2-R cells further corroborated this evidence, as the acquisition of resistance to JQ1 is associated to the compensatory increase in the expression of proteins belonging to cholesterol metabolism, and to the restoration of cell proliferation rate.

Overall, these data suggest that BET inhibition reduces lipid content by hampering SREBP-2 expression and processing, which result in a reduced expression of target genes involved in lipid biosynthesis, uptake and intracellular trafficking.

Despite the fact that more efforts should be done in order to better clarify the specific contribution of each BET protein in the regulation of lipid metabolism, this work provides the proof of principle that epigenetic pathways, influenced by BET protein activity, represent novel physiological mechanisms to control lipid homeostasis. These findings may also set the basis for designing innovative therapeutic approaches aimed at ameliorating the functional outcomes of several disorders characterized by lipid unbalances.

4. Materials and Methods

4.1. Cell Cultures and Generation of HepG2 Resistant to JQ1

N1E-115 neuroblastoma cells were cultured at 5% CO2 in DMEM medium at high glucose, containing 10% (*v/v*) foetal calf serum, L -glutamine (2 mM), and added with penicillin/streptomycin solution. Cells were then seeded at 50% confluency and were induced to differentiate for four hours by adding 2% dimethyl sulfoxide. Differentiated N1E-115 cells were treated with JQ1 (0.1 µM) or vehicle (DMSO, dilution 1:1000 in cell culture media) for 48 hours.

Human foetal foreskin fibroblasts (HFFF2) were grown at 5% CO2 in DMEM (high glucose) additioned with 10% foetal calf serum, L -glutamine (2 mM), and penicillin/streptomycin solution. HFFF2 (60% confluency) were then treated with 0.4 µM JQ1 for 48 hours. Cells treated with vehicle (DMSO, dilution 1:1000 in cell culture media) served as control.

HepG2 cells were routinely cultured at 5% CO2 in DMEM medium at high glucose, containing 10% (*v/v*) foetal calf serum, L -glutamine (2 mM), and added with penicillin/streptomycin solution. Cells were passaged every 3 days and medium changed every 2–3 days. All the experiments were performed with cell confluency of 60%–70%. DMSO stock of 4 mM JQ1 was diluted 1:1000 in cell culture media to obtain the final concentration of 0.4 μM JQ1. Experiments were then performed 48 hours after JQ1 stimulation. Cells treated with vehicle (DMSO, dilution 1:1000 in cell culture media) served as control. The generation of HepG2 cells resistant to JQ1 was carried out by slightly modifying the protocol provided by Gobbi and colleagues [54]. HepG2 resistant cells were selected by continuously applying increasing doses of JQ1, starting from 50 nM and enhancing the drug concentration every 1–2 weeks for 2 months of total treatment. Surviving JQ1-resistant cells were then maintained at 0.4 μM JQ1 throughout. For cell proliferation experiments, chemicals were purchased from Sigma-Aldrich and used at the following concentrations: JQ1 (0.4 μM), mevalonate (100 μM), cholesterol (50 μM), simvastatin (1 μM), 25-hydroxycholesterol (20 μM).

4.2. Oil Red O Staining and Quantification

For the evaluation of lipid content, HepG2 cells were seeded in 12-well plates. Briefly, after 48 hours of JQ1 treatment, cells were fixed in paraformaldehyde (4% solution) for 10 minutes and gently rinsed twice with PBS. Sixty per cent isopropanol was added to fixed cells for 5 minutes, and then washed 2 times with distilled water. Cells were stained with 1 mL of the Oil Red O solution (Sigma-Aldrich, Milan, Italy) for 15 minutes at room temperature with continuous gentle shaking. Subsequently, wells were then rinsed 3 times with distilled water, until no excess stain was seen. Stained cells were visualized in brightfield using an Olympus BX 51 microscope (Olympus Italia, Segrate, Italy), equipped with a Leica DFC 420 camera (Leica Microsystems, Milan, Italy). Electronic images were captured using a Leica Application Suite version 3.5.0 system (Leica Microsystems, Milan, Italy). For relative Oil Red O accumulation by spectrophotometry, after incubation with Oil Red O solution, stained cells were washed 3 times with distilled water, and the dye was eluted by the addition of 1 mL isopropanol for 15 minutes, with gentle shaking. A total of 100 μL of the eluted dye was removed from each sample and were transferred to a clean 96-well plate for reading the absorbance at 540 nm.

4.3. Filipin Staining

Filipin staining was performed as previously described [55]. Filipin staining was performed using Filipin complex (Sigma-Aldrich, catalog #F9765). Filipin stock solution (10mg/mL in PBS) was freshly prepared before the use. Cells were fixed in paraformaldehyde (4% solution) for 10 minutes and rinsed 3 times with PBS. Cells were stained with 1mL Filipin working solution (0.05 mg/mL in PBS) for 2 hours at room temperature, in the dark. Next, wells were rinsed 3 times with PBS to remove excess dye. Finally, cells were viewed in PBS through a fluorescence microscope using a UV filter set (340–380 nm excitation). Images were acquired at 20× magnification using an Axio Imager Z2 (Carl Zeiss, Jena, Germany) equipped with a charge-coupled device (CCD) camera controlled by the ISIS software (MetaSystems, Milano, Italy). Filipin quantification was calculated as the mean fluorescence intensity per cell area by using ImageJ Software for Windows.

4.4. Lysate Preparation and Western Blot Analysis

HepG2 lysate was performed as already reported [56]. Forty-eight hours after treatment, cells were lysed in 80 μL lysis buffer (0.25 M Tris pH 6.8, 10% SDS, phosphatase and protease inhibitor cocktails) by sonication (duty cycle 20%, output 3). Samples were then centrifuged at 10,000g for 10 min to remove cell debris. Protein concentration was assessed by the method of Lowry. Subsequently, Laemmli buffer was added to HepG2 lysates, and samples were boiled for 3 min before loading to the sodium dodecyl sulfate polyacrylamide gel electrophoresis (SDS-PAGE) for subsequent western blot analysis.

Western blot experiments were performed by slightly modifying the previously described protocol [8]. Briefly, proteins (30 µg) from HepG2 lysates were resolved on SDS-PAGE at 40 mA (constant current) for 60 min. Subsequently, proteins were transferred onto nitrocellulose membrane by using Trans-Blot Turbo Transfer System (Bio-Rad Laboratories, Milan, Italy). The nitrocellulose membrane was incubated at room temperature with 5% fat-free milk in Tris-buffered saline (0.138 M NaCl, 0.027 M KCl, 0.025 M Tris-HCl, and 0.05% Tween-20, pH 6.8), and probed at 4 °C overnight with the following primary antibodies: phopsho-HMGCR (Merck Millipore, #09-356, dilution 1:1000), HMGCR (Abcam, ab242315, dilution 1:1000), phospho-ACC (Sigma-Aldrich, SAB4503851, dilution 1:500), ACC (Sigma-Aldrich, SAB4501396, dilution 1:500), SR-B1 (Abcam, Cambridge, UK, ab52629, dilution 1:2000), LDLr (Abcam, ab30532, dilution 1:1000), NPC1 (Novus Biologicals, NB400-148, dilution 1:1000), TMEM97 (Novus Biologicals, Centennial, CO, USA, NBP1-30436, dilution 1:1000), SREBP-2 (Abcam, ab30682, dilution 1:1000), SREBP-1 (Santa Cruz Biotechnology, sc-8984, dilution 1:1000), alpha-Tubulin (Sigma-Aldrich, T6199, dilution 1:10000), vinculin (Sigma-Aldrich, V9131, dilution 1:20.000). Subsequently, membranes were probed for 1 hour with horseradish peroxidase conjugated secondary IgG antibodies (Bio-Rad Laboratories, Milan, Italy). Protein-antibody immunocomplexes onto nitrocellulose were visualized by using clarity ECL Western blotting (Bio-Rad Laboratories, Milan, Italy, #1705061), and chemiluminescence acquisition was carried out through ChemiDoc MP system (Bio-Rad Laboratories). Western blotting images were analyzed by ImageJ (National Institutes of Health, Bethesda, MD, USA) software for Windows. All samples were normalized for protein loading with alpha-Tubulin (chosen as a housekeeping protein). Recorded values were derived from the ratio between arbitrary units obtained from the protein band and the respective housekeeping protein.

4.5. Immunofluorescence Staining

Immunofluorescence of HepG2 cells was performed by following the previously described protocol [17]. Cells were fixed in paraformaldehyde (4% in PBS) and incubated overnight with appropriate antibodies: HMGCR (Abcam, ab242315, dilution 1:100), SREBP-2 (Abcam, ab30682, dilution 1:100), SREBP-1 (Santa Cruz Biotechnology, Dallas, TX, USA, sc-8984, dilution 1:100), SR-B1 (Abcam, ab52629, dilution 1:100), LDLr (Santa Cruz Biotechnology, sc-11824, dilution 1:200), anti-Perilipin-2 (anti-Plin2) antibody (R&D Systems, #MAB76341, dilution 1:100). After incubation with primary antibodies, fixed cells were probed for 1 hour at room temperature with donkey anti-goat secondary antibody Alexa Fluor 488 (ThermoFisher Scientific, Milan, Italy, A-11055), goat anti-rabbit secondary antibody Alexa Fluor 555 (ThermoFisher Scientific, A27039) and goat anti-rabbit secondary antibody Alexa Fluor 488 (ThermoFisher Scientific, A-11008). Coverslips were mounted with Vectashield Antifade mounting medium with DAPI (Vector, H-1200) to visualize nuclear staining. The samples were examined at confocal microscopy (TCS SP8; Leica, Wetzlar, Germany). Images were captured using Leica TCS SP8 equipped with a 40 × 1.40–0.60 NA HCX Plan Apo oil BL objective at RT and Leica LAS X Software.

4.6. Statistical Analysis

All the results are expressed as mean ± standard deviation (SD). Normal distribution of the data was assessed by applying the Shapiro–Wilk test. Unpaired Student's t test was performed to compare means between two experimental groups. When comparing three or more experimental groups, one-way analysis of variance (ANOVA) was carried out, followed by Tukey's or Dunnett's post hoc. $p < 0.05$ was considered to indicate a statistically significant difference. Statistical analysis and graph editing were performed using GRAPHPAD INSTAT3 (GraphPad, La Jolla, CA, USA) for Windows.

Author Contributions: Conceptualization: M.S.; data curation: B.C., C.T., F.B., M.S. and V.P.; formal analysis: C.T. and M.S.; funding acquisition: V.P.; investigation: B.C., C.T., F.B., M.C. and M.S.; methodology: C.T., M.C. and M.S.; project administration: M.S.; resources: F.B., M.S., S.D.B. and V.P.; software: F.B., M.S. and S.D.B.; supervision: M.S.; validation: C.T., M.C. and M.S.; visualization: M.S.; writing—original draft: M.S.; writing—review and editing: C.T., B.C., F.B., G.C., M.C., S.D.B. and V.P. All authors have read and agree to the published version of the manuscript.

Funding: This research was funded by PRIN-MIUR, Contract grant number: 2015SHM58M_004 to VP, Departments of Excellence, 2017, legge 232/2016, art.1, comma 314–337 awarded to Dept. of Science, University Roma Tre, Rome, Italy for 2018–2022.

Conflicts of Interest: The authors declare no conflict of interest.

Abbreviations

ACC	Acyl Coenzyme A carboxylase
BET	Bromodomain and extra-terminal domain
HDAC	histone deacetylases
HDL	high density lipoprotein
HMGCR	3-hydroxy-3-methylglutaryl Coenzyme A reductase
LDL	low density lipoprotein
LDLr	low density lipoprotein receptor
NPC1	Niemann-Pick type C1
ROS	reactive oxygen species
SR-B1	scavenger receptor class B type 1
SREBPs	sterol regulatory element binding proteins
TMEM97	transmembrane protein 97

References

1. Wakil, S.J.; Abu-Elheiga, L.A. Fatty acid metabolism: target for metabolic syndrome. *J. Lipid Res.* **2009**, *50*, S138–S143. [CrossRef]

2. Papackova, Z.; Cahová, M. Fatty Acid Signaling: The New Function of Intracellular Lipases. *Int. J. Mol. Sci.* **2015**, *16*, 3831–3855. [CrossRef]

3. Trapani, L.; Segatto, M.; Pallottini, V. Regulation and deregulation of cholesterol homeostasis: The liver as a metabolic "power station". *World J. Hepatol.* **2012**, *4*, 184–190. [CrossRef] [PubMed]

4. Chen, H.W. Role of cholesterol metabolism in cell growth. *Fed. Proc.* **1984**, *43*, 126–130. [PubMed]

5. Yao, C.H.; Fowle-Grider, R.; Mahieu, N.G.; Liu, G.Y.; Chen, Y.J.; Wang, R.; Singh, M.; Potter, G.S.; Gross, R.W.; Schaefer, J.; et al. Exogenous Fatty Acids Are the Preferred Source of Membrane Lipids in Proliferating Fibroblasts. *Cell Chem. Biol.* **2016**, *23*, 483–493. [CrossRef] [PubMed]

6. Nohturfft, A.; Zhang, S.C. Coordination of Lipid Metabolism in Membrane Biogenesis. *Annu. Rev. Cell Dev. Boil.* **2009**, *25*, 539–566. [CrossRef] [PubMed]

7. Donaldson, W.E. Regulation of fatty acid synthesis. *Fed. Proc.* **1979**, *38*, 2617–2621.

8. Segatto, M.; Manduca, A.; Lecis, C.; Rosso, P.; Jozwiak, A.; Swiezewska, E.; Moreno, S.; Trezza, V.; Pallottini, V. Simvastatin treatment highlights a new role for the isoprenoid/cholesterol biosynthetic pathway in the modulation of emotional reactivity and cognitive performance in rats. *Neuropsychopharmacology* **2014**, *39*, 841–854. [CrossRef]

9. Segatto, M.; Trapani, L.; Lecis, C.; Pallottini, V. Regulation of cholesterol biosynthetic pathway in different regions of the rat central nervous system. *Acta Physiol.* **2012**, *206*, 62–71. [CrossRef]

10. Eberlé, D.; Hegarty, B.; Bossard, P.; Ferré, P.; Foufelle, F. SREBP transcription factors: master regulators of lipid homeostasis. *Biochimie* **2004**, *86*, 839–848. [CrossRef]

11. Shen, W.J.; Azhar, S.; Kraemer, F.B. SR-B1: A Unique Multifunctional Receptor for Cholesterol Influx and Efflux. *Annu. Rev. Physiol.* **2018**, *80*, 95–116. [CrossRef] [PubMed]

12. Cartocci, V.; Segatto, M.; Di Tunno, I.; Leone, S.; Pfrieger, F.W.; Pallottini, V. Modulation of the Isoprenoid/Cholesterol Biosynthetic Pathway During Neuronal Differentiation In Vitro. *J. Cell. Biochem.* **2016**, *117*, 2036–2044. [CrossRef] [PubMed]

13. Cartocci, V.; Catallo, M.; Tempestilli, M.; Segatto, M.; Pfrieger, F.W.; Bronzuoli, M.R.; Scuderi, C.; Servadio, M.; Trezza, V.; Pallottini, V. Altered Brain Cholesterol/Isoprenoid Metabolism in a Rat Model of Autism Spectrum Disorders. *Neuroscience* **2018**, *372*, 27–37. [CrossRef] [PubMed]

14. Trapani, L.; Segatto, M.; Simeoni, V.; Balducci, V.; Dhawan, A.; Parmar, V.S.; Prasad, A.K.; Saso, L.; Incerpi, S.; Pallottini, V. Short- and long-term regulation of 3-hydroxy 3-methylglutaryl coenzyme A reductase by a 4-methylcoumarin. *Biochimie* **2011**, *93*, 1165–1171. [CrossRef] [PubMed]

15. Schulze, H.; Sandhoff, K. Lysosomal Lipid Storage Diseases. *Cold Spring Harb. Perspect. Boil.* **2011**, *3*, a004804. [CrossRef]

16. Wang, C.-Y.; Filippakopoulos, P. Beating the odds: BETs in disease. *Trends Biochem. Sci.* **2015**, *40*, 468–479. [CrossRef]

17. Segatto, M.; Fittipaldi, R.; Pin, F.; Sartori, R.; Ko, K.D.; Zare, H.; Fenizia, C.; Zanchettin, G.; Pierobon, E.S.; Hatakeyama, S.; et al. Epigenetic targeting of bromodomain protein BRD4 counteracts cancer cachexia and prolongs survival. *Nat. Commun.* **2017**, *8*, 1707. [CrossRef]

18. Andrieu, G.P.; Shafran, J.S.; Deeney, J.T.; Bharadwaj, K.R.; Rangarajan, A.; Denis, G.V. BET proteins in abnormal metabolism, inflammation, and the breast cancer microenvironment. *J. Leukoc. Boil.* **2018**, *104*, 265–274. [CrossRef]

19. Stathis, A.; Bertoni, F. BET Proteins as Targets for Anticancer Treatment. *Cancer Discov.* **2018**, *8*, 24–36. [CrossRef]

20. Klein, K. Bromodomain protein inhibition: a novel therapeutic strategy in rheumatic diseases. *RMD Open* **2018**, *4*, e000744. [CrossRef]

21. Hussong, M.; Börno, S.T.; Kerick, M.; Wunderlich, A.; Franz, A.; Sültmann, H.; Timmermann, B.; Lehrach, H.; Hirsch-Kauffmann, M.; Schweiger, M.R. The bromodomain protein BRD4 regulates the KEAP1/NRF2-dependent oxidative stress response. *Cell Death Dis.* **2014**, *5*, e1195. [CrossRef] [PubMed]

22. Siebel, A.L.; Trinh, S.K.; Formosa, M.F.; Mundra, P.A.; Natoli, A.K.; Reddy-Luthmoodoo, M.; Huynh, K.; Khan, A.A.; Carey, A.L.; Van Hall, G.; et al. Effects of the BET-inhibitor, RVX-208 on the HDL lipidome and glucose metabolism in individuals with prediabetes: A randomized controlled trial. *Metabolism* **2016**, *65*, 904–914. [CrossRef] [PubMed]

23. Sakamaki, J.I.; Wilkinson, S.; Hahn, M.; Tasdemir, N.; O'Prey, J.; Clark, W.; Hedley, A.; Nixon, C.; Long, J.S.; New, M.; et al. Bromodomain Protein BRD4 Is a Transcriptional Repressor of Autophagy and Lysosomal Function. *Mol. Cell* **2017**, *66*, 517–532.e9. [CrossRef] [PubMed]

24. Li, F.; Yang, C.; Zhang, H.-B.; Ma, J.; Jia, J.; Tang, X.; Zeng, J.; Chong, T.; Wang, X.; He, D.; et al. BET inhibitor JQ1 suppresses cell proliferation via inducing autophagy and activating LKB1/AMPK in bladder cancer cells. *Cancer Med.* **2019**, *8*, 4792–4805. [CrossRef]

25. Gilham, D.; Wasiak, S.; Tsujikawa, L.M.; Halliday, C.; Norek, K.; Patel, R.G.; Kulikowski, E.; Johansson, J.; Sweeney, M.; Wong, N.C. RVX-208, a BET-inhibitor for treating atherosclerotic cardiovascular disease, raises ApoA-I/HDL and represses pathways that contribute to cardiovascular disease. *Atherosclerosis* **2016**, *247*, 48–57. [CrossRef]

26. Zhang, P.; Wang, D.; Zhao, Y.; Ren, S.; Gao, K.; Ye, Z.; Wang, S.; Pan, C.W.; Zhu, Y.; Yan, Y.; et al. Intrinsic BET inhibitor resistance in SPOP-mutated prostate cancer is mediated by BET protein stabilization and AKT-mTORC1 activation. *Nat. Med.* **2017**, *23*, 1055–1062. [CrossRef]

27. Filippakopoulos, P.; Qi, J.; Picaud, S.; Shen, Y.; Smith, W.B.; Fedorov, O.; Morse, E.M.; Keates, T.; Hickman, T.T.; Felletar, I.; et al. Selective inhibition of BET bromodomains. *Nature* **2010**, *468*, 1067–1073. [CrossRef]

28. Leng, E.; Xiao, Y.; Mo, Z.; Li, Y.; Zhang, Y.; Deng, X.; Zhou, M.; Zhou, C.; He, Z.; He, J.; et al. Synergistic effect of phytochemicals on cholesterol metabolism and lipid accumulation in HepG2 cells. *BMC Complement. Altern. Med.* **2018**, *18*, 122. [CrossRef]

29. Mi, Y.; Tan, D.; He, Y.; Zhou, X.; Zhou, Q.; Ji, S. Melatonin Modulates lipid Metabolism in HepG2 Cells Cultured in High Concentrations of Oleic Acid: AMPK Pathway Activation may Play an Important Role. *Cell Biophys.* **2018**, *76*, 463–470. [CrossRef]

30. Nagarajan, S.R.; Paul-Heng, M.; Krycer, J.R.; Fazakerley, D.J.; Sharland, A.F.; Hoy, A.J. Lipid and glucose metabolism in hepatocyte cell lines and primary mouse hepatocytes: a comprehensive resource for in vitro studies of hepatic metabolism. *Am. J. Physiol. Metab.* **2019**, *316*, E578–E589. [CrossRef]

31. Mehlem, A.; Hagberg, C.E.; Muhl, L.; Eriksson, U.; Falkevall, A. Imaging of neutral lipids by oil red O for analyzing the metabolic status in health and disease. *Nat. Protoc.* **2013**, *8*, 1149–1154. [CrossRef] [PubMed]

32. Goldstein, J.L.; Brown, M.S. Regulation of low-density lipoprotein receptors: implications for pathogenesis and therapy of hypercholesterolemia and atherosclerosis. *Circulation* **1987**, *76*, 504–507. [CrossRef] [PubMed]

33. Yang, X.-P.; Amar, M.J.; Vaisman, B.; Bocharov, A.V.; Vishnyakova, T.G.; Freeman, L.A.; Kurlander, R.J.; Patterson, A.P.; Becker, L.C.; Remaley, A.T. Scavenger receptor-BI is a receptor for lipoprotein(a). *J. Lipid Res.* **2013**, *54*, 2450–2457. [CrossRef] [PubMed]

34. Wojtanik, K.M.; Liscum, L. The Transport of Low Density Lipoprotein-derived Cholesterol to the Plasma Membrane Is Defective in NPC1 Cells. *J. Boil. Chem.* **2003**, *278*, 14850–14856. [CrossRef] [PubMed]

35. Peake, K.B.; Vance, J.E. Defective cholesterol trafficking in Niemann-Pick C-deficient cells. *FEBS Lett.* **2010**, *584*, 2731–2739. [CrossRef] [PubMed]

36. Ebrahimi-Fakhari, D.; Wahlster, L.; Bartz, F.; Werenbeck-Ueding, J.; Praggastis, M.; Zhang, J.; Joggerst-Thomalla, B.; Theiss, S.; Grimm, D.; Ory, D.S.; et al. Reduction of TMEM97 increases NPC1 protein levels and restores cholesterol trafficking in Niemann-pick type C1 disease cells. *Hum. Mol. Genet.* **2016**, *25*, 3588–3599. [CrossRef]

37. Pesiri, V.; Totta, P.; Segatto, M.; Bianchi, F.; Pallottini, V.; Marino, M.; Acconcia, F. Estrogen receptor α L429 and A430 regulate 17β-estradiol-induced cell proliferation via CREB1. *Cell. Signal.* **2015**, *27*, 2380–2388. [CrossRef]

38. Stock, C.J.W.; Michaeloudes, C.; Leoni, P.; Durham, A.L.; Mumby, S.; Wells, A.U.; Chung, K.F.; Adcock, I.M.; Renzoni, E.A.; Lindahl, G.E. Bromodomain and Extraterminal (BET) Protein Inhibition Restores Redox Balance and Inhibits Myofibroblast Activation. *Biomed. Res. Int.* **2019**, *2019*, 1484736. [CrossRef]

39. Roberts, T.C.; Etxaniz, U.; Dall'Agnese, A.; Wu, S.-Y.; Chiang, C.-M.; Brennan, P.E.; Wood, M.J.A.; Puri, P.L. BRD3 and BRD4 BET Bromodomain Proteins Differentially Regulate Skeletal Myogenesis. *Sci. Rep.* **2017**, *7*, 1–16. [CrossRef]

40. Lu, H.; Talbot, S.; Robertson, K.A.; Watterson, S.; Forster, T.; Roy, U.; Ghazal, P. Rapid proteasomal elimination of 3-hydroxy-3-methylglutaryl-CoA reductase by interferon-γ in primary macrophages requires endogenous 25-hydroxycholesterol synthesis. *Steroids* **2015**, *99*, 219–229. [CrossRef]

41. Segatto, M.; Tonini, C.; Pfrieger, F.W.; Trezza, V.; Pallottini, V. Loss of Mevalonate/Cholesterol Homeostasis in the Brain: A Focus on Autism Spectrum Disorder and Rett Syndrome. *Int. J. Mol. Sci.* **2019**, *20*, 3317. [CrossRef]

42. Pahan, K. Lipid-lowering drugs. *Cell. Mol. Life Sci.* **2006**, *63*, 1165–1178. [CrossRef]

43. Pipalia, N.H.; Cosner, C.C.; Huang, A.; Chatterjee, A.; Bourbon, P.; Farley, N.; Helquist, P.; Wiest, O.; Maxfield, F.R. Histone deacetylase inhibitor treatment dramatically reduces cholesterol accumulation in Niemann-Pick type C1 mutant human fibroblasts. *Proc. Natl. Acad. Sci. USA* **2011**, *108*, 5620–5625. [CrossRef]

44. Meaney, S. Epigenetic regulation of cholesterol homeostasis. *Front. Genet.* **2014**, *5*, 311. [CrossRef]

45. Gaur, V.; Connor, T.; Sanigorski, A.; Martin, S.D.; Bruce, C.R.; Henstridge, D.C.; Bond, S.T.; McEwen, K.A.; Kerr-Bayles, L.; Ashton, T.D.; et al. Disruption of the Class IIa HDAC Corepressor Complex Increases Energy Expenditure and Lipid Oxidation. *Cell Rep.* **2016**, *16*, 2802–2810. [CrossRef]

46. Lin, Z.; Bishop, K.S.; Sutherland, H.; Marlow, G.; Murray, P.; Denny, W.A.; Ferguson, L.R. A quinazoline-based HDAC inhibitor affects gene expression pathways involved in cholesterol biosynthesis and mevalonate in prostate cancer cells. *Mol. BioSyst.* **2016**, *12*, 839–849. [CrossRef]

47. Ferrari, A.; Fiorino, E.; Giudici, M.; Gilardi, F.; Galmozzi, A.; Mitro, N.; Cermenati, G.; Godio, C.; Caruso, D.; De Fabiani, E.; et al. Linking epigenetics to lipid metabolism: Focus on histone deacetylases. *Mol. Membr. Boil.* **2012**, *29*, 257–266. [CrossRef]

48. Bartz, F.; Kern, L.; Erz, D.; Zhu, M.; Gilbert, D.; Meinhof, T.; Wirkner, U.; Erfle, H.; Muckenthaler, M.; Pepperkok, R.; et al. Identification of Cholesterol-Regulating Genes by Targeted RNAi Screening. *Cell Metab.* **2009**, *10*, 63–75. [CrossRef]

49. Sasaki, M.; Terao, Y.; Ayaori, M.; Uto-Kondo, H.; Iizuka, M.; Yogo, M.; Hagisawa, K.; Takiguchi, S.; Yakushiji, E.; Nakaya, K.; et al. Hepatic overexpression of idol increases circulating protein convertase subtilisin/kexin type 9 in mice and hamsters via dual mechanisms: sterol regulatory element-binding protein 2 and low-density lipoprotein receptor-dependent pathways. *Arterioscler. Thromb. Vasc. Biol.* **2014**, *34*, 1171–1178. [CrossRef]

50. Kim, Y.-M.; Shin, H.-T.; Seo, Y.-H.; Byun, H.-O.; Yoon, S.-H.; Lee, I.-K.; Hyun, N.-H.; Chung, H.-Y.; Yoon, G. Sterol Regulatory Element-binding Protein (SREBP)-1-mediated Lipogenesis Is Involved in Cell Senescence*. *J. Boil. Chem.* **2010**, *285*, 29069–29077. [CrossRef]

51. Segatto, M.; Trapani, L.; Di Tunno, I.; Sticozzi, C.; Valacchi, G.; Hayek, J.; Pallottini, V. Cholesterol Metabolism Is Altered in Rett Syndrome: A Study on Plasma and Primary Cultured Fibroblasts Derived from Patients. *PLoS ONE* **2014**, *9*, e104834. [CrossRef]

52. Wen, Y.-A.; Xiong, X.; Zaytseva, Y.Y.; Napier, D.L.; Vallee, E.; Li, A.T.; Wang, C.; Weiss, H.L.; Evers, B.M.; Gao, T. Downregulation of SREBP inhibits tumor growth and initiation by altering cellular metabolism in colon cancer. *Cell Death Dis.* **2018**, *9*, 265. [CrossRef]

53. Shi, J.; Vakoc, C.R.; Junwei, S. The mechanisms behind the therapeutic activity of BET bromodomain inhibition. *Mol. Cell* **2014**, *54*, 728–736. [CrossRef]
54. Gobbi, G.; Donati, B.; Valle, I.F.D.; Reggiani, F.; Torricelli, F.; Remondini, D.; Castellani, G.; Ambrosetti, D.C.; Ciarrocchi, A.; Sancisi, V. The Hippo pathway modulates resistance to BET proteins inhibitors in lung cancer cells. *Oncogene* **2019**, *38*, 6801–6817. [CrossRef]
55. Göritz, C.; Thiebaut, R.; Tessier, L.-H.; Nieweg, K.; Moehle, C.; Buard, I.; Dupont, J.-L.; Schurgers, L.J.; Schmitz, G.; Pfrieger, F.W. Glia-induced neuronal differentiation by transcriptional regulation. *Glia* **2007**, *55*, 1108–1122. [CrossRef]
56. Rocha, J.; Trapani, L.; Segatto, M.; Rosa, P.; Nogueira, C.; Zeni, G.; Pallottini, V. Molecular Effects of Diphenyl Diselenide on Cholesterol and Glucose Cell Metabolism. *Curr. Med. Chem.* **2013**, *20*, 4426–4434. [CrossRef]

International Journal of
Molecular Sciences

Article

Epinephrine Infiltration of Adipose Tissue Impacts MCF7 Breast Cancer Cells and Total Lipid Content

Pierre Avril [1,†], Luciano Vidal [1,†], Sophie Barille-Nion [2], Louis-Romée Le Nail [1], Françoise Redini [1], Pierre Layrolle [1], Michelle Pinault [3], Stéphane Chevalier [3], Pierre Perrot [1,4,*] and Valérie Trichet [1]

[1] INSERM, Université de Nantes, UMR1238, Phy-Os, Sarcomes osseux et remodelage des tissus calcifiés, F-44035 Nantes, France; pierre.avril.44@gmail.com (P.A.); luciano.vidal@univ-nantes.fr (L.V.); lrlenail@hotmail.com (L.-R.L.N.); francoise.redini@univ-nantes.fr (F.R.); pierre.layrolle@univ-nantes.fr (P.L.); valerie.trichet@univ-nantes.fr (V.T.)

[2] CRCINA, INSERM, Université d'Angers, Université de Nantes, F-44035 Nantes, France; sophie.barille@univ-nantes.fr

[3] INSERM Université de Tours, UMR1069, Nutrition, Croissance et Cancer, F-37032 Tours, France; michelle.pinault@univ-tours.fr (M.P.); stephane.chevalier@univ-tours.fr (S.C.)

[4] CHU de Nantes, Service de Chirurgie Plastique et des Brûlés, F-44035 Nantes, France

[*] Correspondence: pierre.perrot@chu-nantes.fr; Tel.: +33-2-40-08-73-02

[†] These authors contributed equally to this work.

Received: 6 October 2019; Accepted: 4 November 2019; Published: 11 November 2019

Abstract: Background: Considering the positive or negative potential effects of adipocytes, depending on their lipid composition, on breast tumor progression, it is important to evaluate whether adipose tissue (AT) harvesting procedures, including epinephrine infiltration, may influence breast cancer progression. Methods: Culture medium conditioned with epinephrine-infiltrated adipose tissue was tested on human Michigan Cancer Foundation-7 (MCF7) breast cancer cells, cultured in monolayer or in oncospheres. Lipid composition was evaluated depending on epinephrine-infiltration for five patients. Epinephrine-infiltrated adipose tissue (EI-AT) or corresponding conditioned medium (EI-CM) were injected into orthotopic breast carcinoma induced in athymic mouse. Results: EI-CM significantly increased the proliferation rate of MCF7 cells Moreover EI-CM induced an output of the quiescent state of MCF7 cells, but it could be either an activator or inhibitor of the epithelial mesenchymal transition as indicated by gene expression changes. EI-CM presented a significantly higher lipid total weight compared with the conditioned medium obtained from non-infiltrated-AT of paired-patients. In vivo, neither the EI-CM or EI-AT injection significantly promoted MCF7-induced tumor growth. Conclusions: Even though conditioned media are widely used to mimic the secretome of cells or tissues, they may produce different effects on tumor progression, which may explain some of the discrepancy observed between in vitro, preclinical and clinical data using AT samples.

Keywords: adipose tissue; breast cancer; epinephrine; breast reconstruction

---

1. Introduction

Adipose tissue (AT) is a biologically active tissue, which releases soluble growth factors (vascular endothelium growth factor, insulin growth factor) inducing tissue revascularization, but also produces hormones (leptin, adiponectin), cytokines (interleukin 6) and insoluble fatty acids, which all interact through complex networks within a tumor microenvironment. Different in vitro and pre-clinical studies have demonstrated that AT including mature adipocytes and stem cells, promotes the proliferation, invasion and survival of breast cancer cells through different secreted factors [1–5].

In contrast specific lipids content in peritumoral AT of breast cancer patients were correlated with therapeutic benefits. Decreased levels of two polyunsaturated n-3 fatty acids (n-3 PUFA), docosahexaenoic and eicosapentaenoice acids (DHA 22:6n-3 and EPA 20:5n-3), in peritumoral AT of women were associated with aggressive multifocal tumors compared to unifocal ones. Moreover it was shown that DHA and EPA decreased resistance of experimental mammary tumors to taxanes, anthracyclines or radio-therapy. These preclinical results are supported by the improved outcome of chemotherapy on metastatic breast cancer that was observed in a phase II clinical study including dietary supplementation with n-3 PUFA [6–11].

In addition, AT transfer does not increase the risk of recurrence of breast cancer as recently suggested by large retrospective clinical studies [12–14]. However one retrospective study observed that patients presenting either ductal or lobular intraepithelial neoplasia, had an increased risk of local events in the group who had undergone lipofilling [15]. The method to harvest AT is one of the most important steps governing the success of AT transplantation. Different methods have been described in the literature studying the variables (fat harvesting technique, infiltration solution, donor site, fat processing, etc.) that influence adipocyte survival and the AT engraftment [16–19]. However, the breast cancer recurrence risk related to the AT harvesting method has not yet been investigated. To harvest fat tissue, some surgical teams use for fat harvesting, an infiltration solution that contains epinephrine, to induce vasoconstriction [20], but it is worth noting that epinephrine may enhance lipolysis in AT [21].

In our study, the ephinephrine-lactated Ringer's infiltrated solution adipose tissue conditioned medium (EI-CM) was tested on proliferative and quiescent human Michigan Cancer Foundation-7 (MCF7) breast cancer cells. The lipid composition of conditioned media of non-infiltrated or epinephrine infiltrated-AT from five donors was investigated. EI-CM and epinephrine-infiltrated adipose tissue (EI-AT) were injected into orthotopic induced breast carcinoma in athymic mouse.

2. Results

2.1. MCF7 Cell Proliferation was Enhanced by Epinephrine-Infiltrated Adipose Tissue Conditioned Medium

Proliferation of MCF7 breast carcinoma cells was analyzed in adherent culture conditions by measuring mitochondrial activity (WST-1 assay) and cell viability was controlled by cell counting with trypan blue staining. As cancer cells lost their adherence to plastic when whole epinephrine-infiltrated adipose tissue (EI-AT) was used, EI-AT was replaced by EI-CM to complement cell culture medium in order to mimic secreted factors by EI-AT. Before being treated, cells were cultured overnight (16 h) in a standard medium without fetal bovine serum (FBS) in order to observe a synchronized response to growth factor stimulation. FBS supplementation (10%) enhanced MCF7 mitochondrial activity up to 240% compared to that without FBS (Figure 1a). Similarly supplementation with 50% epinephrine-infiltrated adipose tissue conditioned medium (EI-CM) and 50% MEM α medium (0% FBS) from three different donors enhanced MCF7 mitochondrial activity from 150 to 200% (Figure 1a). The increases of mitochondrial activity were correlated with an increase in cell number by trypan bleu counting. Moreover, independent experiments performed with 50% or 25% of EI-CM of patients n°1 to 3 showed similar increases of MCF7 cell proliferation. The inhibition of the ERK1/2 signaling pathway using 5 μM UO126 induced a 30% decrease of the MCF7 proliferation with EI-CM, whereas it did not change the MCF7 proliferation with 10% FBS (Figure 1b). These results indicate that EI-CM increases MCF7 cell proliferation at least partially through the ERK1/2 signaling pathway.

Cell distribution in each cell-cycle phases was observed by flow cytometry after DNA staining with propidium iodide. During culture without FBS, at least half of the MCF7 cells were in G_0/G_1 phase (54% in Figure 1c, top panel). FBS treatment decreased the proportion of cells in G_0/G_1 phase by half and increased the proportion of cells preparing their mitosis and those replicating their DNA (Figure 1c, middle panel). When MCF7 cells were treated with 25% EI-CM (Figure 1c, low panel), the proportion of cells in G_0/G_1 phase was also reduced by half compared to 0% FBS culture condition. With 25% EI-CM, a higher increase in cells in G_2/M phase was observed compared to 10% FBS (plus

20% versus plus 11%) whereas the increase of cell proportion in S phase was weaker than with 10% FBS (plus 4% versus plus 15%). These results indicate that EI-CM complementation induced cell-cycle activation in MCF7 cells allowing cells to reach the G_2/M phase faster than FBS complementation.

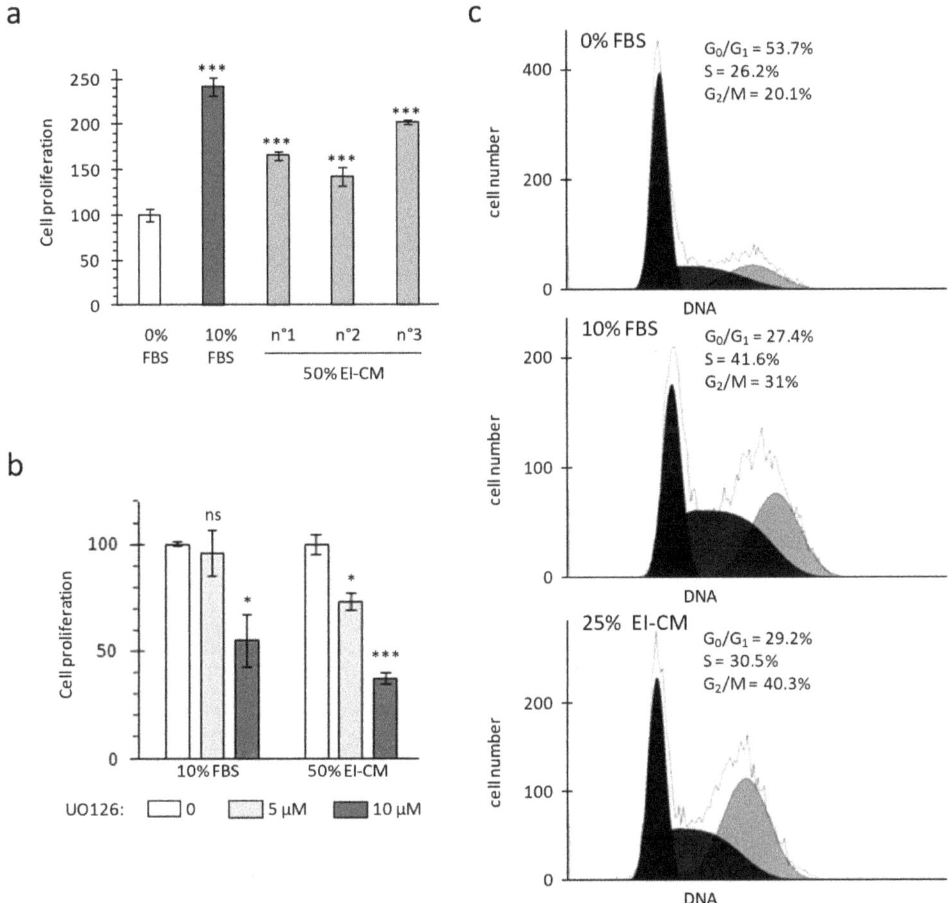

Figure 1. Michigan Cancer Foundation-7 (MCF7) cell growth with ELR solution-infiltrated adipose tissue conditioned medium (EI-CM). MCF7 cells were cultured during 24 h in a medium without any growth factor (0% FBS) or supplemented with fetal bovine serum (10% FBS) or EI-CM (25 to 50%) and MEM with 0%FBS. (**a**) Histogram shows the mitochondrial activity of MCF7 cells, measured by WST-1 assay. EI-CM were derived from 3 different donors (n°1 to n°3). Results are the means of 3 wells and are presented as a percentage of 0% FBS value with standard deviations. Statistically significant differences are indicated in comparison with 0% FBS (***: $p < 0.0001$). Each patient EI-CM was tested in 3 independent experiments. (**b**) Mitochondrial activity of MCF7 cells measured by WST-1 assay. Cells were cultured for 24 h with or without 5 or 10 µM of ERK inhibitor UO126. Results are the means of 3 wells and are presented as a percentage of condition without UO126 with standard deviations. Statistically significant differences are indicated in comparison with 0 UO126 (*: $p < 0.05$; ***: $p < 0.0001$). Two independent experiments were performed. (**c**) Histograms show the distribution of MCF7 cells in cell-cycle phases following DNA detection by flow cytometry. Because only 2–3% of cells were identified in the subG_0 phase, only the proportion of cells in the G_0/G_1, S and G_2/M phases are indicated.

2.2. MCF7 Cell Quiescence was Increased by Sphereoid Culture and Reduced by Epinephrine-Infiltrated Adipose Tissue Conditioned Medium

Cell culture under anchorage-independent conditions induces carcinoma cells to form spheres and to undergo epithelial mesenchymal transition (EMT) which may correlate with a more invading phenotype such as carcinoma stem cells [22,23]. From MCF7 spheres, messenger ribonucleic nucleic acids (mRNAs) were isolated for relative gene expression analysis after three days in culture. Five genes *MYC, CD44, TWIST1, TWIST2* and *SNAI1* (official symbols and full gene names presented in Table 1) which are activated in breast carcinoma stem cells and during EMT exhibited a higher expression in MCF7 cells cultured as spheroids (3-D) compared to that in MCF7 cells cultured in monolayer (2-D) (Figure 2a). In accordance with EMT, E-cadherin gene (*CDH1*) expression was lower in MCF7 spheroids than in monolayers. EI-CM treatment of 3-D cultured MCF7 cells increased expression of *MYC* and *TWIST2* while it decreased that of *CD44* and *SNAI1*, suggesting that the effect of EI-CM on EMT in MCF7 cells could be either as an activator or inhibitor of the EMT.

Tumor recurrence can be explained by the persistence of cancer cells in a quiescent/non-dividing stage that enables them to escape to chemotherapy agents during treatment, whereas microenvironment changes may later activate a cellular switch towards cell division. Quiescent cells corresponding to cells in G_0 phase do not express the Ki-67 protein which is strictly associated with dividing cells in the G_1, S, G_2 or M phases [24]. Under anchorage-independent conditions (3-D), 26.9% of MCF7 cells were in G_0 phase (Figure 2b, left panel). In 3-D culture conditions, 10% FBS complementation did not change cell distribution (Figure 2b, middle panel). In contrast, EI-CM complementation of 3-D cultured MCF7 cells induced a decrease of cell proportion in G_0 phase (13.9%; Figure 2b, right panel).

Immunohistochemistry (IHC) targeting the Ki-67 protein was performed on spheroids formed by MCF7 cells cultured under anchorage-independent conditions (Figure 2c). Consistent with flow cytometry, 25% EI-CM treatment induced an increase of the Ki-67 positive cell proportion (+25%; Figure 2d left panel). Despite cell cycle activation by EI-CM, sphere number and volume (Figure 2d, right panel) were not significantly different between 1% FBS and 25% EI-CM complementation. Altogether, this indicates that EI-CM enabled G_0 to G_1 phase transition of MCF7 cells.

Table 1. List of genes analyzed by real time RT-PCR: Genes are presented with official gene symbols and corresponding full name. Forward and reverse primer sequences used to perform the analyses are indicated.

Official Symbol	Official Full Name	Reverse Primer
HPTR1	Hypoxanthine PhosphoRibosyl Transferase 1	CGAGCAAGACGTTCAGTCCT
CD44	Cluster of Differentiation 44	CGGCAGGTTATATTCAAATCG
TWIST1	Twist-related protein 1	TGCAGAGGTGTGAGGATGGTGC
TWIST2	Twist-related protein 2	AGAAGGTCTGGCAATGGCAGCA
SNAI1	Snail family transcriptional repressor 1	CAGCAGGTGGGCCTGGTCGTA
CDH1	Cadherin 1	CCAGCGGCCCCTTCACAGTC
MYC	Myelocytomatosis viral oncogene homolog	GATCCAGACTCTGACCTTTTGC

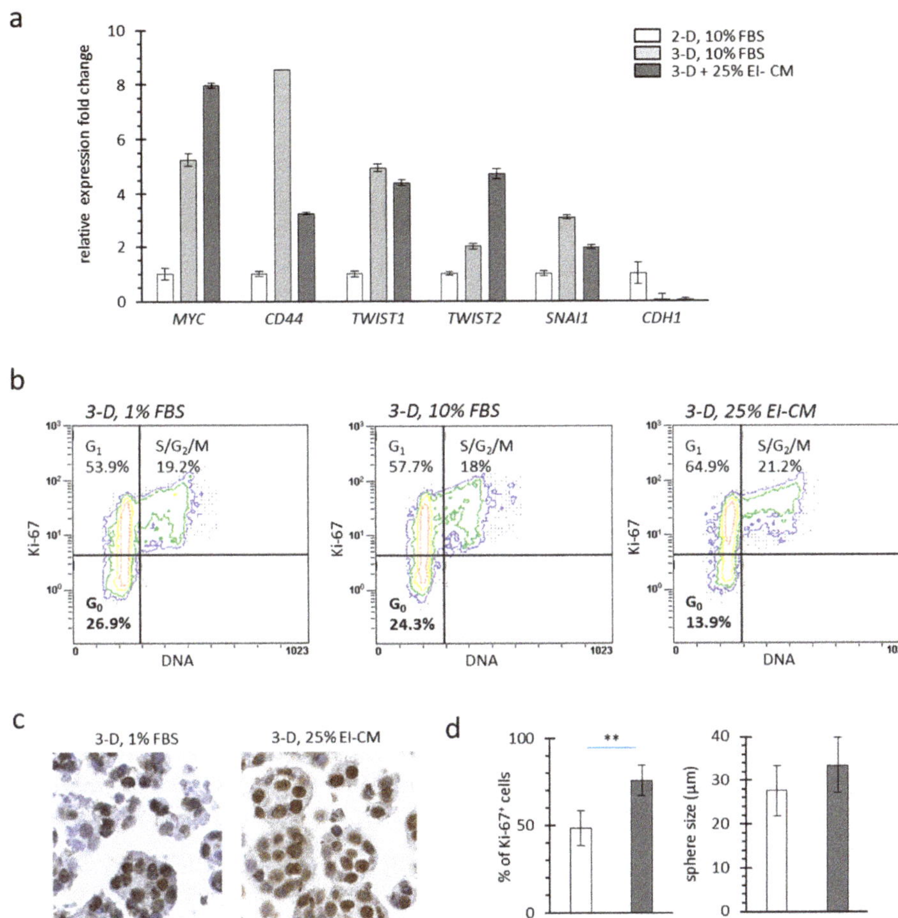

Figure 2. Quiescence of MCF7 cells with ELR solution-infiltrated adipose tissue conditioned medium (EI-CM). (**a**) Relative expression fold changes are presented for mRNA of MCF7 cells cultured either in monolayer (2-D) or in sphere (non-anchorage conditions, 3-D) during 3 days and treated 48 h in a medium supplemented with 1 or 10% FBS or with 25% EI-CM. Expression change of those 5 genes has been correlated to epithelial mesenchymal transition leading to invading phenotype of carcinoma cells. Full gene names and symbols are indicated in Table 1. Means of 3 samples are presented with standard deviations. Significant differences are not indicated as this experiment was not repeated. (**b**) Dot plots show the distribution in cell cycle phases (G_0, G_1 and $S/G_2/M$) following DNA and Ki-67 detection in MCF7 cells cultured in spheres (3-D). Because only 2–3% of cells were identified in subG₀ phase, only the proportion of cells in G_0, G_1, and $S/G_2/M$ phases are indicated. (**c**) Representative images of IHC detection of Ki-67 on MCF7 spheres (3-D). Magnification is indicated. (**d**) Histograms show the Ki-67 index (left panel) and the mean diameter (right panel) of MCF7 spheres. In the left panel, percentages were counted on 6 representative regions for each treatment after Ki-67 detection by IHC. In the right panel, mean diameter was calculated on 90 different spheres for each condition. Statistically significant differences are indicated in comparison with 1% FBS (**: $p < 0.001$). Three independent experiments were performed.

2.3. Epinephrine Infiltration Changed Lipid Content and Proliferative Effect of Adipose Tissue Conditioned Medium

For patients n°4 to 10, two AT samples were successively collected: a first one without epinephrine infiltration and a second one following infiltration with the epinephrine-lactated Ringer's solution (ELR) and were used to obtain, respectively, NI-CM and EI-CM. As previously observed with EI-CM from patients n°1 to 3, the CM obtained from EI-AT samples had an increased proliferative effect on MCF7 cells, whereas their counterpart CM obtained with non-infiltrated AT did not (Figure 3a, left panel). As shown in Figure 3a (right panel), the ELR by itself had no effect on MCF7 cell proliferation. Because epinephrine infiltration may modify the metabolite composition of the AT samples through lipolysis enhanced by beta-adrenergic receptor activation, the comparison of lipid contents between NI-CM and EI-CM was performed to identify potential molecular mediators leading to EI-CM pro-proliferative effects. We compared the fatty acid content of AT-CM samples that were obtained from 5 donor sites either non-injected (NI) or ephinephrine-lactated Ringer's infiltrated (EI). EI-CM showed a higher total lipid content compared to NI-CM of the corresponding donors (Figure 3b). This result suggested that EI-CM and NI-CM may present a different lipid content; however, the percentages of saturated, mono-unsaturated or polyunsaturated fatty acids (PUFAs n-3 and n-6) were similar (Figure 3c) and there were no statistically significant differences in individual fatty acid between EI-CM and NI-CM samples which were derived from 5 donor sites either infiltrated or non-infiltrated with ELR.

2.4. Injection of Epinephrine-Infiltrated Adipose Tissue or Corresponding Conditioned Medium into MCF7 Tumor in Mice

We were able to compare EI-AT and EI-CM injection in a preclinical model of breast carcinoma. Orthotopic breast carcinoma were induced in athymic mice by intraductal injection of MCF7 cells and a single injection of either PBS, EI-CM or EI-AT was performed at the tumor site after 90 days when tumors were detectable (>70 mm^3).

PBS-treated group (Figure 4b, top panel) showed a slow tumor development, reaching a mean volume of 200 mm^3 at day 160. Human Ki-67 protein detection confirmed the presence of tumor cells in mammary ducts (Figure 4a, top panel) as well as in surrounding adipose and connective structures (Figure 4a, middle panel). These observations may indicate that the tumor first grew within mammary ducts before invading the rest of mammary fat pad, as a ductal carcinoma in situ that would have turned invasive.

Two of six tumors in the EI-CM-injected group showed faster development compared to tumors of the PBS-injected group (Figure 4b, top and middle panels, respectively); however differences between the tumor volume means of these two groups were not significant at day 160. Tumor growth was more similar between PBS-and EI-AT-injected groups (Figure 4b, top and low panels, respectively) than between PBS- and EI-CM-injected groups. However the percentages of Ki-67 positive cells ranging from 18 to 26% were not significantly different between EI-AT-, EI-CM- and PBS-treated groups as determined after human Ki-67 protein immunohistochemical staining on tumor samples (Figure 4a). These results indicate that EI-CM may have slightly (but not significantly) promoted MCF7 tumor growth while corresponding whole EI-AT may not have modified breast tumor growth.

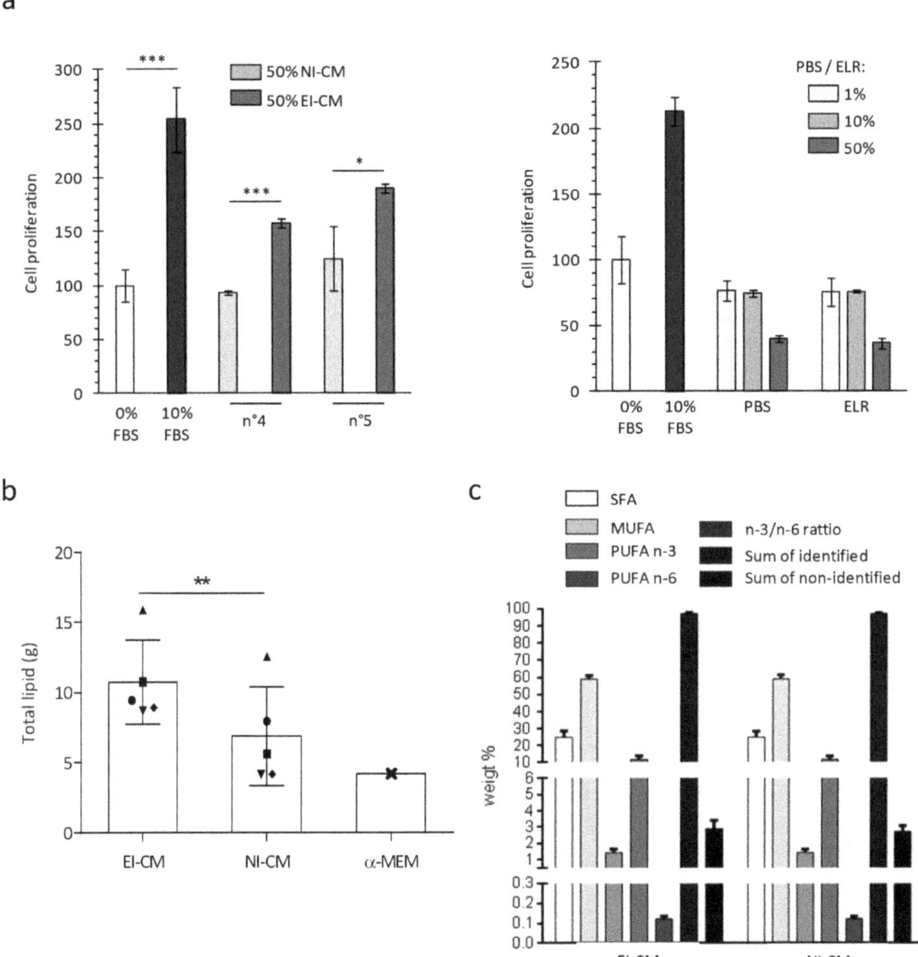

Figure 3. MCF7 cell growth and lipid composition depending on ELR solution infiltration of harvested adipose tissue. (**a**) Histogram shows the mitochondrial activity of MCF7 cells measured after 24 h. For the left panel, cells were cultured in medium without FBS (0% FBS) or supplemented with 10% FBS or with 50% EI-CM and 50% MEM α with 0%FBS. AT samples were obtained from 2 different donors (n°4 and n°5) and were initially not-infiltrated (NI) or infiltrated with ELR (EI). For right panel, cells were cultured in a medium without FBS (0% FBS) or supplemented with 10% FBS, PBS or ELR, representing 1 to 50% of the total volume. *: $p < 0.05$; ***: $p < 0.0001$. (**b**) Histogram shows the total lipid amount detected in conditioned medium from infiltrated with ELR (EI-CM) or not-infiltrated AT (NI-CM) for 5 patients (n°6 to n°10) who are represented by a distinct geometric forms. Lipid amount is indicated in standard culture medium without FBS (MEM α). **: $p = 0.0045$ paired t-test. (**c**) Histogram shows the weight % of fatty acids derived from 5 donors either infiltrated or non-infiltrated with ELR. Saturated, mono-unsaturated or polyunsaturated fatty acids (SFA, MUFA or PUFA) were measured in a conditioned medium of epinephrine lactated Ringer's solution-infiltrated or non-infiltrated adipose tissue (EI-CM or NI-CM).

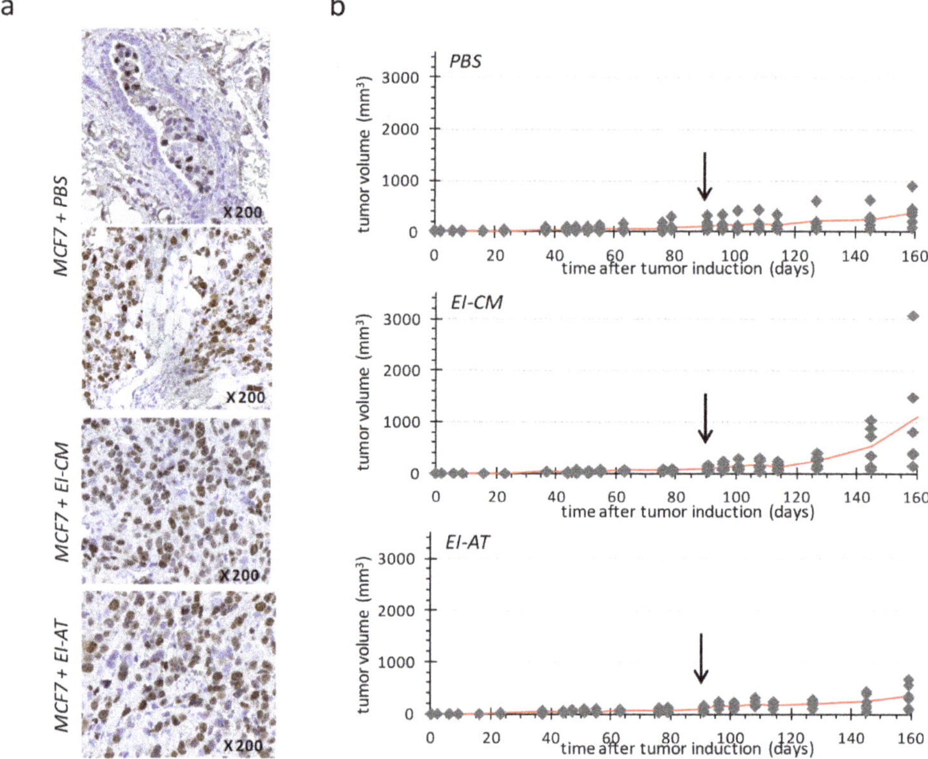

Figure 4. Single injection of either EI-AT or EI-CM in MCF7 tumor induced in athymic mouse. (**a**) Immunohistochemical staining of human Ki-67 protein in MCF7 tumors injected with PBS, EI-CM or EI-AT. Observation of mammary duct (top panel) and mammary fat pad (middle panel). (**b**) Evolution of tumor volume is reported for each group of mice ($n = 6$); the mean tumor volume is represented by a red line. The arrows indicate time of the intra-tumor injection of either PBS (top panel), or EI-CM (middle panel group) or EI-AT (low panel). At days 145 and 160, no significant difference between groups was detected by unpaired nonparametric method.

3. Discussion

AT transplantation has become an increasingly common technique in aesthetic breast augmentation, in non-oncological and in oncological breast reconstruction [25]. Low complication rates, readily available donor sites with low donor-site morbidity and an aesthetic benefit are some of the advantages of the AT transfer. Although AT transplantation has proven effective in breast reconstruction, safety concerns have been raised regarding its use in patients with a history of breast cancer [26–28]. At present, large cohort retrospective studies or systematic literature reviews and meta-analyses suggest that AT transplantation is safe with no increase of cancer recurrence risk for breast cancer patients after treatments [29–33]. However AT harvesting procedures have been poorly described in these clinical studies, whereas the lack of standardized protocols for harvesting may explain unpredictable clinical outcomes with AT engraftment [34–36].

Ephinephrine-lactated Ringer's solution is often used to infiltrate AT before harvesting, in order to reduce bleeding, but it may modify the metabolite composition, especially cholesterol and fatty acids, of AT samples since catecholamines induce lipolysis in adipocytes [37–39]. Interestingly, polyunsaturated n-3 fatty acids in peritumoral AT of breast cancer patients may have beneficial effects on the disease progression.

In this study, we sought to understand how soluble factors secreted by EI-AT may influence the proliferation and quiescent state of breast cancer cells. EI-AT secreted factors that were collected in the conditioned culture medium induced a significant increase in the proliferation rate (150% to 200%) of MCF7 breast cancer cells, while non-infiltrated AT soluble factors did not. EI-AT secreted factors increased MCF7 cell proliferation at least partially through the extracellular-regulated kinase (ERK) 1/2 signaling pathway. In a previous study using similar EI-CM samples, multi cytokine assay has identified interleukin 6 (IL-6) and leptin as molecular candidates to induce increase of osteosarcoma cell proliferation; however neither IL-6 nor leptin have been able to mimic the pro-proliferative effects of EI-CM. By in vitro and preclinical studies, Danilo C. et al. have shown that cholesteryl ester via its cellular receptor (scavenger receptor class B type I, SR-BI) increase breast cancer cell proliferation through the phosphatidylinositol 3-kinase (PI3K)/protein kinase B (AKT) pathway but not through the mitogen-activated protein kinase (MAPK)/ERK1/2 pathway [40]. Despite the important role of ERK1/2 in the proliferation of breast cancer cells in vitro, activation of ERK1/2 was not associated with enhanced proliferation and invasion of 148 clinical mammary carcinomas [41].

During clinical procedures, EI-AT transplantation is never performed in a tumor site with proliferative cancer cells. Plastic surgery is performed following a cancer-free period and at worst, the primary tumor site may contain quiescent/dormant cancer cells. The quiescent state of breast cancer cells in vitro was induced by culture under anchorage-independent conditions, using methylcellulose in the culture medium. We observed that EI-CM induced an output of the quiescent state of MCF7 cells when maintained in non-adherent spheres: 14% of cells in G_0 phase with EI-CM compared to 27% or 24% in 1% and 10% FBS supplementation, respectively. Such 3-D culture conditions induced a slight change from epithelial towards mesenchymal phenotypes of MCF7 cells, as suggested by *MYC*, *CD44*, *TWIST1/2* and *SNAI1* expression increase associated with a decrease of *CDH1* expression. We observed that EI-CM did not enhance such potential EMT in MCF7 cells. In this study, we did not test EI-CM effect on the migration of breast cancer cells and we did not use primary breast cancer stem cells which are of high interest in the progression, treatment resistance and recurrence. Originally, Charvet H.J. et al. have used one breast cancer cell line derived from one out of 10 patient specimens, and not a purchased banked cell line. Charvet H.J. et al. showed a 10-fold migration increase of primary breast cancer cells when cocultured with adipose-derived stem cells isolated from the same patient.

To conduct in vitro assays, an AT-conditioned medium (AT-CM) is usually used in the literature, instead of the whole AT sample itself which would be injected into a patient. AT-CM contains AT secreted and soluble factors including growth factors, cytokines and free fatty acids, but no adipocytes or adipose-derived stem cells. Dirat B. et al. have demonstrated that adipocytes obtained from breast AT during tumorectomy, exhibit a loss of lipid content, an expression increase of proinflammatory cytokines and the ability to increase invasive capacities of breast cancer cell lines.

Conditioned media derived from epinephrine-infiltrated AT showed a pro-proliferative effect on breast cancer cells and significantly higher lipid contents compared to non-infiltrated AT of corresponding patients. However, the percentages of saturated, mono-unsaturated or polyunsaturated fatty acids were similar in EI- or NI-CM. Recently, Wang Y.Y. et al. showed that free fatty acids released from adipocytes were incorporated into breast cancer cells as triglycerides in lipid droplets and that saturated fatty acids but not unsaturated ones were increased in cocultured cells [42]. Additionally, they demonstrated that lipolysis in adipocytes was induced by tumor cell secretions, but was not induced by catecholamines. In our study, only patients with a standard body mass index ranging from 20 to 22 were included. Because obesity is clearly related to a higher risk of cancer [43], including breast cancer risk, it would be of interest to compare the total lipid contents of EI-CM derived from obese and lean patients.

Epinephrine-infiltrated adipose tissue-conditioned medium (EI-CM) and whole epinephrine-infiltrated adipose tissue (EI-AT) of the same patient were compared following a single injection within breast carcinomas induced in athymic mice by intraductal injection of MCF7 cells. Interestingly, carcinoma growth was slow in this preclinical model; tumors were visible only 90 days after MCF7 cell injection, despite implantation of pellet delivering 17β-estradiol, and tumor volumes were less than 400 mm^3 two weeks after EI-AT or EI-CM single injections. A single injection of EI-CM may have slightly but not significantly increased MCF7 tumor growth compared to a single PBS injection. The untranslation of the EI-CM cellular effects to in vivo effects may be due to reversed or transient effects that have not been tested in our in vitro study, or due to neutralization through molecular interactions with physiological liquids. In contrast, corresponding whole EI-AT injection did not modify breast tumor growth, in agreement with clinical studies which show that AT transfer did not increase tumor recurrence and then, may have no effect on quiescent tumor cells.

MCF-7 cell line which is a luminal A subtype of breast cancer expressing estrogen and progesterone receptors, is not the more aggressive and invasive model. It will be of high interest to test a model with a higher metastatic potential such as the MDA-MB-231 cell line, a basal subtype of triple negative breast cancer. We observed that EI-CM of patients n°1 to 3 induced 50% increase in the proliferation rate of MDA-MB-231 cells in culture (data not shown), but we did not investigate further this cell line as we were not able to establish an adequate in vivo model using it. A low tumor incidence (50%) with high variability of tumor size was obtained using MDA-MB-231 cell injection in nude mice.

In conclusion, epinephrine-infiltration of AT induces secretion of factors including lipids. This may contribute to the pro-proliferative effect and output of the quiescent state that were observed in vitro on MCF7 breast cancer cells. Moreover, such epinephrine-induced secreted factors seem to increase the in vivo growth of MCF7-induced tumor in mice. However, a single injection of whole epinephrine-infiltrated AT did not increase the slow progression of MCF7-induced tumor in mice, revealing a discrepancy between the effects of AT-secreted soluble factors in the conditioned medium and the whole AT sample which would be injected into a patient. The proportion of polyunsaturated fatty acids was not modified by the epinephrine infiltration, despite the significant increase in secreted lipids by EI-AT. The results of the EI-AT presented here do not call into question the safety of AT transplantation, however it would be of interest to compare cancer recurrences in breast cancer patients following the transplantation of AT harvested with or without epinephrine infiltration.

4. Materials and Methods

4.1. Adipose Tissue (AT)

AT samples: Human adipose tissue (AT) was obtained from abdominal liposuction during plastic surgery at the University Hospital of Nantes. Donors (patients n°1 to 10) with no significant medical history gave informed consent for the use of surplus AT sample for anonymized unlinked research, as validated by the "Comité de Protection des Personnes des Pays de la Loire" and by the "Ministère de la Recherche" (Art. L. 1245-2 of the French public health code, Law no. 2004-800 of 6 August 2004, Official Journal of 7 August 2004) with declaration to the "Commission Nationale de l'Informatique et des Libertés". Ten AT samples were collected using a 12-gauge, 12-hole cannula (Khouri Harvester) connected to a 10 mL Luer-Lock syringe after infiltration with 0.1% epinephrine lactated Ringer's solution as performed in our department to reduce bleeding. All patients had a body mass index ranging from 20 to 22. For five patients, we obtained both epinephrine infiltrated (EI) and non-infiltrated (NI) samples. Samples were centrifuged at 3000 rpm with a 9.5 cm radius fixed angle rotor for 1 min (Medilite™, Thermo Fisher Scientific, Illkirch, France) at room temperature. After centrifugation, the samples were separated into 3 layers: the upper one composed of oil, the middle one composed of the adipose tissue and the bottom one with blood and infiltration solution. Only middle layers corresponding to AT samples were collected.

AT-Conditioned Medium (AT-CM): AT samples were placed in cell culture inserts (pore size 3 μm; Becton Dickinson, Le Pont de Claix, France) with Minimum Essential Medium alpha (Gibco® MEM α; Life technologies, St Aubin, France) with nucleosides and 1 g/L D-Glucose (MEM) under serum-free conditions. After 24 h, inserts with AT were removed and AT-CM was collected and frozen at minus 20 °C.

4.2. Culture Conditions

MCF7 cells were initially derived from a human breast adenocarcinoma ATCC number HTB-22, (ATCC, Manassas, VA, USA). They are a luminal subtype and express estrogen, progesterone and glucocorticoid receptors. MEM α medium was supplemented with 10% fetal bovine serum (FBS) and used to culture cells at 37 °C in a humidified atmosphere (5% CO_2/95% air). For culture under anchorage-independent conditions, 1 mL MEM α medium was supplemented with 1.05% of methylcellulose (R&D Systems, Lille, France) and 1% FBS, and was seeded with 1×10^5 cells into a well of 24-well plate for 3 days. Then 0.5 mL MEM α medium supplemented with 1% or 10% FBS or with 25% EI-CM were added for 2 days. Complete FBS starving (0%) was avoided in 3-D culture to maintain a low proportion (<5%) of cells in subG0 phase.

4.3. Cell Viability

Three thousand cells per well were cultured into 96-well plates with medium supplemented with FBS, CM or chemical inhibitor of ERK1/2 phosphorylation, UO126 (R&D Systems). After 24 h, WST-1 reagent (Roche Diagnostics, Meylan, France) was added to each well for 2 h at 37 °C. Then absorbance was read at 450 nm.

4.4. Cell Cycle Analysis

MCF7 cells were obtained and analyzed as previously described [44]. Briefly, DNA was stained in ethanol-fixed cells with propidium iodide (50 µg/mL; Sigma-Aldrich, Lyon, France) and Ki-67 protein was eventually detected using a FITC-coupled mouse anti-human Ki-67 antibody (Becton-Dickinson, Le pont de Claix, France). Cell fluorescence was measured by flow cytometry (Cytomics FC500; Beckman Coulter, Villepinte, France). Cell-cycle distribution was analyzed for 20,000 events using MultiCycle AV Software, Windows version (Phoenix Flow System, San Diego, CA, USA) to obtain histograms of cell repartition in each cell-cycle phase and CXP Analysis software version 2.2 (Beckman Coulter, Villepinte, France) to obtain dot-plots of DNA/Ki-67 double-staining.

4.5. Reverse Transcription and Quantitative PCR

Gene expression was observed as previously described [45], after RNA extraction (NucleoSpin RNA II; Machery-Nagel, Düren, Germany), reverse transcription with ThermoScript RT (Invitrogen Life Technologies, Villebon sur Yvette, France) and cDNA amplification using the IQ SYBR Green Supermix (Bio-Rad, Marne la Coquette, France). For quantitative analysis, the iCycler iQ Real-time PCR Detection system (Bio-Rad), was used to calculate relative fold change of gene expression, following the delta delta Ct method [46]. The *HPRT1* reference gene was used for normalization. Primer sequences with corresponding gene symbol and name are indicated in Table 1.

4.6. Breast Carcinoma Model

Eight-week-old female athymic mice (NMRI nu/nu) were obtained from Elevages Janvier (Le Genest St Isle, France). They were housed under pathogen-free conditions at the Experimental Therapy Unit (Faculty of Medicine, Nantes, France). The experimental protocol was approved by the regional committee on animal ethics (CEEA.PdL.06) and the Minister of Agriculture (Authorisation number: 9634) and was conducted following the guidelines "Charte nationale portant sur l'éthique de l'expérimentation animale" of the French ethical committee. The mice were anaesthetized by inhalation of an isoflurane-air mix (2% for induction and 0.5% for maintenance, 1 L/min) before injection with 2×10^6 MCF7 cells in 30 µL of Matrigel (R&D Systems) diluted in phosphate buffered saline (PBS 50%) into the 4th left mammary duct. At the time of cell injection, a pellet delivering 17β-estradiol

(Innovative Research of America, Sarasota, FL, USA) was subcutaneously implanted between the neck and the left shoulder. Formula (l^2xL)/2, where l and L represent the smallest and largest diameter respectively, was used to calculate the tumor volume [47].

4.7. Histology Analysis

Tumor samples were fixed in 4% buffered paraformaldehyde (PFA) for 48 h, while sphere samples were fixed in 4% PFA for 15. Three μm-thick sections of tumors or spheres embedded in paraffin were dewaxed, rehydrated and then treated with 3% H_2O_2 for 15 min at room temperature. Human Ki-67 immunohistochemistry detection was then performed with a mouse monoclonal anti-human Ki-67 (MIB-1 clone; Dako, Les Ulis, France) and revealed with a biotinylated goat anti-mouse Immunoglobulin G secondary antibody and Streptavidin-Horse Radish Peroxydase complexes (Dako) that were observed following an incubation with 3,3′-Diaminobenzidine (DAB, Dako). Nuclei were counterstained with a Gill-Haematoxylin solution. ImageJ software (NIH, Bethesda, MD, USA) was used to calculate the proportion of Ki-67-positive cells, from counting >15,000 nuclei in 6 sections of each tumor sample or >5000 nuclei in 6 sections of each sphere sample.

4.8. Lipid Analysis

Fatty acid composition analysis using gas chromatography: AT-CM were frozen in liquid nitrogen before total lipid extraction according to the FOLCH method with chloroform-methanol 2:1 (*v/v*). Extracts from AT-CM were washed with saline and separated into two phases. The chloroform phase transferred to a new tube was evaporated. Lipid extracts were resuspended with 200 μl of chloroform-methanol 2:1 (*v/v*). Triglycerids (TG) were separated on silica gel TLC plates (LK5, 20*20; Merck St Quentin, Yvelines, France) for thin layer chromatography. After spot scraping, TG were collected and treated as fatty acids methyl esters (FAME) for gas chromatography analysis.

Derivatization of fatty acids was performed with 14% boron trifluoride (in methanol), which resulted in the formation of methyl esters. The derivatization mixture was incubated and shaken for 30 min at 100 °C. Finally, FAME were extracted twice with hexane and then evaporated to dryness. Batch samples were analzsed with a gas chromatograph (GC-2010plus; Shimadzu Scientific instruments, Noisiel, France) through a capillary column (SGE BPX70 GC Capillary Columns; Chromoptic SAS, Courtaboeuf, France). A hydrogen carrier gas was maintained at 120 kPa. Oven temperature was set to 60 °C to 220 °C and flame ionization detector temperature at 280 °C for fatty acid detection. FAME identification was done by comparing relative retention times of samples to those obtained for pure standard mixtures (Supelco 37 fatty acid methyl Ester mix; Sigma Aldrich). The relative amount of each fatty acid (saturated, mono-unsaturated or polyunsaturated fatty acid) was quantified by integrating the baseline peak divided by the peak area corresponding to all fatty acids, using the GC solutions software (Shimadzu Scientific instruments, Noisiel, France).

4.9. Statistical Analysis

Microsoft Excel software (Redmond, WA, USA) was used. In vitro experiments results were analyzed following the analysis of variance t-test. In vivo experimentation results were analyzed with the unpaired nonparametric method and Dunn's multiple comparisons following the Kruskal-Wallis test.

Author Contributions: Conceptualization, S.B.-N., P.P. and V.T.; methodology, P.A. and M.P.; validation, L.-R.L.N., P.P. and V.T.; formal analysis, P.A. and V.T.; writing—original draft preparation, L.V., S.C. and V.T.; writing—review and editing, L.-R.L.N., V.T. and P.P.; project administration, V.T.; funding acquisition, F.R., P.L. and V.T.

Funding: This work was partly supported by INSERM, the University of Nantes and Cancéropole Grand Ouest (AOE2016).

Acknowledgments: The authors wish to thank Brennan M for the manuscript revision, Guiho R and Bitteau K for their help with in vivo experiments and Charrier C for the histology experiments.

Conflicts of Interest: The authors declare no conflicts of interest. The funders had no role in the design of the study; in the collection, analyses, or interpretation of data; in the writing of the manuscript, or in the decision to publish the results.

References

1. Wolfson, B.; Eades, G.; Zhou, Q. Adipocyte activation of cancer stem cell signaling in breast cancer. *World J. Biol. Chem.* **2015**, *6*, 39–47. [CrossRef]

2. D'Esposito, V.; Liguoro, D.; Ambrosio, M.R.; Collina, F.; Cantile, M.; Spinelli, R.; Raciti, G.A.; Miele, C.; Valentino, R.; Campiglia, P.; et al. Adipose microenvironment promotes triple negative breast cancer cell invasiveness and dissemination by producing CCL5. *Oncotarget* **2016**, *7*, 24495–24509.

3. Nieman, K.M.; Romero, I.L.; Van Houten, B.; Lengyel, E. Adipose tissue and adipocytes supports tumorigenesis and metastasis. *Biochim. Biophys. Acta* **2013**, *1831*, 1533–1541. [CrossRef]

4. Iyengar, P.; Combs, T.P.; Shah, S.J.; Gouon-Evans, V.; Pollard, J.W.; Albanese, C.; Flanagan, L.; Tenniswood, M.P.; Guha, C.; Lisanti, M.P.; et al. Adipocyte-secreted factors synergistically promote mammary tumorigenesis through induction of anti-apoptotic transcriptional programs and proto-oncogene stabilization. *Oncogene* **2003**, *22*, 6408–6423. [CrossRef]

5. Dirat, B.; Bochet, L.; Dabek, M.; Daviaud, D.; Dauvillier, S.; Majed, B.; Wang, Y.Y.; Meulle, B.; Salles, B.; Gonidec, S.L.; et al. Cancer-associated adipocytes exhibit an activated phenotype and contribute to breast cancer invasion. *Cancer Res.* **2011**, *71*, 2455–2465. [CrossRef] [PubMed]

6. Kornfeld, S.; Goupille, C.; Vibet, S.; Chevalier, S.; Pinet, A.; Lebeau, J.; Tranquart, F.; Bougnoux, P.; Martel, E.; Maurin, A.; et al. Reducing endothelial NOS activation and interstitial fluid pressure with n-3 PUFA offset tumor chemoresistance. *Carcinogenesis* **2012**, *33*, 260–267. [CrossRef]

7. Ouldamer, L.; Goupille, C.; Vildé, A.; Arbion, F.; Body, G.; Chevalier, S.; Cottier, J.P.; Bougnoux, P. N-3 Polyunsaturated Fatty Acids of Marine Origin and Multifocality in Human Breast Cancer. *PLoS ONE* **2016**, *11*, e0147148. [CrossRef]

8. Chauvin, L.; Goupille, C.; Blanc, C.; Pinault, M.; Domingo, I.; Guimaraes, C.; Bougnoux, P.; Chevalier, S.; Maheo, K. Long chain n-3 polyunsaturated fatty acids increase the efficacy of docetaxel in mammary cancer cells by downregulating Akt and PKCε/δ-induced ERK pathways. *Biochim. Biophys. Acta* **2016**, *1861*, 380–390. [CrossRef] [PubMed]

9. Biondo, P.D.; Brindley, D.N.; Sawyer, M.B.; Field, C.J. The potential for treatment with dietary long-chain polyunsaturated n-3 fatty acids during chemotherapy. *J. Nutr. Biochem.* **2008**, *19*, 787–796. [CrossRef]

10. Calviello, G.; Serini, S.; Piccioni, E.; Pessina, G. Antineoplastic effects of n-3 polyunsaturated fatty acids in combination with drugs and radiotherapy: Preventive and therapeutic strategies. *Nutr. Cancer* **2009**, *61*, 287–301. [CrossRef]

11. Bougnoux, P.; Hajjaji, N.; Maheo, K.; Couet, C.; Chevalier, S. Fatty acids and breast cancer: Sensitization to treatments and prevention of metastatic re-growth. *Prog. Lipid. Res.* **2010**, *49*, 76–86. [CrossRef]

12. Kronowitz, S.J.; Mandujano, C.C.; Liu, J.; Kuerer, H.M.; Smith, B.; Garvey, P.; Jagsi, R.; Hsu, L.; Hanson, S.; Valero, V. Lipofilling of the breast does not increase the risk of recurrence of breast cancer: A matched controlled study. *Plast. Reconstr. Surg.* **2016**, *137*, 385–393. [CrossRef]

13. Charvet, H.J.; Orbay, H.; Wong, M.S.; Sahar, D.E. The oncologic safety of breast fat grafting and contradictions between basic science and clinical studies: A systematic review of the recent literature. *Ann. Plast. Surg.* **2015**, *75*, 471–479. [CrossRef]

14. Gentile, P.; Casella, D.; Palma, E.; Calabrese, C. Engineered fat graft enhanced with adipose-derived stromal vascular fraction cells for regenerative medicine: Clinical, histological and instrumental evaluation in breast reconstruction. *J. Clin. Med.* **2019**, *8*, 504. [CrossRef]

15. Petit, J.Y.; Rietjens, M.; Botteri, E.; Rotmensz, N.; Bertolini, F.; Curigliano, G.; Rey, P.; Garusi, F.; De Lorenzi, S.; Martella, S.; et al. Evaluation of fat grafting safety in patients with intraepithelial neoplasia: A matched-cohort study. *Ann. Oncol. Off. J. Eur. Soc. Med. Oncol.* **2013**, *24*, 1479–1484. [CrossRef]

16. Geissler, P.J.; Davis, K.; Roostaeian, J.; Unger, J.; Huang, J.; Rohrich, R.J. Improving fat transfer viability: The role of aging, body mass index, and harvest site. *Plast. Reconstr. Surg.* **2014**, *134*, 227–232. [CrossRef]

17. Gentile, P.; De Angelis, B.; Di Pietro, V.; Amorosi, V.; Scioli, M.G.; Orlandi, A.; Cervelli, V. Gentle Is Better: The original "Gentle Technique" for fat placement in breast lipofilling. *J. Cutan. Aesthet. Surg.* **2018**, *11*, 120–126. [CrossRef] [PubMed]

18. Gentile, P.; Orlandi, A.; Scioli, M.G.; Di Pasquali, C.; Bocchini, I.; Curcio, C.B.; Floris, M.; Fiaschetti, V.; Floris, R.; Cervell, V. A comparative translational study: The combined use of enhanced stromal vascular fraction and platelet-rich plasma improves fat grafting maintenance in breast reconstruction. *Stem Cells Transl. Med.* **2012**, *1*, 341–351. [CrossRef]

19. Gentile, P.; Scioli, M.G.; Orlandi, A.; Cervelli, V. Breast reconstruction with enhanced stromal vascular fraction fat grafting: What is the best method? *Plast. Reconstr. Surg. Glob. Open* **2015**, *3*, e406. [CrossRef]

20. Hamza, A.; Lohsiriwat, V.; Rietjens, M. Lipofilling in breast cancer surgery. *Gland Surg.* **2013**, *2*, 7–14. [PubMed]

21. Large, V.; Hellström, L.; Reynisdottir, S.; Lönnqvist, F.; Eriksson, P.; Lannfelt, L.; Arner, P. Human beta-2 adrenoceptor gene polymorphisms are highly frequent in obesity and associate with altered adipocyte beta-2 adrenoceptor function. *J. Clin. Investig.* **1997**, *100*, 3005–3013. [CrossRef] [PubMed]

22. Girard, Y.K.; Wang, C.; Ravi, S.; Howell, M.C.; Mallela, J.; Alibrahim, M.; Green, R.; Hellemann, G.; Mohapatra, S.S.; Mohapatra, S. A 3D fibrous scaffold inducing tumoroids: A platform for anticancer drug development. *PLoS ONE* **2013**, *8*, e75345. [CrossRef]

23. Al-Hajj, M.; Wicha, M.S.; Benito-Hernandez, A.; Morrison, S.J.; Clarke, M.F. Prospective identification of tumorigenic breast cancer cells. *Proc. Natl. Acad. Sci. USA* **2003**, *100*, 3983–3988. [CrossRef]

24. Cuylen, S.; Blaukopf, C.; Politi, A.Z.; Müller-Reichert, T.; Neumann, B.; Poser, I.; Ellenberg, J.; Hyman, A.A.; Gerlich, D.W. Ki-67 acts as a biological surfactant to disperse mitotic chromosomes. *Nature* **2016**, *535*, 308–312. [CrossRef]

25. Niddam, J.; Vidal, L.; Hersant, B.; Meningaud, J.P. Primary Fat Grafting to the Pectoralis Muscle during Latissimus Dorsi Breast Reconstruction. *Plast. Reconstr. Surg. Glob. Open* **2016**, *4*, 1059. [CrossRef]

26. Lohsiriwat, V.; Curigliano, G.; Rietjens, M.; Goldhirsch, A.; Petit, J.Y. Autologous fat transplantation in patients with breast cancer: «silencing» or "fueling" cancer recurrence? *Breast Edinb. Scotl.* **2011**, *20*, 351–357. [CrossRef] [PubMed]

27. Charvet, H.J.; Orbay, H.; Harrison, L.; Devi, K.; Sahar, D.E. In vitro effects of adipose-derived stem cells on breast cancer cells harvested from the same patient. *Ann. Plast. Surg.* **2016**, *76*, S241–S245. [CrossRef]

28. Fraser, J.K.; Hedrick, M.H.; Cohen, S.R. Oncologic risks of autologous fat grafting to the breast. *Aesthet. Surg. J.* **2011**, *31*, 68–75. [CrossRef]

29. Silva-Vergara, C.; Fontdevila, J.; Weshahy, O.; Yuste, M.; Descarrega, J.; Grande, L. Breast cancer recurrence is not increased with lipofilling reconstruction: A case-controlled study. *Ann. Plast. Surg.* **2017**, *79*, 243–248. [CrossRef] [PubMed]

30. Largo, R.D.; Tchang, L.A.H.; Mele, V.; Scherberich, A.; Harder, Y.; Wettstein, R.; Schaefer, D. Efficacy, safety and complications of autologous fat grafting to healthy breast tissue: A systematic review. *J. Plast. Reconstr. Aesthetic Surg.* **2014**, *67*, 437–448. [CrossRef]

31. Cohen, O.; Lam, G.; Karp, N.; Choi, M. Determining the oncologic safety of autologous fat grafting as a reconstructive modality: An institutional review of breast cancer recurrence rates and surgical outcomes. *Plast. Reconstr. Surg.* **2017**, *140*, 382e–392e. [CrossRef]

32. Wazir, U.; El Hage Chehade, H.; Headon, H.; Oteifa, M.; Kasem, A.; Mokbel, K. Oncological safety of lipofilling in patients with breast cancer: A meta-analysis and update on clinical practice. *Anticancer Res.* **2016**, *36*, 4521–4528. [CrossRef]

33. Gigli, S.; Amabile, M.I.; Pastena, F.D.; De Luca, A.; Gulia, C.; Manganaro, L.; Monti, M.; Ballesio, L. Lipofilling outcomes mimicking breast cancer recurrence: Case report and update of the literature. *Anticancer. Res.* **2017**, *37*, 5395–5398.

34. Suszynski, T.M.; Sieber, D.A.; Van Beek, A.L.; Cunningham, B.L. Characterization of adipose tissue for autologous fat grafting. *Aesthet. Surg. J.* **2015**, *35*, 194–203. [CrossRef]

35. Spear, S.L.; Coles, C.N.; Leung, B.K.; Gitlin, M.; Parekh, M.; Macarios, D. The safety, effectiveness, and efficiency of autologous fat grafting in breast surgery. *Plast. Reconstr. Surg. Glob. Open* **2016**, *4*, e827. [CrossRef]

36. Waked, K.; Colle, J.; Doornaert, M.; Cocquyt, V.; Blondeel, P. Systematic review: The oncological safety of adipose fat transfer after breast cancer surgery. *Breast Edinb. Scotl.* **2017**, *31*, 128–136. [CrossRef]

37. Lafontan, M.; Berlan, M. Fat cell adrenergic receptors and the control of white and brown fat cell function. *J. Lipid Res.* **1993**, *34*, 1057–1091.

38. Bougnères, P.; Stunff, C.L.; Pecqueur, C.; Pinglier, E.; Adnot, P.; Ricquier, D. In vivo resistance of lipolysis to epinephrine. A new feature of childhood onset obesity. *J. Clin. Investig.* **1997**, *99*, 2568–2573. [CrossRef]

39. Jocken, J.W.E.; Blaak, E.E. Catecholamine-induced lipolysis in adipose tissue and skeletal muscle in obesity. *Physiol. Behav.* **2008**, *94*, 219–230. [CrossRef]

40. Danilo, C.; Gutierrez-Pajares, J.L.; Mainieri, M.A.; Mercier, I.; Lisanti, M.P.; Frank, P.G. Scavenger receptor class B type I regulates cellular cholesterol metabolism and cell signaling associated with breast cancer development. *Breast Cancer Res.* **2013**, *15*, R87. [CrossRef]

41. Milde-Langosch, K.; Bamberger, A.M.; Rieck, G.; Grund, D.; Hemminger, G.; Müller, V.; Longing, T. Expression and prognostic relevance of activated extracellular-regulated kinases (ERK1/2) in breast cancer. *Br. J. Cancer* **2005**, *92*, 2206–2215. [CrossRef]

42. Wang, Y.Y.; Attané, C.; Milhas, D.; Dirat, B.; Dauvillier, S.; Guerard, A.; Gilhodes, G.; Lazar, I.; Alet, N.; Laurent, V.; et al. Mammary adipocytes stimulate breast cancer invasion through metabolic remodeling of tumor cells. *JCI. Insight.* **2017**, *2*, e87489. [CrossRef]

43. Basen-Engquist, K.; Chang, M. Obesity and cancer risk: Recent review and evidence. *Curr. Oncol. Rep.* **2011**, *13*, 71–76. [CrossRef]

44. Chipoy, C.; Brounais, B.; Trichet, V.; Battaglia, S.; Berreur, M.; Oliver, L.; Juin, P.; Redini, F.; Heymann, D.; Blanchard, F. Sensitization of osteosarcoma cells to apoptosis by oncostatin M depends on STAT5 and p53. *Oncogene* **2007**, *26*, 6653–6664. [CrossRef]

45. Avril, P.; Le Nail, L.R.; Brennan, M.Á.; Rosset, P.; De Pinieux, G.; Layrolle, P.; Layrolle, P.; Heymann, D.; Trichet, V.; Perrot, P. Mesenchymal stem cells increase proliferation but do not change quiescent state of osteosarcoma cells: Potential implications according to the tumor resection status. *J. Bone Oncol.* **2015**, *5*, 5–14. [CrossRef]

46. Livak, K.J.; Schmittgen, T.D. Analysis of relative gene expression data using real-time quantitative PCR and the 2(-Delta Delta C(T)) Method. *Methods* **2001**, *25*, 402–408. [CrossRef]

47. Gernapudi, R.; Yao, Y.; Zhang, Y.; Wolfson, B.; Roy, S.; Duru, N.; Eades, G.; Yang, P.; Zhou, Q. Targeting exosomes from preadipocytes inhibits preadipocyte to cancer stem cell signaling in early-stage breast cancer. *Breast Cancer Res. Treat.* **2015**, *150*, 685–695. [CrossRef]

Review

In Search for Genes Related to Atherosclerosis and Dyslipidemia Using Animal Models

Anastasia V. Poznyak [1], Andrey V. Grechko [2], Reinhard Wetzker [3] and Alexander N. Orekhov [4,5,†,*]

[1]   Institute for Atherosclerosis Research, Skolkovo Innovative Center, 121609 Moscow, Russia; tehhy_85@mail.ru
[2]   Federal Scientific Clinical Center for Resuscitation and Rehabilitation, 109240 Moscow, Russia; noo@fnkcrr.ru
[3]   Department of Anesthesiology and Intensive Care Medicine, Jena University Hospital, 07743 Jena, Germany; reinhard.wetzker@uni-jena.de
[4]   Laboratory of Angiopathology, Institute of General Pathology and Pathophysiology, 125315 Moscow, Russia
[5]   Institute of Human Morphology, 117418 Moscow, Russia
*   Correspondence: a.h.opexob@gmail.com; Tel.: +7-903-169-08-66
†   Current address: Institute of General pathology and Pathophysiology, Baltiyskaya str. 8, 125315 Moscow, Russia.

Received: 14 January 2020; Accepted: 17 March 2020; Published: 18 March 2020

Abstract: Atherosclerosis is a multifactorial chronic disease that affects large arteries and may lead to fatal consequences. According to current understanding, inflammation and lipid accumulation are the two key mechanisms of atherosclerosis development. Animal models based on genetically modified mice have been developed to investigate these aspects. One such model is low-density lipoprotein (LDL) receptor knockout (KO) mice ($ldlr^{-/-}$), which are characterized by a moderate increase of plasma LDL cholesterol levels. Another widely used genetically modified mouse strain is apolipoprotein-E KO mice ($apoE^{-/-}$) that lacks the primary lipoprotein required for the uptake of lipoproteins through the hepatic receptors, leading to even greater plasma cholesterol increase than in $ldlr^{-/-}$ mice. These and other animal models allowed for conducting genetic studies, such as genome-wide association studies, microarrays, and genotyping methods, which helped identifying more than 100 mutations that contribute to atherosclerosis development. However, translation of the results obtained in animal models for human situations was slow and challenging. At the same time, genetic studies conducted in humans were limited by low sample sizes and high heterogeneity in predictive subclinical phenotypes. In this review, we summarize the current knowledge on the use of KO mice for identification of genes implicated in atherosclerosis and provide a list of genes involved in atherosclerosis-associated inflammatory pathways and their brief characteristics. Moreover, we discuss the approaches for candidate gene search in animals and humans and discuss the progress made in the field of epigenetic studies that appear to be promising for identification of novel biomarkers and therapeutic targets.

Keywords: atherosclerosis; mutations; epigenetics

---

1. Introduction

Atherosclerosis is a chronic vascular disease that affects large and small arteries and is characterized by the development of lipid-rich plaques in the vascular wall. The growing plaques reduce the vascular lumen and, in the case of so-called unstable plaques, can trigger thrombogenesis on the surface. The resulting ischemia can lead to fatal consequences if it affects a vital organ, such as heart [1]. During the last few decades, a significant progress has been made in the understanding of atherosclerosis pathogenesis. However, no effective treatment for the disease has been developed so far [2]. The greatest challenge is the lack of effective treatment approaches, which is the consequence of our incomplete understanding of the processes that underlie the pathogenesis. Genetic base is one of the most important features that are involved in the disease development.

Currently, the most widely used anti-atherosclerosis drugs are statins [3]. The aim of statin therapy is reducing the blood level of low-density lipoprotein (LDL) cholesterol, which is a well-known pro-atherogenic agent. It was demonstrated that statin therapy reduces the risk of cardiovascular events by approximately one third [4]. However, statin therapy alone cannot be regarded as effective treatment of atherosclerosis. Increasing the plasma levels of high-density lipoprotein (HDL), which has anti-atherogenic properties, is another promising approach. Accumulating evidence shows that high levels of HDL cholesterol can inhibit atherosclerosis progression. To date, several therapeutic agents that increase HDL levels are known, including niacin, fibrates, and statins. Among the recently developed medications are apoA-I-phospholipid complexes, human apoA-I and apoA-I-mimetic peptides, and inhibitors of cholesterol ester transfer protein, which have reached the level of clinical trials. However, the inhibitor of cholesterol ester transport torcetrapib failed to demonstrate a clinical benefit. The search for new HDL-rising therapeutic agents currently continues [5,6].

Inflammation is known to play a key role in atherosclerosis initiation and development. Ever since the concept of atherosclerosis being a chronic inflammatory condition was established, the search for anti-inflammatory agents that may be effective against atherosclerosis is ongoing. However, the complexity of signaling cascades regulating the activities of immune cells in atherosclerotic lesions makes their regulation challenging. Moreover, anti-inflammatory therapy of atherosclerosis should target lesion-specific inflammatory processes while leaving the normal immune response uncompromised [7]. Genes related to the inflammation are considered as possibly implicated in the atherogenesis, as well as genes related to the lipid metabolism.

The search for genetic determinants of atherosclerosis has been ongoing for decades. Currently, this line of research is focused on identifying the candidate genes implicated in known atherogenesis pathways and conducting association studies to evaluate their roles in the pathology development. This approach resulted in establishing the pro-atherogenic roles of numerous genes [8]. In parallel, genome-wide linkage studies were carried out to identify atherogenesis-regulating quantitative trait loci (QTL). This approach appears to be promising for identifying new atherosclerosis-related genes in an unbiased manner [9]. Finally, studies of epigenetic modifications associated with atherosclerosis present further opportunities for possible therapeutic intervention. Since the whole genome sequence information was available for humans and mice, including haplotype information, it became possible to conduct genome-wide association studies with relatively high speed.

The search for genes responsible for atherosclerosis has several challenges. First, human genome-wide association studies (GWAS) that represent a powerful modern tool for such analyses do not take into account the impact of the environmental factors and molecular processes within a particular tissue. These factors can, however, be followed using murine models. Thus, there exist currently more than 100 different strains of atherosclerotic mice with various properties' combinations. This diversity also complicates the interpretation of findings on different models [10]. Another challenge is the difficulty of comparison of human and murine genes potentially related to atherosclerosis. Genes that were found in mouse QTL studies can be further tested in human association studies to check whether they are associated with atherosclerosis [11].

2. Methodology of Genetic Studies of Atherosclerosis

Every study begins with choosing the most suitable model. Despite the fact that traditionally used models of atherosclerosis greatly improved our knowledge about atherogenesis, especially in its early stages, there is currently a need for novel, more advanced models. One of the challenging areas of research is studying thrombosis associated with atherosclerotic plaque rupture, which is difficult to model in animals.

Moreover, there is a need for mouse models deficient for apoE and LDL receptors with a genetic background other than C57BL/6. However, genetic differences are present even between different substrains of C57BL/6 mice. For example, Almodovar et al. identified fourteen single nucleotide polymorphisms (SNPs) resulting in differences between the two most popular lines from Jackson Laboratory and Taconic [12]. Moreover, the Nnt deletion on chromosome 13 was shown to be different between these two lines. At the phenotypic level, these differences resulted in variability of adiposity between the substrains that were relevant for atherosclerosis research. Therefore, differences in the genetic background within one line may cause reproducibility problems and also affect the testing parameters in an unpredictable way.

It is clear by now that the atherosclerosis is a polygenic disorder, and it is necessary to use models with various genetic backgrounds to identify the genetic features that are responsible for the disease development with confidence confidently. Several works have been conducted to assess the genetic variability between mouse strains. Grainger et al. performed an investigation on intercrosses of BALB/cJ (BALB) and SM/J (SM) apolipoprotein E-deficient (*apoE*$^{-/-}$) mice to identify chromosomal regions harboring genes contributing to carotid atherosclerosis. With the use of QTL analysis and bioinformatics tools, they found out that there are five significant QTL, among which the one on the chromosome 12 had the highest LOD score. Potential candidate genes were listed: *Arhgap5, Akap6, Mipol1, Clec14a, Fancm, Nin, Dact1, Rtn1*, and *Slc38a6* [13]. Another study, performed by Wang et al., engaged NOD (non-obese diabetic) mice. They found that the knockdown of *ldlr* or *apoE* in NOD mice did not lead to atherosclerosis development. By contrast, C57BL/6 mice with the ApoE deficiency developed the disease when fed with a high-fat diet. Moreover, simultaneous knockdown of both *ldlr* and *apoE* resulted in a severe atherosclerosis [14].

However, the large majority of the mouse models of atherosclerosis are currently created on the C57BL/6 genetic background. This limits their mapping power and coverage of allelic diversity. QTL analysis on F2 intercrosses of SM/J- *apoE*$^{-/-}$ and BALB/cJ- *apoE*$^{-/-}$ mice revealed that *tnfaip3* was the most potent causal gene. These results show the importance of the use of different strains in the studies aimed at identification of the genetic causes of the atherosclerosis [15].

Creation of mouse models of atherosclerosis allows for revealing new atherosclerosis-related QTL [11]. Such identification may, in its turn, identify previously unknown atherogenesis pathways and new potential therapeutic targets. Although human and mouse orthologs of atherosclerosis-related genes do not always coincide, a search can be conducted for other players of the identified pathway that may prove to be useful in that regard.

Inflammation and lipid metabolism deregulation represent good entry points for searching for relevant atherosclerosis-related genes. However, due to the crucial role of many of these genes for reproduction and survival, they have been subject to great selective pressure and therefore their pro-inflammatory polymorphisms can be found easily. To date, six genes involved into the inflammatory response were identified: *ALOX5AP* [16] and *MEF2A* [17] in human linkage studies of myocardial infarction, *Alox5* [18] and *Tnfsf4* [19] in mouse linkage studies of atherosclerosis and *LTA* [20] and *PSMA6* [21] in human genome-wide association studies of myocardial infarction (Table 1) [17–20,22,23]. The obtained results indicate that genetic predisposition to inflammation may account for a considerable part of variance of atherosclerosis incidence in populations [9].

Table 1. Brief summary on six important genes potentially involved in atherosclerosis development.

Name	Gene ID	Link to Atherosclerosis	Normal Function	Identification (Organism, Method)	Reference
ALOX5AP	241	Genetic variants are potentially contributed to the higher coronary heart disease risk	Part of leukotriene biosynthesis pathway	Human, haplotype association study	[22]
MEF2A	4205	MEF2A signaling pathway is involved in pathogenesis of familial CAD and MI	Myocyte-specific transcription factor	Human, linkage study	[17]
Alox5	11689	homozygote 5LO null mice develop smaller atherosclerotic lesions	Part of leukotriene biosynthesis pathway	Mouse, linkage study	[18]
Tnfsf4	22164	blocking the OX-40/OX40L interaction reduced atherogenesis	OX40 ligand	Mouse, linkage study	[19]
LTA	4049	Genetic variants are potentially contributed to the higher MI risk	Lymphocyte cytokine	Human, GWAS	[20]
PSMA6	5687	PSMA6 rs_1048990 polymorphism may contribute to MI susceptibility in type 2 diabetes	-	Human, GWAS	[23]

CAD: coronary artery disease; MI: myocardial infarction; GWAS: genome-wide association study.

Despite the progress made, identification of atherosclerosis-related genes by means of linkage studies remains difficult. One of the challenges is the complexity of atherosclerosis pathogenesis: the involvement of different cell types in the pathological process, the heterogeneity of genes implicated in the process, and the fact that each of them by themselves may have only a small effect, and the role of the environmental factors. As a result, most genetic variants that were identified in genome-wide association studies have not been previously captured in QTL. Moreover, genes responsible for a certain trait tend to cluster within the chromosome and interact with each other, making positional cloning difficult.

It has been demonstrated that not all genes that influence the phenotype of the QTL may be detected by positional cloning [18]. Other QTL genes can be found by breaking down congenic regions to smaller chunks and analyzing them separately. Multiple causal QTL genes can be identified by association studies conducted for genes in the QTL peak, also called "peak wide mapping" [24]. The chance of identifying human atherosclerosis genes can potentially be increased by using the mouse–human comparative genomic approach. Whether a particular gene is associated with the increased risk of atherosclerosis can be determined by candidate gene association studies that are, however, limited by the high frequency of false-positive results. Moreover, this approach for gene searching is biased, since only genes that are suspected of being related to atherosclerosis are tested. Even the established associations provide little information on whether and how the polymorphisms are linked to the altered gene functions. Genome-wide association studies were quite successful in detecting unexpected candidate atherosclerosis-related genes, but could not be used widely enough due to high costs and complexity of the procedures.

Microarray and genotypic methods offer the possibility of quantifying the expression of numerous (up to thousands) of genes simultaneously in an unbiased way. These methods allow for locating QTLs for phenotypes based on the expression data (so called expression QTLs, or eQTLs). To identify candidate genes for further studies, a combination of clinical QTL with eQTL can be used [25].

Bone marrow transplantation can be considered as an additional approach to identifying the pro-atherosclerotic roles of specific genes in mice. For instance, this method allowed for demonstrating the impact of hematopoietic-expressed genes on atherosclerosis [26].

Each of the approaches listed above has its limitations and advantages. The most promising strategy for finding new genes that are responsible for atherosclerotic development appears to be combining more than one approach in a single study.

3. Genetic Aspects of Atherosclerotic Disease in Mice

Mouse is the most commonly used model animal for studying human diseases, including atherosclerosis. First attempts to create a suitable mouse model of the disease were based on dietary modifications. Wild type mice that are fed with a regular chow diet do not develop atherosclerosis spontaneously. For that purpose, special diets have been developed, such as high fat (15%), cholesterol (1.25%), and cholate (0.5% cholic acid) diet, or "Paigen Diet". This diet is commonly used to induce atherosclerosis lesions in wild type animals. The diet is characterized by a reduced conversion of cholesterol into bile acid, and impaired cholesterol clearance, which leads to increased plasma cholesterol. Moreover, cholate-rich diet alters the expression of several genes that are responsible for lipid metabolism and inflammatory response [27].

More recently, the development of new genetic tools allowed for creating genetically modified mouse models of human atherosclerosis that allowed for better accuracy in reproducing human atherosclerotic lesions. First, mice deficient for low density lipoprotein receptor ($ldlr^{-/-}$) were created. These animals lack the primary receptor responsible for LDL cholesterol (LDL-C) uptake, which results in its increased plasma concentration. Further increase of LDL-C in these animals is achieved by feeding them with a Western-type diet (WTD). Another approach is to cross the $ldlr^{-/-}$ mice with another mutant strain, such as *apobec1* knock-out mice. Mice deficient for apolipoprotein-E ($apoE^{-/-}$) represent another atherosclerosis model, which is currently widely used. In these animals, circulating lipoprotein contains no primary ligand that is used for facilitating the lipoprotein uptake through hepatic receptors. This results in a plasma cholesterol increase that is more prominent than in $ldlr^{-/-}$ mice, with spontaneous development of atherosclerotic lesions in animals fed with a regular diet and further increase of the disease induced by the WTD [28].

Mouse models of atherosclerosis proved to be very helpful for genetic studies of the disease. Over 100 mutations involved in the pathogenesis of atherosclerosis have been identified in mice. Many of these genes may be involved in common pathways in atherosclerosis development [29]. However, only a few of these identified genes were demonstrated to play a causative role in human coronary artery disease. A possible explanation for this is the high selection pressure on these genes that prevents loss-of-function mutations to spread in the human population. Another reason is the differences in the pathogenesis mechanisms of atherosclerosis in humans and mice. Such differences may include many aspects, starting from the size, structure, and elasticity of the arteries and ending with distinct mechanisms of complicated plaque formation, which are difficult to reproduce in mice. Moreover, atherosclerotic disease affects different arteries in humans and mice, with coronary, carotid, and cerebral arteries being the most important in humans, with aorta and proximal large vessels being the first affected in mice [30].

The translation of the genetic findings from mouse models to human satiation was slow and unconvincing. This resulted not only from the limitations of mouse models, but also from the insufficient development of human genetics to identify significant genetic associations with atherosclerotic disease.

4. Candidate Gene Approaches in Mice

Two main approaches have been used for candidate atherosclerosis-related gene testing in mice. The first method, a candidate gene loss-of-function approach, involves the creation of genetically-modified animals deficient for a specific gene with subsequent induction of atherosclerosis either by crossing the mouse line of interest with $ldlr^{-/-}$ or $apoE^{-/-}$ mice or by using the atherosclerosis-inducing diet (Figure 1). This approach allowed identifying associations of more than 100 genes with atherosclerotic disease in mice. About 60% of these genes were initially studied in mice with $apoE^{-/-}$ genetic background, around 25% in mice with $ldlr^{-/-}$ background, and 10% in animals with diet-induced disease. Many of the tested candidate genes had no or only minimal effect on atherosclerosis development. Some genes only had an effect in a certain genetic background; for instance, $icam1^{-/-}$ had an effect only in $apoE^{-/-}$ (but not $ldlr^{-/-}$) mice. A QTL analysis allowed mapping 43 significant loci, of which 19 were identified in $apoE^{-/-}$ animals, 12 in $ldlr^{-/-}$ animals, and 12 in the diet-induced model [15].

The gene of arachidonate 5-lipoxygenase (*alox5*) was shown to be linked to atherosclerosis in mice, since animals deficient for this gene developed atherosclerotic phenotypes. However, genetic studies in humans have not confirmed the association of *ALOX5* gene with coronary artery atherosclerosis. At the same time, another human atherosclerosis-associated gene, *CXCL12*, could be mapped to the orthologous mouse chromosome Chr6, at a distance of less than 1 Mb from *alox5*, indicative of a potential contribution of *cxcl12* polymorphisms to the Artles QTL [31].

Another gene potentially causally related to atherosclerosis is *tnfsf4*, which was identified in Ath1, one of the first atherosclerosis-associated QTLs mapped in mice. This gene, also known as Ox401, is a member of the tumor necrosis ligand superfamily. Mice deficient for this gene demonstrated decreased atherosclerosis progression. Moreover, studies in humans have revealed SNP that were related to myocardial infarction with a marginal significance [32], which was, however, not confirmed by other studies in humans [33]. The human ortholog *TNFSF4* has no known association locus in its proximity. A disintegrin and metalloproteinase 17 (*adam17*) has been proposed as another gene causally related to atherosclerosis, but the effect of its knock-out remains to be investigated [34].

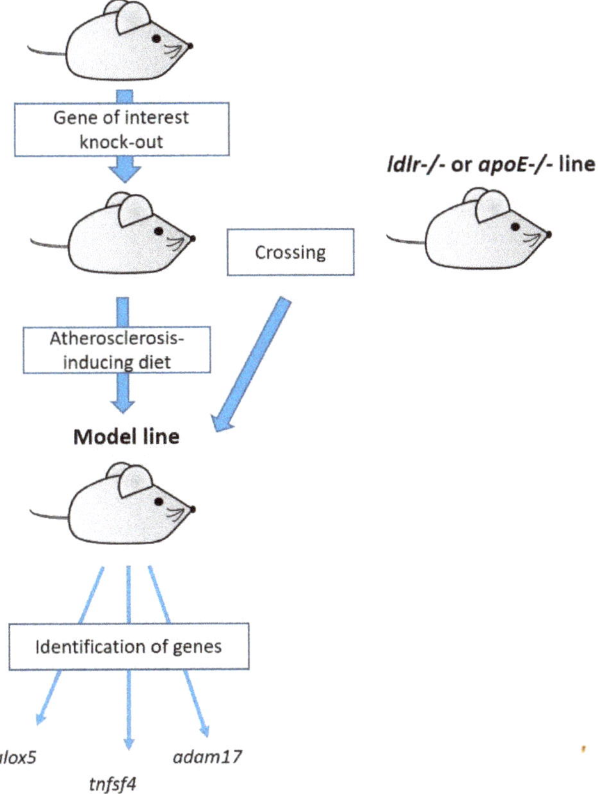

Figure 1. The scheme of a candidate loss-of-function approach and the atherosclerosis-related genes it helped to identify.

5. Genetics of Atherosclerosis in Humans

Atherosclerosis is a multifactorial disease involving a range of biological mechanisms, from lipid metabolism disruption to inflammation. It is therefore unlikely to pinpoint a unique gene that is responsible for disease initiation. However, a range of causal genes acting together can probably be identified and used for finding potential therapeutic targets. In humans, numerous genes have been shown to be important for the individual's susceptibility to atherosclerosis. Genetic variants and mutations that increase the risk of atherosclerosis and related cardiovascular diseases were identified. If each of these variants would have relatively little effect in the overall balance, their combination with each other and with the genetic background may have pronounced effects. Genetic studies in humans are challenging because of the genotype complexity, limited sample sizes, and heterogeneity of the observed phenotypes.

6. Candidate Gene Studies in Humans

Candidate gene studies allow for revealing associations of individual gene polymorphisms with atherosclerosis. To date, several genetic loci that are likely to play a role in the disease pathogenesis have been revealed [35]. However, a large part of candidate gene studies has not yielded convincing results or reported associations were weak and impossible to confirm. Some progress was achieved after development of a cardiovascular gene-centric 50K SNP array, which allowed for revealing several new genes with a significant association with coronary artery disease [36].

In general, genetic markers are less powerful predictors of atherosclerosis than traditional risk factors, such as gender, age, or behavioral factors. However, more detailed studies, including studies on twins, have revealed the existence of genetic susceptibility to atherosclerosis. In rare cases, such as familial hypercholesterolemia (FH), atherosclerosis can be inherited following Mendelian laws. In FH, mutations in the *LDLR* gene lead to a prominent increase of plasma LDL cholesterol level, which strongly increases susceptibility to atherosclerosis [37]. The genetic variants leading to FH-associated atherosclerosis have been studied using a QTL approach in several large families and by studying a single large family under a Mendelian inheritance hypothesis. QTL mapping in humans encounters the same challenge as it does in mice: the large size of the linkage regions that hinders the causal genes identification [9]. This may explain the small number of the identified atherosclerosis-associated genetic loci that could be confirmed by several studies.

The identified atherosclerosis-associated candidate genes include lipoprotein receptor-related protein 6 (*LRP6*) at 12q13.2, arachidonate 5-lipoxygenase-activating protein (*ALOX5AP*) at 13q12-13, and myocyte enhancer factor 2A (*MEF2A*) at 15q26.3 [38]. These genes have not been identified in studies on mouse models. Further studies using a genome-wide association approach have identified new loci outside of the original confidence intervals for *LRP6* and *ALOX5AP*, indicating that these may be distinct loci. At the same time, *MEF2A* may be linked to a broader region on chromosome 15 that includes another gene, *ADAMTS7* [29].

Genome-wide association studies, or GWAS, represent a powerful tool that makes possible simultaneous investigation of millions of polymorphisms to reveal their association with a certain phenotype within a large population [39]. Studies using GWAS have revealed several human genetic loci strongly associated with coronary artery disease and myocardial infarction [40]. Other examples of large GWAS are the CARDIoGRAM consortium that has analyzed data from 14 different GWAS for a total of 140,000 patients, and reported 13 new coronary artery disease-associated loci and the C4D consortium that investigated 70,000 patients from South Asia and Europe and identified four new loci [41,42]. Apart from the loci mentioned above, numerous other loci were identified that may be significant ($p < 0.05$, but higher than the significance threshold currently used for GWAS, which is 5×10^{-8}). It is possible that further research will demonstrate significance of some of these loci in a larger meta-analysis.

GWAS helped to reveal the association between several genes implicated in triglyceride metabolism and cardiovascular disease, including *APOA5* and *APOC3* [43]. Rare *APOA5* mutations were shown to be associated with enhanced plasma triglyceride levels and, at the same time, the elevated risk of coronary artery disease [44]. By contrast, rare loss-of-function mutations in the *APOC3* gene were demonstrated to lower both the plasma triglyceride levels and the cardiovascular risk [45]. These findings make the *APOC3* and *APOA5* the promising lipid metabolism-related target for future development of atherosclerosis and dyslipidemia treatment approaches.

Loss-of-function variants of another gene related to the triglyceride metabolism, *ANGPTL3*, were revealed to be atheroprotective due to their association with the decrease in plasma levels of triglycerides, LDL cholesterol, and HDL cholesterol [46]. Rare variants of *HSD17B13* were found to be significantly associated with triglycerides and HDL in white individuals with type 2 diabetes. This provided an explanation for the observed lipid variation in response to fenofibrate treatment in individuals with type 2 diabetes treated with statins [47].

7. Epigenetic Factors

Epigenetic factors play important roles in many human diseases, including atherosclerosis. They are increasingly recognized as disease modifiers and potential therapeutic targets. Among the known epigenetic factors are DNA methylation, histone modification, and the effects of various non-coding RNAs. Studying of epigenetic factors is challenging because of their complexity and dynamic nature, but also by the limited access to in vivo material and tissue heterogeneity, since epigenetic modifications are often cell type-specific [48]. These limitations make single cell analysis technologies. Single cell

RNA sequencing allowed revealing disease stage-specific markers with subsequent isolation of specific cell population suitable for epigenetic profiling. Combined with modern computational strategies for data analysis, these approaches can help with revealing gene regulatory networks for a particular cell type that are present in atherosclerotic plaques [49].

In contrast to genetic, epigenetic modifications have a dynamic nature. They can be influenced by the environmental stimuli that can modify the cell phenotype, gene expression patterns, and the expression and regulation of the transcription factors [50]. For instance, it was demonstrated in mice that transplantation of macrophages into a new tissue environment changed the epigenetic profiles and gene expression patters and rendered the transplanted cells phenotypically similar to tissue-resident macrophages [51].

7.1. DNA Methylation

DNA methylation is performed by methyltransferases (DNMT) that add methyl groups to the 5′ position of cytosine rings in the CpG dinucleotides. Insufficient DNA methylation leads to gene activation, and hypermethylation to gene silencing [52]. Recent studies revealed significant levels of DNA hydroxymethylation within genes characterized by active transcription, as well as in their enhancer regions [53].

Recent studies have demonstrated that, in both humans and mice, overall DNA hypermethylation of CpG islands is observed in atherosclerosis [54]. Moreover, genome-wide DNA methylation sequencing revealed a positive correlation between DNA methylation status and atherosclerotic lesion grade [55]. It is therefore likely that DNA methylation plays a role in atherosclerosis development, and therefore a DNA methylation status can potentially be used as an atherosclerosis biomarker.

7.2. Histone Modification

Histones are highly alkaline proteins that play a major role in nucleosome formation. Histone modifications are represented by a complex of covalent post-translational modifications, such as phosphorylation, methylation, and acetylation. Histones can also be ubiquitinated and SUMOylated [56]. Chromatin-remodeling complexes are capable to control histone modifications in a dynamic way [57]. The process is carried out by 'writer' and 'eraser' complexes that introduce or remove covalent modifications of lysine or, less frequently, arginine residues of histone proteins.

7.3. Long Non-Coding RNAs

Long non-coding RNAs (lncRNAs) are commonly defined as RNA transcripts shorter than 200 nucleotides that do not encode a functional protein. LncRNAs are polyadenylated, contain only 2–3 exons, and are typically spliced [58]. Most of them are transcribed by polymerase II. LncRNAs are associated with the regulations of numerous cellular processes in mammals, such as chromatin remodeling, chromatin modification, dosage compensation effect, genomic imprinting, and others [59]. These molecules do not act by themselves, but form complexes and execute their regulatory functions through interacting with proteins. LncRNAs were shown to interact with different enzymes and complexes and thus affect DNA methylation, histone methylation, and acetylation. They also are able to regulate the transcription by the interaction with transcriptional factors. The implication of lncRNAs in post-transcriptional regulation is based on their ability to interact with splicing factors and proteins and subsequently regulate mRNA alternative splicing, and splicing factors can also directly regulate lncRNA alternative splicing [59].

However, lncRNA are better to define not as non-coding, but as likely to be non-coding, because it is impossible to verify the absence of coding properties with certainty. The most significant piece of criteria used to distinguish whether the transcript encodes a peptide is the lack of open reading frames (ORFs). The first attempt to identify lncRNAs and to summarize the data available from studies in mice was the FANTOM project [60]. It used the cDNA cloning with the subsequent Sanger sequencing.

Although comprehensive data on lncRNAs remain to be collected, there is growing evidence that lncRNAs play a role in epigenetic and/or transcriptional processes through recruiting chromatin modifying and transcriptional factors to DNA among other mechanisms [61]. It was also shown that lncRNAs can participate in post-transcriptional regulators by controlling translation, splicing, and mRNA stability [62].

In cardiovascular diseases, several lncRNAs have been identified as epigenetic regulators playing a role in the pathological processes. For instance, the antisense non-coding RNA in the Ink4 locus (ANRIL) that is transcribed from the human 9p21.3 locus, the expression of which is strongly associated with the incidence of coronary artery disease [63,64]. It was shown that ANRIL could promote proliferation of human vascular smooth muscle cells through recruiting repressive PcG protein complexes to the cell cycle inhibitor genes *CDKN2A/B* [65]. In the endothelial cells, ANRIL was shown to be induced by NF-κB signaling. In this cell type, it can induce the expression of pro-inflammatory genes IL-6 and IL-8 through recruiting the transcription factor YY1 [66].

Another lncRNA that was shown to be involved in atherosclerosis development is lincRNA-p21. In atherosclerotic plaques from $apoE^{-/-}$ atherosclerosis mouse model and from human coronary arteries, this lncRNA was found to be downregulated. Moreover, knock-down of lincRNA-p21 was demonstrated to increase the neointima growth in a mouse model of carotid artery vascular injury. This process was partially dependent on p53-dependent apoptotic genes regulation in vascular smooth muscle cells [67]. Finally, several lncRNAs were also found to be involved in such processes as inflammation and innate immunity regulation, which makes them relevant for atherosclerosis research [68,69]. Future studies are likely to add to the growing list of lncRNAs implicated in atherosclerosis and evaluate their potential as biomarkers or even therapeutic targets.

8. Conclusions

The complexity of atherosclerosis pathogenesis makes it impossible to identify a precise set of genes responsible for the disease development. Instead, multiple genes have been identified that can participate in different stages of atherosclerosis progress. Finding genetic determinants for atherosclerosis can crucially improve not only general understanding of underlying processes, but also an individualized medication. The same symptom was formed by environmental factors, genetic predisposition or their combinations in different patients, and these patients may call for different therapies.

There are three main challenges impeding studies in the field of finding genetic and epigenetic determinants for atherosclerosis. The first one is to establish a suitable model because of the complexities in translating results obtained from classical models to humans and because of serious limitations of using human material. The second challenge is to create an economically viable and unbiased approach that can be widely used in order to make data be able to be reproduced and standardized. Despite the general understanding of how to target the particular gene in the case of therapeutic strategy, the further elaboration of a genetically-based treatment or preventive approach for atherosclerosis also remains challenging.

Funding: This work was supported by the Russian Science Foundation (Grant #19-15-00010).

Conflicts of Interest: The authors declare no conflict of interest. The funders had no role in the design of the study; in the collection, analyses, or interpretation of data; in the writing of the manuscript, or in the decision to publish the results.

References

1. Martinez, M.S.; Garcia, A.; Luzardo, E.; Chavez-Castillo, M.; Olivar, L.C.; Salazar, J.; Velasco, M.I.; Rojas Quintero, J.J.; Bermudez, V. Energetic metabolism in cardiomyocytes: Molecular bases of heart ischemia and arrhythmogenesis. *Vessel Plus* **2017**, *1*, 130–141. [CrossRef]

2. Gowdar, S.; Syal, S.; Chabra, L. Probable protective role of diabetes mellitus in takosubo cardiomyopathy: A review. *Vessel Plus* **2017**, *1*, 129–136.

3. Kazi, D.S.; Penko, J.M.; Bibbins-Domingo, K. Statins for Primary Prevention of Cardiovascular Disease: Review of Evidence and Recommendations for Clinical Practice. *Med. Clin. N. Am.* **2017**, *101*, 689–699. [CrossRef] [PubMed]

4. Todd, J.; Farmer, J.A. Optimal low-density lipoprotein levels: Evidence from epidemiology and clinical trials. *Curr. Atheroscler. Rep.* **2006**, *8*, 157–162. [CrossRef]

5. Brousseau, M.E. Emerging role of high-density lipoprotein in the prevention of cardiovascular disease. *Drug Discov. Today* **2005**, *10*, 1095–1101. [CrossRef]

6. Rollins, J.; Chen, Y.; Paigen, B.; Wang, X. In search of new targets for plasma high-density lipoprotein cholesterol levels: Promise of human-mouse comparative genomics. *Trends Cardiovasc. Med.* **2006**, *16*, 220–234. [CrossRef]

7. Hansson, G.K.; Libby, P. The immune response in atherosclerosis: A double-edged sword. *Nat. Rev. Immunol.* **2006**, *6*, 508–519. [CrossRef]

8. Arnett, D.K.; Baird, A.E.; Barkley, R.A.; Basson, C.T.; Boerwinkle, E.; Ganesh, S.K.; Herrington, D.M.; Hong, Y.; Jaquish, C.; McDermott, D.A.; et al. American Heart Association Council on Epidemiology and Prevention; American Heart Association Stroke Council; Functional Genomics and Translational Biology Interdisciplinary Working Group. *Funct. Genom. Transl. Biol. Interdiscip. Work. Group* **2007**, *115*, 2878–2901.

9. Chen, Y.; Rollins, J.; Paigen, B.; Wang, X. Genetic and genomic insights into the molecular basis of atherosclerosis. *Cell Metab.* **2007**, *6*, 164–179. [CrossRef]

10. Bennett, B.J.; Davis, R.C.; Civelek, M.; Orozco, L.; Wu, J.; Qi, H.; Pan, C.; Packard, R.R.; Eskin, E.; Yan, M.; et al. Genetic Architecture of Atherosclerosis in Mice: A Systems Genetics Analysis of Common Inbred Strains. *PLoS Genet.* **2015**, *11*, e1005711. [CrossRef]

11. Wang, X.; Ishimori, N.; Korstanje, R.; Rollins, J.; Paigen, B. Identifying novel genes for atherosclerosis through mouse-human comparative genetics. *Am. J. Hum. Genet.* **2005**, *77*, 1–15. [CrossRef] [PubMed]

12. Almodovar, A.J.; Luther, R.J.; Stonebrook, C.L.; Wood, P.A. Genomic structure and genetic drift in C57BL/6 congenic metabolic mutant mice. *Mol. Genet. Metab.* **2013**, *110*, 396–400. [CrossRef] [PubMed]

13. Grainger, A.T.; Jones, M.B.; Chen, M.H.; Shi, W. Polygenic Control of Carotid Atherosclerosis in a BALB/cJ × SM/J Intercross and a Combined Cross Involving Multiple Mouse Strains. *G3* **2017**, *7*, 731–739. [CrossRef] [PubMed]

14. Wang, X.; Huang, R.; Zhang, L.; Li, S.; Luo, J.; Gu, Y.; Chen, Z.; Zheng, Q.; Chao, T.; Zheng, W.; et al. A severe atherosclerosis mouse model on the resistant NOD background. *Dis. Model. Mech.* **2018**, *11*, dmm033852. [CrossRef]

15. Garrett, N.E., 3rd; Grainger, A.T.; Li, J.; Chen, M.H.; Shi, W. Genetic analysis of a mouse cross implicates an anti-inflammatory gene in control of atherosclerosis susceptibility. *Mamm. Genome* **2017**, *28*, 90–99. [CrossRef]

16. Helgadottir, A.; Manolescu, A.; Thorleifsson, G.; Gretarsdottir, S.; Jonsdottir, H.; Thorsteinsdottir, U.; Samani, N.J.; Gudmundsson, G.; Grant, S.F.; Thorgeirsson, G.; et al. The gene encoding 5-lipoxygenase activating protein confers risk of myocardial infarction and stroke. *Nat. Genet.* **2004**, *36*, 233–239. [CrossRef]

17. Wang, L.; Fan, C.; Topol, S.E.; Topol, E.J.; Wang, Q. Mutation of MEF2A in an inherited disorder with features of coronary artery disease. *Science* **2003**, *302*, 1578–1581. [CrossRef]

18. Ghazalpour, A.; Wang, X.; Lusis, A.J.; Mehrabian, M. Complex inheritance of the 5-lipoxygenase locus influencing atherosclerosis in mice. *Genetics* **2006**, *173*, 943–951. [CrossRef]

19. van Wanrooij, E.J.; van Puijvelde, G.H.; de Vos, P.; Yagita, H.; van Berkel, T.J.; Kuiper, J. Interruption of the Tnfrsf4/Tnfsf4 (OX40/OX40L) pathway attenuates atherogenesis in low-density lipoprotein receptor-deficient mice. *Arterioscler. Thrombs. Vasc. Biol.* **2007**, *27*, 204–210. [CrossRef]

20. Ozaki, K.; Ohnishi, Y.; Iida, A.; Sekine, A.; Yamada, R.; Tsunoda, T.; Sato, H.; Sato, H.; Hori, M.; Nakamura, Y.; et al. Functional SNPs in the lymphotoxin-alpha gene that are associated with susceptibility to myocardial infarction. *Nat. Genet.* **2002**, *32*, 650–654. [CrossRef]

21. Shiffman, D.; Ellis, S.G.; Rowland, C.M.; Malloy, M.J.; Luke, M.M.; Iakoubova, O.A.; Pullinger, C.R.; Cassano, J.; Aouizerat, B.E.; Fenwick, R.G.; et al. Identification of four gene variants associated with myocardial infarction. *Am. J. Hum. Genet.* **2005**, *77*, 596–605. [CrossRef] [PubMed]

22. van der Net, J.B.; Versmissen, J.; Oosterveer, D.M.; Defesche, J.C.; Yazdanpanah, M.; Aouizerat, B.E.; Steyerberg, E.W.; Malloy, M.J.; Pullinger, C.R.; Kane, J.P.; et al. Arachidonate 5-lipoxygenase-activating protein (ALOX5AP) gene and coronary heart disease risk in familial hypercholesterolemia. *Atherosclerosis* **2009**, *203*, 472–478. [CrossRef] [PubMed]

23. Barbieri, M.; Marfella, R.; Rizzo, M.R.; Boccardi, V.; Siniscalchi, M.; Schiattarella, C.; Siciliano, S.; Lemme, P.; Paolisso, G. The -8 UTR C/G polymorphism of PSMA6 gene is associated with susceptibility to myocardial infarction in type 2 diabetic patients. *Atherosclerosis* **2008**, *201*, 117–123. [CrossRef] [PubMed]

24. Wang, L.; Hauser, E.R.; Shah, S.H.; Pericak-Vance, M.A.; Haynes, C.; Crosslin, D.; Harris, M.; Nelson, S.; Hale, A.B.; Granger, C.B.; et al. Peakwide mapping on chromosome 3q13 identifies the kalirin gene as a novel candidate gene for coronary artery disease. *Am. J. Hum. Genet.* **2007**, *80*, 650–663. [CrossRef] [PubMed]

25. Hubner, N.; Wallace, C.A.; Zimdahl, H.; Petretto, E.; Schulz, H.; Maciver, F.; Mueller, M.; Hummel, O.; Monti, J.; Zidek, V.; et al. Integrated transcriptional profiling and linkage analysis for identification of genes underlying disease. *Nat. Genet.* **2005**, *37*, 243–253. [CrossRef] [PubMed]

26. Lieu, H.D.; Withycombe, S.K.; Walker, Q.; Rong, J.X.; Walzem, R.L.; Wong, J.S.; Hamilton, R.L.; Fisher, E.A.; Young, S.G. Eliminating atherogenesis in mice by switching off hepatic lipoprotein secretion. *Circulation* **2003**, *107*, 1315–1321. [CrossRef]

27. Xu, G.; Zhao, L.; Fuchs, M. Differences between hepatic and biliary lipid metabolism and secretion in genetically gallstone-susceptible and gallstone-resistant mice. *Chin. Med. J.* **2002**, *115*, 1292–1295.

28. Getz, G.S.; Reardon, C.A. Diet and murine atherosclerosis. *Arterioscler. Thrombs. Vasc. Biol.* **2006**, *26*, 242–249. [CrossRef]

29. Stylianou, I.M.; Bauer, R.C.; Reilly, M.P.; Rader, D.J. Genetic basis of atherosclerosis: Insights from mice and humans. *Circ. Res.* **2012**, *110*, 337–355. [CrossRef]

30. Libby, P.; Ridker, P.M.; Hansson, G.K. Progress and challenges in translating the biology of atherosclerosis. *Nature* **2011**, *473*, 317–325. [CrossRef]

31. Stein, O.; Thiery, J.; Stein, Y. Is there a genetic basis for resistance to atherosclerosis? *Atherosclerosis* **2002**, *160*, 1–10. [CrossRef]

32. Wang, X.; Ria, M.; Kelmenson, P.M.; Eriksson, P.; Higgins, D.C.; Samnegård, A.; Petros, C.; Rollins, J.; Bennet, A.M.; Wiman, B.; et al. Positional identification of TNFSF4, encoding OX40 ligand, as a gene that influences atherosclerosis susceptibility. *Nat. Genet.* **2005**, *37*, 365–372. [CrossRef] [PubMed]

33. Koch, W.; Hoppmann, P.; Mueller, J.C.; Schömig, A.; Kastrati, A. Lack of support for association between common variation in TNFSF4 and myocardial infarction in a German population. *Nat. Genet.* **2008**, *40*, 1386–1387. [CrossRef] [PubMed]

34. Holdt, L.M.; Thiery, J.; Breslow, J.L.; Teupser, D. Increased ADAM17 mRNA expression and activity is associated with atherosclerosis resistance in LDL-receptor deficient mice. *Arterioscler. Thrombs. Vasc. Biol.* **2008**, *28*, 1097–1103. [CrossRef]

35. Cohen, J.C.; Boerwinkle, E.; Mosley, T.H., Jr.; Hobbs, H.H. Sequence variations in PCSK9, low LDL, and protection against coronary heart disease. *N. Engl. J. Med.* **2006**, *354*, 1264–1272. [CrossRef]

36. Keating, B.J.; Tischfield, S.; Murray, S.S.; Bhangale, T.; Price, T.S.; Glessner, J.T.; Galver, L.; Barrett, J.C.; Grant, S.F.; Farlow, D.N.; et al. Concept, design and implementation of a cardiovascular gene-centric 50 k SNP array for large-scale genomic association studies. *PLoS ONE* **2008**, *3*, e3583. [CrossRef]

37. Hobbs, H.H.; Russell, D.W.; Brown, M.S.; Goldstein, J.L. The LDL receptor locus in familial hypercholesterolemia: Mutational analysis of a membrane protein. *Ann. Rev. Genet.* **1990**, *24*, 133–170. [CrossRef]

38. Mani, A.; Radhakrishnan, J.; Wang, H.; Mani, A.; Mani, M.A.; Nelson-Williams, C.; Carew, K.S.; Mane, S.; Najmabadi, H.; Wu, D.; et al. LRP6 mutation in a family with early coronary disease and metabolic risk factors. *Science* **2007**, *315*, 1278–1282. [CrossRef]

39. McCarthy, M.I.; Abecasis, G.R.; Cardon, L.R.; Goldstein, D.B.; Little, J.; Ioannidis, J.P.; Hirschhorn, J.N. Genome-wide association studies for complex traits: Consensus, uncertainty and challenges. *Nat. Rev. Genet.* **2008**, *9*, 356–369. [CrossRef]

40. Wellcome Trust Case Control Consortium. Genome-wide association study of 14,000 cases of seven common diseases and 3000 shared controls. *Nature* **2007**, *447*, 661–678. [CrossRef]

41. Schunkert, H.; König, I.R.; Kathiresan, S.; Reilly, M.P.; Assimes, T.L.; Holm, H.; Preuss, M.; Stewart, A.F.; Barbalic, M.; Gieger, C.; et al. Large-scale association analysis identifies 13 new susceptibility loci for coronary artery disease. *Nat. Genet.* **2011**, *43*, 333–338. [CrossRef] [PubMed]

42. Coronary Artery Disease (C4D) Genetics Consortium. A genome-wide association study in Europeans and South Asians identifies five new loci for coronary artery disease. *Nat. Genet.* **2011**, *43*, 339–344. [CrossRef] [PubMed]

43. Kessler, T.; Vilne, B.; Schunkert, H. The impact of genome-wide association studies on the pathophysiology and therapy of cardiovascular disease. *EMBO Mol. Med.* **2016**, *8*, 688–701. [CrossRef] [PubMed]

44. Do, R.; Stitziel, N.O.; Won, H.H.; Jørgensen, A.B.; Duga, S.; Angelica Merlini, P.; Kiezun, A.; Farrall, M.; Goel, A.; Zuk, O.; et al. Exome sequencing identifies rare LDLR and APOA5 alleles conferring risk for myocardial infarction. *Nature* **2015**, *518*, 102–106. [CrossRef] [PubMed]

45. TG and HDL Working Group of the Exome Sequencing Project, National Heart, Lung, and Blood Institute; Crosby, J.; Peloso, G.M.; Auer, P.L.; Crosslin, D.R.; Stitziel, N.O.; Lange, L.A.; Lu, Y.; Tang, Z.Z.; Zhang, H.; et al. Loss-of-function mutations in APOC3, triglycerides, and coronary disease. *N. Engl. J. Med.* **2014**, *371*, 22–31.

46. Dewey, F.E.; Gusarova, V.; Dunbar, R.L.; O'Dushlaine, C.; Schurmann, C.; Gottesman, O.; McCarthy, S.; Van Hout, C.V.; Bruse, S.; Dansky, H.M.; et al. Genetic and Pharmacologic Inactivation of ANGPTL3 and Cardiovascular Disease. *N. Engl. J. Med.* **2017**, *377*, 211–221. [CrossRef] [PubMed]

47. Rotroff, D.M.; Pijut, S.S.; Marvel, S.W.; Jack, J.R.; Havener, T.M.; Pujol, A.; Schluter, A.; Graf, G.A.; Ginsberg, H.N.; Shah, H.S.; et al. Genetic Variants in HSD17B3, SMAD3, and IPO11 Impact Circulating Lipids in Response to Fenofibrate in Individuals with Type 2 Diabetes. *Clin. Pharmacol. Ther.* **2018**, *103*, 712–721. [CrossRef]

48. Jia, S.J.; Gao, K.Q.; Zhao, M. Epigenetic regulation in monocyte/macrophage: A key player during atherosclerosis. *Cardiovasc. Ther.* **2017**, *355*, e12262. [CrossRef]

49. Mostafavi, S.; Yoshida, H.; Moodley, D.; LeBoité, H.; Rothamel, K.; Raj, T.; Ye, C.J.; Chevrier, N.; Zhang, S.Y.; Feng, T.; et al. Immunological Genome Project Consortium. Parsing the Interferon Transcriptional Network and Its Disease Associations. *Cell* **2016**, *164*, 564–578. [CrossRef]

50. Domcke, S.; Bardet, A.F.; Adrian Ginno, P.; Hartl, D.; Burger, L.; Schübeler, D. Competition between DNA methylation and transcription factor determines binding of NRF1. *Nature* **2015**, *528*, 575–579. [CrossRef]

51. Lavin, Y.; Winter, D.; Blecher-Gonen, R.; David, E.; Keren-Shaul, H.; Merad, M.; Jung, S.; Amit, I. Tissue-resident macrophage enhancer landscapes are shaped by the local microenvironment. *Cell* **2014**, *159*, 1312–1326. [CrossRef]

52. Tirado-Magallanes, R.; Rebbani, K.; Lim, R.; Pradhan, S.; Benoukraf, T. Whole genome DNA methylation: Beyond genes silencing. *Oncotarget* **2017**, *8*, 5629–5637. [CrossRef] [PubMed]

53. Greco, C.M.; Kunderfranco, P.; Rubino, M.; Larcher, V.; Carullo, P.; Anselmo, A.; Kurz, K.; Carell, T.; Angius, A.; Latronico, M.V.; et al. DNA hydroxymethylation controls cardiomyocyte gene expression in development and hypertrophy. *Nat. Commun.* **2016**, *7*, 12418. [CrossRef] [PubMed]

54. Zaina, S.; Heyn, H.; Carmona, F.J.; Varol, N.; Sayols, S.; Condom, E.; Ramírez-Ruz, J.; Gomez, A.; Gonçalves, I.; Moran, S.; et al. DNA methylation map of human atherosclerosis. *Circ. Cardiovasc. Genet.* **2014**, *7*, 692–700. [CrossRef] [PubMed]

55. Valencia-Morales Mdel, P.; Zaina, S.; Heyn, H.; Carmona, F.J.; Varol, N.; Sayols, S.; Condom, E.; Ramírez-Ruz, J.; Gomez, A.; Moran, S.; et al. The DNA methylation drift of the atherosclerotic aorta increases with lesion progression. *BMC Med. Genom.* **2015**, *8*, 7. [CrossRef] [PubMed]

56. Pons, D.; de Vries, F.R.; van den Elsen, P.J.; Heijmans, B.T.; Quax, P.H.; Jukema, J.W. Epigenetic histone acetylation modifiers in vascular remodelling: New targets for therapy in cardiovascular disease. *Eur. Heart J.* **2009**, *30*, 266–277. [CrossRef] [PubMed]

57. Helin, K.; Dhanak, D. Chromatin proteins and modifications as drug targets. *Nature* **2013**, *502*, 480–488. [CrossRef]

58. Ulitsky, I.; Bartel, D.P. lncRNAs: Genomics, evolution, and mechanisms. *Cell* **2013**, *154*, 26–46. [CrossRef]

59. Zhang, X.; Wang, W.; Zhu, W.; Dong, J.; Cheng, Y.; Yin, Z.; Shen, F. Mechanisms and Functions of Long Non-Coding RNAs at Multiple Regulatory Levels. *Int. J. Mol. Sci.* **2019**, *20*, 5573. [CrossRef]

60. Okazaki, Y.; Furuno, M.; Kasukawa, T.; Adachi, J.; Bono, H.; Kondo, S.; Nikaido, I.; Osato, N.; Saito, R.; Suzuki, H.; et al. Analysis of the mouse transcriptome based on functional annotation of 60,770 full-length cDNAs. *Nature* **2002**, *420*, 563–573.

61. Khyzha, N.; Alizada, A.; Wilson, M.D.; Fish, J.E. Epigenetics of Atherosclerosis: Emerging Mechanisms and Methods. *Trends Mol. Med.* **2017**, *23*, 332–347. [CrossRef] [PubMed]

62. Batista, P.J.; Chang, H.Y. Long noncoding RNAs: Cellular address codes in development and disease. *Cell* **2013**, *152*, 1298–1307. [CrossRef] [PubMed]

63. Helgadottir, A.; Thorleifsson, G.; Manolescu, A.; Gretarsdottir, S.; Blondal, T.; Jonasdottir, A.; Jonasdottir, A.; Sigurdsson, A.; Baker, A.; Palsson, A.; et al. A common variant on chromosome 9p21 affects the risk of myocardial infarction. *Science* **2007**, *316*, 1491–1493. [CrossRef] [PubMed]
64. McPherson, R.; Pertsemlidis, A.; Kavaslar, N.; Stewart, A.; Roberts, R.; Cox, D.R.; Hinds, D.A.; Pennacchio, L.A.; Tybjaerg-Hansen, A.; Folsom, A.R.; et al. A common allele on chromosome 9 associated with coronary heart disease. *Science* **2007**, *316*, 1488–1491. [CrossRef] [PubMed]
65. Holdt, L.M.; Hoffmann, S.; Sass, K.; Langenberger, D.; Scholz, M.; Krohn, K.; Finstermeier, K.; Stahringer, A.; Wilfert, W.; Beutner, F.; et al. Alu elements in ANRIL non-coding RNA at chromosome 9p21 modulate atherogenic cell functions through trans-regulation of gene networks. *PLoS Genet.* **2013**, *9*, e1003588. [CrossRef] [PubMed]
66. Zhou, X.; Han, X.; Wittfeldt, A.; Sun, J.; Liu, C.; Wang, X.; Gan, L.M.; Cao, H.; Liang, Z. Long non-coding RNA ANRIL regulates inflammatory responses as a novel component of NF-κB pathway. *RNA Biol.* **2016**, *13*, 98–108. [CrossRef]
67. Wu, G.; Cai, J.; Han, Y.; Chen, J.; Huang, Z.P.; Chen, C.; Cai, Y.; Huang, H.; Yang, Y.; Liu, Y.; et al. LincRNA-p21 regulates neointima formation, vascular smooth muscle cell proliferation, apoptosis, and atherosclerosis by enhancing p53 activity. *Circulation* **2014**, *130*, 1452–1465. [CrossRef]
68. Krawczyk, M.; Emerson, B.M. p50-associated COX-2 extragenic RNA (PACER) activates COX-2 gene expression by occluding repressive NF-κB complexes. *Elife* **2014**, *3*, e01776. [CrossRef]
69. Atianand, M.K.; Hu, W.; Satpathy, A.T.; Shen, Y.; Ricci, E.P.; Alvarez-Dominguez, J.R.; Bhatta, A.; Schattgen, S.A.; McGowan, J.D.; Blin, J.; et al. A Long Noncoding RNA lincRNA-EPS Acts as a Transcriptional Brake to Restrain Inflammation. *Cell* **2016**, *165*, 1672–1685. [CrossRef]

International Journal of
Molecular Sciences

Review

The Diabetes Mellitus–Atherosclerosis Connection: The Role of Lipid and Glucose Metabolism and Chronic Inflammation

Anastasia Poznyak [1], Andrey V. Grechko [2], Paolo Poggio [3], Veronika A. Myasoedova [3] Valentina Alfieri [3] and Alexander N. Orekhov [1,4,5,*]

[1]   Institute for Atherosclerosis Research, Skolkovo Innovative Center, 121609 Moscow, Russia; tehhy_85@mail.ru
[2]   Federal Scientific Clinical Center for Resuscitation and Rehabilitation, 109240 Moscow, Russia; noo@fnkcrr.ru
[3]   Unit for the Study of Aortic, Valvular and Coronary Pathologies, Centro Cardiologico Monzino IRCCS, 20138 Milano, Italy; Paolo.Poggio@ccfm.it (P.P.); veronika.myasoedova@gmail.com (V.A.M.); valentina.alfieri@ccfm.it (V.A.)
[4]   Laboratory of Angiopathology, Institute of General Pathology and Pathophysiology, 125315 Moscow, Russia
[5]   Institute of Human Morphology, 117418 Moscow, Russia
*   Correspondence: a.h.opexob@gmail.com; Tel.: +7-903-169-0866

Received: 19 February 2020; Accepted: 4 March 2020; Published: 6 March 2020

Abstract: Diabetes mellitus comprises a group of carbohydrate metabolism disorders that share a common main feature of chronic hyperglycemia that results from defects of insulin secretion, insulin action, or both. Insulin is an important anabolic hormone, and its deficiency leads to various metabolic abnormalities in proteins, lipids, and carbohydrates. Atherosclerosis develops as a result of a multistep process ultimately leading to cardiovascular disease associated with high morbidity and mortality. Alteration of lipid metabolism is a risk factor and characteristic feature of atherosclerosis. Possible links between the two chronic disorders depending on altered metabolic pathways have been investigated in numerous studies. It was shown that both types of diabetes mellitus can actually induce atherosclerosis development or further accelerate its progression. Elevated glucose level, dyslipidemia, and other metabolic alterations that accompany the disease development are tightly involved in the pathogenesis of atherosclerosis at almost every step of the atherogenic process. Chronic inflammation is currently considered as one of the key factors in atherosclerosis development and is present starting from the earliest stages of the pathology initiation. It may also be regarded as one of the possible links between atherosclerosis and diabetes mellitus. However, the data available so far do not allow for developing effective anti-inflammatory therapeutic strategies that would stop atherosclerotic lesion progression or induce lesion reduction. In this review, we summarize the main aspects of diabetes mellitus that possibly affect the atherogenic process and its relationship with chronic inflammation. We also discuss the established pathophysiological features that link atherosclerosis and diabetes mellitus, such as oxidative stress, altered protein kinase signaling, and the role of certain miRNA and epigenetic modifications.

Keywords: atherosclerosis; diabetes mellitus; cardiovascular disease; chronic inflammation; lipid metabolism; hyperglycemia

---

1. Introduction

Atherosclerosis is a widespread chronic inflammatory disorder of the arterial wall that often leads to disability and even death. At its final stages, atherosclerosis manifests itself as a lesion of the intimal layer of the arterial wall and accumulation of plaques. Subsequent erosion or rupture of atherosclerotic plaques triggers thrombotic events that can potentially be fatal. Decades of intensive research made it clear that atherosclerosis has complex pathogenesis, the main components of which are lipid accumulation and chronic inflammation in the arterial wall [1]. Atherosclerosis is classically associated with altered lipid metabolism and hypercholesterolemia [2]. An elevated level of circulating modified low-density lipoprotein (LDL) is a known risk factor of cardiovascular diseases [3]. However, the disease pathogenesis appears to be more complex than lipid metabolism changes and involves multiple factors, the most prominent of which is inflammation [4].

The chain of pathological events that leads to atherosclerosis development is believed to be initiated by local endothelial dysfunction, which may be caused by blood flow turbulence near the sites of artery bends or bifurcations. The blood vessel endothelium responds to the mechanical stress with activation that subsequently leads to the recruitment of circulating immune cells. Circulating monocytes adhere to the injured area of the arterial wall and penetrate inside, differentiating into macrophages that actively participate in lipid uptake through phagocytosis and give rise to foam cells that are abundantly present in atherosclerotic plaques [5].

Detailed study of atherosclerotic lesion development is complicated by the fact that the process may differ considerably in humans and available model animals [6,7]. However, the main outlines of the process could be established. The early stage of atherosclerotic lesion development is known as "fatty streak", an area in the vascular wall that is characterized by intracellular lipid accumulation by foam cells, which also contains vascular smooth muscle cells (VSMCs) and T lymphocytes. Fatty streaks can further progress to atherosclerotic lesions if chronic injury of the endothelium persists. In the growing lesions, intracellular lipid accumulation involves several cell types. The recruited macrophages internalize LDL particles via phagocytosis and contribute to the local production of inflammation mediators. Resident intimal cells also actively participate in this process. The stellar-shaped macrovascular pericytes form a three-dimensional cellular network in the subendothelial layer of the intima, forming contacts with each other and endothelial cells and ensuring tissue homeostasis. This network is disrupted in atherosclerotic plaques due to pericyte phenotypic changes, leading to a loss of intercellular contacts and increased production of extracellular matrix components [8]. VSMCs involved in the pathological process can also undergo a phenotypic switch, possibly acquiring proliferative and secretory properties [9]. At the later stages of the disease development, plaques can acquire a stable fibrous cap that separates them from the vessel milieu. Destabilization of the plaque occurs through depletion and rupture of the fibrous cap facilitated by matrix metalloproteinases (MMPs) that provoke extracellular matrix degradation. Macrophages and other inflammatory cells serve as important sources of these enzymes in the plaque [10]. The mechanisms responsible for plaque erosion need to be further investigated. These processes are especially difficult to model in atherosclerotic animals [6]. Inflammatory events, such as local platelet-mediated neutrophil activation, release of myeloperoxidase, toll-like receptor (TLR-2) signaling, and neutrophil-mediated injury, appear to play their roles in this process [11].

Atherosclerotic plaques can reduce the lumen of the blood vessel, leading to ischemia and metabolic changes in the alimented tissues [12]. Even more dangerous is thrombogenesis induced by unstable plaques and, in some cases, on the surface of undamaged plaques, which can often lead to fatal consequences [13].

2. Diabetes Mellitus

Diabetes mellitus is a group of disorders of carbohydrate metabolism, whose main feature is chronic hyperglycemia that results from defects of insulin secretion, insulin action, or a combination of those. Metabolic abnormalities observed in diabetes can be caused by the low level of insulin production and/or insulin resistance of the target tissues. The disease affects primarily skeletal muscles and adipose tissue, but also liver, at the level of insulin receptors, the signal transduction system, and/or effector enzymes or genes [14]. Symptoms of hyperglycemia include polyuria, polydipsia, weight loss, sometimes accompanied by polyphagia, and blurred vision. It can also be accompanied by growth impairment and susceptibility to certain infections. The direct life-threatening consequences of uncontrolled diabetes are hyperglycemia with ketoacidosis or the nonketotic hyperosmolar syndrome [15]. However, some patients, mostly with type 2 diabetes, can remain asymptomatic during the early years of the disease.

2.1. Type 1 Diabetes

Type 1 diabetes (T1D) is caused by the autoimmune destruction of insulin-producing pancreatic β-cells [16]. The classic trio of T1D symptoms is polydipsia, polyphagia, and polyuria. Most often the disease is diagnosed in children and adolescents, who usually demonstrate the abovementioned combination of symptoms and a marked hyperglycemia that necessitates lifelong exogenous insulin replacement. The study of T1D pathogenesis was mostly based on two animal models of the disease: the nonobese diabetic mouse and the BioBreeding-diabetes-prone rat, both of which are characterized by progressive T-cell-mediated destruction of β-cells [17]. However, the differences between rodent models and the human situation limited the transferability of the obtained results. In humans, autoantibodies were present in 70–80% of patients at the time of diagnosis [18]. Immunosuppressive and immunointerventive approaches for preventing T1D did not result in preservation of β-cell function or acted only temporally [19,20].

Patients with T1D develop pancreatic lesions that can lead to acute pancreatitis with leukocyte accumulation [17]. The disease affects both exocrine and endocrine components of the pancreas.

2.2. Type 2 Diabetes

Increasing prevalence of type 2 diabetes mellitus (T2D), which now affects more than 370 million people, is a result of the worldwide increased incidence of obesity. Diabetes is diagnosed from fasting and two-hour glucose levels following a standardized oral glucose load. Prediabetes is often determined as the distinction between impaired fasting glucose and/or impaired glucose tolerance [21]. However, T2D should probably be regarded as a continuum of disease stages with increasing severity, in which the degree of plasma glucose increase relies on the magnitude of β-cell deficiency. Insulin resistance is already well established when impaired glucose tolerance is present and the increase in glucose, even across the normal range, is due to a continuous decline in β-cell function [22]. The disease was shown to be heritable in some cases, and individuals with first-degree relatives affected by diabetes are at increased risk of its development [23]. It was demonstrated that β-cell function is heritable [24] and that β-cell function crucially determines glucose intolerance and T2D in different racial and ethnic groups [25]. It is currently known that the pathogenesis of T2D is heterogeneous, and processes other than insulin resistance and β-cell dysfunction are involved in its development.

3. Diabetes Mellitus and Atherosclerosis: Pathophysiological Conjunction

Diabetes mellitus and atherosclerosis appear to be connected through several pathological pathways. Increased risk and accelerated development of atherosclerosis have been shown in studies on diabetic patients. For instance, several studies have reported the early development of atherosclerosis in adolescents and children with T1D [26,27]. Among the factors explaining such acceleration, dyslipidemia with increased levels of atherogenic LDL, hyperglycemia, oxidative stress, and increased inflammation have been proposed (Figure 1).

3.1. Dyslipidemia in Diabetic Patients

One of the most studied links between diabetes and atherosclerosis is the level of small dense LDL (sdLDL), which is a known risk factor of atherosclerosis. Although LDL is currently recognized as the main source of intracellular lipid accumulation in the plaque, native LDL particles do not cause prominent lipid accumulation in cultured cells; hence, they do not possess atherogenicity. It is atherogenic modification of LDL, which alters the physical–chemical characteristics of LDL particles, that triggers massive lipid accumulation [3]. A cascade of multiple modifications of LDL has been proposed as a plausible model for LDL atherogenic modification in the blood. According to this model, an LDL particle first becomes desialylated, which is followed by the increase of particle density, decrease of its size, and acquisition of negative charge [28]. Such particles can be isolated based on their physical properties. Very-low-density LDL (VLDL) is another subfraction of altered LDL described in humans. Small dense lipoprotein particles are more susceptible to oxidation because of the reduced antioxidant content and altered lipid composition. It is likely that oxidation takes place at the last stages of atherogenic LDL modification [29,30].

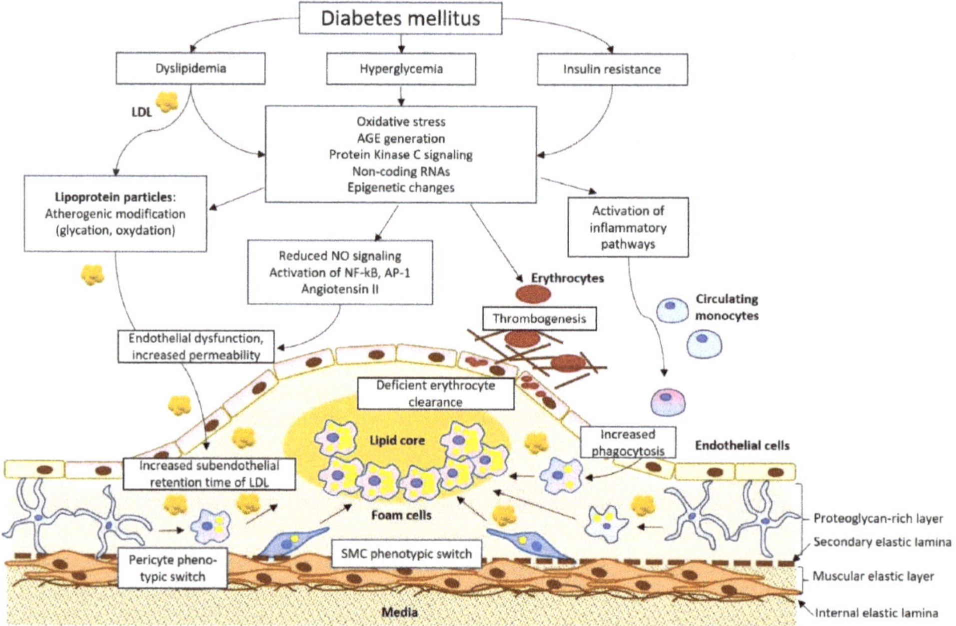

Figure 1. A simplified scheme of the pathophysiological connection of diabetes mellitus and atherosclerosis. Dyslipidemia, hyperglycemia, and insulin resistance result in a spectrum of physiological changes, including the formation of atherogenic low-density lipoprotein (LDL), advanced glycation end products (AGE), and activation of pro-inflammatory signaling that impact different cell types of the arterial wall, resulting in atherosclerotic lesion development. SMC, smooth muscular cells.

After penetration into the subendothelial space at the atherosclerotic lesion site, modified LDL particles reside there for a longer time due to interaction with proteoglycans and therefore have increased chances of being internalized by the lesion cells. Moreover, modified LDL has a lower affinity to LDL receptor (LDLR) and is therefore internalized mainly through unspecific phagocytosis, which leads to intracellular cholesterol accumulation rather than normal degradation of lipoprotein particles. These processes result in the formation of foam cells with the cytoplasm filled with accumulated lipid droplets [31].

Studies conducted in a murine model of diabetes clearly demonstrated the significance of LDL modification for the increase of the subendothelial retention time of lipoprotein particles. The authors extracted the LDL fraction from the blood of T1D patients and healthy controls, injected it into diabetic mice, and measured its retention in the atherosclerosis-prone areas of the arterial wall. It turned out that retention of LDL obtained from T1D patients was more than fourfold higher than that of LDL obtained from control subjects [32].

A cross-sectional study revealed altered lipoprotein levels in young adolescents with T1D and T2D. Increased concentrations of apolipoprotein B, sdLDL, and LDL-cholesterol were not highly prevalent in young subjects with T1D but had increased prevalence in subjects with T2D [33]. Importantly, studies that did not assess atherogenic LDL subfractions and only looked at the total level of circulating LDL may have missed this risk factor in diabetic patients, since the latter may remain within the normal range even when sdLDL is abnormally elevated.

Epidemiological studies indirectly demonstrated the effect of T1D- and T2D-associated metabolic changes on atherosclerosis development. For instance, the Diabetes Control and Complications Trial in patients with T1D and the United Kingdom Prospective Diabetes Study in patients with T2D demonstrated that patients on the conventional treatment who did not achieve adequate blood glucose control were at higher risk of vascular complications than those receiving intensive treatment to ensure strict glucose control [34,35].

Diabetes-associated dyslipidemia received much attention during recent years and became the subject of numerous review papers [36,37]. Alteration of the blood lipid profile in diabetes is linked to elevated hepatic production of triglyceride-rich lipoproteins, leading to the increased formation of atherogenic VLDL. This alteration can partially be corrected by insulin treatment.

3.2. The Role of Hyperglycemia and Advanced Glycation End-Products in Atherosclerosis

The risk of diabetic cardiovascular complications after exposure to high glucose levels for a certain period of time is called "metabolic memory" or "legacy effect". One of the possible mechanisms of this effect is the formation of advanced glycation end-products (AGE), which occurs when the blood glucose level is high. These compounds are not easily metabolized, accumulating in patients with a long history of inadequate blood glucose control. Such accumulation may accelerate the progression of vascular disease in diabetic patients. Numerous studies have established the link between inadequately controlled blood glucose level and microvascular complications of diabetes, such as renal and retinal symptoms. However, the relationship between elevated blood glucose and atherosclerosis of large arteries appears to be less straightforward. Direct pro-atherogenic effects of glucose levels on the cell types typically present in atherosclerotic lesions could not be demonstrated [38]. It remains possible that elevated glucose acts primarily on tissues, including liver or adipose tissue, and the effect on atherosclerotic lesion cells is mediated by altered signaling from these tissues. An elevated intracellular glucose level increases the flux through cellular metabolic pathways, such as the mitochondrial electron transport system, which may result in reactive oxygen species (ROS) overproduction. Moreover, glucose metabolites can induce pro-inflammatory responses through activation of protein kinase C-beta and aldose reductase [39].

Another possibility is that elevated glucose acts primarily through extracellular mechanisms, for example, by inducing glycation and glycoxidation of proteins, resulting in AGE formation. AGE accumulate in diabetic patients when the blood glucose level is elevated and appear to play an important role in atherosclerosis development. These molecules influence endothelium activation and surface expression of adhesion molecules, thereby promoting the adhesion and entrance of monocytes/macrophages into the subendothelial space during the initial stages of plaque formation. Moreover, these molecules enhance cytokine release by macrophages, thereby maintaining a pro-inflammatory context within the developing plaque. Another mechanism is glycation of LDL particles, which can be regarded as one of the atherogenic modifications of LDL. It was also shown that AGE may inhibit reverse cholesterol transport by reducing the expression of ATP-binding membrane cassette transporters A1 and G1 (ABCA1 and ABCG1) on monocytes, to enhance vasoconstriction by increasing endothelin-1 levels, and to reduce vasodilation by decreasing nitric oxide levels. Finally, AGE participate in the modification of extracellular matrix molecules, which also promotes atherosclerotic lesion development [40,41]. Modification of the extracellular matrix proteins by excessive glycation promotes interaction with AGE receptor RAGE on macrophages, endothelial cells, VSMCs, and other cell types. Such interaction results in pro-inflammatory effects and increased intracellular ROS generation [42].

In mouse models of atherosclerosis, such as apolipoprotein E-deficient (*apoE*$^{-/-}$) mice with chemically induced diabetes, RAGE deficiency was shown to alleviate atherosclerotic lesion development [43]. These findings open the possibility of using RAGE inhibition for reducing atherosclerosis development in diabetic patients.

The direct role of AGE in stimulating the expression of scavenger receptors and promoting phagocytosis has been revealed in a recent study [44]. In this study, AGE-modified bovine serum albumin induced morphological changes in cultured murine macrophages, increasing their phagocytic activity. This effect was attenuated by a fucose-containing sulphated polysaccharide, fucoidan, which has known anti-inflammatory properties. Diabetes is known to be associated with a pro-inflammatory state, which is discussed below in more detail. It is possible that enhanced glucose uptake by lesional cells is promoted by the pro-inflammatory signaling and increased phagocytic activity by lesion macrophages rather than the direct effect of hyperglycemia.

Further insights on possible links between hyperglycemia and atherosclerosis came from animal studies. Studies of hyperglycemia's effect on vascular lesions in the *apoE*$^{-/-}$ mouse model revealed that advanced lesions appear in hyperglycemic mice earlier than they do in normoglycemic controls. Moreover, accelerated atherogenesis was observed earlier than any detectable divergence in the plasma lipid parameters in normoglycemic mice [45]. A new model of hyperglycemia-accelerated atherosclerosis was created by crossing *apoE*$^{-/-}$ or LDLR-deficient mouse strains with mice carrying a point mutation in the gene encoding insulin (Ins2$^{+/Akita}$:*apoE*$^{-/-}$ mice) [45]. These animals were characterized by spontaneous development of diabetes and atherosclerosis, presenting with insulin deficiency, hypercholesterolemia (predominantly through LDL-cholesterol increase), and accelerated formation of atherosclerotic plaques while kept on a regular chow diet. The authors reported deficient lipoprotein clearance through lipolysis-stimulated lipoprotein receptors and altered lipoprotein composition. This animal model was expected to be useful for studying atherosclerosis in the context of T1D and testing possible therapeutic approaches. For instance, Ins2$^{+/Akita}$:*apoE*$^{-/-}$ mice were used to demonstrate the beneficial effect of leptin on atherosclerotic plaque progression [46].

Excessive glycation may also play a role at later stages of atherosclerosis development. As demonstrated in a recent study, glycation of erythrocytes in T2D patients may promote their internalization by the endothelial cells via phagocytosis, which impairs endothelial function. This process is likely to contribute to unstable plaque development with subsequent thrombosis in patients with T2D and atherosclerosis [47].

The level of AGE may also be used for diagnostic purposes to assess the risk of atherosclerosis development and vascular complications in diabetic patients. In a recent study, measurement of skin AGE levels through autofluorescence (AF) in Japanese T1D patients and their gender- and age-matched healthy controls demonstrated the increased AF in diabetes that appeared to be an independent risk factor for carotid atherosclerosis [48].

3.3. The Role of Oxidative Stress

Diabetes is known to be associated with both increased ROS production and reduced activity of antioxidant systems [49]. Studies in vitro have demonstrated that increased ROS production is linked to hyperglycemia [50]. Further studies in animals have revealed the involvement of NADPH oxidase family protein Nox1, which was up-regulated in diabetic mice. Knockdown of this protein alleviated atherosclerosis progression in such animals [51]. The role of oxidative stress in diabetes-associated atherosclerosis was confirmed in experiments on *apoE*$^{-/-}$ mice deficient for one of the main regulators of antioxidant enzymes, glutathione peroxidase 1 (Gpx1). Upon diabetes induction with streptozotocin, animals that were also deficient for Gpx1 had accelerated atherogenesis, with increased plaque size, macrophage infiltration, and increased expression of inflammatory markers, while restoration of Gpx1 reduced atherogenesis [52]. Overall, vascular ROS increase appears to be closely related to atherosclerosis in the diabetic context, and antioxidant therapies may still be considered for the management of the disease, although more selective approaches are needed to achieve relevant results with antioxidant drugs [40].

3.4. The Role of Protein Kinase C (PKC) Activation

Protein kinase C (PKC) is one of the key protein kinases mediating the cellular signaling pathway, which responds to cytokines, growth factors, and other messenger molecules [53]. Increased glucose uptake by vascular cells results in increased synthesis of diacylglycerol, which is an activator of PKC. Enhanced PKC activation can also result in response to oxidative stress [54]. Increased vascular PKC activation was confirmed in animal models of diabetes. Enhanced PKC signaling has numerous pro-atherogenic effects, including reduced production of NO and impaired vasodilation, endothelial dysfunction and increased permeability, and increased production of cytokines and extracellular matrix [42]. The complexity of intracellular signaling cascades activated by PKC makes it difficult to pinpoint the exact mechanism of its pro-atherogenic effect. However, studies in *apoE*$^{-/-}$ mice have shown that chemical or genetic inhibition of PKCβ led to reduced formation of atherosclerotic lesions [55]. Another study conducted in *apoE*$^{-/-}$ and in *apoE*$^{-/-}$ and *PKCβ*$^{-/-}$ double knock-out mice with chemically induced diabetes showed that increased PKCβ activation was linked to accelerated atherosclerosis development through the induction of CD11c and pro-inflammatory activation of macrophages. Correspondingly, inhibition of PKCβ reduced the size of atherosclerotic lesions [56]. These findings indicate that PKC may be considered as a potential therapeutic target.

3.5. Diabetes-Associated Chronic Inflammation and Atherosclerosis

Chronic inflammation is a known feature that is common to both atherosclerosis and diabetes mellitus. Atherosclerosis is currently regarded as a chronic inflammatory condition. In patients with T2D, increased activity of inflammasomes and elevated levels of nucleotide-binding oligomerization domain-like receptor 3 (NLRP3) were demonstrated, together with increased levels of pro-inflammatory cytokines interleukin (IL)-1β and IL-18 [57]. One of the direct links between atherosclerosis and diabetes identified within the inflammatory pathways is neutrophil extracellular trap activation, or NETosis, a special type of cell death of macrophages, during which the cells release chromatin into the extracellular space to trap and kill bacteria. This process is known to be elevated in chronic sterile inflammation and autoimmune conditions where it contributes to pathology development [42]. Increased levels of NETosis markers were found in patients with T2D [58]. Moreover, it was shown that NETosis may be enhanced in hyperglycemic conditions [59]. The possible role of enhanced NETosis in atherosclerosis development was confirmed in animal models. The atherosclerotic *apoE*$^{-/-}$ mice that also lacked neutrophil elastase and proteinase-3 necessary for NETosis had reduced atherosclerotic lesion formation compared with single knock-out animals [60].

Naturally, an active search for anti-inflammatory drugs that could reduce the risk of atherosclerotic cardiovascular disease in diabetic patients was conducted during recent years [54]. Among the anti-inflammatory drugs used to treat diabetic patients are salicylates, which were shown to reduce the glucose level while being effective for cardiovascular disease prevention and reducing the risk of thrombosis [61]. The use of inflammatory cytokine inhibitors appeared to be a promising approach for reducing cardiovascular risk in diabetic patients. It was shown that canakinumab, a monoclonal antibody that binds and neutralizes IL-1β, significantly reduced markers of inflammation in patients with controlled diabetes mellitus and high cardiovascular risk, but had no major effect on LDL-cholesterol [62]. Another study showed that canakinumab had a similar beneficial effect for reducing cardiovascular risk in patients with and without diabetes, but had no effect on de novo diabetes incidence [63]. The search for effective anti-inflammatory therapies reducing atherosclerosis in diabetic patients continues [64].

3.6. The Role of Circulating Non-Coding RNAs

Non-coding RNAs were shown to be implicated in numerous human disorders and are currently regarded as possible biomarkers and disease modifiers. Advances in genetic methods allowed non-coding RNAs to be studied in more detail and revealed their associations with pathogenic processes. MicroRNA (miRNA) are short RNA fragments that can inhibit the expression of certain genes at the mRNA level. These RNA fragments can be produced by multiple cell types and tissues and can be found circulating in the blood either free or confined in membrane microvesicles. Accumulating evidence highlights miRNAs as important possible biomarkers. However, the complexity of the miRNA landscape associated with human diseases, including diabetes, is so high that it is probably more promising to study miRNA signatures (combinations of multiple miRNA) rather than single miRNA types [65].

In humans, more than 2500 miRNAs have been identified, and several of them were shown to play a role in diabetes mellitus pathogenesis. In particular, multiple miRNAs were found to be involved in the development of microvascular complications of diabetes [66]. Among the miRNA which are relevant for atherosclerosis as well as diabetes, miR-146 and miR-126 have received much attention. miR-146a and miR-146b play an important role in the endothelial cells, where their expression is induced by inflammatory cytokine signaling and serves as a negative feedback loop to control inflammatory endothelial activation [67]. Therefore, regulation of these miRNAs is likely to be implicated at the initial stages of atherosclerotic lesion development in diabetic patients. miR-126 expression was shown to be a risk factor of T2D development, and this miRNA played a protective role in a mouse model of atherosclerosis [68,69].

Another important miRNA, miR-378a, was shown to play an important role in the regulation of metabolism, including energy and glucose homeostasis [70]. A very recent study implicated this miRNA in atherosclerosis development. The authors showed that miRNA-378a targets signal regulatory protein alpha (SIRPa), thereby regulating phagocytosis and polarization of macrophages. Moreover, the level of this miRNA was reduced in the aorta of *apoE*$^{-/-}$ mice in comparison to controls, highlighting its important role in the regulation of atherosclerosis-associated processes [71].

Another non-coding RNA that likely plays a role in diabetes-associated atherosclerosis is long non-coding RNA Dnm3os (dynamin 3 opposite strand). This RNA was shown to be increased in macrophages from diabetic mice, including *apoE*$^{-/-}$ diabetic mice, as well as in monocytes from T2D patients. Overexpression of this RNA promoted inflammatory gene expression and phagocytosis by macrophages and led to chromatin epigenetic changes, further promoting the inflammatory response [72]. More studies are needed to identify and characterize relevant non-coding RNAs that may have detrimental or beneficial effects on cardiovascular risk in diabetic patients.

3.7. The Role of Epigenetic Mosdification

Both persistent and temporal hyperglycemic exposure were shown to influence several significant cellular signaling pathways, including PKC activation and oxidative stress, described above, and transforming growth factor (TGF)-β-SMAD-MAPK signaling [73,74]. Moreover, hyperglycemia was shown to enhance the flux into the polyol and hexosamine pathways and increase the formation of AGE that are also associated with alterations in signaling pathways [75,76]. All these various effects make hyperglycemia a prominent risk factor for diabetic complications and vascular events. Chromatin changes play an important regulative role in the establishing of the link between glycemia and vascular complications.

Hyperglycemia is associated with the range of chromatin modifications that affect the genetic signature of vascular endothelial cells. For example, a genome-wide sequencing study of aortic endothelial cells exposed to a high level of glucose revealed histone H3K9/K14 hyperacetylation patterns that were inversely associated with DNA methylation in CpG clusters. This finding correlates with the activation of transcription of pathways linked to atherogenic effects and vascular diseases. The study demonstrated that hyperglycemia can induce epigenetic changes in the vascular endothelium that are relevant for atherosclerosis development, thereby providing another link between diabetes and atherosclerosis pathogenesis [77].

Transient hyperglycemia also triggers mono-methylation of H3 histones at lysine 4 (H3K4m1) and other histone lysine modifications. H3K4m1 was shown to be written by the Set7 lysine methyltransferase. The observed changes at the promoter of the *RELA* gene encoding the NF-κB-p65 subunit persisted for 5–6 days after the cells were returned to a normoglycemic state [78,79]. Thus, cytokines, chemokines, and adhesion molecules are affected by hyperglycemia through the regulation of one of the crucial pro-inflammatory transcription factors related to vascular and metabolic complications, among which is atherosclerosis [80]. Among these molecules, vascular cell adhesion molecule 1 (VCAM-1), which promotes adhesion of monocytes to the arterial endothelial cells, and monocyte chemoattractant protein 1 (MCP-1), responsible for macrophage infiltration, should be highlighted [81]. The enhanced expression of both NF-κB-dependent MCP-1 and VCAM-1 and genes encoding NFκB-p65 itself was observed in aortas of apolipoprotein A knockdown mice that were previously exposed to hyperglycemia.

The concluding result of these hyperglycemia-induced changes is the transcriptional activation of genes that are related to endothelial dysfunction [82]. Acetylation, as well as hyperacetylation, is also possible and can lead to the enhanced expression of the following genes relevant for atherosclerotic lesion development at different stages through the inflammatory response and extracellular matrix degradation: *MCP-1, MMP10, ICAM, HMOX1,* and *SLC7A11* [76].

Further research identified the role of Set7/9, which possibly coactivates NF-κB transcriptional activity in monocytes in response to inflammation through the activation of the H3K4me promotor, and the analogous effect was observed in endothelial cells in response to hyperglycemia [80,81]. Moreover, the induction of Set7-mediated up-regulation of *HMOX1* was shown to be beyond the methylation of histones but linked to hyperglycemia [78]. It was also shown that early intrusive monitoring of the glycemic profile in diabetic patients can play an important role in preventing vascular complications. This indicates the important role of hyperglycemia in the long-term outlook, leading to a phenomenon called "metabolic memory" [82].

All the aforementioned findings suggest numerous links between hyperglycemia, epigenetic changes, and cardiovascular risk that cannot be ignored. However, there are still many blind spots in the understanding of the underlying molecular mechanisms and their connections.

4. Conclusions

Both types of diabetes mellitus have been shown to be independent risk factors for accelerated atherosclerosis development. It is now clear that the pathogenesis of diabetes mellitus and atherosclerosis are closely linked, but the mechanisms and molecular interactions of this linkage are still under discussion. Among the known pathological mechanisms connecting diabetes and atherosclerosis are dyslipidemia, hyperglycemia with AGE production, increased oxidative stress, and inflammation. Despite the continuing search for novel therapeutic approaches, few medications have shown strong beneficial effects with regard to reducing the risk of atherosclerosis development in the specific population of diabetic patients. Adequate glycemic control and reduction of known risk factors remain the most frequently used strategies for protecting such patients. More studies are needed to reveal the exact signaling mechanisms of diabetes-associated macrovascular damage and to identify specific therapeutic targets.

Author Contributions: A.N.O., V.A.M., V.A. and P.P. conceptualized the manuscript, A.P. wrote the manuscript text, V.A.M., A.V.G. and A.N.O. reviewed the text, A.V.G. and A.N.O. obtained funding. All authors have read and agreed to the published version of the manuscript.

Funding: This work was supported by the Russian Science Foundation (Grant #18-15-00254).

Conflicts of Interest: The authors declare no conflict of interest.

References

1. Falk, E. Pathogenesis of atherosclerosis. *J. Am. Coll. Cardiol.* **2006**, *47* (Suppl. 8), C7. [CrossRef]
2. Miname, M.H.; Santos, R.D. Reducing cardiovascular risk in patients with familial hypercholesterolemia: Risk prediction and lipid management. *Prog. Cardiovasc. Dis.* **2019**, *62*, 414–422. [CrossRef] [PubMed]

3.   Summerhill, V.I.; Grechko, A.V.; Yet, S.F.; Sobenin, I.A.; Orekhov, A.N. The atherogenic role of circulating modified lipids in atherosclerosis. *Int. J. Mol. Sci.* **2019**, *20*, 3561. [CrossRef] [PubMed]

4.   Taleb, S. Inflammation in atherosclerosis. *Arch. Cardiovasc. Dis.* **2016**, *109*, 708–715. [CrossRef] [PubMed]

5.   Cheng, F.; Torzewski, M.; Degreif, A.; Rossmann, H.; Canisius, A.; Lackner, K.J. Impact of glutathione peroxidase-1 deficiency on macrophage foam cell formation and proliferation: Implications for atherogenesis. *PLoS ONE* **2013**, *8*, e72063. [CrossRef]

6.   Schwartz, S.M.; Galis, Z.S.; Rosenfeld, M.E.; Falk, E. Plaque rupture in humans and mice. *Arterioscler. Thromb. Vasc. Biol.* **2007**, *27*, 705–713. [CrossRef]

7.   Aliev, G.; Castellani, R.J.; Petersen, R.B.; Burnstock, G.; Perry, G.; Smith, M.A. Pathobiology of familial hypercholesterolemic atherosclerosis. *J. Submicrosc. Cytol. Pathol.* **2004**, *36*, 225–240.

8.   Rekhter, M.D.; Andreeva, E.R.; Mironov, A.A.; Orekhov, A.N. Three-dimensional cytoarchitecture of normal and atherosclerotic intima of human aorta. *Am. J. Pathol.* **1991**, *138*, 569–580.

9.   Xepapadaki, E.; Zvintzou, E.; Kalogeropoulou, C.; Filou, S.; Kypreos, K.E. The Antioxidant Function of HDL in Atherosclerosis. *Angiology* **2020**, *71*, 112–121. [CrossRef]

10. Shah, P.K. Molecular mechanisms of plaque instability. *Curr. Opin. Lipidol.* **2007**, *18*, 492–499. [CrossRef]

11. Folco, E.J.; Mawson, T.L.; Vromman, A.; Bernardes-Souza, B.; Franck, G.; Persson, O.; Nakamura, M.; Newton, G.; Luscinskas, F.W.; Libby, P. Neutrophil Extracellular Traps Induce Endothelial Cell Activation and Tissue Factor Production through Interleukin-1α and Cathepsin G. *Arterioscler. Thromb. Vasc. Biol.* **2018**, *38*, 1901–1912. [CrossRef] [PubMed]

12. Martínez, M.S.; García, A.; Luzardo, E.; Chávez-Castillo, M.; Olivar, L.C.; Salazar, J.; Velasco, M.; Rojas Quintero, J.J.; Bermúdez, V. Energetic metabolism in cardiomyocytes: Molecular basis of heart ischemia and arrhythmogenesis. *Vessel Plus* **2017**, *1*, 130–141. [CrossRef]

13. Shah, P.K. Inflammation, infection and atherosclerosis. *Trends Cardiovasc. Med.* **2019**, *28*, 468–472. [CrossRef] [PubMed]

14. Kharroubi, A.T.; Darwish, H.M. Diabetes mellitus: The epidemic of the century. *World J. Diabetes.* **2015**, *6*, 850–867. [CrossRef] [PubMed]

15. American Diabetes Association. Diagnosis and classification of diabetes mellitus. *Diabetes Care.* **2010**, *33* (Suppl. 1), S62–S69. [CrossRef]

16. Bluestone, J.A.; Herold, K.; Eisenbarth, G. Genetics, pathogenesis and clinical interventions in type 1 diabetes. *Nature* **2010**, *464*, 1293–1300. [CrossRef]

17. Roep, B.O.; Atkinson, M.; von Herrath, M. Satisfaction (not) guaranteed: Re-evaluating the use of animal models of type 1 diabetes. *Nat. Rev. Immunol.* **2004**, *4*, 989–997. [CrossRef]

18. Notkins, A.L.; Lernmark, A. Autoimmune type 1 diabetes: Resolved and unresolved issues. *J. Clin. Investig.* **2001**, *108*, 1247–1252. [CrossRef]

19. Keymeulen, B.; Walter, M.; Mathieu, C.; Kaufman, L.; Gorus, F.; Hilbrands, R.; Vandemeulebroucke, E.; Van de Velde, U.; Crenier, L.; De Block, C.; et al. Four-year metabolic outcome of a randomized controlled CD3-antibody trial in recent-onset type 1 diabetic patients depends on their age and baseline residual beta cell mass. *Diabetologia* **2010**, *53*, 614–623. [CrossRef]

20. Wherrett, D.K.; Bundy, B.; Becker, D.J.; DiMeglio, L.A.; Gitelman, S.E.; Goland, R.; Gottlieb, P.A.; Greenbaum, C.J.; Herold, K.C.; Marks, J.B.; et al. Type 1 Diabetes TrialNet GAD Study Group. Antigen-based therapy with glutamic acid decarboxylase (GAD) vaccine in patients with recent-onset type 1 diabetes: A randomised double-blind trial. *Lancet* **2011**, *378*, 319–327. [CrossRef]

21. Genuth, S.; Alberti, K.G.; Bennett, P.; Buse, J.; Defronzo, R.; Kahn, R.; Kitzmiller, J.; Knowler, W.C.; Lebovitz, H.; Lernmark, A.; et al. Expert Committee on the Diagnosis and Classification of Diabetes Mellitus. Follow-up report on the diagnosis of diabetes mellitus. *Diabetes Care* **2003**, *26*, 3160–3167. [PubMed]

22. Stancáková, A.; Javorský, M.; Kuulasmaa, T.; Haffner, S.M.; Kuusisto, J.; Laakso, M. Changes in insulin sensitivity and insulin release in relation to glycemia and glucose tolerance in 6414 Finnish men. *Diabetes* **2009**, *58*, 1212–1221. [CrossRef] [PubMed]

23. Cnop, M.; Vidal, J.; Hull, R.L.; Utzschneider, K.M.; Carr, D.B.; Schraw, T.; Scherer, P.E.; Boyko, E.J.; Fujimoto, W.Y.; Kahn, S.E. Progressive loss of beta-cell function leads to worsening glucose tolerance in first-degree relatives of subjects with type 2 diabetes. *Diabetes Care* **2007**, *30*, 677–682. [CrossRef] [PubMed]

24. Elbein, S.C.; Hasstedt, S.J.; Wegner, K.; Kahn, S.E. Heritability of pancreatic beta-cell function among nondiabetic members of Caucasian familial type 2 diabetic kindreds. *J. Clin. Endocrinol. Metab.* **1999**, *84*, 1398–1403. [PubMed]

25. Jensen, C.C.; Cnop, M.; Hull, R.L.; Fujimoto, W.Y.; Kahn, S.E. American Diabetes Association GENNID Study Group. Beta-cell function is a major contributor to oral glucose tolerance in high-risk relatives of four ethnic groups in the U.S. *Diabetes* **2002**, *51*, 2170–2178. [CrossRef] [PubMed]

26. Dahl-Jorgensen, K.; Larsen, J.R.; Hanssen, K.F. Atherosclerosis in childhood and adolescent type 1 diabetes: Early disease, early treatment? *Diabetologia* **2005**, *48*, 1445–1453. [CrossRef] [PubMed]

27. Orchard, T.J.; Costacou, T.; Kretowski, A.; Nesto, R.W. Type 1 diabetes and coronary artery disease. *Diabetes Care* **2006**, *29*, 2528–2538. [CrossRef]

28. Berneis, K.K.; Krauss, R.M. Metabolic origins and clinical significance of LDL heterogeneity. *J. Lipid Res.* **2002**, *43*, 1363–1379. [CrossRef]

29. Ivanova, E.A.; Bobryshev, Y.V.; Orekhov, A.N. LDL electronegativity index: A potential noved index for predicting cardiovascular disease. *Vasc. Health Risk Manag.* **2015**, *11*, 525–532.

30. Ivanova, E.A.; Myasoedova, V.A.; Melnichenko, A.A.; Grechko, A.V.; Orekhov, A.N. Small dense low-density lipoprotein as biomarker for atherosclerotic diseases. *Oxid Med. Cell Longev.* **2017**, *2017*, 1273042. [CrossRef]

31. Maguire, E.M.; Pearce, S.W.A.; Xiao, Q. Foam cell formation: A new target for fighting atherosclerosis and cardiovascular disease. *Vasc. Pharmacol.* **2019**, *112*, 54–71. [CrossRef] [PubMed]

32. Hagensen, M.K.; Mortensen, M.B.; Kjolby, M.; Palmfeldt, J.; Bentzon, J.F.; Gregersen, S. Increased retention of LDL from type 1 diabetic patients in atherosclerosis-prone areas of the murine arterial wall. *Atherosclerosis* **2019**, *286*, 156–162. [CrossRef] [PubMed]

33. Albers, J.J.; Marcovina, S.M.; Imperatore, G.; Snively, B.M.; Stafford, J.; Fujimoto, W.Y.; Mayer-Davis, E.J.; Petitti, D.B.; Pihoker, C.; Dolan, L.; et al. Prevalence and Determinants of Elevated Apolipoprotein B and Dense Low-Density Lipoprotein in Youths with Type 1 and Type 2 Diabetes. *J. Clin. Endocrinol. Metab.* **2008**, *93*, 735–742. [CrossRef] [PubMed]

34. Diabetes Control and Complications Trial (DCCT)/Epidemiology of Diabetes Interventions and Complications (EDIC) Study Research Group. Intensive Diabetes Treatment and Cardiovascular Outcomes in Type 1 Diabetes: The DCCT/EDIC Study 30-Year Follow-up. *Diabetes Care* **2016**, *39*, 686–693. [CrossRef] [PubMed]

35. Holman, R.R.; Paul, S.K.; Bethel, M.A.; Matthews, D.R.; Neil, H.A. 10-year follow-up of intensive glucose control in type 2 diabetes. *N. Engl. J. Med.* **2008**, *359*, 1577–1589. [CrossRef] [PubMed]

36. Hirano, T. Pathophysiology of diabetic dyslipidemia. *J. Atheroscler. Thromb.* **2018**, *25*, 771–782. [CrossRef]

37. Taskinen, M.R.; Borén, J. New insights into the pathophysiology of dyslipidemia in type 2 diabetes. *Atherosclerosis* **2015**, *239*, 483–495. [CrossRef]

38. Bornfeldt, K.E. Does Elevated Glucose Promote Atherosclerosis? Pros and Cons. *Circ. Res.* **2016**, *119*, 190–193. [CrossRef]

39. Tabit, C.E.; Shenouda, S.M.; Holbrook, M.; Fetterman, J.L.; Kiani, S.; Frame, A.A.; Kluge, M.A.; Held, A.; Dohadwala, M.M.; Gokce, N.; et al. Protein kinase C-β contributes to impaired endothelial insulin signaling in humans with diabetes mellitus. *Circulation* **2013**, *127*, 86–95. [CrossRef]

40. Katakami, N. Mechanism of Development of Atherosclerosis and Cardiovascular Disease in Diabetes Mellitus. *J. Atheroscler. Thromb.* **2018**, *25*, 27–39. [CrossRef]

41. Khan, M.I.; Pichna, B.A.; Shi, Y.; Bowes, A.J.; Werstuck, G.H. Evidence supporting a role for endoplasmic reticulum stress in the development of atherosclerosis in a hyperglycaemic mouse model. *Antioxid. Redox Signal.* **2009**, *11*, 2289–2298. [CrossRef] [PubMed]

42. Zeadin, M.G.; Petlura, C.I.; Werstuck, G.H. Molecular mechanisms linking diabetes to the accelerated development of atherosclerosis. *Can. J. Diabetes* **2013**, *37*, 345–350. [CrossRef] [PubMed]

43. Soro-Paavonen, A.; Watson, A.M.; Li, J.; Paavonen, K.; Koitka, A.; Calkin, A.C.; Barit, D.; Coughlan, M.T.; Drew, B.G.; Lancaster, G.I.; et al. Receptor for advanced glycation end products (RAGE) deficiency attenuates the development of atherosclerosis in diabetes. *Diabetes* **2008**, *57*, 2461–2469. [CrossRef] [PubMed]

44. Hamasaki, S.; Kobori, T.; Yamazaki, Y.; Kitaura, A.; Niwa, A.; Nishinaka, T.; Nishibori, M.; Mori, S.; Nakao, S.; Takahashi, H. Effects of scavenger receptors-1 class A stimulation on macrophage morphology and highly modified advanced glycation end product-protein phagocytosis. *Sci. Rep.* **2018**, *8*, 5901. [CrossRef] [PubMed]

45. Jun, J.Y.; Ma, Z.; Pyla, R.; Segar, L. Spontaneously diabetic Ins2+/Akita:apoE-deficient mice exhibit exaggerated hypercholesterolemia and atherosclerosis. *Am. J. Physiol. Endocrinol. Metab.* **2011**, *301*, E145–E154. [CrossRef] [PubMed]

46. Jun, J.Y.; Ma, Z.; Pyla, R.; Segar, L. Leptin treatment inhibits the progression of atherosclerosis by attenuating hypercholesterolemia in type 1 diabetic Ins2+/Akita:apoE-/- mice. *Atherosclerosis* **2012**, *225*, 341–347. [CrossRef]

47. Catan, A.; Turpin, C.; Diotel, N.; Patche, J.; Guerin-Dubourg, A.; Debussche, X.; Bourdon, E.; Ah-You, N.; Le Moullec, N.; Besnard, M.; et al. Aging and glycation promote erythrocyte phagocytosis by human endothelial cells: Potential impact in atherothrombosis under diabetic conditions. *Atherosclerosis* **2019**, *291*, 87–98. [CrossRef]

48. Osawa, S.; Katakami, N.; Kuroda, A.; Takahara, M.; Sakamoto, F.; Kawamori, D.; Matsuoka, T.; Matsuhisa, M.; Shimomura, I. Skin Autofluorescence is Associated with Early-stage Atherosclerosis in Patients with Type 1 Diabetes. *J. Atheroscler. Thromb.* **2017**, *24*, 312–326. [CrossRef]

49. Di Marco, E.; Jha, J.C.; Sharma, A.; Wilkinson-Berka, J.L.; Jandeleit-Dahm, K.A.; de Haan, J.B. Are reactive oxygen species still the basis for diabetic complications? *Clin. Sci.* **2015**, *129*, 199–216. [CrossRef]

50. Giacco, F.; Brownlee, M. Oxidative stress and diabetic complications. *Circ. Res.* **2010**, *107*, 1058–1070. [CrossRef]

51. Gray, S.P.; Di Marco, E.; Okabe, J.; Szyndralewiez, C.; Heitz, F.; Montezano, A.C.; de Haan, J.B.; Koulis, C.; El-Osta, A.; Andrews, K.L.; et al. NADPH oxidase 1 plays a key role in diabetes mellitus-accelerated atherosclerosis. *Circulation* **2013**, *127*, 1888–1902. [CrossRef]

52. Chew, P.; Yuen, D.Y.; Stefanovic, N.; Pete, J.; Coughlan, M.T.; Jandeleit-Dahm, K.A.; Thomas, M.C.; Rosenfeldt, F.; Cooper, M.E.; de Haan, J.B. Antiatherosclerotic and renoprotective effects of ebselen in the diabetic apolipoprotein E/GPx1-double knockout mouse. *Diabetes* **2010**, *59*, 3198–3207. [CrossRef] [PubMed]

53. Newton, A.C. Regulation of the ABC kinases by phosphorylation: Protein kinase C as a paradigm. *Biochem. J.* **2008**, *88*, 1341–1378. [CrossRef] [PubMed]

54. Nishikawa, T.; Edelstein, D.; Du, X.L.; Yamagishi, S.; Matsumura, T.; Kaneda, Y.; Yorek, M.A.; Beebe, D.; Oates, P.J.; Hammes, H.P.; et al. Normalizing mitochondrial superoxide production blocks three pathways of hyperglycaemic damage. *Nature* **2000**, *404*, 787–790. [CrossRef] [PubMed]

55. Harja, E.; Chang, J.S.; Lu, Y.; Leitges, M.; Zou, Y.S.; Schmidt, A.M.; Yan, S.F. Mice deficient in PKCbeta and apolipoprotein E display decreased atherosclerosis. *FASEB J.* **2009**, *23*, 1081–1091. [CrossRef]

56. Kong, L.; Shen, X.; Lin, L.; Leitges, M.; Rosarion, R.; Zou, Y.S.; Yan, S.F. PKCβ promotes vascular inflammation and acceleration of atherosclerosis in diabetic ApoE null mice. *Arterioscler. Thromb. Vasc. Biol.* **2013**, *33*, 1779–1787. [CrossRef]

57. Lee, H.M.; Kim, J.J.; Kim, H.J.; Shong, M.; Ku, B.J.; Jo, E.K. Upregulated NLRP3 inflammasome activation in patients with type 2 diabetes. *Diabetes* **2013**, *62*, 194–204. [CrossRef]

58. Menegazzo, L.; Ciciliot, S.; Poncina, N.; Mazzucato, M.; Persano, M.; Bonora, B.; Albiero, M.; Vigili de Kreutzenberg, S.; Avogaro, A.; Fadini, G.P. NETosis is induced by high glucose and associated with type 2 diabetes. *Acta Diabetol.* **2015**, *52*, 497–503. [CrossRef]

59. Joshi, M.B.; Lad, A.; Bharath Prasad, A.S.; Balakrishnan, A.; Ramachandra, L.; Satyamoorthy, K. High glucose modulates IL-6 mediated immune homeostasis through impeding neutrophil extracellular trap formation. *FEBS Lett.* **2013**, *587*, 2241–2246. [CrossRef]

60. Nahrendorf, M.; Swirski, F.K. Immunology. Neutrophil-macrophage communication in inflammation and atherosclerosis. *Science* **2015**, *349*, 237–238. [CrossRef]

61. Tsalamandris, S.; Antonopoulos, A.S.; Oikonomou, E.; Papamikroulis, G.A.; Vogiatzi, G.; Papaioannou, S.; Deftereos, S.; Tousoulis, D. The role of inflammation in diabetes: Current concepts and future perspectives. *Eur. Cardiol.* **2019**, *14*, 50–59. [CrossRef] [PubMed]

62. Goldfine, A.B.; Conlin, P.R.; Halperin, F.; Koska, J.; Permana, P.; Schwenke, D.; Shoelson, S.E.; Reaven, P.D. A randomised trial of salsalate for insulin resistance and cardiovascular risk factors in persons with abnormal glucose tolerance. *Diabetologia* **2013**, *56*, 714–723. [CrossRef] [PubMed]

63. Ridker, P.M.; Howard, C.P.; Walter, V.; Everett, B.; Libby, P.; Hensen, J.; Thuren, T.; CANTOS Pilot Investigative Group. Effects of interleukin-1beta inhibition with canakinumab on hemoglobin A1c, lipids, C-reactive protein, interleukin-6, and fibrinogen: A phase IIb randomized, placebo-controlled trial. *Circulation* **2012**, *126*, 2739–2748. [CrossRef] [PubMed]

64. Everett, B.M.; Donath, M.Y.; Pradhan, A.D.; Thuren, T.; Pais, P.; Nicolau, J.C.; Glynn, R.J.; Libby, P.; Ridker, P.M. Anti-inflammatory therapy with canakinumab for the prevention and management of diabetes. *J. Am. Coll. Cardiol.* **2018**, *71*, 2392–2401. [CrossRef]

65. Vasu, S.; Kumano, K.; Darden, C.M.; Rahman, I.; Lawrence, M.C.; Naziruddin, B. MicroRNA signatures as future biomarkers for diagnosis of diabetes states. *Cells* **2019**, *8*, 1533. [CrossRef] [PubMed]

66. Barutta, F.; Bellini, S.; Mastrocola, R.; Bruno, G.; Gruden, G. Micro RNA and microvascular complications of diabetes. *Int. J. Endocrinol.* **2018**, *2018*, 6890501. [CrossRef]

67. Cheng, H.S.; Sivachandran, N.; Lau, A.; Boudreau, E.; Zhao, J.L.; Baltimore, D.; Delgado-Olguin, P.; Cybulsky, M.I.; Fish, J.E. MicroRNA-146 represses endothelial activation by inhibiting pro-inflammatory pathways. *EMBO Mol. Med.* **2013**, *5*, 1017–1034. [CrossRef]

68. Zernecke, A.; Bidzhekov, K.; Noels, H.; Shagdarsuren, E.; Gan, L.; Denecke, B.; Hristov, M.; Koppel, T.; Jahantigh, M.N.; Lutgens, E.; et al. Selivery of microRNA-126 by apoptotic bodies induces CXCL12-dependent vascular protection. *Sci. Signal.* **2009**, *2*, ra81. [CrossRef]

69. Machado, I.F.; Teodoro, J.S.; Palmeira, C.M.; Rolo, A.P. miR-378a: A new emerging microRNA in metabolism. *Cell. Mol. Life Sci.* **2019**, in press. [CrossRef]

70. Chen, W.; Li, X.; Wang, J.; Song, N.; Zhu, A.; Jia, L. miR-378a Modulates Macrophage Phagocytosis and Differentiation through Targeting CD47-SIRPα Axis in Atherosclerosis. *Scand. J. Immunol.* **2019**, *90*, e12766. [CrossRef]

71. Das, S.; Reddy, M.A.; Senapati, P.; Stapleton, K.; Lanting, L.; Wang, M.; Amaram, V.; Ganguly, R.; Zhang, L.; Devaraj, S.; et al. Diabetes Mellitus-Induced Long Noncoding RNA Dnm3os Regulates Macrophage Functions and Inflammation via Nuclear Mechanisms. *Arterioscler. Thromb. Vasc. Biol.* **2018**, *38*, 1806–1820. [CrossRef] [PubMed]

72. Clempus, R.E.; Griendling, K.K. Reactive oxygen species signaling in vascular smooth muscle cells. *Cardiovasc. Res.* **2006**, *71*, 216–225. [CrossRef] [PubMed]

73. Kanwar, Y.S.; Wada, J.; Sun, L.; Xie, P.; Wallner, E.I.; Chen, S.; Chugh, S.; Danesh, F.R. Diabetic nephropathy: Mechanisms of renal disease progression. *Exp. Biol. Med.* **2008**, *233*, 4–11. [CrossRef] [PubMed]

74. Brownlee, M. Biochemistry and molecular cell biology of diabetic complications. *Nature* **2001**, *414*, 813–820. [CrossRef]

75. Kim, W.; Hudson, B.I.; Moser, B.; Guo, J.; Rong, L.L.; Lu, Y.; Qu, W.; Lalla, E.; Lerner, S.; Chen, Y.; et al. Receptor for advanced glycation end products and its ligands: A journey from the complications of diabetes to its pathogenesis. *Ann. N. Y. Acad. Sci.* **2005**, *1043*, 553–561. [CrossRef]

76. Pirola, L.; Balcerczyk, A.; Tothill, R.W.; Haviv, I.; Kaspi, A.; Lunke, S.; Ziemann, M.; Karagiannis, T.; Tonna, S.; Kowalczyk, A.; et al. Genome-wide analysis distinguishes hyperglycemia regulated epigenetic signatures of primary vascular cells. *Genome Res.* **2011**, *21*, 1601–1615. [CrossRef]

77. Brasacchio, D.; Okabe, J.; Tikellis, C.; Balcerczyk, A.; George, P.; Baker, E.K.; Calkin, A.C.; Brownlee, M.; Cooper, M.E.; El-Osta, A. Hyperglycemia induces a dynamic cooperativity of histone methylase and demethylase enzymes associated with gene-activating epigenetic marks that coexist on the lysine tail. *Diabetes* **2009**, *58*, 1229–1236. [CrossRef]

78. Okabe, J.; Orlowski, C.; Balcerczyk, A.; Tikellis, C.; Thomas, M.C.; Cooper, M.E.; El-Osta, A. Distinguishing hyperglycemic changes by Set7 in vascular endothelial cells. *Circ. Res.* **2012**, *110*, 1067–1076. [CrossRef]

79. Keating, S.T.; Plutzky, J.; El-Osta, A. Epigenetic Changes in Diabetes and Cardiovascular Risk. *Circ. Res.* **2016**, *118*, 1706–1722. [CrossRef]

80. El-Osta, A.; Brasacchio, D.; Yao, D.; Pocai, A.; Jones, P.L.; Roeder, R.G.; Cooper, M.E.; Brownlee, M. Transient high glucose causes persistent epigenetic changes and altered gene expression during subsequent normoglycemia. *J. Exp. Med.* **2008**, *205*, 2409–2417, Erratum in: *J. Exp. Med.* **2008**, *205*, 2683. [CrossRef]

81. De Rosa, S.; Arcidiacono, B.; Chiefari, E.; Brunetti, A.; Indolfi, C.; Foti, D.P. Type 2 Diabetes Mellitus and Cardiovascular Disease: Genetic and Epigenetic Links. *Front. Endocrinol.* **2018**, *9*, 2. [CrossRef] [PubMed]

82. Ceriello, A. The emerging challenge in diabetes: The "metabolic memory". *Vasc. Pharmacol.* **2012**, *57*, 133–138. [CrossRef] [PubMed]

International Journal of
Molecular Sciences

Review

The Role of High-Density Lipoproteins in Endothelial Cell Metabolism and Diabetes-Impaired Angiogenesis

Khalia R. Primer [1,2,3], Peter J. Psaltis [1,2], Joanne T.M. Tan [1,2] and Christina A. Bursill [1,2,3,*]

1   Faculty of Health and Medical Sciences, University of Adelaide, Adelaide, SA 5000, Australia;
    khalia.primer@sahmri.com (K.R.P.); peter.psaltis@sahmri.com (P.J.P.); joanne.tan@sahmri.com (J.T.M.T.)
2   Vascular Research Centre, South Australian Health and Medical Research Centre,
    Adelaide, SA 5000, Australia
3   Centre for Nanoscale Biophotonics, Adelaide, SA 5000, Australia
*   Correspondence: Christina.Bursill@sahmri.com; Tel.: +61-881-284-788

Received: 16 April 2020; Accepted: 18 May 2020; Published: 21 May 2020

Abstract: Diabetes mellitus affects millions of people worldwide and is associated with devastating vascular complications. A number of these complications, such as impaired wound healing and poor coronary collateral circulation, are characterised by impaired ischaemia-driven angiogenesis. There is increasing evidence that high-density lipoproteins (HDL) can rescue diabetes-impaired angiogenesis through a number of mechanisms, including the modulation of endothelial cell metabolic reprogramming. Endothelial cell metabolic reprogramming in response to tissue ischaemia is a driver of angiogenesis and is dysregulated by diabetes. Specifically, diabetes impairs pathways that allow endothelial cells to upregulate glycolysis in response to hypoxia adequately and impairs suppression of mitochondrial respiration. HDL rescues the impairment of the central hypoxia signalling pathway, which regulates these metabolic changes, and this may underpin several of its known pro-angiogenic effects. This review discusses the current understanding of endothelial cell metabolism and how diabetes leads to its dysregulation whilst examining the various positive effects of HDL on endothelial cell function.

Keywords: diabetes mellitus; angiogenesis; high-density lipoprotein; endothelial cell; metabolism; metabolic reprogramming

---

1. Introduction

Diabetes mellitus is one of the most debilitating and prevalent diseases in the world and imposes a significant health and economic burden upon the global community. As of 2017, diabetes affects 425 million people worldwide. This number is expected to reach 629 million by 2045 [1]. The social and financial burden of the disease is staggering. Worldwide healthcare expenditure tops 850 billion USD, and this is predicted to leap to 958 billion USD by 2045. The vascular complications associated with diabetes are extremely diverse. Diabetes is an established risk factor for cardiovascular disease and contributes to multiple vascular complications, including neuropathy, retinopathy, nephropathy, impaired wound healing, and poor outcomes related to myocardial infarction (MI). All of these are associated with dysregulated angiogenesis. Successful wound repair and coronary collateral vessel formation post-MI are specifically dependent on appropriate pro-angiogenic responses to ischaemia, and there is emerging evidence that this is regulated by endothelial cell metabolism. Despite advances in therapeutic strategies for these vascular complications, many patients do not respond effectively to current treatments. This represents an unmet clinical need for the development of new therapies that rescue impaired ischaemia-driven angiogenesis in diabetes. Accumulating evidence supports high-density lipoproteins (HDL) as an exciting new therapeutic option. HDL can promote in vitro angiogenesis functions in high glucose conditions and augment ischaemia-driven angiogenesis in diabetic pre-clinical models [2,3]. This review will discuss in detail the multiple mechanisms for the pro-angiogenic effects of HDL in diabetes, including the emerging evidence for a central role in the correction of endothelial cell metabolism.

2. High-Density Lipoproteins

High-density lipoproteins (HDL) are highly heterogenous particles composed of an outer layer of apolipoproteins (apo) and phospholipids, surrounding a core of esterified cholesterol (Figure 1). ApoA-I is the predominant protein moiety in HDL and is thought to impart the many biological properties of HDL. HDL cholesterol (HDL-C) refers to the HDL particles in the bloodstream that are specifically carrying cholesterol. This is distinct from 'HDL', which refers to the particle itself, and which can have specific effects independent of cholesterol efflux and can be modified by various disease milieu [4]. It is necessary to distinguish between these two forms of notation, as this highlights the breadth of effects had by HDL. Additionally, investigating the role of HDL as a discrete particle independent of cholesterol efflux may identify reasons why HDL-C-raising therapies have demonstrated limited clinical benefit [4].

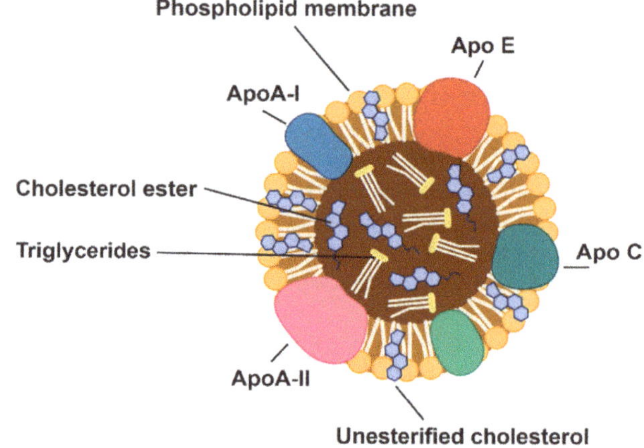

Figure 1. The structure of a high-density lipoprotein particle. Apo, apolipoprotein.

HDL primarily mediates reverse cholesterol transport by carrying cholesterol from peripheral tissues, including from macrophages in atherosclerotic plaques, to the liver for metabolism and excretion. This is a critical function that is a key contributor to the cardioprotective properties of HDL. HDL effluxes cholesterol by interacting with two cholesterol transporters ABCA1 and ABCG1 and the scavenger receptor SR-BI. ABCA1 interacts specifically with lipid-free apoA-I. Once apoA-I has acquired lipid to become a discoidal particle, it can then interact with ABCG1 and SR-BI. ABCA1 and ABCG1 contribute to the majority of the cholesterol efflux exchange but are also linked to a number of downstream signalling pathways, some of which are associated with angiogenesis. ABCG1, for example, mediates downstream signalling events that lead to elevated endothelial nitric oxide synthase (eNOS) activity, an important promoter of neovascularisation. This pathway is dependent on cholesterol efflux as it reduces the intracellular concentration of oxysterol 7-ketocholesterol [5]. Of the three cell-surface proteins that interact with HDL, SR-BI mediates the most downstream signalling pathways. These are either dependent or independent of cholesterol efflux. SR-BI is distinct to the other cholesterol transporters in that cholesterol can be exchanged bidirectionally between the cell and the HDL particle. Many of the signalling pathways downstream of SR-BI overlap significantly with those that regulate angiogenesis, including the phosphoinositide 3-kinase

(PI3K)/Akt and MAPK pathways that will be discussed in more detail later in this review [6]. For the most part, apoA-I is believed to be the key component that mediates the multiple effects of HDL. The only other reported component of HDL that stimulates significant vascular biological effects is a sphingolipid, which interacts with the S1P1 receptor and activates downstream signalling events that are very similar to those downstream of SR-BI in endothelial cells and regulates angiogenesis [6]. It should be noted, however, that sphingolipid makes up only a tiny proportion of the HDL particle, and it has been suggested that its physiological relevance within HDL is likely to be limited.

HDL has been implicated in numerous diseases. These include rheumatoid arthritis (RA), a chronic inflammatory disorder of the joints that causes gradual destruction of the bones, which becomes increasingly painful for a patient. It has been reported that patients with RA have overall decreased HDL-C levels, but increased amounts of dysfunctional and pro-inflammatory HDL [7]. Similarly, metabolic syndrome is associated with decreased HDL-C levels, which is accompanied by obesity, hypertriglyceridaemia, insulin resistance, and impaired glucose tolerance [7,8]. HDL has also long been studied extensively in the context of vascular disease. Strong inverse relationships have been established between low levels of circulating HDL and the incidence of cardiovascular disease (CVD) [7–9]. However, the active, positive effect that HDL has on the cardiovascular system is complex and remains to be fully characterized.

3. Diabetes-Impaired Angiogenesis

Impaired ischaemia-driven angiogenesis is a hallmark of certain diabetic vascular complications, such as impaired wound healing and poor recovery following myocardial infarction. Angiogenesis is the protrusion of new vessels from existing ones into a tissue to form a mature vascular network that can re-oxygenate the ischaemic site. Across all diabetic vascular complications associated with impaired ischaemia-driven angiogenesis, epidemiological data show that increased HDL-C levels are associated with improved outcomes, or lower severity or incidence of the complication [7,10–12].

3.1. Wound Healing

Diabetic patients are particularly susceptible to developing persistent wounds that heal slowly and expose the patient to infection and further harm. Combined with other diabetic complications, such as peripheral neuropathy, which reduces nerve function and sensation in the extremities, impaired wound healing can lead to the development of diabetic foot ulcers (DFU). The pathophysiology of DFU is complex and multifactorial, and no therapy currently exists, which adequately addresses these complexities to achieve a clinical benefit. The current treatment strategy for DFU primarily involves the management of the patient's co-morbidities that may contribute to ulcer development, such as glycaemic status or physical fitness. Additionally, regular dressing changes, management of infection, offloading and debridement of the wound tissue are employed to enhance the likelihood of a DFU healing correctly [13,14]. Outside of these standard-of-care approaches, there are very few effective therapies that actively improve the wound healing process.

Angiogenesis is a central component of wound healing. One reason for this is that neovessels deliver oxygen and nutrients to promote new tissue formation during the proliferative phase of healing [15]. Diabetic wounds exhibit decreased capillary density and vascularity due to insufficient angiogenesis [15,16]. Specifically, it has been shown that in chronic diabetic wounds, the angiogenic response normally induced by hypoxia is dysregulated [15,17]. Consistent with this, agents that inhibit angiogenesis, such as TNP-470 and SU5416, have been shown to inhibit wound repair [18,19]. Therefore, promoting wound angiogenesis presents a strategy to accelerate wound closure. This has been confirmed in murine wound healing studies using pro-angiogenic agents, such as vascular endothelial growth factor A (VEGFA) [20].

Whilst limited, there is epidemiological evidence that circulating HDL levels are inversely associated with the risk of DFU and amputation. For example, it has been reported that lower HDL levels are associated with increased DFU severity as classified by Wagner's severity level [10], and are also subsequently associated with an increased incidence of lower extremity amputation and wound-related death [11]. This suggests that higher HDL levels may be protective against the pathophysiological development of chronic diabetic wounds.

3.2. Recovery Following Myocardial Infarction

In the context of myocardial infarction, ischaemia-driven angiogenesis is a critical component of coronary collateral circulation development. Chronic imbalances in myocardial oxygen supply and demand can lead to tissue ischaemia, which induces adaptive neovascularisation to increase oxygen supply to the myocardium. The extent to which a patient possesses coronary collateral circulation plays a critical role in determining how vulnerable they may be to athero-occlusive disease in the coronary arteries. A well-developed coronary collateral circulation is associated with improved cardiac function and survival following myocardial infarction (MI), as well as improved prognosis in the context of stable chronic coronary disease. Similar to wound healing, patients with diabetes have poorer coronary collateral circulation, which is associated with much worse outcomes following MI and revascularisation procedures [21–23].

The relationship between HDL and myocardial infarction has also been investigated. A study in the early 1980s examined the prognostic significance of HDL-C levels after patients had recovered from an acute MI. This study, conducted with data from the Coronary Drug Project, determined that low levels of HDL-C were associated with increased mortality following an acute MI [24]. Since then, additional studies have found supporting evidence that low levels of HDL-C or a high low-density lipoprotein cholesterol (LDL-C)/HDL-C ratio are common in patients with acute MI [25] and that these patients have a significantly higher in-hospital mortality rate [25,26] and cardiac mortality rate [26,27].

There is also an association between HDL and coronary collateral circulation development. In a retrospective cross-sectional study, Hasan et al. determined that low levels of HDL cholesterol were an independent predictor of poor coronary collateral circulation [12]. These results suggest a role for HDL levels in the pathophysiology of MI.

3.3. Dysregulated Angiogenesis in Diabetes

Within the complex pathology of diabetes mellitus, there are also a number of complications in which angiogenesis is upregulated or otherwise dysregulated. These include diabetic retinopathy (DR), which can lead to blindness, and diabetic nephropathy (DN), which can cause significant damage to the kidneys. The role of angiogenesis in these complications is very different from the role of impaired ischaemia-driven angiogenesis in wound healing and MI.

DR is characterised by two clinical stages, non-proliferative DR and proliferative DR (PDR). PDR represents a more advanced disease stage and is characterised by abnormal, increased neovascularisation of the retina. During this stage, the new abnormal vessels may leak into the vitreous fluid of the eye and cause serious impairment to the patient's vision. Mechanistically, this increase in dysfunctional vessel formation is initially caused by retinal ischaemia which induces the VEGF pathway. This pathology is also heavily reliant on inflammatory pathways. Furthermore, current therapies for PDR are primarily anti-angiogenic, whilst therapies for impaired wound healing and myocardial ischaemia are pro-angiogenic [28].

DN presents another completely distinct pathology associated with dysregulated angiogenesis. DN is one of the leading causes of end-stage kidney disease in the world, and its development is complex. Excess hyperglycaemia-induced reactive oxygen species (ROS) cause damage to the kidney's glomeruli, leading to excretion of albumin in the urine. This damage is not underpinned by excess or inadequate angiogenesis, but rather a dysfunction of the endothelial cells, podocytes, and mesangial cells which mediate glomerular filtration barrier function [29].

The differences between these complex pathologies highlight the breadth of effects diabetes has on the vasculature and demonstrates why the development of clinically-effective therapies is difficult. This review will focus on impaired ischaemia-driven angiogenesis, and specifically on the metabolism of endothelial cells found in cardiac vessels or associated with wound healing in the peripheral vasculature. We will discuss the disturbances caused by diabetes and the current evidence for how HDL regulates or corrects these pathways.

4. Physiological Angiogenesis and Its Regulation by HDL

Physiological ischaemia-driven angiogenesis is complex and regulated by many different factors. Endothelial cells are the main vessel-forming cells in angiogenesis and exist in three sub-types; migratory tip cells, proliferative stalk cells, and quiescent phalanx cells [30]. These sub-types are influenced by signalling molecules in a dynamic fashion to regulate cellular metabolism and behaviour, such as migration and proliferation.

Under normal conditions, endothelial cells are quiescent. This can be interrupted by a number of physiological factors to induce the shift to the migratory and proliferative phenotype critical for angiogenesis. One of these factors is hypoxia signalling. Hypoxia signalling is characterised by a decrease in the activity of the prolyl hydroxylase domain (PHD) proteins, which ordinarily utilise oxygen to target the hypoxia-inducible factors (HIF) for degradation [31]. To achieve this, the PHD proteins hydroxylate HIF subunits, which are then recognised by the von Hippel–Lindau (VHL) protein of the E3 ubiquitin ligase complex. The HIF subunits are then quickly degraded through the proteasomal degradation pathway [31]. Under conditions of low oxygen, the decrease in PHD protein activity allows for the HIF subunit HIF-1α to accumulate, rather than be degraded. HIF-1α then travels from the cytosol of a cell to the nucleus and, in combination with other subunits, initiates transcription of key angiogenic genes [32,33]. Potentially the most important of these is VEGFA, which is secreted by cells to create a signalling molecule concentration gradient. This gradient serves to induce endothelial cell proliferation and migration and leads them to the origin of hypoxia, forming new blood vessels as they travel [3,34]. In addition to this, fibroblast growth factors (FGFs) are also increased in response to hypoxia and contribute to increased endothelial cell proliferation in a similar way [35,36].

Other key contributors to this process are endothelial progenitor cells (EPCs). EPCs circulate in the bloodstream and respond to pro-angiogenic signals to differentiate into mature endothelial cells and contribute to neovascularization [37]. Combined, HIF-1α and VEGFA reprogram critical aspects of endothelial cell function to ensure their survival in hypoxia and the creation of new blood vessels.

Some HIF-1α-independent pathways and transcription factors have also been found to be critically important in angiogenesis. One of these is the forkhead box O (FOXO) transcription factor, FOXO1. FOXO1 is an effector of the PI3K/Akt pathway, and its nuclear activity can be inhibited by Akt-mediated phosphorylation. Wilhelm et al. demonstrated that FOXO1 acts as an enforcer of endothelial cell quiescence and that its deletion in mice causes an uncontrolled increase in vessel sprouting. Contrastingly, its overexpression severely restricted angiogenesis and led to vessel thinning [38].

Another important transcription coactivator is peroxisome proliferator-activated receptor gamma coactivator 1-alpha (PGC-1α). PGC-1α is well recognised as a central modulator of cellular metabolism and mitochondrial biogenesis. In muscle cells, PGC-1α has also been shown to potently upregulate VEGF under hypoxic conditions in a mechanism that is independent of HIF-1α. Arany et al. demonstrated this by exposing cultured muscle cells to a low oxygen environment. This induced the expression of PGC-1α, which then coactivated oestrogen-related receptor-α (ERR-α) on the *VEGF* promoter, leading to increased expression of VEGF and promotion of capillary density in skeletal muscle. This identified a role for PGC-1α in the regulation of angiogenic signalling, potentially in connection with its role in metabolism [39]. FOXO1 and PGC-1α are predominantly recognised as modulators of metabolism but are also clearly associated with angiogenesis.

HDL has been shown to regulate ischaemia-induced angiogenesis in a number of ways. An early study showed that in a murine model of hindlimb ischaemia, infusions of reconstituted HDL (rHDL, apoA-I complexed with phospholipid) increased the number of EPCs which homed to the ischaemic limb, resulting in improved reperfusion [40]. Mechanistically, in vitro studies demonstrated that rHDL promoted the differentiation of EPCs through the PI3K/Akt pathway [40]. Furthermore, studies conducted in patients with type 2 diabetes mellitus found that infusions of rHDL improved vascular function [41], and increased the number of circulating EPCs [42].

Studies have also examined the effects of rHDL on the HIF-1α/VEGFA hypoxia signalling pathway, which initiates angiogenesis. Neovascularisation and blood flow reperfusion were increased by apoA-I infusions in the murine hindlimb ischaemia model. This study also examined the effect of rHDL on endothelial cell function in vitro and found that rHDL enhanced hypoxia-stimulated migration, proliferation, and tubulogenesis. It was then determined that these effects were mediated by augmentation of HIF-1α, VEGFA and VEGF receptor 2 (VEGFR2) [43]. It was demonstrated by a second study that these effects were due to changes in the post-translational modulation of HIF-1α induced by rHDL. In this study, Tan et al. found that through an interaction with scavenger receptor B I (SR-BI) and the PI3K/Akt pathway, rHDL increased the expression of the E3 ubiquitin ligases Siah1 and Siah2, which are responsible for targeting the PHD proteins for degradation. This resulted in decreased PHD protein activity, allowing HIF-1α to accumulate and promote transcription of VEGFA [2] (Figure 2).

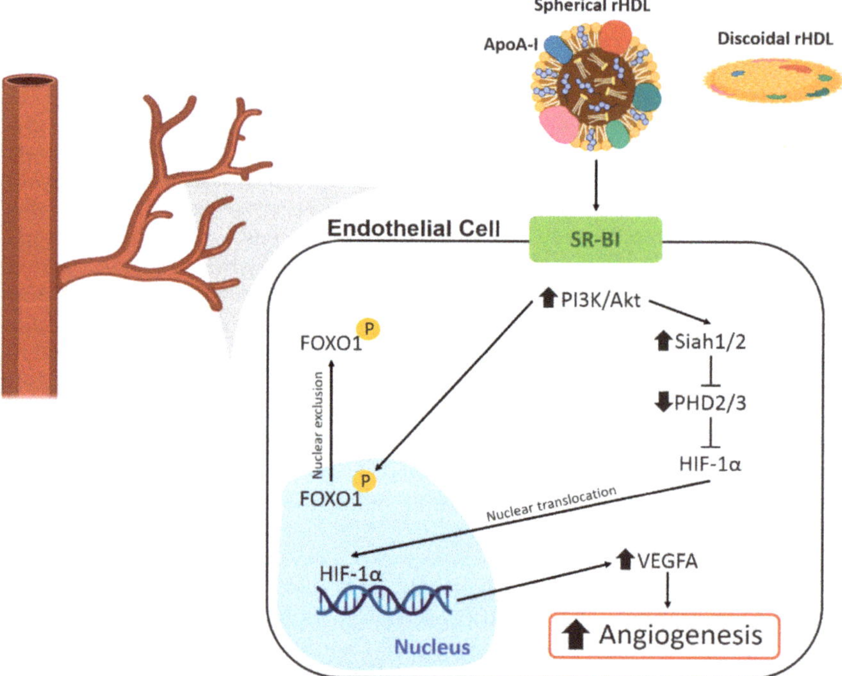

Figure 2. A schematic of the known effects of high-density lipoproteins (HDL) on angiogenic signalling pathways in endothelial cells. FOXO1, Forkhead Box O1; HIF-1α, hypoxia-inducible factor 1α; PHD, prolyl hydroxylase domain; PI3K, phosphoinositide 3-kinase; SR-BI, scavenger receptor B I.

There is also evidence that HDL regulates FOXO1 to positively effect endothelial cell function. A study by Theofilatos et al. demonstrated that treatment of human aortic endothelial cells with rHDL led to increased phosphorylation of FOXO1 by Akt, followed by its exclusion from the nucleus [44] (Figure 2). This resulted in increased expression of angiopoietin-like 4 (ANGPTL4), which is a potent pro-angiogenic molecule [45].

The exact mechanism by which HDL achieves these various salutary effects is complex and may be due to multiple factors. The effects on the HIF-1α/VEGFA signalling pathways have been shown to be dependent on an interaction between the rHDL particle and scavenger receptor class B type 1

(SR-BI) [3]. SR-BI plays a role in reverse cholesterol transport, and its presence is essential for rHDL to elicit its pro-angiogenic effects [3]. However, the role of cholesterol efflux was not directly addressed in these studies. Additionally, several studies referred to in this review have demonstrated positive effects on the vasculature or EPCs with only the apoA-I protein moiety of HDL. This indicates that there may be multiple mechanisms by which HDL and its apoA-I component are able to elicit positive effects on endothelial cell function.

Taken together, these studies highlight that rHDL regulates multiple aspects of pro-angiogenic signalling pathways to achieve a positive effect or 'rescue' of diabetes-impaired angiogenesis. However, given the complexity of diabetes and the associated vascular complications, it is necessary to fully understand the breadth of effects had by rHDL to develop it as a potential therapeutic agent. The HIF-1α/VEGFA hypoxia signalling pathway is well-characterised and considered the main driver of ischaemia-induced angiogenesis. HDL has been implicated in multiple steps of this pathway. The response elicited by this pathway in endothelial cells is complex, and an orchestration of many different factors is required for the cells to respond to hypoxia and participate in angiogenesis correctly. Endothelial cell metabolism and the pathways which reprogram it in response to hypoxia are fast emerging as drivers of angiogenesis in their own right. HDL is also emerging as a regulator of this, supporting its already diverse effects on endothelial cells and angiogenesis.

5. Endothelial Cell Metabolism

To support their wide range of critical functions, endothelial cells possess unique metabolic capabilities. Without metabolism there can be no energy production or replenishment of nutrient pools, and without this, an endothelial cell cannot fulfil any of its necessary functions. To understand how we can target endothelial cell metabolism to have a positive impact on diabetes-impaired angiogenesis, we must first understand its regulatory mechanisms (Figure 3).

5.1. Glycolysis

Glycolysis describes a chain of reactions by which one molecule of glucose is broken down into two molecules of pyruvate. Glycolysis occurs in the cytosol, does not require oxygen, and can, therefore, occur in the presence or absence of oxygen. There are many steps in this chain of reactions, one of which is ATP-producing. Glycolysis produces approximately two molecules of ATP for every molecule of glucose. There are also many metabolic intermediates produced as glycolysis progresses, and these have their own roles in the regulation of endothelial cell function, which will be discussed later.

Active endothelial cells rely on glycolysis for the bulk of their ATP production, though this may seem counter-intuitive, given their ordinarily easy access to oxygen from the bloodstream [30]. However, the sheer quantity of oxygen that endothelial cells are exposed to could easily lead the cells to experience oxidative stress if they did not restrict flux through pathways that consume oxygen. Furthermore, reducing reliance on oxygen-consuming pathways ensures that endothelial cells are always primed to function in hypoxia, which is useful for revascularisation of ischaemic tissues. This is in contrast to many other cell types for which full oxidation of glucose through mitochondrial respiration is the most efficient method of ATP production.

Upon encountering hypoxia, endothelial cells undergo a substantial metabolic shift. It is essential that endothelial cells further upregulate glycolysis to account for the increased energy demands of proliferation and migration, whilst simultaneously keeping oxygen consumption low to avoid oxidative stress. To achieve this, VEGF and FGF signalling support an increase in glycolytic flux. This means that the rate of glucose breakdown through glycolysis increases, producing more ATP, more glycolytic intermediates, and more pyruvate. This section will discuss the key roles of VEGF and FGFs in regulating glycolytic flux.

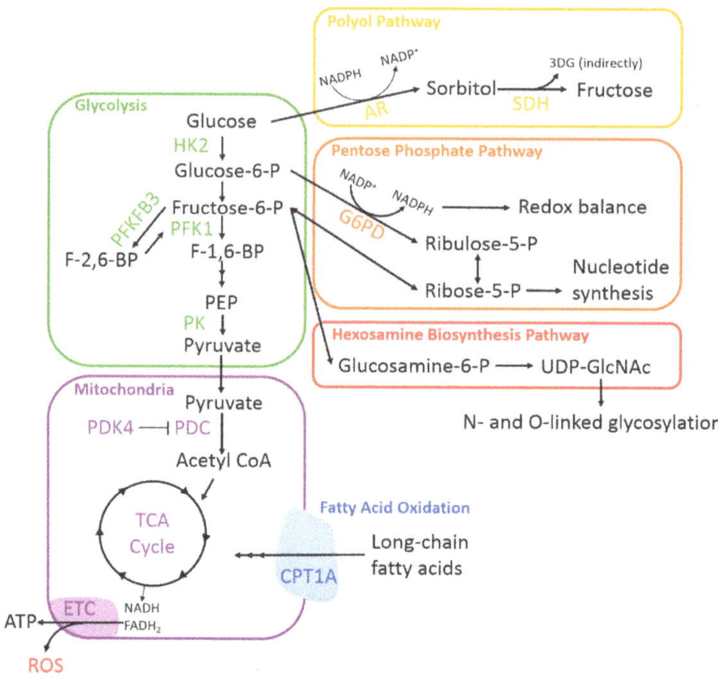

Figure 3. Diagram of endothelial cell metabolism. A simplified diagram of endothelial cell metabolism depicting the known metabolic pathways and their rate-limiting enzymes. 3DG, 3-deoxyglucosone; AR, aldose reductase; CPT1A, carnitine palmitoyltransferase 1A; ETC, electron transport chain; G6PD, glucose-6-phosphate dehydrogenase; HK2, hexokinase 2; NADPH, nicotinamide adenine dinucleotide phosphate; PDC, pyruvate dehydrogenase complex; PDK4, pyruvate dehydrogenase kinase 4; PFK1, phosphofructokinase 1; PFKFB3, phosphofructokinase-2/fructose-2,6-bisphosphatase isoform 3; PK, pyruvate kinase; SDH, sorbitol dehydrogenase; UDP-GlcNAc, uridine diphosphate N-acetylglucosamine.

First, the increase in glycolytic flux has been found to be mediated, at least in part, by an interaction between VEGF and 6-phosphofructo-2-kinase/fructose-2,6-biphosphatase 3 (PFKFB3) [30]. An important rate-limiting step in glycolysis is the conversion of fructose-6-phosphate (F6P) to fructose-1,6-bisphosphate (F1,6P$_2$) by 6-phosphofructo-1-kinase (PFK-1). PFKFB3 synthesises fructose-2,6-bisphosphate (F2,6P$_2$), a strong activator of PFK-1. Therefore, the activity of PFKFB3 represents an avenue for potent upregulation of glycolysis. De Bock et al. have shown that PFKFB3 knockdown using short hairpin RNA (shRNA) reduced glycolytic flux by 35% in both microvascular and arterial endothelial cells, indicating that PFKFB3 is important for maintaining adequate levels of glycolytic flux in endothelial cells. Importantly, it was shown that VEGF increases both PFKFB3 expression and glycolysis. With respect to the effect on angiogenesis, knockdown of PFKFB3 decreased vessel sprouting in endothelial cell spheroids. Further investigation of this in vivo revealed that mice with endothelial cells deficient in PFKFB3 displayed significant defects in retinal blood vessel development as well as decreased vascular area in the hindbrain [30].

Other glycolytic enzymes have also been implicated in the regulation of endothelial cell angiogenic functions. One of these is hexokinase 2 (HK2), which phosphorylates a molecule of glucose to produce glucose-6-phosphate. This is the first step in glycolysis and is also rate-limiting. Yu et al. investigated the relationship between FGF2 and HK2 in endothelial cells with the aim of delineating the mechanism by which FGF2 stimulates angiogenesis. This group found that mouse embryos deficient in endothelial FGF receptor 1 (FGFR1) exhibited reduced vessel branching in the skin. To elucidate this mechanism, human umbilical vein endothelial cells (HUVECs) were treated with FGF, which significantly enhanced glycolysis and HK2 expression. Knockdown of HK2 significantly reduced glycolysis, whilst adenoviral-mediated overexpression increased glycolysis. This study identified a relationship between FGF and HK2 in endothelial cells and determined that HK2-mediated glycolysis was essential for angiogenesis [36].

Pyruvate kinase (PK) catalyses the conversion of phosphoenolpyruvate to pyruvate, which is the final rate-limiting step in glycolysis. When PK activity is low, upstream glycolytic intermediates may accumulate and be shunted to various side-pathways. Endothelial cells predominantly express the PKM2 isoform of this enzyme, and this was examined in the context of angiogenesis by Kim et al. This group knocked down PKM2 using siRNA and observed significant suppression of cell proliferation, migration, and tubule formation in vitro. This was associated with a decrease in extracellular acidification, which represents the completion of the glycolytic pathway. Furthermore, endothelial cell-specific deletion of PKM2 in mice was associated with a significant reduction in retinal vessel density and branching. Finally, PKM2 was found to achieve its effects on the proliferation via inhibition of p53, causing a blockade of the cell cycle. Although this role of PKM2 was found to be independent of its enzymatic activity, this study nevertheless highlights the close relationship between the integrity of metabolic pathways and the angiogenic capacity of endothelial cells [46].

Together, this research highlights the importance of glycolysis in angiogenesis and identifies key regulatory mechanisms that may be implicated in diabetes-impaired angiogenesis (Figure 3).

5.2. Mitochondrial Respiration and Hypoxia Tolerance

It has also been shown that endothelial cells maintain very low levels of mitochondrial respiration relative to their rate of glycolysis [30]. Mitochondrial respiration, which encompasses both the tricarboxylic acid (TCA) cycle and oxidative phosphorylation, ordinarily consumes a high quantity of oxygen and subsequently contributes to the production of reactive oxygen species (ROS). Suppressing oxidative phosphorylation may be necessary for endothelial cells to reduce ROS production and maintain redox homeostasis.

Recently, a key regulatory step has been identified, which may be critical for adequate suppression of mitochondrial respiration in endothelial cells. The pyruvate–dehydrogenase complex (PDC) catalyses the conversion of pyruvate to acetyl coenzyme A (CoA) so that it may condense with oxaloacetate to enter the TCA cycle and contribute to mitochondrial respiration [47]. The pyruvate dehydrogenase lipoamide kinases (PDK) of which there are four subtypes, PDK1-4, are particularly important as they inhibit the PDC. The PDKs inhibit the PDC by phosphorylating three serine residues in the E1α subunit of the PDC in a tissue-specific manner [47,48]. This decreases the amount of acetyl CoA produced from glucose-derived pyruvate, which fuels the TCA cycle, thereby reducing the rate of respiration.

The importance of PDK4 in hypoxia-induced metabolic reprogramming was established by Aragones et al. when they demonstrated that myofibers deficient in PHD1, which normally degrades HIF proteins, exhibited elevated levels of PDK4. These myofibers were significantly more tolerant of hypoxia, consumed less oxygen, and were experiencing less oxidative stress [49]. Whilst Aragones et al. did observe that this protection against oxidative stress was independent of increased angiogenesis, it does demonstrate a direct mechanistic link between hypoxia signalling and PDK4. Consistent with this, our group has demonstrated that PDK4 was increased in human coronary artery endothelial cells (HCAECs) exposed to hypoxia. Furthermore, PDK4 knockdown by short interfering RNA (siRNA) significantly impaired endothelial cell tubule formation and migration in vitro [50]. This indicated that suppression of glucose-driven mitochondrial respiration might be essential for maintaining the endothelial cell capacity for angiogenesis, potentially by decreasing mitochondrial production of ROS or allowing for compensatory increases in fatty acid oxidation (Figure 3).

5.3. Fatty Acid Oxidation

Fatty acid oxidation (FAO) represents an alternate avenue for the production of acetyl CoA from long-chain fatty acids and bypasses the PDC. This is achieved by direct production of acetyl CoA from long-chain fatty acids, which can then condense with oxaloacetate to form citrate and fuel the TCA cycle. In endothelial cells, specifically, the main mechanism which regulates FAO involves AMP-activated protein kinase (AMPK), which is activated by hypoxia [51]. AMPK indirectly activates carnitine palmitoyl transferase 1A (CPT1A), which shuttles long-chain fatty acids into the mitochondria and represents a rate-limiting step of FAO [52].

Whilst FAO can be used for the generation of ATP through mitochondrial respiration, endothelial cells rely predominantly on glycolysis for energy production. Recent research does, however, point to FAO as an essential contributor to endothelial cell functions beyond energy production.

Kalucka et al. examined the metabolism of quiescent endothelial cells. This group found that quiescent endothelial cells upregulated FAO three times more than proliferative endothelial cells and that this was primarily in support of nicotinamide adenine dinucleotide phosphate (NADPH) regeneration for redox homeostasis. Blocking FAO in these cells induced significant oxidative stress, whilst supplementing them with acetate rescued this. This demonstrates that FAO is essential for maintaining redox balance in quiescent endothelial cells [53].

In contrast, Schoors et al. investigated FAO in proliferative endothelial cells undergoing vessel sprouting, again with a focus on identifying an alternative role for the process beyond energy production. Schoors et al. knocked down CPT1A and found that this impaired vessel sprouting in endothelial cell spheroids due to decreased cell proliferation. CPT1A knockdown did not lower ATP levels, and only increased ROS levels by approximately 20%, a level thought to have a positive effect on endothelial cell proliferation. It was then determined that acetyl CoA from FAO was cycling through the TCA cycle and contributing to de novo synthesis of deoxyribonucleotides in endothelial cells, and that knockdown of CPT1A reduced this and subsequently impaired de novo DNA synthesis. This culminated in the impairment of endothelial cell proliferation and overall vessel sprouting [54]. This contrasts with the role of FAO in maintaining redox balance in quiescent endothelial cells but nevertheless suggests that FAO is important for correct endothelial cell function and angiogenesis.

5.4. The Pentose Phosphate Pathway

Glycolytic side pathways are also critical for endothelial cell function. These involve various glycolytic intermediates and have a wide range of functions that can be either positive or negative for an active endothelial cell. Thus far, the pentose phosphate pathway (PPP) has been demonstrated to be the most important of these side pathways. The PPP utilises glucose-6-phosphate from glycolysis in two different branching pathways. The oxidative branch (oxPPP) comprises an irreversible reaction which generates NADPH and ribose-5-phosphate (R5P), whilst the non-oxidative branch produces only R5P [55]. NADPH is critical for redox homeostasis as it allows for the conversion of oxidised glutathione (GSSG) to its reduced form (GSH). Reduced glutathione is an essential antioxidant which neutralises reactive oxygen species [56]. Therefore, this mechanism is likely to play an important role in maintaining the redox balance of an endothelial cell during hypoxia-driven angiogenesis. Additionally, the R5P produced by this pathway is essential for the synthesis of nucleotides, which likely supports the increased proliferation required during vessel sprouting. The rate of the oxPPP is determined by glucose-6-phosphate dehydrogenase (G6PD). The importance of this pathway and its close relationship with the rate of glycolysis implicates it in diabetes-impaired angiogenesis.

6. Endothelial Cell Metabolism and Diabetes-Impaired Angiogenesis

The prevailing hypothesis regarding diabetes-impaired angiogenesis is that hyperglycaemia causes increased production of reactive oxygen species (ROS), which leads to general endothelial cell dysfunction. However, this is only one aspect of the complex pathology. Diabetic hyperglycaemia also negatively affects individual transcription factors, signalling molecules, and metabolic enzymes, which contribute to endothelial cell dysfunction and impaired angiogenesis. The HIF-1α/VEGFA hypoxia signalling pathway is a clear example of this. Furthermore, ROS production is closely tied to cellular metabolism, as we have discussed previously. Therefore, changes in cellular metabolism elicited by diabetes potentially underpin the increase in ROS production. Mitochondrial respiration is a major source of ROS, and several glycolytic side pathways contribute to cellular redox balance. The relationships between these pathways are complex, and it is difficult to tell from where the negative effects on angiogenesis stem.

6.1. Diabetes Impairs Central Metabolic Pathways

Angiogenesis is initiated by the HIF-1α/VEGFA hypoxia signalling pathway, which has a myriad of complex downstream effects that orchestrate the vessel sprouting and maturation process. It has been consistently demonstrated that hyperglycaemia or in vitro high glucose conditions negatively affect the expression of HIF-1α and its translocation to the nucleus [3]. This effect was found to be due to a high glucose-induced increase in the PHD proteins, which tag HIF-1α for degradation. Increased degradation of HIF-1α means it cannot travel to the nucleus and induce the transcription of pro-angiogenic genes. This, therefore, also impairs the hypoxia-induced expression of VEGFA and is subsequently associated with the impairment of angiogenesis under high glucose conditions [3].

These negative effects on the central HIF-1α/VEGFA signalling pathway are also associated with downstream impairments to hypoxia-induced metabolic reprogramming. VEGFA is known to increase PFKFB3 expression in endothelial cells to support adequate upregulation of glycolysis in hypoxia. The effect of hyperglycaemia on PFKFB3 was tested by Rudnicki et al., who demonstrated that high glucose reduced the mRNA levels of *Pfkfb3* in endothelial cells isolated from mice [57] (Figure 4).

Suppression of mitochondrial respiration is also a critical aspect of metabolic reprogramming in response to hypoxia [49]. The expression of PDK4, which inactivates the PDC in endothelial cells, is known to be increased in response to hypoxia [58]. Recent studies by our group show that, importantly, this induction is impaired when the endothelial cells are exposed to high glucose, indicating a failure of the metabolic reprogramming response. These changes were also associated with a marked increase in mitochondrial respiration, which is a known source of ROS, as well as impaired in vitro angiogenesis [50].

PGC-1α is an important mediator of mitochondrial biogenesis and metabolism and is significantly affected by diabetes. Sawada et al. aimed to fully characterise the effect of diabetes on PGC-1α and determine how this affected endothelial cell function and angiogenesis. First, they demonstrated that PGC-1α was significantly elevated in endothelial cells from the heart or lung tissue from several different diabetic mouse models [59]. This effect was replicated in vitro with cultured primary endothelial cells exposed to high glucose. Furthermore, endothelial cells overexpressing PGC-1α exhibited significantly impaired migration, as demonstrated with a transwell assay. Mechanistically, PGC-1α was found to inhibit Akt phosphorylation of eNOS, which is essential for angiogenesis [59]. This study identified a strong link between the modulation of metabolism and control of angiogenesis, as well as highlighting the negative effects of diabetes on this mechanism.

Combined, these negative effects of diabetes affect central metabolic pathways and mediators, which normally support angiogenesis.

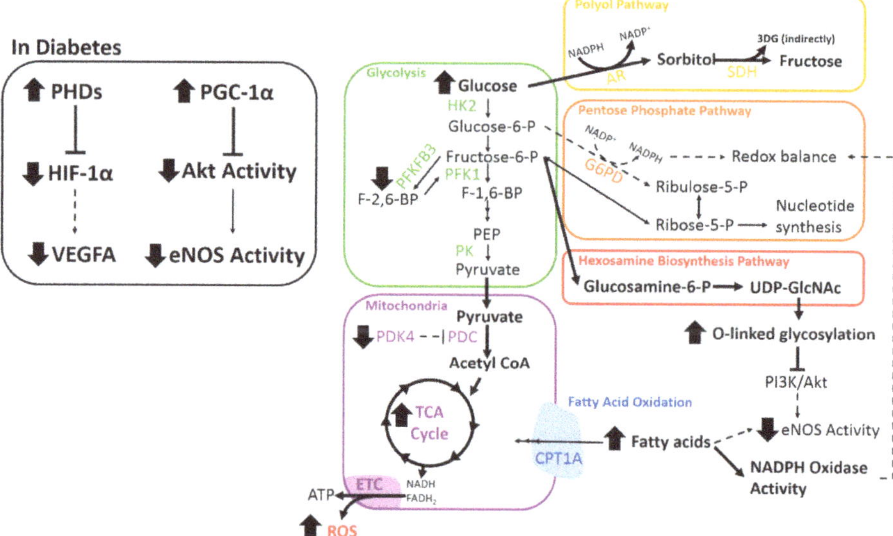

Figure 4. The effects of diabetes on endothelial cell metabolism. A diagram depicting the changes in endothelial cell metabolism elicited by diabetes. Bold lines indicate an increase, dashed lines indicate a decrease. 3DG, 3-deoxyglucosone; AR, aldose reductase; CPT1A, carnitine palmitoyltransferase 1A; ETC, electron transport chain; eNOS, endothelial nitric oxide synthase; G6PD, glucose-6-phosphate dehydrogenase; HIF-1α, hypoxia-inducible factor 1α; HK2, hexokinase 2; NADPH, nicotinamide adenine dinucleotide phosphate; PDC, pyruvate dehydrogenase complex; PDK4, pyruvate dehydrogenase kinase 4; PFK1, phosphofructokinase 1; PFKFB3, phosphofructokinase-2/fructose-2,6-bisphosphatase isoform 3; PHDs, prolyl hydroxylase domain proteins; PI3K, phosphoinositide 3-kinase; PK, pyruvate kinase; SDH, sorbitol dehydrogenase; UDP-GlcNAc, uridine diphosphate N-acetylglucosamine; VEGF, vascular endothelial growth factor. Bold arrows indicate an increase, dashed arrows indicate a decrease.

6.2. Glycolytic Side-Pathways

In addition to its negative effects on hypoxia signalling, hyperglycaemia also increases ROS production through a number of metabolic pathways.

The oxidative branch of the pentose phosphate pathway has been examined in this context, and it is well established that high glucose inhibits the rate-limiting enzyme of the oxPPP, glucose-6-phosphate dehydrogenase (G6PD) [60]. This blocks production of NADPH, which is essential for converting oxidised glutathione (GSSG) to its reduced form (GSH), thereby reducing the ability of the cells to adequately neutralise ROS [60]. This mechanism severely impairs the ability of endothelial cells to maintain redox balance in diabetes.

Once a cell begins to accumulate ROS through this mechanism, an additional cascade of events begins. Excess ROS damages a cell's DNA, which can lead to aberrant transcription of various proteins. One specific protein that is aberrantly activated by ROS-induced DNA damage in endothelial cells is poly-ADP-ribose polymerase 1 (PARP1) [61]. PARP1 post-translationally modifies proteins to inactivate them, and one relevant target is the glycolytic enzyme glyceraldehyde-3-phosphate dehydrogenase (GAPDH) [61–63]. When GAPDH is inactivated in this manner, it causes a blockade in the glycolytic pathway which allows for the accumulation of intermediates that cannot flow further through glycolysis. With the known effect of hyperglycaemia on G6PD keeping these intermediates from being shunted to a useful pathway, such as the oxPPP, the intermediates instead increase flux through upstream pathways which are detrimental to cellular function and known to be increased in diabetic endothelial cells: the polyol pathway and hexosamine biosynthesis pathway (HBP).

In diabetes, excess glucose is shunted into the polyol pathway and converted to sorbitol by aldose reductase, a reaction which uses NADPH [64]. This increased usage of NADPH may further exhaust a cell's ability to replenish the antioxidant GSH and exacerbate oxidative stress. Sorbitol is then converted to fructose by sorbitol dehydrogenase. This reaction also indirectly generates 3-deoxyglucosone (3DG) through the hydrolysis of the intermediate fructose-3-phosphate [65]. 3DG is a highly reactive compound known to contribute to the production of advanced glycated end-products (AGEs) [66,67]. AGEs are proteins or lipids that have been post-translationally glycated and are implicated in many aspects of diabetes-induced EC dysfunction. They have been shown to specifically inhibit angiogenesis by increasing extracellular matrix degradation by matrix metalloproteinases [68,69]. To conclude, increased flux through the polyol pathway caused by hyperglycaemia exacerbates oxidative stress and contributes to impaired angiogenesis.

One of the upstream glycolytic metabolites blocked by GAPDH inactivation is fructose-6-phosphate (F6P). F6P enters the hexosamine biosynthesis pathway and is converted to uridine diphosphate N-acetylglucosamine (UDP-GlcNAc), which is an essential substrate for N-linked and O-linked glycosylation of proteins. This post-translational modification of proteins is normally very important for normal cellular function. However, under hyperglycaemic conditions, excess F6P produced from glucose leads to aberrantly increased O-linked glycosylation of target proteins. Federici et al. examined this pathway in endothelial cells and found that O-linked glycosylation was significantly increased in HCAECs exposed to high glucose conditions [70]. This was found to impair PI3K/Akt signalling, which was then associated with a decrease in eNOS phosphorylation by Akt. Phosphorylation of eNOS increases the production of nitric oxide, which is pro-angiogenic [71,72]. A similar study by Luo et al. supported these findings by demonstrating that O-GlcNAc levels were increased in a murine streptozotocin model of diabetes and that this was indeed associated with impaired angiogenesis in an aortic ring assay. The group also showed that increased O-GlcNAc leads to reduced endothelial cell migration and tubule formation in vitro. Importantly, overexpression of O-GlcNAcase, a protein that reverses O-linked glycosylation, rescued migration and tubule formation of endothelial cells [72]. This indicates that increased flux through the HBP caused by hyperglycaemia impairs the PI3K/Akt signalling pathway, which has negative consequences for angiogenesis.

6.3. Diabetes and Fatty Acid Oxidation

The role of FAO in diabetes is complex and varies across tissue and cell types. It has been determined that FAO is essential for angiogenesis through its support of endothelial cell proliferation and biomass synthesis [54] Additionally, FAO is essential for the maintenance of quiescence and redox homeostasis in quiescent endothelial cells [53]. One additional aspect to consider is that whilst diabetes is characterised by hyperglycaemia, it is also associated with high circulating levels of free fatty acids (FFAs). Several studies have examined the effect of elevated FFAs on general endothelial function. Steinberg et al. investigated this in healthy patients with experimentally increased FFA levels and observed that the increase in FFAs was associated with endothelial dysfunction [73]. In a subsequent study, the same research group demonstrated that elevation of FFAs impaired endothelial insulin-mediated vasodilation and endothelial nitric oxide production [74]. This result was also demonstrated by Vigili et al., who showed that elevated FFAs impaired endothelium-dependent vasodilation [75]. One study has examined the effect of high glucose and elevated FFAs on ROS production in endothelial cells in vitro. Inoguchi et al. found that both high glucose and elevated FFAs increased superoxide production by NADPH oxidases in a protein kinase C-dependent manner [76]. This has some relevance for angiogenesis, given what is known about the necessity to regulate ROS production in a hypoxic environment and highlights the importance of the oxPPP in generating NADPH for maintaining redox balance. Whilst these effects do not necessarily directly reflect changes in angiogenesis, they do provide some insight into the general endothelial dysfunction caused by FFA elevation in patients with diabetes. It is necessary to investigate these effects further to understand how angiogenesis may be affected by elevated FFAs, and whether diabetes negatively affects FAO in endothelial cells.

These studies demonstrate that the effects of diabetes on endothelial cell metabolism are wide-ranging, complex, and significantly impair various angiogenic functions (Figure 4).

7. Emerging Role of HDL in Mechanisms of Endothelial Cell Metabolic Reprogramming and Diabetes-Impaired Angiogenesis

Earlier in this review, we highlighted the various positive effects of HDL on the cardiovascular system and angiogenesis in response to ischaemia and in diabetes. We have also established the importance of endothelial cell metabolic reprogramming as a driver of angiogenesis, and the various negative effects had upon its regulation by diabetes. Multiple lines of evidence point to a role for HDL in the regulation of endothelial cell metabolism in the context of diabetes-impaired angiogenesis.

7.1. HDL and Hypoxia Signalling

There is much evidence supporting the concept that HDL augments diabetes-impaired HIF-1α activity and VEGFA expression, resulting in the rescue of angiogenesis both in vitro and in vivo.

Specifically, Tan et al. found that rHDL rescued diabetes-impaired angiogenesis in two murine models of hindlimb ischaemia and wound healing. rHDL was shown to stabilise HIF-1α in high glucose by suppressing the PHD2 and three proteins, leading to an increase in the expression of VEGFA and rescuing high glucose-impaired tubulogenesis in vitro [3].

VEGF is one of the most important downstream targets of HIF-1α, and its expression is known to be impaired in diabetes [3]. The relationship between VEGF and glycolytic protein PFKFB3 has been well documented and is also essential for angiogenesis [30]. PFKFB3 expression is also impaired under high glucose conditions, which may contribute to negative effects on angiogenesis. rHDL is known to rescue diabetes-impaired expression of VEGFA [3], and this may also lead to the support of increased glycolytic flux in hypoxia and underpin some of the known pro-angiogenic activity of rHDL.

Furthermore, our research group has recently determined that PDK4, which inhibits mitochondrial respiration, is a target of rHDL. In human coronary artery endothelial cells, we have demonstrated that PDK4 expression is increased in hypoxia, but that this induction is impaired by exposure to high glucose conditions [50]. This aligns with the impairment to HIF-1α signalling seen under diabetic conditions in previous studies. Pre-incubation of the endothelial cells with rHDL restored the induction of PDK4 under high glucose and hypoxic conditions. When the oxygen consumption of the cells was examined as a measure of mitochondrial respiration, it was found that high glucose significantly increased oxygen consumption, but that pre-incubation with rHDL was able to return these levels to the baseline. These effects were also associated with the rescue of high glucose-impaired endothelial cell migration and tubulogenesis under hypoxic conditions [50]. We, therefore, propose that rHDL augments metabolic reprogramming via the upregulation of PDK4, decreasing oxygen consumption and potentially protecting the cells against the development of oxidative stress in hypoxia.

7.2. HDL and PI3K/Akt Signalling

Diabetes also affects endothelial cell metabolism via impairment of PI3K/Akt signalling. Hyperglycaemia causes the increased flux of fructose-6-phosphate through the hexosamine biosynthesis pathway, leading to dysregulation of O-linked glycosylation. One of the targets of this dysregulated glycosylation is the PI3K/Akt signalling pathway. This signalling pathway controls a number of critical mechanisms, including phosphorylation of eNOS by Akt. In diabetes, this pathway was significantly impaired in in vitro and in vivo models, and was associated with impaired angiogenesis [72]. A number of studies have demonstrated that rHDL can increase PI3K/Akt signalling; therefore, this is another pathway by which rHDL may support endothelial cell function and rescue angiogenesis in diabetes. One study demonstrated that intravenous injections of rHDL promoted differentiation of endothelial progenitor cells by increasing phosphorylation of Akt by PI3K in peripheral mononuclear cells. This was associated with augmentation of blood flow and increased capillary density in ischaemic limbs [40]. Another study demonstrated that the pre-incubation of endothelial cells with rHDL significantly increased protein expression of both PI3K and Akt under high glucose conditions [3]. This was subsequently associated with a significant increase in phosphorylation of eNOS by Akt and the rescue of high glucose-impaired tubulogenesis in vitro [3]. Furthermore, specific inhibition of PI3K/Akt with LY294002 in endothelial cells attenuated the augmentation of in vitro tubulogenesis, and HIF-1α and VEGFA by rHDL in response to hypoxia [2]. Finally, it has been demonstrated that rHDL initiates the nuclear exclusion of FOXO1 by increasing its phosphorylation by Akt, which has a number of positive effects on endothelial cell function [44].

These results demonstrate that high-density lipoproteins affect multiple aspects of endothelial cell metabolism to achieve a pro-angiogenic effect under diabetic conditions (Figure 5). Several potential downstream targets of HDL have also been identified and represent avenues for further study.

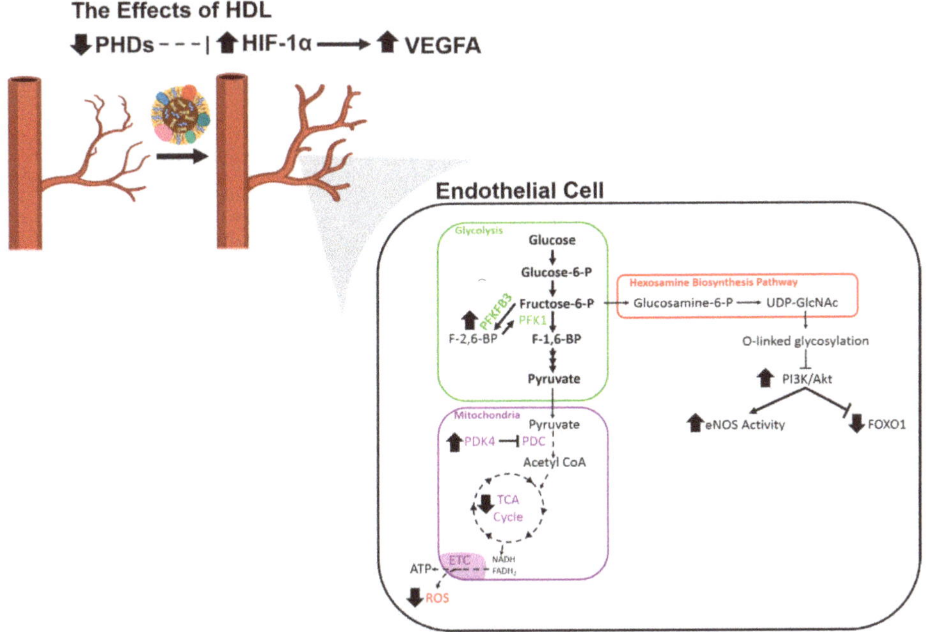

Figure 5. The known and proposed effects of HDL on endothelial cell metabolism. A diagram depicting the changes in endothelial cell metabolism caused by HDL under conditions of high glucose and hypoxia. Bold lines indicate an increase, dashed lines indicate a decrease. ETC, electron transport chain; eNOS, endothelial nitric oxide synthase; FOXO1, forkhead box O1; HIF-1α, hypoxia-inducible factor 1α; PDC, pyruvate dehydrogenase complex; PDK4, pyruvate dehydrogenase kinase 4; PHDs, prolyl hydroxylase domain proteins; PI3K, phosphoinositide 3-kinase; UDP-GlcNAc, uridine diphosphate N-acetylglucosamine; VEGF, vascular endothelial growth factor.

8. Conclusions

The dysregulated angiogenesis associated with diabetes mellitus presents a complex contribution to the pathology of diabetic vascular complications and the associated mortality. Therapies to correct this dysregulation are severely lacking, as is our understanding of the complexities of this process. Endothelial cell metabolism is a central driver of angiogenesis, but many aspects of metabolic reprogramming are negatively affected by diabetes, though our knowledge of this is still developing. HDL continues to represent a promising avenue for therapeutic development, as its multiple protective functions culminate in the rescue of diabetes-impaired angiogenesis and improved healing of chronic diabetic wounds. However, due to the relative lack of clinical success with HDL-raising therapies, further research must be conducted to fully elucidate the mechanisms involved in these processes and identify the best approach for harnessing the protective effects of HDL.

Author Contributions: K.R.P. researched and wrote the manuscript, P.J.P. and J.T.M.T. provided scientific advice and edited the manuscript, C.A.B. oversaw the conceptualization and writing process, and edited the manuscript. All authors have read and agreed to the published version of the manuscript.

Acknowledgments: Research in the authors' laboratory is currently supported by the Heart Foundation Lin Huddleston Fellowship (C.A.B.).

Conflicts of Interest: The authors declare no conflict of interest.

Abbreviations

3DG	3-deoxyglucosone
AGE	Advanced glycation end products
AMPK	AMP-activated protein kinase
ANGPTL4	Angiopoietin-like 4
Apo	Apolipoprotein
CoA	Coenzyme A
CPT1A	Carnitine Palmitoyltransferase 1A
CVD	Cardiovascular disease
DFU	Diabetic foot ulcer
DN	Diabetic nephropathy
DR	Diabetic retinopathy
eNOS	Endothelial nitric oxide synthase
EPC	Endothelial progenitor cell
ERRα	Oestrogen-related receptor α
F1,6P2	Fructose-1,6-bisphosphate
F2,6P2	Fructose-2,6-bisphosphate
F6P	Fructose-6-phosphate
FAO	Fatty acid oxidation
FFA	Free fatty acid
FGF	Fibroblast growth factor
FGFR1	Fibroblast growth factor receptor 1
FOXO	Forkhead box O
G6PD	Glucose-6-phosphate dehydrogenase
GAPDH	Glyceraldehyde-3-phosphate dehydrogenase
GSH	Reduced glutathione
GSSG	Oxidised glutathione
HBP	Hexosamine biosynthesis pathway
HCAEC	Human coronary artery endothelial cell
HDL	High-density lipoprotein
HDL-C	High-density lipoprotein cholesterol
HIF	Hypoxia-inducible factor
HIF-1α	Hypoxia-inducible factor 1α
HK2	Hexokinase 2
HUVEC	Human umbilical vein endothelial cell
LDL	Low-density lipoprotein
LDL-C	Low-density lipoprotein cholesterol
MAPK	Mitogen-activated protein kinase
MI	Myocardial infarction
NADPH	Nicotinamide adenine dinucleotide phosphate

PARP1	Poly (ADP-Ribose) Polymerase 1
PDC	Pyruvate dehydrogenase complex
PDK	Pyruvate dehydrogenase kinase
PDK4	Pyruvate dehydrogenase kinase 4
PDR	Proliferative diabetic retinopathy
PFK-1	Phosphofructokinase-1
PFKFB3	6-phosphofructo-2-kinase/fructose-2,6-biphosphatase 3
PGC-1α	Peroxisome proliferator-activated receptor gamma co-activator 1α
PHD	Prolyl hydroxylase domain
PI3K	Phosphoinositide 3-kinases
PK	Pyruvate kinase
PPP	Pentose phosphate pathway
R5P	Ribose-5-phosphate
RA	Rheumatoid arthritis
rHDL	Reconstitute high-density lipoprotein
ROS	Reactive oxygen species
shRNA	Short hairpin RNA
siRNA	Small interfering RNA
SR-BI	Scavenger receptor BI
TCA	Tricarboxylic acid
UDP-GlcNAc	Uridine diphosphate N-acetylglucosamine
VEGFA	Vascular endothelial growth factor A
VEGFR2	Vascular endothelial growth factor receptor 2
VHL	Von Hippel-Lindau

References

1. IDF. *IDF Diabetes Atlas*, 8th ed.; IDF: Brussels, Belgium, 2017.
2. Tan, J.T.M.; Prosser, H.C.; Vanags, L.Z.; Monger, S.A.; Ng, M.K.C.; Bursill, C.A. High-density lipoproteins augment hypoxia-induced angiogenesis via regulation of post-translational modulation of hypoxia-inducible factor 1α. *FASEB J.* **2013**, *28*, 206–217. [CrossRef] [PubMed]
3. Tan, J.T.M.; Prosser, H.C.; Dunn, L.L.; Vanags, L.Z.; Ridiandries, A.; Tsatralis, T.; Leece, L.; Clayton, Z.; Yuen, S.C.G.; Robertson, S.; et al. High Density Lipoproteins Rescue Diabetes-Impaired Angiogenesis via Scavenger Receptor Class B Type I. *Diabetes* **2016**, *65*, 3091–3103. [CrossRef] [PubMed]
4. Davidson, W.S. HDL-C vs HDL-P: How Changing One Letter Could Make a Difference in Understanding the Role of High-Density Lipoprotein in Disease. *Clin. Chem.* **2014**, *60*, e1–e3. [CrossRef] [PubMed]
5. Terasaka, N.; Yu, S.; Yvan-Charvet, L.; Wang, N.; Mzhavia, N.; Langlois, R.; Pagler, T.; Li, R.; Welch, C.L.; Goldberg, I.J.; et al. ABCG1 and HDL protect against endothelial dysfunction in mice fed a high-cholesterol diet. *J. Clin. Investig.* **2008**, *118*, 3701–3713. [CrossRef] [PubMed]
6. Prosser, H.C.; Ng, M.K.C.; Bursill, C.A. The role of cholesterol efflux in mechanisms of endothelial protection by HDL. *Curr. Opin. Lipidol.* **2012**, *23*, 182–189. [CrossRef] [PubMed]
7. Cho, K.H. High-Density Lipoproteins as Biomarkers and Therapeutic Tools. In *Impacts of Lifestyle, Diseases, and Environmental Stressors on HDL*; Springer: Berlin/Heidelberg, Germany, 2019; Volume 1.
8. Grundy, S.M. Pre-Diabetes, Metabolic Syndrome, and Cardiovascular Risk. *J. Am. Coll. Cardiol.* **2012**, *59*, 635–643. [CrossRef]
9. Ahmed, H.; Miller, M.; Nasir, K.; McEvoy, J.W.; Herrington, D.; Blumenthal, R.S.; Blaha, M.J. Primary Low Level of High-Density Lipoprotein Cholesterol and Risks of Coronary Heart Disease, Cardiovascular Disease, and Death: Results From the Multi-Ethnic Study of Atherosclerosis. *Am. J. Epidemiol.* **2016**, *183*, 875–883. [CrossRef]
10. Kartono, T.; Mallapasi, M.N.; Mulawardi, M.; Laidding, S.R.; Aminyoto, M.; Prihantono, P. Correlation of HDL cholesterol serum and Wagner's severity level of diabetic foot ulcers. *Int. J. Res. Med. Sci.* **2017**, *5*, 5129. [CrossRef]

11. Ikura, K.; Hanai, K.; Shinjyo, T.; Uchigata, Y. HDL cholesterol as a predictor for the incidence of lower extremity amputation and wound-related death in patients with diabetic foot ulcers. *Atheroscler* **2015**, *239*, 465–469. [CrossRef]

12. Kadi, H.; Ozyurt, H.; Ceyhan, K.; Koc, F.; Celik, A.; Burucu, T. The Relationship between High-Density Lipoprotein Cholesterol and Coronary Collateral Circulation in Patients with Coronary Artery Disease. *J. Investig. Med.* **2012**, *60*, 808–812. [CrossRef]

13. Prashanth, V.; Rayman, G.; Ketan, D.; Driver, V.; Hartemann, A.; Londahl, M.; Piaggesi, A.; Apelqvist, J.; Attinger, C.; Game, F. Effectiveness of interventions to enhance healing of chronic foot ulcers in diabetes: A systematic review. *Diabetes Metab. Res. Rev.* **2019**, *36*, 3284.

14. Rayman, G.; Vas, P.; Dhatariya, K.; Driver, V.; Hartemann, A.; Londahl, M.; Piaggesi, A.; Apelqvist, J.; Attinger, C.; Game, F.; et al. Guidelines on use of interventions to enhance healing of chronic foot ulcers in diabetes (IWGDF 2019 update). *Diabetes/Metab. Res. Rev.* **2020**, *36*, e3283. [CrossRef] [PubMed]

15. Okonkwo, U.A.; DiPietro, L.A. Diabetes and Wound Angiogenesis. *Int. J. Mol. Sci.* **2017**, *18*, 1419. [CrossRef] [PubMed]

16. Lin, C.-J.; Lan, Y.-M.; Ou, M.-Q.; Ji, L.-Q.; Lin, S.-D. Expression of miR-217 and HIF-1α/VEGF pathway in patients with diabetic foot ulcer and its effect on angiogenesis of diabetic foot ulcer rats. *J. Endocrinol. Investig.* **2019**, *42*, 1307–1317. [CrossRef] [PubMed]

17. Brem, H.; Tomic-Canic, M. Cellular and molecular basis of wound healing in diabetes. *J. Clin. Investig.* **2007**, *117*, 1219–1222. [CrossRef]

18. Klein, S.A.; Bond, S.J.; Gupta, S.; Yacoub, O.A.; Anderson, G.L. Angiogenesis Inhibitor TNP-470 Inhibits Murine Cutaneous Wound Healing. *J. Surg. Res.* **1999**, *82*, 268–274. [CrossRef]

19. Roman, C.D.; Choy, H.; Nanney, L.; Riordan, C.; Parman, K.; Johnson, D.; Beauchamp, R.D. Vascular Endothelial Growth Factor-Mediated Angiogenesis Inhibition and Postoperative Wound Healing in Rats. *J. Surg. Res.* **2002**, *105*, 43–47. [CrossRef]

20. Galiano, R.D.; Tepper, O.M.; Pelo, C.R.; Bhatt, K.A.; Callaghan, M.; Bastidas, N.; Bunting, S.; Steinmetz, H.G.; Gurtner, G.C. Topical Vascular Endothelial Growth Factor Accelerates Diabetic Wound Healing through Increased Angiogenesis and by Mobilizing and Recruiting Bone Marrow-Derived Cells. *Am. J. Pathol.* **2004**, *164*, 1935–1947. [CrossRef]

21. Sasmaz, H.; Yilmaz, M.B.; Madanmohan, T.; Nandeesha, H.; Pavithran, P. Coronary Collaterals in Obese Patients: Impact of Metabolic Syndrome. *Angiology* **2008**, *60*, 164–168. [CrossRef]

22. Brackbill, M.L.; Sytsma, C.S.; Sykes, K. Perioperative Outcomes of Coronary Artery Bypass Grafting: Effects of Metabolic Syndrome and Patient's Sex. *Am. J. Crit. Care* **2009**, *18*, 468–473. [CrossRef]

23. Hoffmann, R.; Stellbrink, E.; Schröder, J.; Gräwe, A.; Vogel, G.; Blindt, R.; Kelm, M.; Radke, P.W. Impact of the Metabolic Syndrome on Angiographic and Clinical Events after Coronary Intervention Using Bare-Metal or Sirolimus-Eluting Stents. *Am. J. Cardiol.* **2007**, *100*, 1347–1352. [CrossRef] [PubMed]

24. Berge, K.G.; Canner, P.L.; Hainline, A. High-density lipoprotein cholesterol and prognosis after myocardial infarction. *Circulation* **1982**, *66*, 1176–1178. [CrossRef] [PubMed]

25. Acharjee, S.; Roe, M.; Amsterdam, E.A.; Holmes, D.N.; Boden, W.E. Relation of Admission High-Density Lipoprotein Cholesterol Level and in-Hospital Mortality in Patients with Acute Non-ST Segment Elevation Myocardial Infarction (from the National Cardiovascular Data Registry). *Am. J. Cardiol.* **2013**, *112*, 1057–1062. [CrossRef] [PubMed]

26. Park, J.Y.; Rha, S.-W.; Elnagar, A.; Choi, B.G.; Im, S.I.; Kim, S.; Na, J.O.; Han, S.; Choi, C.U.; Lim, H.E.; et al. The impact of low level of high-density lipoprotein cholesterol on 6-month angiographic and 2-year clinical outcomes in acute myocardial infarction patients undergoing primary percutaneous coronary intervention. *J. Am. Coll. Cardiol.* **2012**, *59*, E18. [CrossRef]

27. Kim, J.S.; Kim, W.; Woo, J.S.; Lee, T.W.; Ihm, C.G.; Kim, Y.G.; Moon, J.Y.; Lee, S.H.; Jeong, M.H.; Jeong, K.H.; et al. The Predictive Role of Serum Triglyceride to High-Density Lipoprotein Cholesterol Ratio According to Renal Function in Patients with Acute Myocardial Infarction. *PLoS ONE* **2016**, *11*, e0165484. [CrossRef]

28. Wang, W.; Lo, A.C.Y. Diabetic Retinopathy: Pathophysiology and Treatments. *Int. J. Mol. Sci.* **2018**, *19*, 1816. [CrossRef]

29. Cao, Z.; Cooper, M.E. Pathogenesis of diabetic nephropathy. *J. Diabetes Investig.* **2011**, *2*, 243–247. [CrossRef]

30. De Bock, K.; Georgiadou, M.; Schoors, S.; Kuchnio, A.; Wong, B.W.; Cantelmo, A.R.; Quaegebeur, A.; Ghesquière, B.; Cauwenberghs, S.; Eelen, G.; et al. Role of PFKFB3-Driven Glycolysis in Vessel Sprouting. *Cell* **2013**, *154*, 651–663. [CrossRef]

31. Appelhoff, R.J.; Tian, Y.-M.; Raval, R.R.; Turley, H.; Harris, A.L.; Pugh, C.W.; Ratcliffe, P.J.; Gleadle, J. Differential Function of the Prolyl Hydroxylases PHD1, PHD2, and PHD3 in the Regulation of Hypoxia-inducible Factor. *J. Biol. Chem.* **2004**, *279*, 38458–38465. [CrossRef]

32. Jiang, B.-H.; Zheng, J.Z.; Leung, S.W.; Roe, R.; Semenza, G.L. Transactivation and Inhibitory Domains of Hypoxia-inducible Factor 1α. *J. Biol. Chem.* **1997**, *272*, 19253–19260. [CrossRef]

33. Forsythe, J.A.; Jiang, B.-H.; Iyer, N.V.; Agani, F.; Leung, S.W.; Koos, R.D.; Semenza, G.L. Activation of vascular endothelial growth factor gene transcription by hypoxia-inducible factor 1. *Mol. Cell. Biol.* **1996**, *16*, 4604–4613. [CrossRef] [PubMed]

34. Yamakawa, M.; Liu, L.X.; Date, T.; Belanger, A.J.; Vincent, K.A.; Akita, G.Y.; Kuriyama, T.; Cheng, S.H.; Gregory, R.J.; Jiang, C. Hypoxia-Inducible Factor-1 Mediates Activation of Cultured Vascular Endothelial Cells by Inducing Multiple Angiogenic Factors. *Circ. Res.* **2003**, *93*, 664–673. [CrossRef] [PubMed]

35. Ornitz, D.M.; Itoh, N. The Fibroblast Growth Factor signaling pathway. *Wiley Interdiscip. Rev. Dev. Biol.* **2015**, *4*, 215–266. [CrossRef]

36. Pengchun, Y.; Kerstin, W.; Dubrac, A.; Tung, J.K.; Alves, T.C.; Fang, J.S.; Xie, Y.; Zhu, J.; Chen, Z.; De Smet, F. FGF-dependent metabolic control of vascular development. *Nature* **2017**, *545*, 224–228.

37. Hristov, M. Endothelial Progenitor Cells Isolation and Characterization. *Trends Cardiovasc. Med.* **2003**, *13*, 201–206. [CrossRef]

38. Wilhelm, K.; Happel, K.; Eelen, G.; Schoors, S.; Oellerich, M.F.; Lim, R.; Zimmermann, B.; Aspalter, I.M.; Franco, C.A.; Boettger, T.; et al. FOXO1 couples metabolic activity and growth state in the vascular endothelium. *Nature* **2016**, *529*, 216–220. [CrossRef] [PubMed]

39. Arany, Z.; Foo, S.-Y.; Ma, Y.; Ruas, J.L.; Bommi-Reddy, A.; Girnun, G.; Cooper, M.; Laznik, D.; Chinsomboon, J.; Rangwala, S.M.; et al. HIF-independent regulation of VEGF and angiogenesis by the transcriptional coactivator PGC-1α. *Nature* **2008**, *451*, 1008–1012. [CrossRef]

40. Sumi, M.; Sata, M.; Miura, S.-I.; Rye, K.-A.; Toya, N.; Kanaoka, Y.; Yanaga, K.; Ohki, T.; Saku, K.; Nagai, R. Reconstituted High-Density Lipoprotein Stimulates Differentiation of Endothelial Progenitor Cells and Enhances Ischemia-Induced Angiogenesis. *Arter. Thromb. Vasc. Biol.* **2007**, *27*, 813–818. [CrossRef]

41. Nieuwdorp, M.; Vergeer, M.; Bisoendial, R.J.; op'T Roodt, J.; Levels, H.; Birjmohun, R.S.; Kuivenhoven, J.A.; Basser, R.; Rabelink, T.J.; Kastelein, J.J.P.; et al. Reconstituted HDL infusion restores endothelial function in patients with type 2 diabetes mellitus. *Diabetologia* **2008**, *51*, 1081–1084. [CrossRef]

42. Van Oostrom, O.; Nieuwdorp, M.; Westerweel, P.; Hoefer, I.; Basser, R.; Stroes, E.; Verhaar, M.C. Reconstituted HDL Increases Circulating Endothelial Progenitor Cells in Patients with Type 2 Diabetes. *Arter. Thromb. Vasc. Biol.* **2007**, *27*, 1864–1865. [CrossRef]

43. Prosser, H.; Ng, M.; Bursill, C. Multifunctional Regulation of Angiogenesis by High Density Lipoproteins. *Hear. Lung Circ.* **2012**, *21*, S45. [CrossRef]

44. Theofilatos, D.; Fotakis, P.; Valanti, E.; Sanoudou, D.; Zannis, V.; Kardassis, D. HDL-apoA-I induces the expression of angiopoietin like 4 (ANGPTL4) in endothelial cells via a PI3K/AKT/FOXO1 signaling pathway. *Metabolism* **2018**, *87*, 36–47. [CrossRef] [PubMed]

45. Chong, H.C.; Chan, J.S.K.; Goh, C.Q.; Gounko, N.V.; Luo, B.; Wang, X.; Foo, S.; Wong, M.T.C.; Choong, C.; Kersten, S.; et al. Angiopoietin-like 4 Stimulates STAT3-mediated iNOS Expression and Enhances Angiogenesis to Accelerate Wound Healing in Diabetic Mice. *Mol. Ther.* **2014**, *22*, 1593–1604. [CrossRef] [PubMed]

46. Kim, B.; Jang, C.; Dharaneeswaran, H.; Li, J.; Bhide, M.; Yang, S.; Li, K.; Arany, Z. Endothelial pyruvate kinase M2 maintains vascular integrity. *J. Clin. Investig.* **2018**, *128*, 4543–4556. [CrossRef] [PubMed]

47. Jeoung, N.H.; Harris, R.A. Role of Pyruvate Dehydrogenase Kinase 4 in Regulation of Blood Glucose Levels. *Korean Diabetes J.* **2010**, *34*, 274–283. [CrossRef]

48. Zhang, Y.; Ma, K.; Sadana, P.; Chowdhury, F.; Gaillard, S.; Wang, F.; McDonnell, N.P.; Unterman, T.G.; Elam, M.B.; Park, E.A. Estrogen-related Receptors Stimulate Pyruvate Dehydrogenase Kinase Isoform 4 Gene Expression. *J. Biol. Chem.* **2006**, *281*, 39897–39906. [CrossRef]

49. Aragonés, J.; Schneider, M.; Van Geyte, K.; Fraisl, P.; Dresselaers, T.; Mazzone, M.; Dirkx, R.; Zacchigna, S.; Lemieux, H.; Jeoung, N.H.; et al. Deficiency or inhibition of oxygen sensor Phd1 induces hypoxia tolerance by reprogramming basal metabolism. *Nat. Genet.* **2008**, *40*, 170–180. [CrossRef]

50. Primer, K.; Solly, E.; Psaltis, P.; Tan, J.; Bursill, C. High-density Lipoproteins Rescue Diabetes-impaired Angiogenesis by Restoring Cellular Metabolic Reprogramming Responses to Hypoxia. *Heart Lung Circ.* **2019**, *28*, 365. [CrossRef]

51. Fisslthaler, B.; Fleming, I. Activation and Signaling by the AMP-Activated Protein Kinase in Endothelial Cells. *Circ. Res.* **2009**, *105*, 114–127. [CrossRef]

52. Dagher, Z.; Ruderman, N.B.; Tornheim, K.; Ido, Y. Acute Regulation of Fatty Acid Oxidation and AMP-Activated Protein Kinase in Human Umbilical Vein Endothelial Cells. *Circ. Res.* **2001**, *88*, 1276–1282. [CrossRef]

53. Kalucka, J.; Bierhansl, L.; Conchinha, N.V.; Missiaen, R.; Elia, I.; Brüning, U.; Scheinok, S.; Treps, L.; Cantelmo, A.R.; Dubois, C.; et al. Quiescent Endothelial Cells Upregulate Fatty Acid β-Oxidation for Vasculoprotection via Redox Homeostasis. *Cell Metab.* **2018**, *28*, 881–894.e13. [CrossRef] [PubMed]

54. Schoors, S.; Bruning, U.; Missiaen, R.; Queiroz, K.C.S.; Borgers, G.; Elia, I.; Zecchin, A.; Cantelmo, A.R.; Christen, S.; Goveia, J.; et al. Fatty acid carbon is essential for dNTP synthesis in endothelial cells. *Nature* **2015**, *520*, 192–197. [CrossRef] [PubMed]

55. Boros, L.G.; Lee, P.; Brandes, J.; Cascante, M.; Muscarella, P.; Schirmer, W.; Melvin, W.S.; Ellison, E. Nonoxidative pentose phosphate pathways and their direct role in ribose synthesis in tumors: Is cancer a disease of cellular glucose metabolism? *Med. Hypotheses* **1998**, *50*, 55–59. [CrossRef]

56. Prasai, P.K.; Shrestha, B.; Orr, A.W.; Pattillo, C.B. Decreases in GSH:GSSG activate vascular endothelial growth factor receptor 2 (VEGFR2) in human aortic endothelial cells. *Redox Biol.* **2018**, *19*, 22–27. [CrossRef]

57. Rudnicki, M.; Abdifarkosh, G.; Nwadozi, E.; Ramos, S.V.; Makki, A.; Sepa-Kishi, D.M.; Ceddia, R.B.; Perry, C.G.; Roudier, E.; Haas, T.L. Endothelial-specific FoxO1 depletion prevents obesity-related disorders by increasing vascular metabolism and growth. *eLife* **2018**, *7*, 7. [CrossRef]

58. Lee, J.H.; Kim, E.-J.; Kim, D.-K.; Lee, J.-M.; Park, S.B.; Lee, I.-K.; Harris, R.A.; Lee, M.-O.; Choi, H.-S. Hypoxia induces PDK4 gene expression through induction of the orphan nuclear receptor ERRgamma. *PLoS ONE* **2012**, *7*, 115–125.

59. Sawada, N.; Jiang, A.; Takizawa, F.; Safdar, A.; Manika, A.; Tesmenitsky, Y.; Kang, K.-T.; Bischoff, J.; Kalwa, H.; Sartoretto, J.L.; et al. Endothelial PGC-1α mediates vascular dysfunction in diabetes. *Cell Metab.* **2014**, *19*, 246–258. [CrossRef]

60. Zhang, Z.; Apse, K.; Pang, J.; Stanton, R.C. High Glucose Inhibits Glucose-6-phosphate Dehydrogenase via cAMP in Aortic Endothelial Cells. *J. Biol. Chem.* **2000**, *275*, 40042–40047. [CrossRef]

61. Du, X.; Matsumura, T.; Edelstein, D.; Rossetti, L.; Zsengeller, Z.; Szabo, C.; Brownlee, M. Inhibition of GAPDH activity by poly(ADP-ribose) polymerase activates three major pathways of hyperglycemic damage in endothelial cells. *J. Clin. Investig.* **2003**, *112*, 1049–1057. [CrossRef]

62. Andrabi, S.A.; Umanah, G.K.E.; Chang, C.; Stevens, D.A.; Karuppagounder, S.S.; Gagné, J.-P.; Poirier, G.G.; Dawson, V.L.; Dawson, V.L. Poly(ADP-ribose) polymerase-dependent energy depletion occurs through inhibition of glycolysis. *Proc. Natl. Acad. Sci. USA* **2014**, *111*, 10209–10214. [CrossRef]

63. Devalaraja-Narashimha, K.; Padanilam, B.J. PARP-1 inhibits glycolysis in ischemic kidneys. *J. Am. Soc. Nephrol.* **2008**, *20*, 95–103. [CrossRef] [PubMed]

64. Steinmetz, P.R.; Balko, C.; Gabbay, K.H. The Sorbitol Pathway and the Complications of Diabetes. *N. Engl. J. Med.* **1973**, *288*, 831–836. [CrossRef] [PubMed]

65. Lal, S.; Randall, W.C.; Taylor, A.H.; Kappler, F.; Walker, M.; Brown, T.; Szwergold, B.S. Fructose-3-phosphate production and polyol pathway metabolism in diabetic rat hearts. *Metabolism* **1997**, *46*, 1333–1338. [CrossRef]

66. Szwergold, B.; Kappler, F.; Brown, T. Identification of fructose 3-phosphate in the lens of diabetic rats. *Science* **1990**, *247*, 451–454. [CrossRef]

67. Hamada, Y.; Araki, N.; Horiuchi, S.; Hotta, N. Role of polyol pathway in nonenzymatic glycation. *Nephrol. Dial. Transplant.* **1996**, *11*, 95–98. [CrossRef]

68. Tamarat, R.; Silvestre, J.-S.; Huijberts, M.; Benessiano, J.; Ebrahimian, T.G.; Duriez, M.; Wautier, M.-P.; Wautier, J.L.; Levy, B. Blockade of advanced glycation end-product formation restores ischemia-induced angiogenesis in diabetic mice. *Proc. Natl. Acad. Sci. USA* **2003**, *100*, 8555–8560. [CrossRef]

69. Matafome, P.; Sena, C.; Seiça, R. Methylglyoxal, obesity, and diabetes. *Endocrine* **2013**, *43*, 472–484. [CrossRef]

70. Federici, M. Insulin-Dependent Activation of Endothelial Nitric Oxide Synthase Is Impaired by O-Linked Glycosylation Modification of Signaling Proteins in Human Coronary Endothelial Cells. *Circulation* **2002**, *106*, 466–472. [CrossRef]
71. Namba, T.; Koike, H.; Murakami, K.; Aoki, M.; Makino, H.; Hashiya, N.; Ogihara, T.; Kaneda, Y.; Kohno, M.; Morishita, R. Angiogenesis Induced by Endothelial Nitric Oxide Synthase Gene Through Vascular Endothelial Growth Factor Expression in a Rat Hindlimb Ischemia Model. *Circulation* **2003**, *108*, 2250–2257. [CrossRef]
72. Luo, B.; Soesanto, Y.; McClain, D. Protein modification by O-linked GlcNAc reduces angiogenesis by inhibiting Akt activity in endothelial cells. *Arter. Thromb. Vasc. Biol.* **2008**, *28*, 651–657. [CrossRef]
73. Steinberg, H.O.; Tarshoby, M.; Monestel, R.; Hook, G.; Cronin, J.; Johnson, A.; Bayazeed, B.; Baron, A.D. Elevated circulating free fatty acid levels impair endothelium-dependent vasodilation. *J. Clin. Investig.* **1997**, *100*, 1230–1239. [CrossRef] [PubMed]
74. Steinberg, H.O.; Paradisi, G.; Hook, G.; Crowder, K.; Cronin, J.; Baron, A.D. Free fatty acid elevation impairs insulin-mediated vasodilation and nitric oxide production. *Diabetes* **2000**, *49*, 1231–1238. [CrossRef] [PubMed]
75. De Kreutzenberg, S.V.; Crepaldi, C.; Marchetto, S.; Calo, L.; Tiengo, A.; Del Prato, S.; Avogaro, A. Plasma Free Fatty Acids and Endothelium-Dependent Vasodilation: Effect of Chain-Length and Cyclooxygenase Inhibition. *J. Clin. Metab.* **2000**, *85*, 793–798. [CrossRef] [PubMed]
76. Inoguchi, T.; Li, P.; Umeda, F.; Yu, H.Y.; Kakimoto, M.; Imamura, M.; Aoki, T.; Etoh, T.; Hashimoto, T.; Naruse, M.; et al. High glucose level and free fatty acid stimulate reactive oxygen species production through protein kinase C–dependent activation of NAD(P)H oxidase in cultured vascular cells. *Diabetes* **2000**, *49*, 1939–1945. [CrossRef] [PubMed]

International Journal of
Molecular Sciences

Article

Adipose Tissue and Brain Metabolic Responses to Western Diet—Is There a Similarity between the Two?

Arianna Mazzoli [1], Maria Stefania Spagnuolo [2], Cristina Gatto [1], Martina Nazzaro [1], Rosa Cancelliere [1], Raffaella Crescenzo [1], Susanna Iossa [1,*] and Luisa Cigliano [1,*]

[1] Department of Biology, University of Naples Federico II, 80134 Naples, Italy; arianna.mazzoli@unina.it (A.M.); crigatto51@gmail.com (C.G.); martinanazzaro52@gmail.com (M.N.); cancelliererosa@gmail.com (R.C.); rcrescen@unina.it (R.C.)

[2] Department of Bio-Agrofood Science, Institute for the Animal Production System in Mediterranean Environment, National Research Council Naples (CNR-ISPAAM), 80147 Naples, Italy; mariastefania.spagnuolo@cnr.it

* Correspondence: susiossa@unina.it (S.I.); luisa.cigliano@unina.it (L.C.)

Received: 13 December 2019; Accepted: 23 January 2020; Published: 25 January 2020

Abstract: Dietary fats and sugars were identified as risk factors for overweight and neurodegeneration, especially in middle-age, an earlier stage of the aging process. Therefore, our aim was to study the metabolic response of both white adipose tissue and brain in middle aged rats fed a typical Western diet (high in saturated fats and fructose, HFF) and verify whether a similarity exists between the two tissues. Specific cyto/adipokines (tumor necrosis factor alpha (TNF-α), adiponectin), critical obesity-inflammatory markers (haptoglobin, lipocalin), and insulin signaling or survival protein network (insulin receptor substrate 1 (IRS), Akt, Erk) were quantified in epididymal white adipose tissue (e-WAT), hippocampus, and frontal cortex. We found a significant increase of TNF-α in both e-WAT and hippocampus of HFF rats, while the expression of haptoglobin and lipocalin was differently affected in the various tissues. Interestingly, adiponectin amount was found significantly reduced in e-WAT, hippocampus, and frontal cortex of HFF rats. Insulin signaling was impaired by HFF diet in e-WAT but not in brain. The above changes were associated with the decrease in brain derived neurotrophic factor (BDNF) and synaptotagmin I and the increase in post-synaptic protein PSD-95 in HFF rats. Overall, our investigation supports for the first time similarities in the response of adipose tissue and brain to Western diet.

Keywords: adipose tissue; hippocampus; frontal cortex; adiponectin; haptoglobin; lipocalin; BDNF; synaptic proteins

1. Introduction

Lifestyle, nutrition and lack of physical exercise is increasing the number of overweight or obese people in the global population. The increased consumption of sugar- and fat-rich industrial foods, which are cheaper and easily available, contributes to the accumulation of peripheral and/or visceral adipose tissue [1]. In the last decade, dietary fats and sugars were also identified as risk factors for cognitive decline and neurodegeneration, through altered brain metabolism, neuroinflammation and neuronal dysfunction [2–6].

The increased risk for obese individuals to develop brain disease might be due to the capacity of adipose tissue to communicate with the brain and impact its function. Therefore, the analysis of specific markers, whose adipose and/or brain levels are affected by diet, might contribute to unveil the intersection between obesity and neurodegeneration. In particular, proinflammatory cytokines, namely tumor necrosis factor alpha (TNFα), are implicated in the development of neuronal insulin resistance [7,8]. Since both cytokines and insulin in the brain regulate synaptic plasticity, learning and memory, neuroinflammation and neuronal insulin resistance have been proposed to be involved in obesity-associated brain impairment [9].

Middle-age, an earlier stage of the aging process, is a life phase in which humans and animals are more prone to develop diet-induced metabolic alterations. In fact, overweight is especially prevalent in the middle-aged population (40.2%) compared to younger or older adults (32.3% and 37.0%, respectively) [10]. In addition, overweight during midlife has been associated to a higher risk of developing cognitive disorders, including Alzheimer's disease [11,12]. Accordingly, we recently found that short-term dietary treatment with a high fat-high fructose diet was able to elicit inflammation and oxidative stress in plasma and brain of middle-aged rats [6,13]. We therefore decided to investigate the possible similarity in the response of white adipose tissue and brain in middle aged rats that were fed a typical Western diet (high in saturated fats and fructose, HFF) for 4 weeks. In particular, specific cyto/adipokines (TNF-α, adiponectin), critical obesity-inflammatory markers (haptoglobin, lipocalin), and insulin signaling or survival protein network (insulin receptor substrate 1 (IRS), Akt, Erk) were measured in both adipose tissue and hippocampus/frontal cortex, two key areas for learning and memory. Moreover, in order to further highlight whether the HFFdiet is associated with an impairment of brain function, we evaluated brain-derived neurotrophic factor (BDNF) and its receptor (Tropomyosin receptor kinase B, TrkB), as well as specific pre- and post-synaptic proteins in the hippocampus and frontal cortex, known to be susceptible to nutritional stimuli.

2. Results

2.1. Metabolic Characterization

Dietary administration of HFF diet elicited caloric hyperphagia in the first and, less markedly, in the second week, while during the third and fourth week the caloric intake was similar in the two groups of rats (Figure 1A). In addition, HFF rats exhibited significantly higher body weight gain, as well as higher body content of epididymal white adipose tissue (e-WAT), both as absolute amount and as % of body weight (Figure 1B).

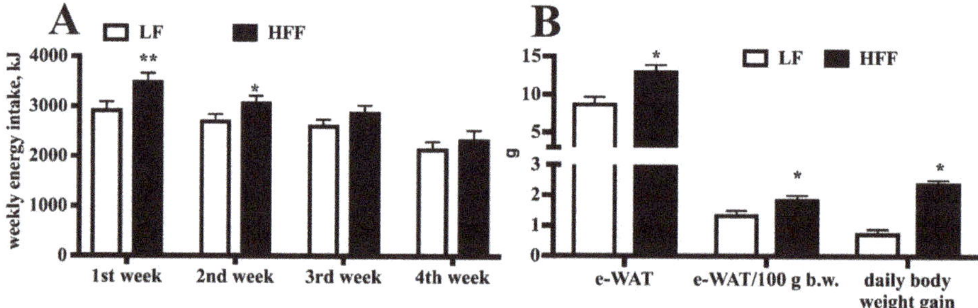

Figure 1. Weekly energy intake (**A**), epididymal white adipose tissue (e-WAT) weight and body weight gain (**B**) in middle-aged rats fed low fat (LF) or high fat-high fructose (HFF) diet for four weeks. Values are the means ± SEM of eight rats. * $p < 0.05$, ** $p < 0.01$ compared to low-fat diet (two-way ANOVA followed by Tukey post-test for energy intake data and two-tailed Student's t-test for e-WAT weight and daily body weight gain data).

Key proteins involved in the regulation of adipocyte function, namely uncoupling protein 2 (UCP2), peroxisome proliferator-activated receptor alpha (PPAR-α), and peroxisome proliferator-activated receptor gamma coactivator 1-alpha (PGC-1α), were found to be downregulated in HFF rats compared to rats fed low fat (LF) diet (Figure 2).

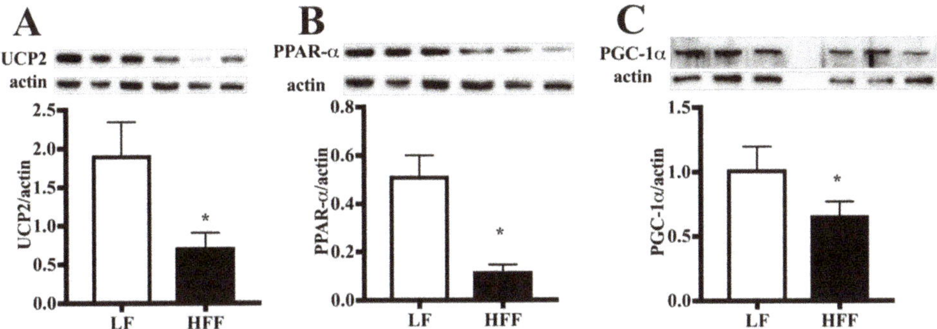

Figure 2. Protein content (with representative western blot) of uncoupling protein 2 (UCP2) (**A**), peroxisome proliferator activated receptor alpha (PPAR-α) (**B**) and peroxisome proliferator-activated receptor gamma coactivator 1-alpha (PGC-1α) (**C**) in epididymal white adipose tissue (e-WAT) from middle-aged rats fed low fat (LF) or high fat-high fructose (HFF) diet for four weeks. Values are the means ± SEM of eight rats. * $p <0.05$ compared to low-fat diet (two-tailed Student's t-test).

2.2. Inflammatory Markers

Since increased intake of fat and fructose is associated with systemic and tissue inflammation [14,15], we wanted to verify whether this response was elicited after our dietary treatment. To this end, we assessed markers of inflammation in plasma, e-WAT and brain areas. TNF-α levels were significantly increased in plasma, e-WAT and hippocampus of HFF rats compared to controls, while no variation was found in frontal cortex (Figure 3). Furthermore, we titrated the level of adiponectin, an adipokine which plays a role in the reduction of oxidative stress and inflammatory cascade [16] and was also reported to exert neuroprotective effect [17]. Adiponectin amount was found significantly reduced in plasma, e-WAT, hippocampus, and frontal cortex of HFF fed rats (Figure 4).

The level of haptoglobin and lipocalin, which are considered markers of both inflammation and adiposity [18,19], was measured. As shown in Figure 5, no diet-associated variation of lipocalin concentration was detected in plasma, hippocampus or cortex, while in e-WAT lipocalin protein content significantly increased in HFF rats. Conversely, haptoglobin level was found increased, following HFF diet, in plasma, frontal cortex, and hippocampus, while no changes were detected in e-WAT (Figure 6).

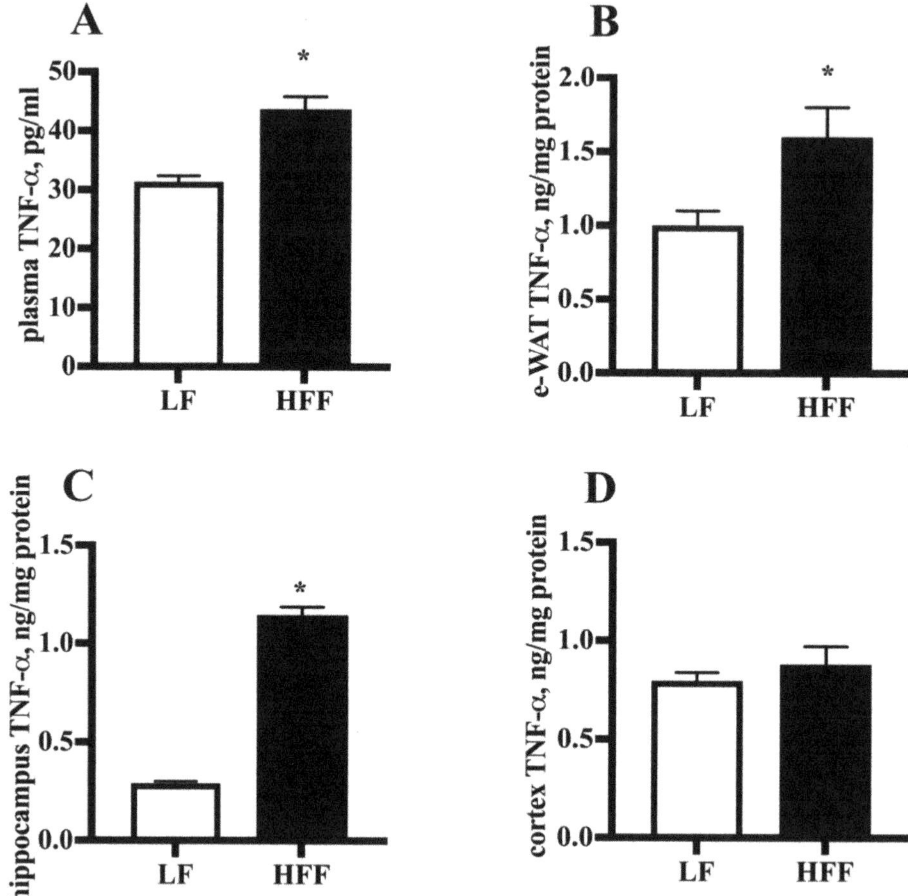

Figure 3. Levels of tumor necrosis factor alpha (TNF-α) detected by enzyme-linked immunosorbent assayin plasma (**A**), epididymal white adipose tissue (e-WAT) (**B**), hippocampus (**C**), and frontal cortex (**D**) from middle-aged rats fed low fat (LF) or high fat-high fructose (HFF) diets for four weeks. Values are the means ± SEM of eight rats. * $p < 0.05$ compared to low-fat diet (two-tailed Student's t-test).

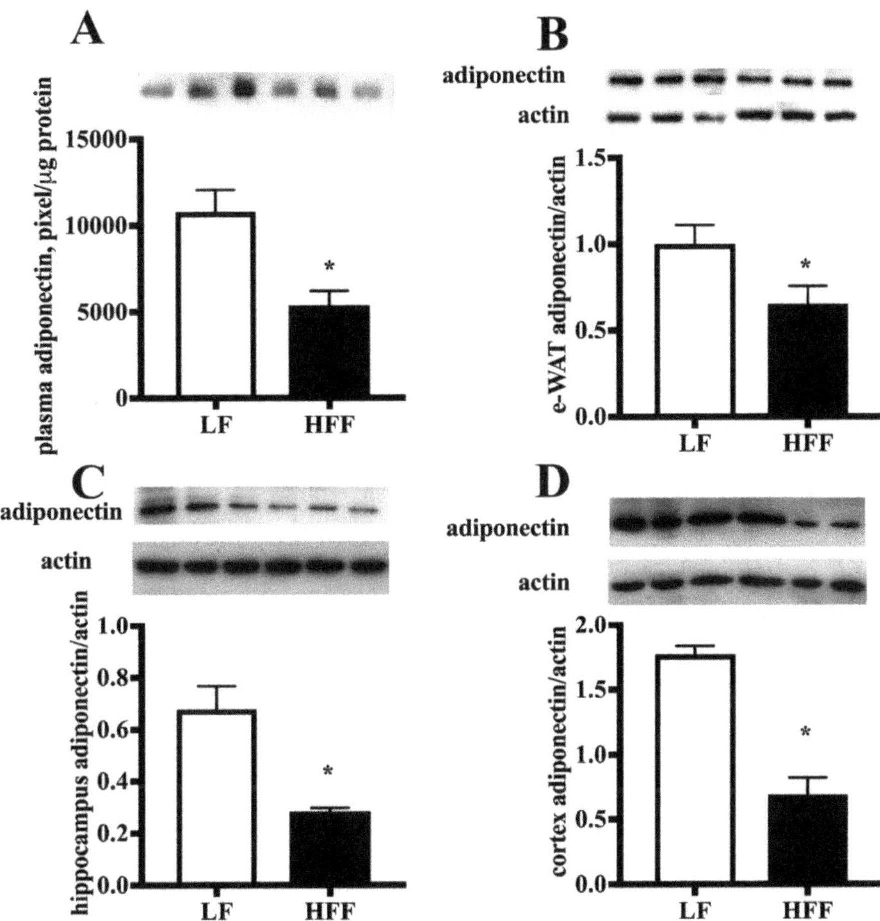

Figure 4. Protein content (with representative western blot) of adiponectin in plasma (**A**), epididymal white adipose tissue (e-WAT) (**B**), hippocampus (**C**), and frontal cortex (**D**) from middle-aged rats fed low fat (LF) or high fat-high fructose (HFF) diets for four weeks. Plasma adiponectin was detected by using 5 μL of sample. Values are the means ± SEM of eight rats. * $p < 0.05$ compared to low-fat diet (two-tailed Student's t-test).

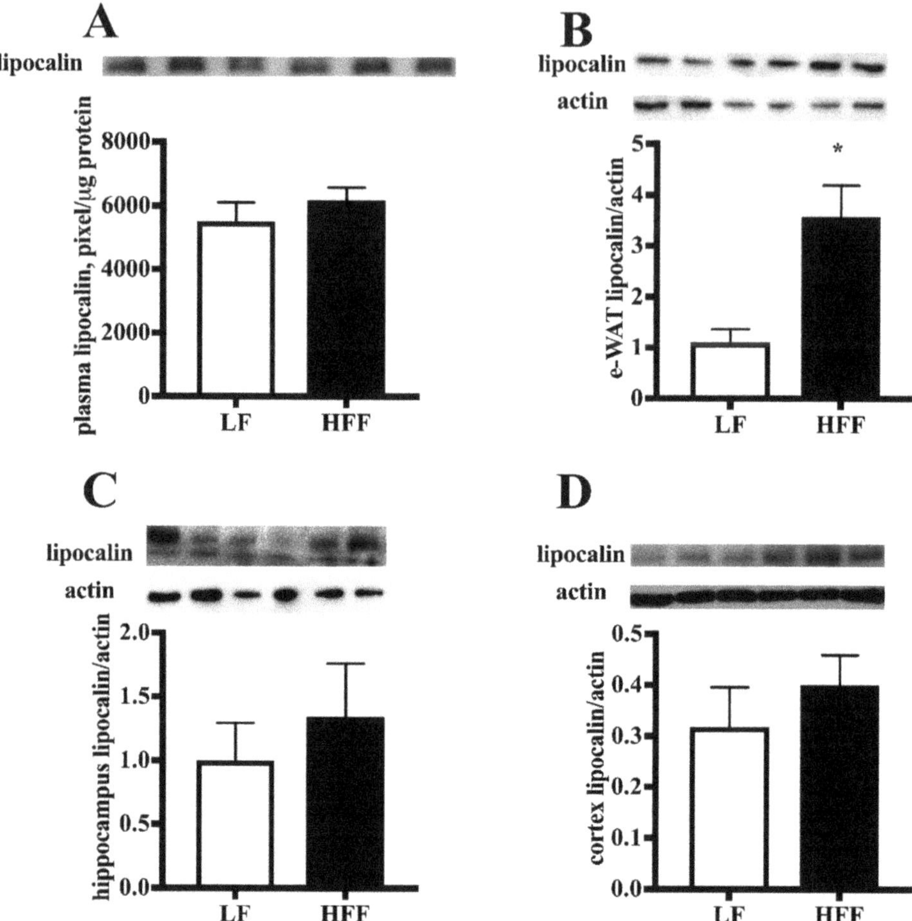

Figure 5. Protein content (with representative western blot) of lipocalin in plasma (**A**), epididymal white adipose tissue (e-WAT) (**B**), hippocampus (**C**), and frontal cortex (**D**) from middle-aged rats fed low fat (LF) or high fat-high fructose (HFF) diet for four weeks. Values are the means ± SEM of eight rats. * $p < 0.05$ compared to low-fat diet (two-tailed Student's t-test).

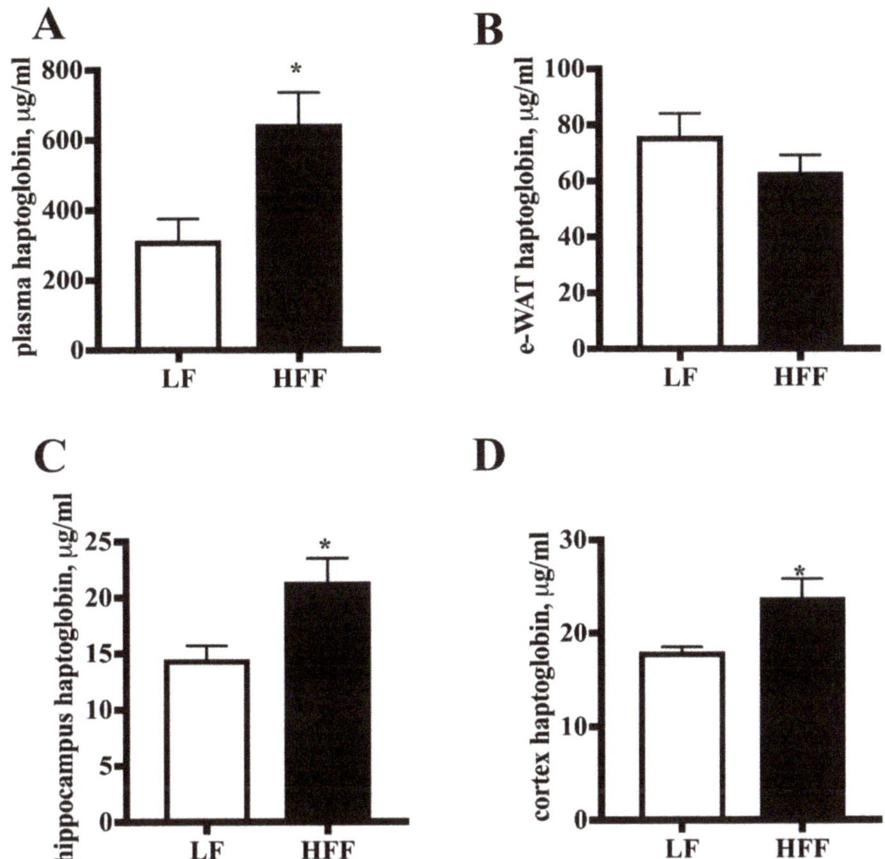

Figure 6. Levels of haptoglobin detected by enzyme-linked immunosorbent assay in plasma (**A**), epididymal white adipose tissue (e-WAT) (**B**), hippocampus (**C**), and frontal cortex (**D**) from middle-aged rats fed low fat (LF) or high fat-high fructose (HFF) diet for four weeks. Values are the means ± SEM of eight rats. * $p < 0.05$ compared to low-fat diet (two-tailed Student's t-test).

2.3. Insulin Signaling

Plasma metabolic profile evidenced that the HFF diet was able to induce systemic insulin resistance, with higher fasting glucose and insulin levels and increased Homeostasis model assessment (HOMA) index (Figure 7A). At the tissue level, insulin signaling was impaired by HFF diet in e-WAT, as shown by a significant decrease in the activatory phosphorylation of IRS (Figure 7B), while this effector was not affected by diet in hippocampus (Figure 7C) and cortex (Figure 7D).

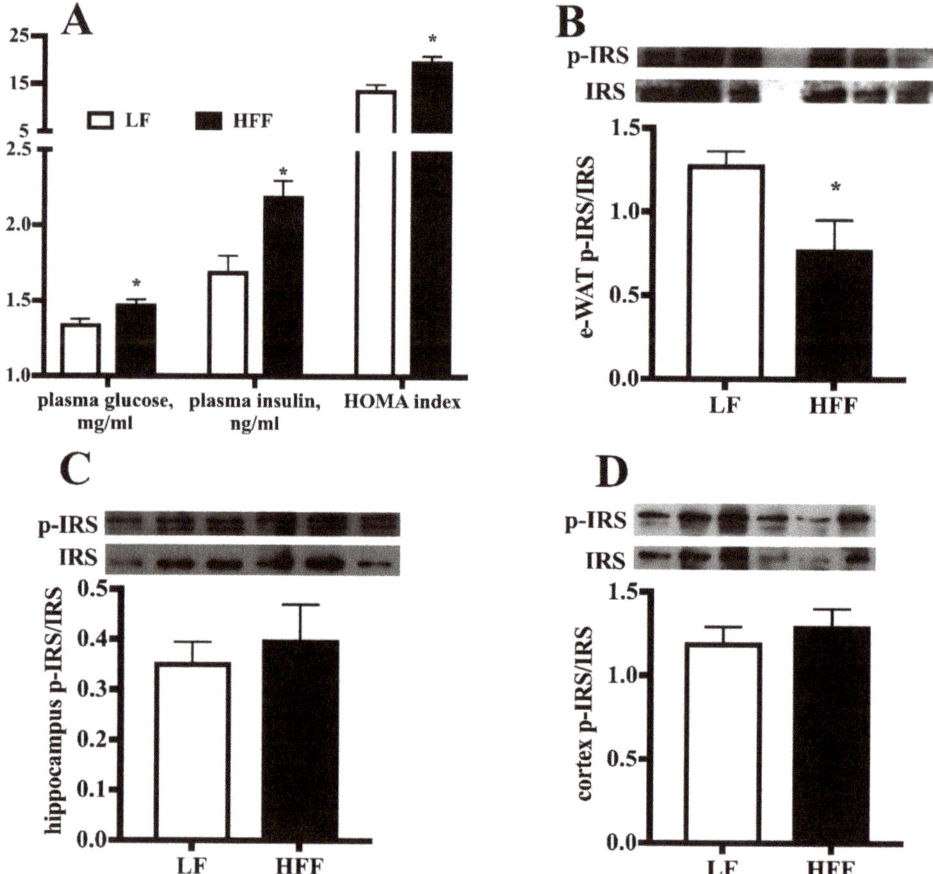

Figure 7. Plasma levels of glucose and insulin and Homeostatic model assessment (HOMA) index (**A**), protein content (with representative western blot) of phosphorylated Insulin Receptor Substrate (p-IRS) in epididymal white adipose tissue (e-WAT) (**B**), hippocampus (**C**), and frontal cortex (**D**) from middle-aged rats fed low fat (LF) or high fat-high fructose (HFF) diet for four weeks. Values are the means ± SEM of eight rats. * $p < 0.05$ compared to low-fat diet (two-tailed Student's t-test).

The analysis of downstream effectors of IRS, namely Akt and Erk, revealed an impairment for both effectors in e-WAT (Figure 8A,B). The activation of the Akt and Erk pathway was found significantly decreased by HFF diet in frontal cortex (Figure 8E,F), while in hippocampus was evident a significant increase in the Erk activation (Figure 8D), with no variation in the Akt pathway (Figure 8C).

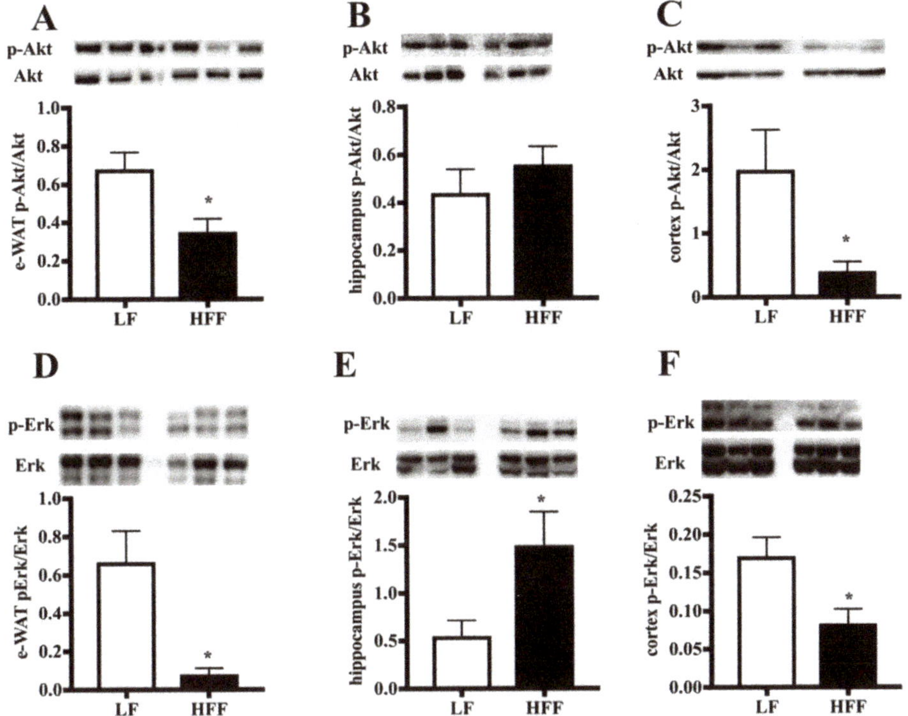

Figure 8. Protein content (with representative western blot) of phosphorylated Akt (p-Akt) and Erk (p-Erk) in epididymal white adipose tissue (e-WAT) (**A,D**), hippocampus (**B,E**), and frontal cortex (**C,F**) from middle-aged rats fed low fat (LF) or high fat-high fructose (HFF) diets for four weeks. Values are the means ± SEM of eight rats. * $p < 0.05$ compared to low-fat diet (two-tailed Student's t-test).

2.4. BDNF and TrkB

BDNF is a key cerebral factor involved in a wide range of neurophysiological processes and has a multipotent impact on brain signaling and synaptic plasticity [20]. The BDNF level was measured in both the hippocampus and frontal cortex, and a significant diet dependent decrease was observed (Figure 9A,C). We also analyzed adipose tissue samples for the presence of BDNF, but we detected only a very faint band (data not shown). At present, we cannot say whether this very faint band reflects a small amount of BDNF produced locally in e-WAT or a small amount coming from the plasma. Further work is required to discriminate between the two options. We further investigated whether the dietary treatment could affect the amount of TrkB receptor, as BDNF activities, such as enhancement of synaptic plasticity, neuroprotection, and stimulation of neuronal fibers growth, are mediated by neurotrophin binding to this receptor [20]. As shown in Figure 9B, TrkB level was significantly lower in the hippocampus of HFF rats, while it was not affected by diet in the frontal cortex (Figure 9D).

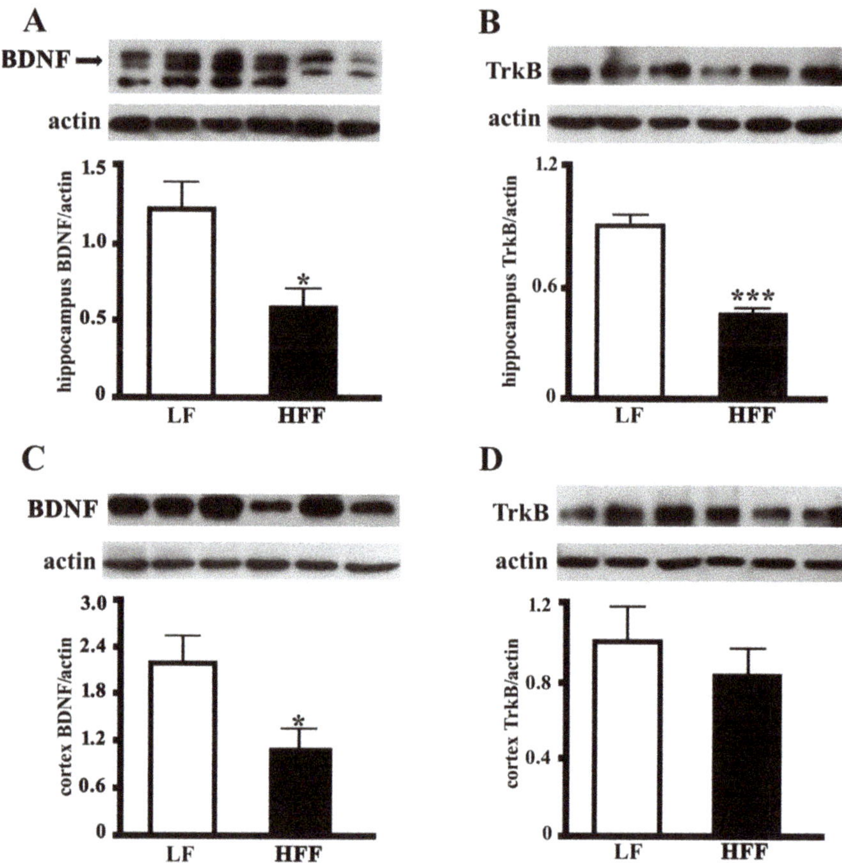

Figure 9. Protein content (with representative western blot) of brain derived neurotrophic factor (BDNF) and tropomyosin receptor kinase b (TrkB) levels in hippocampus (**A**,**B**), and frontal cortex (**C**,**D**) from middle-aged rats fed low fat (LF) or high fat-high fructose (HFF) diets for four weeks. Values are the means ± SEM of eight rats. * $p < 0.05$ compared to low-fat diet (two-tailed Student's t-test).

2.5. Synaptic Proteins in the Hippocampus and Frontal Cortex

In order to clarify the impact of HFF diet on key markers of synaptic function, the levels of presynaptic proteins synaptophysin, synapsin I and synaptotagmin I, and post synaptic density protein 95 (PSD-95) were measured in hippocampus and frontal cortex of HFF and control rats. Synaptophysin and synapsin I are involved in synaptic growth, as they play key roles in synapse formation, maturation, and maintenance [21]. Synaptotagmin I is a major calcium sensor for transmitter release at central synapse and is also crucial for clamping synaptic vesicle fusion in mammalian neurons [22]. HFF diet did not affect synapsin I or synaptophysin level, while it was associated with a marked decrease of synaptotagmin both in the hippocampus (Figure 10) and the frontal cortex (Figure 11). In the next step, we examined the level of PSD-95, a key protein for the function of neurotransmitter receptors [23]. Interestingly, we found that the expression PSD-95 was significantly increased in hippocampus (Figure 10) and frontal cortex (Figure 11) of HFF rats.

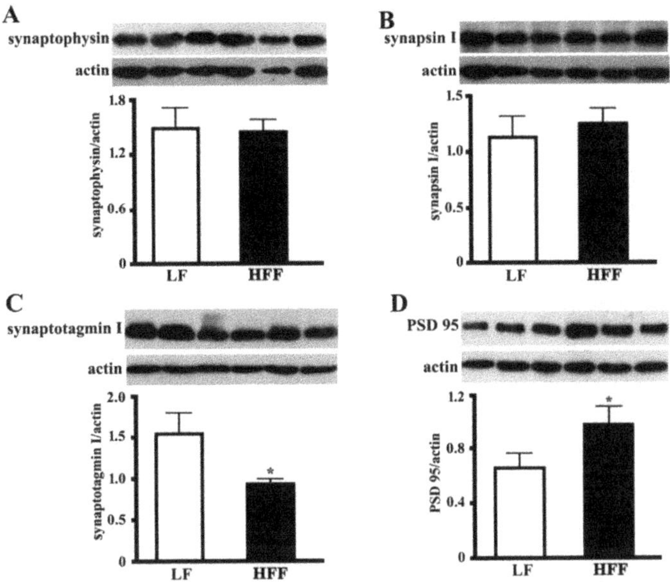

Figure 10. Protein content (with representative western blot) of presynaptic proteins synaptophysin (**A**), synapsin I (**B**) and synaptotagmin I (**C**), and postsynaptic protein PSD-95 (**D**) in the hippocampus from middle-aged rats fed low fat (LF) or high fat-high fructose (HFF) diets for four weeks. Values are the means ± SEM of eight rats. * $p < 0.05$ compared to low-fat diet (two-tailed Student's t-test).

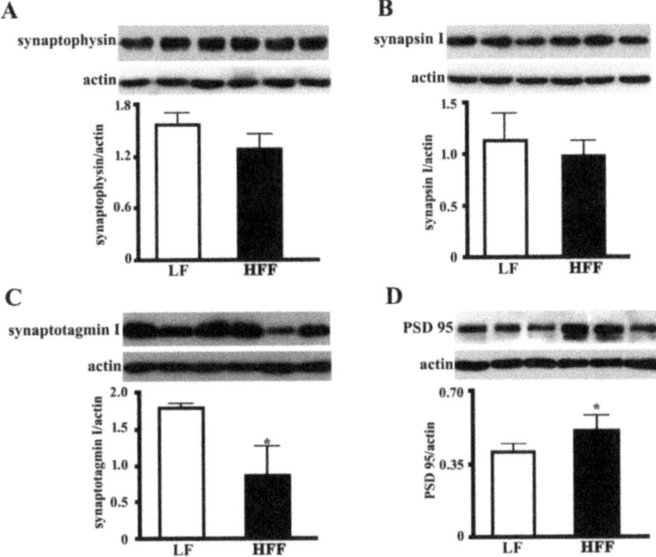

Figure 11. Protein content (with representative western blot) of presynaptic proteins synaptophysin (**A**), synapsin I (**B**) and synaptotagmin I (**C**), and postsynaptic protein PSD-95 (**D**) in the frontal cortex from middle-aged rats fed low fat (LF) or high fat-high fructose (HFF) diets for four weeks. Values are the means ± SEM of eight rats. * $p < 0.05$ compared to low-fat diet (two-tailed Student's t-test).

3. Discussion

The risk of brain dysfunction has been shown to be increased in overweight or obese subjects [24–26]. The underlying mechanisms are not precisely defined, but obesity has been associated with several processes related to the acceleration of aging, including oxidative stress and inflammation [27,28]. The link between obesity and brain disease can be further clarified by studying the alteration of adipose tissue physiology, concomitantly with brain modifications in these conditions. In fact, it has been suggested that the molecules released or produced by the adipose tissue could be the molecular link between obesity and brain dysfunction [29]. Indeed, expansion of adipose tissue leads to local inflammation and release of cytokines and adipokines into the systemic circulation, which in turn could contribute to the pathogenesis of brain disorders [30,31]. Since this issue has not been investigate in depth, we sought to investigate the effect of a short-term Western diet on white adipose tissue functional markers and their possible similarity with changes in two critical brain areas for learning and memory, namely, the hippocampus and frontal cortex.

Metabolic phenotyping of middle-aged rats evidenced the deleterious effects of HFF diet, with the increase in the absolute and relative weight of e-WAT. This adipose depot is among the largest and most readily accessible fat pads in the rat [32], and has been shown to reach its maximum expansion earlier (after 4 weeks of high fat feeding) than other adipose tissue depots [33]. The expansion of e-WAT in HFF rats well correlates with the increased caloric intake during the first two weeks and with the downregulation in the content of key proteins involved in the regulation of adipocyte function, namely UCP2, PPAR-α, and PGC-1α. In fact, PPARα and PGC-1α play a central role in the metabolic regulation in adipose tissue [34,35]. In addition, UCP2 may influence systemic metabolism by regulating the release of adipokines by this tissue [36]. Interestingly, we recently reported a similar decrease in UCP2, PPAR-α, and PGC-1α in brain regions of middle-aged rats fed the same Western diet [6].

HFF diet elicited whole-body insulin resistance. The same metabolic impairment was evident also in e-WAT, where downstream effectors of insulin signaling (IRS, Akt, and Erk) were less activated in HFF rats, thus suggesting a condition of insulin resistance in this tissue.

Systemic inflammation was also found in HFF rats, as demonstrated by the increase in plasma levels of both TNF-α and haptoglobin, which is one of the most represented acute-phase proteins. Similarly, e-WAT from HFF rats exhibited increased TNF-α content, in agreement with the frequent association between insulin resistance and inflammation [37]. The above increase is of relevance and could be at the basis of increased lipocalin content found in e-WAT from these animals. In fact, lipocalin expression in adipocytes is regulated by obesity and TNF-α and it can, in turn, induce insulin resistance [38]. In line with the systemic and adipose inflammatory status, the here presented results show that a condition of brain inflammation occurs in response to the HFF diet. As a matter of fact, a diet-associated increase in TNF-α was observed in hippocampus, and haptoglobin level was found higher both in hippocampus and frontal cortex of HFF rats. It is worth mentioning that no data are available thus far on diet-associated changes of haptoglobin in brain of middle-aged rats. The increase in brain haptoglobin might represent a protective mechanism against the condition of enhanced oxidative stress found in hippocampus and frontal cortex [6], due to its well-known antioxidant activity [39,40].

The above inflammatory status might contribute to the onset of brain insulin resistance, however, interestingly, we did not find an alteration in IRS activation. It is therefore possible that HFF diet, administered for only 4 weeks, is able to induce adipose tissue insulin resistance, as well as brain inflammation and oxidative stress [6], but does not trigger impaired response to insulin in brain. Of note, we recently found a decrease in the protein content of insulin degrading enzyme in hippocampus and cortex of middle-aged rats fed an HFF diet (data not shown). Thus, the impairment in the pathway of degradation of insulin in the brain could act as a potentiating factor on the action of this hormone.

The increased activation of the Erk pathway in the hippocampus could be envisaged as marker of damage of this brain area. In fact, Erk activation seems to play an active role in several models of neuronal death, such as hyperglycaemia-mediated neuronal damage [41] or β-amyloid-induced neuronal death [42]. On the other hand, frontal cortex is differently affected, and the decrease in Akt and Erk pathways is likely due to the impairment in other mechanisms, including neurotrophin signaling, since both Akt and Erk are at the crossroad of several intracellular pathways.

The analysis of changes in adiponectin shows a downregulation in e-WAT. This result is in good agreement with lower protein levels of UCP2 and PPAR-α, since it has been shown that adiponectin expression is regulated via adipose PPAR-α [43], and UCP2 controls adiponectin gene expression in adipose tissue [40].

A diet-associated reduction of adiponectin was also observed in both hippocampus and frontal cortex. The origin of adiponectin found in the brain has been debated. Adiponectin can enter the brain by passing through the blood–brain barrier [44]. Hence, the peripheral adipose tissue could be the source of brain adiponectin, even though it has been reported that adiponectin is also produced inside the brain [45]. It is worth mentioning that, beside its role in improving insulin sensitivity and modulating lipid and carbohydrate metabolism, adiponectin was reported to influence neurogenesis, hippocampal synapses and synaptic plasticity [46–48]. Furthermore, there is strong evidence supporting the neuroprotective effects of adiponectin in cell culture and animal models [49,50]. Therefore, adiponectin decrease might underlie alteration in brain functioning, as suggested by the reduction in BDNF, a further critical marker of brain functioning, playing a key role in modulation of adult neurogenesis [51] and synaptic function [52]. Our result is in agreement with previous studies showing a downregulation of BDNF signaling both in middle-aged [53] and adolescent [54] rats exposed to a Western diet. Accordingly, we found a significant decrease in the amount of synaptotagmin I in both hippocampus and frontal cortex of middle aged HFF rats. As known, synaptotagmin I participates in the regulation of synaptic vesicle exocytosis, acting for clamping synaptic vesicle fusion in mammalian neurons [22]. Despite a lower level of BDNF in both hippocampus and frontal cortex of the HFF group, no differences in the other markers, synaptophysin and synapsin I, were detected. Whether this is due to the fact that BDNF levels are still enough to maintain downstream targets or to the activation of other signaling pathways remains to be elucidated. Intriguingly, a significant increase in the post-synaptic critical protein PSD-95 was found in both hippocampus and frontal cortex, that could represent a compensatory mechanism against HFF diet-induced presynaptic alterations.

4. Materials and Methods

4.1. Materials

Bovine serum albumin fraction V (BSA), rabbit anti-human haptoglobin, salts, and buffers were purchased from Sigma-Aldrich (St. Louis, MO, USA). The dye reagent for protein titration was from Bio-Rad (Hercules, CA, USA), and the polyvinylidene difluoride (PVDF) membrane was from GE Healthcare (Milan, Italy). Horseradish peroxidase (HRP)-conjugated secondary antibodies were from Immunoreagent, (Raleigh, NC, USA) (goat anti-rabbit or goat anti-mouse) or from Sigma-Aldrich (St. Louis, MO, USA) (rabbit anti-goat). Fuji Super RX 100 films were from Laboratorio Elettronico Di Precisione (Naples, Italy).

4.2. Experimental Design

Male Sprague-Dawley rats were purchased from Charles River (Calco, Como, Italy) and used for the experiments. All rats were caged singly in a temperature-controlled room (23 ± 1 °C) with a 12-h light/dark cycle (06.30–18.30). Treatment, housing, and euthanasia of animals met the guidelines set by the Italian Health Ministry. All experimental procedures involving animals were approved by the "Comitato Etico-Scientifico per la Sperimentazione Animale" of the University of Naples Federico II (260/2015-PR).

For the experiments, we used middle-aged rats (11 months old), that were divided in two groups, each composed of eight rats, that were fed an HFF or LF diet for 4 weeks. The composition of the two diets is reported in Table 1. During the dietary treatment, body weight and food and water intake were monitored daily. At the end of the experimental period, the animals were anaesthetized with sodium Tiopental (40 mg/kg body weight) and euthanized by decapitation, and blood, epididymal white adipose tissue (e-WAT), hippocampus, and frontal cortex were harvested. In particular, hippocampus and frontal cortex were dissected as previously published [6,55]. Samples were then snap frozen in liquid nitrogen and stored at −80 °C for subsequent analyses.

Table 1. Diet composition.

	Low Fat	High Fat-High Fructose
Component, g/1000 g		
Standard chow [a]	395.3	231.5
Sunflower oil	19.3	19.3
Casein	59.7	133.3
Water	175.7	175.4
AIN-93 Mineral mix	11.4	11.4
AIN-93 Vitamin mix	3.2	3.2
Choline	0.7	0.7
Methionine	0.9	0.9
Cornstarch	333.8	——-
Butter	——-	129.8
Fructose	——-	294.6
Energy content and composition		
ME content, kJ/g [b]	11.2	14.9
Lipids, J/100 J	10.5	39.3
Proteins, J/100 J	19.9	19.8
Complex carbohydrates, J/100 J	63.9	7.5
Simple sugars, J/100 J	5.7	33.4

[a] 4RF21, Mucedola, Italy; [b] Estimated by computation using values (kJ/g) for energy content as follows: protein 16.736, lipid 37.656, and carbohydrate 16.736. ME = metabolizable energy.

4.3. Metabolic Analyses

The blood samples were centrifuged at 1400× g for 8 min at 4 °C. After centrifugation at 1400× g for 8 min at 4 °C, plasma was isolated and stored at −20 °C until used for determination of substrates and hormones. Plasma glucose concentration was measured by a colorimetric enzymatic method (Pokler Italia, Pontecagnano, Italy). Plasma insulin concentration was measured using an enzyme-linked immunosorbent assay (ELISA) kit (Diametra, Segrate, Italy) in a single assay to avoid interassay variations. HOMA index was calculated as follows: (Glucose (mg/dL) × Insulin (mU/L))/405) [56].

4.4. Markers of Inflammation in Plasma, e-WAT, Frontal Cortexok, and Hippocampus

TNF-α concentrations were determined using a rat-specific enzyme linked immunosorbent assay (R&D Systems, Minneapolisok, MN, USA) according to the manufacturer's instruction.

Haptoglobin concentration in plasma, adipose tissue, frontal cortex, and hippocampus samples was measured by ELISA. Samples were diluted (plasma = 1:9000–1:70,000; adipose tissue, 1:1000–1:30,000; frontal cortex and hippocampus, 1:1000–1:30,000) with coating buffer (7 mM Na_2CO_3, 17 mM $NaHCO_3$, 1.5 mM NaN_3, pH 9.6), and aliquots (50 μL) were then incubated in the wells of a microtitre plate (Immuno MaxiSorp; overnight, 4 °C). Washing and blocking were carried out as previously reported [57], then, the wells were incubated (1 h, 37 °C) with 50 μL of rabbit anti-human haptoglobin (1:500 in 130 mM NaCl, 20 mM Tris-HCl, 0.05% Tween, pH 7.4, containing 0.25% BSA), followed by 60 μl of HRP-coniugated secondary antibody (1:5000 dilution). Peroxidase-catalyzed color development from o-phenylenediamine was measured at 492 nm.

4.5. Western Blotting

Proteins were extracted from e-WAT by diluting tissue samples 1:1 with lysis buffer (20.0 mmol/L Tris, pH 8, 5% glycerol, 138 mM NaCl, 2.7 mM KCl, 1% NP-40, 5 mM ethylenediaminetetraacetic acid, 5% protease inhibitor cocktail, 1% phosphatase inhibitor cocktail). Homogenates were centrifuged (15,000× g, 15 min at 4 °C) and the supernatants were then collected. Aliquots of 20 µg were used for electrophoresis.

Proteins were extracted from hippocampus and frontal cortex by homogenizing aliquots (about 40 mg) of frozen tissues in six volumes (w/v) of cold buffer, as previously published [58]. Homogenates were centrifuged (14,000× g, 45 min, 4 °C), and supernatants were then collected. Aliquots of 40 µg were used for electrophoresis.

All the plasma samples were adjusted to protein concentration of 10 µg/µL and 5 µL were used for electrophoresis.

Samples were fractionated by electrophoresis on 12.5% (to quantify BDNF, synaptophysin, synaptotagmin I, adiponectin, lipocalin), or 10% (to quantify synapsin I, PSD-95, TrkB, p-Akt, p-Erk, p-IRS, PGC-1α, PPAR-α, UCP2) polyacrylamide gel, under denaturing and reducing conditions. After electrophoresis, proteins were blotted onto PVDF membrane, essentially as previously reported [6].

The membranes were pre-blocked (1 h, 37 °C) in PBS, 3% bovine albumin serum, 0.3% Tween 20 (blocking PBS), for p-Akt, p-Erk, p-IRS, lipocalin, PGC-1α, PPAR-α, and UCP2 detection, or in 130 mM NaCl, 20 mM Tris-HCl, pH 7.4, 0.05% Tween 20 (T-TBS) containing 5% non-fat milk (blocking TBS) for all other markers. Membranes were then incubated overnight at 4 °C with antibodies at the appropriate dilutions (see Table S1)

Membranes were washed and then incubated for 1 h with the appropriate HRP-conjugated secondary antibodies. For p-Akt, p-Erk, lipocalin, and PPAR-α, the membranes were washed and incubated at room temperature with a chemiluminescent substrate, Immobilon (Millipore Corporation, Billerica, MA 01821, USA). For all the other markers, detection was carried out using the Excellent Chemiluminescent detection Kit (ElabScience, distributed by Microtech, Naples, Italy).

Quantitative densitometry of the bands was carried out by analyzing chemidoc images or digital images of X-ray films exposed to immunostained membranes using Image Lab Software (Biorad Laboratories S.r.l., Segrate (MI)—Italy).

4.6. Statistical Analysis

Data were expressed as mean values ± SEM. The program GraphPad Prism 8 (GraphPad Software, San Diego, CA, USA) was used to perform statistical analysis by applying two-tailed, unpaired, Student's t-test or two-way ANOVA followed by Tukey post-test. $p < 0.05$ was considered significant.

5. Conclusions

Overall, the general picture of this study (summarized in Figure 12), in middle aged rodent model, shows that the responses to the HFF diet are similar in adipose tissue and brain tissue for several (UCP2, PPAR-α, PGC-1α, TNF-α, adiponectin), but not all the measured parameters (i.e., insulin resistance onset in e-WAT but not brain, or lipocalin and haptoglobin, whose level is not modified in brain or e-WAT, respectively). Whether this similarity is due to crosstalk between the two areas or they are interconnected by a third player is currently unknown and represents an important issue for further studies.

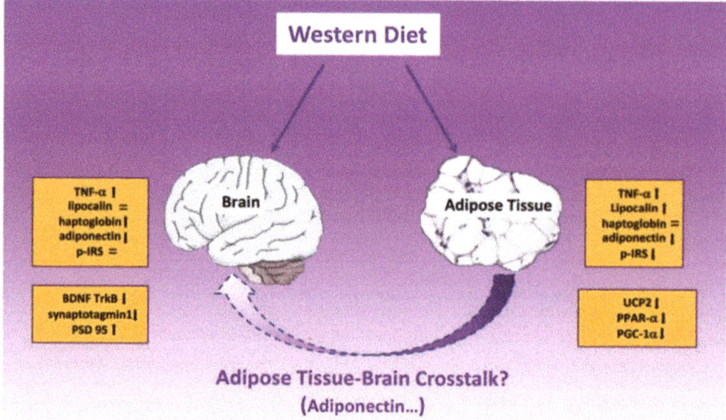

Figure 12. Summary of the modifications induced by high fat-high fructose (western) diet in brain (hippocampus and cortex) and white adipose tissue from middle-aged rats. UCP2=uncoupling protein 2 PPAR-α=peroxisome proliferator activated receptor alpha, PGC-1α=peroxisome proliferator-activated receptor gamma coactivator 1-alpha, TNF-α= tumor necrosis factor alpha, p-IRS=phosphorylated insulin receptor substrate, BDNF=brain derived neurotrophic factor, TrkB=tropomyosin receptor kinase b.

Supplementary Materials: Supplementary materials can be found at http://www.mdpi.com/1422-0067/21/3/786/s1. Table S1:List of antibodies used for western blot with provider, ref. no. and appropriate dilution.

Author Contributions: Conceptualization, S.I. and L.C.; Experiments, A.M., M.S.S., C.G., M.N., R.C. (Rosa Cancelliere), and R.C. (Raffaella Crescenzo); Analysis of data, A.M., M.S.S., C.G., M.N., R.C. (Rosa Cancelliere), R.C. (Raffaella Crescenzo), S.I., and L.C.; Writing—Original draft preparation, S.I., L.C., A.M. and M.S.S.; Writing—Review and editing, all the authors; Funding acquisition, S.I. and L.C. All authors have read and agreed to the published version of the manuscript.

Funding: This work was supported by a grant from the University of Naples Federico II—Ricerca Dip 2017 and by a FIRB—Futuro in Ricerca grant (RBFR12QW4I_004) from the Italian Ministry of Education, University and Research (MIUR).

Acknowledgments: The authors wish to thank Emilia de Santis for skillful management of animal house.

Conflicts of Interest: The authors declare no conflict of interest.

References

1. Hill, J.O.; Wyatt, H.R.; Peters, J.C. Energy balance and obesity. *Circulation* **2012**, *126*, 126–132. [CrossRef] [PubMed]
2. Luppino, F.; de Wit, L.; Bouvy, P.F.; Et, A. Overweight, obesity, and depression: A systematic review and meta-analysis of longitudinal studies. *Arch. Gen. Psychiatry* **2010**, *67*, 220–229. [CrossRef] [PubMed]
3. Gustafson, D.R. Adiposity and cognitive decline: Underlying mechanisms. *J. Alzheimers Dis.* **2012**, *30*, 203–217. [CrossRef] [PubMed]
4. Arshad, N.; Lin, T.S.; Yahaya, M.F. Metabolic syndrome and its effect on the brain: Possible mechanism. *CNS Neurol. Disord. Drug Targets* **2018**, *17*, 595–603. [CrossRef]
5. Cigliano, L.; Spagnuolo, M.S.; Crescenzo, R.; Cancelliere, R.; Iannotta, L.; Mazzoli, A.; Liverini, G.; Iossa, S. Short-term fructose feeding induces inflammation and oxidative stress in the hippocampus of young and adult rats. *Mol. Neurobiol.* **2018**, *55*, 2869–2883. [CrossRef]
6. Crescenzo, R.; Spagnuolo, M.S.; Cancelliere, R.; Iannotta, L.; Mazzoli, A.; Gatto, C.; Iossa, S.; Cigliano, L. Effect of Initial Aging and High-Fat/High-Fructose Diet on Mitochondrial Bioenergetics and Oxidative Status in Rat Brain. *Mol. Neurobiol.* **2019**, *56*, 7651–7663. [CrossRef]

7. Bomfim, T.R.; Forny-Germano, L.; Sathler, L.B.; Brito-Moreira, J.; Houzel, J.C.; Decker, H.; Silverman, M.A.; Kazi, H.; Melo, H.M.; McClean, P.L.; et al. An anti-diabetes agent protects the mouse brain from defective insulin signaling caused by Alzheimer's disease–associated Ab oligomers. *J. Clin. Invest.* **2012**, *122*, 1339–1353. [CrossRef]

8. Lourenco, M.V.; Clarke, J.R.; Frozza, R.L.; Bomfim, T.R.; Forny-Germano, L.; Batistam, A.F.; Sathler, L.B.; Brito-Moreira, J.; Amaral, O.B.; Silva, C.A.; et al. TNF-α mediates PKR-dependent memory impairment and brain IRS-1 inhibition induced by Alzheimer's b-amyloid oligomers in mice and monkeys. *Cell Metab.* **2013**, *18*, 831–843. [CrossRef]

9. De Felice, F.G.; Ferreira, S.T. Inflammation, defective insulin signaling, and mitochondrial dysfunction as common molecular denominators connecting type 2 diabetes to Alzheimer disease. *Diabetes* **2014**, *63*, 2262–2272. [CrossRef]

10. Ogden, C.L.; Carroll, M.D.; Fryar, C.D.; Flegal, K.M. Prevalence of obesity among adults and youth: United States, 2011–2014. *NCHS Data Briefs* **2015**, *219*, 1–8.

11. Kivipelto, M.; Ngandu, T.; Fratiglioni, L.; Viitanen, M.; Kåreholt, I.; Winblad, B.; Helkala, E.L.; Tuomilehto, J.; Soininen, H.; Nissinen, A. Obesity and vascular risk factors at midlife and the risk of dementia and Alzheimer disease. *Arch. Neurol.* **2005**, *62*, 1556–1560. [CrossRef] [PubMed]

12. Whitmer, R.A.; Gunderson, E.P.; Barrett-Connor, E.; Quesenberry, C.P.; Yae, K. Obesity in middle age and future risk of dementia: A 27-year longitudinal population-based study. *BMJ* **2005**, *330*, 1360. [CrossRef] [PubMed]

13. Mazzoli, A.; Crescenzo, R.; Cigliano, L.; Spagnuolo, M.S.; Cancelliere, R.; Gatto, C.; Iossa, S. Early hepatic oxidative stress and mitochondrial changes following western diet in middle aged rats. *Nutrients* **2019**, *11*, 2670. [CrossRef] [PubMed]

14. Della Corte, K.W.; Perrar, I.; Penczynski, K.J.; Schwingshackl, L.; Herder, C.; Buyken, A.E. Effect of dietary sugar intake on biomarkers of subclinical inflammation: A systematic review and meta-analysis of intervention studies. *Nutrients* **2018**, *10*, 606. [CrossRef]

15. Duan, Y.; Zeng, L.; Zheng, C.; Song, B.; Li, F.; Kong, X.; Xu, K. Inflammatory links between high fat diets and diseases. *Front Immunol.* **2018**, *9*, 2649. [CrossRef]

16. Esmaili, S.; Xu, A.; George, J. The multifaceted and controversial immunometabolic actions of adiponectin. *Trends Endocrinol. Metab.* **2014**, *25*, 444–451. [CrossRef]

17. Bloemer, J.; Pinky, P.D.; Govindarajulu, M.; Hong, H.; Judd, R.; Amin, R.H.; Moore, T.; Dhanasekaran, M.; Reed, M.N.; Suppiramaniam, V. Role of adiponectin in central nervous system disorders. *Neural. Plast.* **2018**, 4593530. [CrossRef]

18. Maffei, M.; Barone, I.; Scabia, G.; Santini, F. The multifaceted haptoglobin in the context of adipose tissue and metabolism. *Endocr. Rev.* **2016**, *37*, 403–416. [CrossRef]

19. Zhang, J.; Wu, Y.; Zhang, Y.; LeRoith, D.; Bernlohr, D.A.; Chen, X. The role of lipocalin 2 in the regulation of inflammation in adipocytes and macrophages. *Mol. Endocrinol.* **2008**, *22*, 1416–1426. [CrossRef]

20. Kowiański, P.; Lietzau, G.; Czuba, E.; Waśkow, M.; Steliga, A.; Moryś, J. BDNF: A key factor with multipotent impact on brain signaling and synaptic plasticity. *Cell Mol. Neurobiol.* **2018**, *38*, 579–593. [CrossRef]

21. Cesca, F.; Baldelli, P.; Valtorta, F.; Benfenati, F. The synapsins: Key actors of synapse function and plasticity. *Prog. Neurobiol.* **2010**, *91*, 313–348. [CrossRef] [PubMed]

22. Courtney, N.A.; Bao, H.; Briguglio, J.S.; Chapman, E.R. Synaptotagmin 1 clamps synaptic vesicle fusion in mammalian neurons independent of complexin. *Nat. Commun.* **2019**, *10*, 4076. [CrossRef] [PubMed]

23. Chen, X.; Nelson, C.D.; Li, X.; Winters, C.A.; Azzam, R.; Sousa, A.A.; Leapman, R.D.; Gainer, H.; Sheng, M.; Reese, T.S. PSD-95 is required to sustain the molecular organization of the postsynaptic density. *J. Neurosci.* **2011**, *31*, 6329–6338. [CrossRef] [PubMed]

24. Anstey, K.J.; Cherbuin, N.; Budge, M.; Young, J. Body mass index in midlife and late life as a risk factor for dementia: A meta-analysis of prospective studies. *Obes. Rev.* **2011**, *12*, e426–e437. [CrossRef]

25. Jauch-Chara, K.; Oltmanns, K.M. Obesity a neuropsychological disease? Systematic review and neuropsychological model. *Prog. Neurobiol.* **2014**, *114*, 4–101. [CrossRef]

26. Xu, W.L.; Atti, A.R.; Gatz, M.; Pedersen, N.L.; Johansson, B.; Fratiglioni, L. Midlife overweight and obesity increase late-life dementia risk: A population-based twin study. *Neurology* **2011**, *76*, 1568–1574. [CrossRef]

27. Dandona, P.; Aljada, A.; Chaudhuri, A.; Mohanty, P.; Garg, R. Metabolic syndrome: A comprehensive perspective based on interactions between obesity, diabetes, and inflammation. *Circulation* **2005**, *111*, 1448–1454. [CrossRef]

28. Bhat, N.R. Linking cardiometabolic disorders to sporadic Alzheimer's disease: A perspective on potential mechanisms and mediators. *J. Neurochem.* **2010**, *115*, 551–562. [CrossRef]

29. Parimisetty, A.; Dorsemans, A.C.; Awada, R.; Ravanan, P.; Diotel, N.; Lefebvre d'Hellencourt, C. Secret talk between adipose tissue and central nervous system via secreted factors—an emerging frontier in the neurodegenerative research. *J. Neuroinflamm.* **2016**, *13*, 67. [CrossRef]

30. Huffman, D.M.; Barzilai, N. Role of visceral adipose tissue in aging. *Biochim. Biophys. Acta Gen. Subj.* **2009**, *1790*, 1117–1123. [CrossRef]

31. Sutinen, E.M.; Pirttilä, T.; Anderson, G.; Salminen, A.; Ojala, J.O. Pro-inflammatory interleukin-18 increases Alzheimer's disease associated amyloid-β production in human neuron-like cells. *J. Neuroinflamm.* **2012**, *9*, 1–14. [CrossRef] [PubMed]

32. Chusyd, D.E.; Wang, D.; Huffman, D.M.; Nagy, T.R. Relationships between rodent white adipose fat pads and human white adipose fat depots. *Front. Nutr.* **2016**, *3*, 10. [CrossRef] [PubMed]

33. Guo, H.; Bazuine, M.; Jin, D.; Huang, M.M.; Cushman, S.W.; Chen, X. Evidence for the regulatory role of lipocalin 2 in high-fat diet-induced adipose tissue remodeling in male mice. *Endocrinology* **2013**, *54*, 3525–3538. [CrossRef] [PubMed]

34. Takahashi, H.; Sanada, K.; Nagai, H.; Li, Y.; Aoki, Y.; Ara, T.; Seno, S.; Matsuda, H.; Yu, R.; Kawada, T.; et al. Over-expression of PPARα in obese mice adipose tissue improves insulin sensitivity. *Biochem. Biophys. Res. Commun.* **2017**, *493*, 108–114. [CrossRef] [PubMed]

35. Cheng, C.F.; Ku, H.C.; Lin, H. PGC-1α as a pivotal factor in lipid and metabolic regulation. *Int. J. Mol. Sci.* **2018**, *19*, 3447. [CrossRef]

36. Chevillotte, E.; Giralt, M.; Miroux, B.; Ricquier, D.; Villarroya, F. Uncoupling protein-2 controls adiponectin gene expression in adipose tissue through the modulation of reactive oxygen species production. *Diabetes* **2007**, *56*, 1042–1050. [CrossRef]

37. Shoelson, S.E.; Herrero, L.; Naaz, A. Obesity, Inflammation, and Insulin Resistance. *Gastroenterol* **2007**, *132*, 2169–2180. [CrossRef]

38. Yan, Q.W.; Yang, Q.; Mody, N.; Graham, T.E.; Hsu, C.H.; Xu, Z.; Houstis, N.E.; Kahn, B.B.; Rosen, E.D. The Adipokine Lipocalin 2 Is Regulated by Obesity and Promotes Insulin Resistance. *Diabetes* **2007**, *56*, 2533–2540. [CrossRef]

39. Salvatore, A.; Cigliano, L.; Bucci, E.M.; Corpillo, D.; Velasco, S.; Carlucci, A.; Pedone, C.; Abrescia, P. Haptoglobin binding to apolipoprotein A-I prevents damage from hydroxyl radicals on its stimulatory activity of the enzyme lecithin-cholesterol acyl-transferase. *Biochemistry* **2007**, *46*, 11158–11168. [CrossRef]

40. Maresca, B.; Spagnuolo, M.S.; Cigliano, L. Haptoglobin modulates beta-amyloid uptake by U-87MG astrocytes cell line. *J. Mol. Neurosci.* **2015**, *56*, 35–47. [CrossRef]

41. Zhang, J.Z.; Jing, L.; Guo, F.Y.; Ma, Y.; Wang, Y.L. Inhibitory effect of ketamine on phosphorylation of the extracellular signal-regulated kinase 1/2 following brain ischemia and reperfusion in rats with hyperglycemia. *Exp. Toxicol. Pathol.* **2007**, *59*, 227–235. [CrossRef] [PubMed]

42. Frasca, G.; Carbonaro, V.; Merlo, S.; Copani, A.; Sortino, M.A. Integrins mediate beta-amyloid-induced cell-cycle activation and neuronal death. *J. Neurosci. Res.* **2008**, *86*, 350–355. [CrossRef] [PubMed]

43. Goto, T.; Lee, J.Y.; Teraminami, A.; Kim, Y.I.; Hirai, S.; Uemura, T.; Inoue, H.; Takahashi, N.; Kawada, T.J. Activation of peroxisome proliferator-activated receptor-alpha stimulates both differentiation and fatty acid oxidation in adipocytes. *Lipid Res.* **2011**, *52*, 873–884. [CrossRef]

44. Thundyil, J.; Pavlovski, D.; Sobey, C.G.; Arumugam, T.V. Adiponectin receptor signalling in the brain. *Br. J. Pharm.* **2012**, *165*, 313–327. [CrossRef] [PubMed]

45. Wilkinson, M.; Brown, R.; Imran, S.A.; Ur, E. Adipokine gene expression in brain and pituitary gland. *Neuroendocrinology* **2007**, *86*, 191–209. [CrossRef] [PubMed]

46. Zhang, D.; Wang, X.; Lu, X.Y. Adiponectin exerts neurotrophic effects on dendritic arborization, spinogenesis, and neurogenesis of the dentate gyrus of male mice. *Endocrinology* **2016**, *157*, 2853–2869. [CrossRef] [PubMed]

47. Zhang, D.; Guo, M.; Zhang, W.; Lu, X.Y. Adiponectin stimulates proliferation of adult hippocampal neural stem/progenitor cells through activation of p38 mitogen-activated protein kinase (p38MAPK)/glycogen synthase kinase 3β (GSK-3β)/β-catenin signaling cascade. *J. Biol. Chem.* **2011**, *286*, 44913–44920. [CrossRef]

48. Yau, S.Y.; Li, A.; Hoo, R.L.; Ching, Y.P.; Christie, B.R.; Lee, T.M.; Xu, A.; So, K.F. Physical exercise-induced hippocampal neurogenesis and antidepressant effects are mediated by the adipocyte hormone adiponectin. *Proc. Natl. Acad Sci. USA* **2014**, *111*, 15810–15815. [CrossRef]

49. Letra, L.; Rodrigues, T.; Matafome, P.; Santana, I.; Seiça, R. Adiponectin and sporadic Alzheimer's disease: Clinical and molecular links. *Front Neuroendocr.* **2017**, *52*, 1–11. [CrossRef]

50. Chan, K.H.; Lam, K.S.; Cheng, O.Y.; Kwan, J.S.; Ho, P.W.; Cheng, K.K.; Chung, S.K.; Ho, J.W.; Guo, V.Y.; Xu, A. Adiponectin is protective against oxidative stress induced cytotoxicity in amyloid-beta neurotoxicity. *PLoS ONE* **2012**, *7*, e52354. [CrossRef]

51. Scharfman, H.; Goodman, J.; Macleod, A.; Phani, S.; Antonelli, C.; Croll, S. Increased neurogenesis and the ectopic granule cells after intrahippocampal BDNF infusion in adult rats. *Exp. Neurol.* **2005**, *192*, 348–356. [CrossRef] [PubMed]

52. Leal, G.; Afonso, P.M.; Salazara, I.L.; Duarte, C.B. Regulation of hippocampal synaptic plasticity by BDNF. *Brain Res.* **2015**, *1621*, 82–101. [CrossRef] [PubMed]

53. Stranahan, A.M.; Norman, E.D.; Lee, K.; Cutler, R.G.; Telljohann, R.S.; Egan, J.M.; Mattson, M.P. Diet-induced insulin resistance impairs hippocampal synaptic plasticity and cognition in middle-aged rats. *Hippocampus* **2008**, *18*, 1085–1088. [CrossRef] [PubMed]

54. Hussain, Y.; Jain, S.K.; Samaiya, P.K. Short-term westernized (HFFD) diet fed in adolescent rats: Effect on glucose homeostasis, hippocampal insulin signaling, apoptosis and related cognitive and recognition memory function. *Behav. Brain Res.* **2019**, *361*, 113–121. [CrossRef] [PubMed]

55. Spagnuolo, M.S.; Bergamo, P.; Crescenzo, R.; Iannotta, L.; Treppiccione, L.; Iossa, S.; Cigliano, L. Brain Nrf2 pathway, autophagy, and synaptic function proteins are modulated by a short-term fructose feeding in young and adult rats. *Nutr. Neurosci.* **2018**. [CrossRef] [PubMed]

56. Cacho, J.; Sevillano, J.; de Castro, J.; Herrera, E.; Ramos, M.P. Validation of simple indexes to assess insulin sensitivity during pregnancy in Wistar and Sprague-Dawley rats. *Am. J. Physiol. Endocrinol. Metab.* **2008**, *295*, E1269–E1276. [CrossRef]

57. Spagnuolo, M.S.; Mollica, M.P.; Maresca, B.; Cavaliere, G.; Cefaliello, C.; Trinchese, G.; Scudiero, R.; Crispino, M.; Cigliano, L. High Fat Diet and Inflammation—Modulation of Haptoglobin Level in Rat Brain. *Front Cell. Neurosci.* **2015**, *9*, 479. [CrossRef]

58. Spagnuolo, M.S.; Maresca, B.; Mollica, M.P.; Cavaliere, G.; Cefaliello, C.; Trinchese, G.; Esposito, M.G.; Scudiero, R.; Crispino, M.; Abrescia, P.; et al. Haptoglobin increases with age in rat hippocampus and modulates Apolipoprotein E mediated cholesterol trafficking in neuroblastoma cell lines. *Front. Cell. Neurosci.* **2014**, *8*, 212. [CrossRef]

International Journal of
Molecular Sciences

Article

High-Cholesterol Diet Decreases the Level of Phosphatidylinositol 4,5-Bisphosphate by Enhancing the Expression of Phospholipase C (PLCβ1) in Rat Brain

Yoon Sun Chun and Sungkwon Chung *

Department of Physiology, Sungkyunkwan University School of Medicine, Suwon, Gyeonggi-do 16419, Korea;
ysun129@skku.edu
* Correspondence: schung@skku.edu; Tel.: +82-31-299-6103

Received: 17 December 2019; Accepted: 8 February 2020; Published: 10 February 2020

Abstract: Cholesterol is a critical component of eukaryotic membranes, where it contributes to regulating transmembrane signaling, cell–cell interaction, and ion transport. Dysregulation of cholesterol levels in the brain may induce neurodegenerative diseases, such as Alzheimer's disease, Parkinson disease, and Huntington disease. We previously reported that augmenting membrane cholesterol level regulates ion channels by decreasing the level of phosphatidylinositol 4,5-bisphosphate (PIP$_2$), which is closely related to β-amyloid (Aβ) production. In addition, cholesterol enrichment decreased PIP$_2$ levels by increasing the expression of the β1 isoform of phospholipase C (PLC) in cultured cells. In this study, we examined the effect of a high-cholesterol diet on phospholipase C (PLCβ1) expression and PIP$_2$ levels in rat brain. PIP$_2$ levels were decreased in the cerebral cortex in rats on a high-cholesterol diet. Levels of PLCβ1 expression correlated with PIP$_2$ levels. However, cholesterol and PIP$_2$ levels were not correlated, suggesting that PIP$_2$ level is regulated by cholesterol via PLCβ1 expression in the brain. Thus, there exists cross talk between cholesterol and PIP$_2$ that could contribute to the pathogenesis of neurodegenerative diseases.

Keywords: phosphatidylinositol 4,5-bisphosphate; phospholipase C; cholesterol; high-cholesterol diet

1. Introduction

Lipid metabolism including cholesterol, oxysterols, fatty acids, and phospholipids is involved in numerous neurodegenerative diseases including Alzheimer's disease [1,2]. Cholesterol is a key component of plasma membrane bilayers, where it affects structural and physical functions including fluidity, curvature, and stiffness. It also regulates the functions of membrane proteins and is involved in transmembrane signaling processes, membrane trafficking, endocytosis, and ion transport [3,4]. Cholesterol also participates in the biosynthesis of bile acids and steroid hormones in the plasma membrane, which in turn play important functional and biological roles as signal transducers [5]. Cholesterol is particularly enriched in the brain, and the central nervous system contains 23% of all cholesterol in the whole human body. Cholesterol in neurons and astrocytes controls synaptic transmission [6]. Thus, changes in cholesterol levels and homeostasis have been associated with brain diseases and neurodegenerative disorders such as Alzheimer's disease (AD), Parkinson's disease (PD), and Huntington's disease [7].

Phosphatidylinositol 4,5-bisphosphate (PIP$_2$) plays a pivotal role in cell membranes, regulating biological functions including signal transduction, membrane trafficking, transporter functions, and ion channels [8,9]. Phospholipase C (PLC), which is grouped into four major families (β, γ, δ, and ε), hydrolyzes PIP$_2$ to produce inositol 1,4,5-trisphosphate and diacylglycerol [10]. These products induce calcium release from intracellular stores and activate phosphorylation of cAMP response element-binding protein and neuronal gene expression in the brain [11,12]. Decrease of PIP_2 in the cell membrane results in impaired long-term potentiation and reduced cognition [13]. Levels of PIP_2 are closely related to the production of β-amyloid peptide (Aβ), a culprit in AD [14]. Many ion channels are regulated by PIP_2, and the hydrolysis of PIP_2 by phospholipase C has been shown to reduce their activities [8]. We demonstrated that augmenting membrane cholesterol levels regulated ion channels by decreasing PIP_2 [15], suggesting that a close relationship between plasma membrane-enriched lipids, cholesterol, and PIP_2. The human ether-a-go-go related gene (HERG) K^+ channel, which is known to be modulated by PIP_2, is inhibited by cholesterol enrichment via increased phospholipase C (PLCβ1) expression [16]. Consistent with this, we showed that increasing cholesterol levels in cultured cells increased PLCβ1 and PLCβ3 expression levels among PLC isoforms (β1, β2, β3, β4, γ1, and γ2) and increased PLCβ1 expression induced the decrease of PIP_2 levels [17]. These results suggest that there may exist cross talk among two plasma membrane-enriched lipids, cholesterol and PIP_2, via expression of PLCβ1. However, these results obtained from cultured cells are not confirmed in the brain yet.

In this study, we tested whether cholesterol enrichment affected PLCβ1 expression and PIP_2 level in brain. We found that PIP_2 levels decreased significantly in the cerebral cortices of rats on a high-cholesterol diet, repeating our previous in vitro result [15]. The high-cholesterol diet slightly increased expression of PLCβ1 but not that of PLCβ3. Levels of PLCβ1 expression correlated with those of PIP_2, whereas levels of cholesterol and PIP_2 did not. These results could suggest that PIP_2 levels are regulated by cholesterol via PLCβ1 expression in the brain.

2. Results

2.1. PIP$_2$ Levels Were Down-Regulated in Rats on a High-Cholesterol Diet

We previously found that enriching membrane cholesterol decreased PIP_2 in cultured cells [14,16]. In order to test whether cholesterol augmentation in the brain affects levels of PIP_2 in vivo, 13-week-old Sprague Dawley (SD) rats were fed with either normal or high-cholesterol diets for 6 weeks [18–21]. We observed that the body weight was not altered in high-cholesterol diet rats compared to normal diet rats (data not shown). First, we tested the levels of free cholesterol in the cerebral cortex. For this purpose, we obtained the membrane fractions from the cerebral cortices and measured the levels of free membrane cholesterol using assay kits. As shown in Figure 1A, the free cholesterol level was 390 ± 29 μg/mL ($n = 10$) in the rats on the normal diet, whereas it increased by 15% to 450 ± 44 μg/mL ($n = 10$) in the rats on the high-cholesterol diet. Cholesterol levels was increased by 15.4% by high-cholesterol diet. However, the difference between the two groups was not statistically significant ($p = 0.24$). This was likely owing to the varying cholesterol levels within each group as shown by scattered data in Figure 1A. Alternatively, the non-significant difference could have been because of the existence of different pools of cholesterol in the central nervous system. We next analyzed the PIP_2 levels in the cerebral cortex membranes using assay kits. As shown in Figure 1B, PIP_2 levels were 5.81 ± 0.34 pM ($n = 10$) in the rats on the normal diet, but they significantly decreased by 12.9% to 5.06 ± 0.26 pM ($n = 10$) in the rats on the high-cholesterol diet. These data indicate that PIP_2 was down-regulated by cholesterol augmentation in vivo.

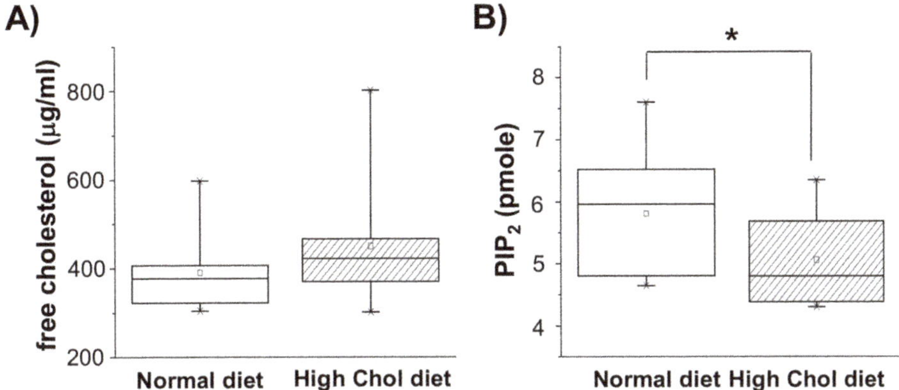

Figure 1. High-cholesterol diet decreased phosphatidylinositol 4,5-bisphosphate (PIP_2) levels in the rat cerebral cortex. Starting at 13 weeks of age, 10 male rats were placed on a high-cholesterol diet and 10 male rats were placed on a normal diet for 6 weeks as described in Materials and Methods. Cerebral cortex was removed and membrane and cytosol fractions were obtained. (**A**) The levels of free cholesterol in the membrane fractions were measured by cholesterol assay kit. Box plots show average cholesterol level in normal diet (open bar) or high-cholesterol diet (hatched bar) groups. The horizontal black lines represent the median of each distribution and the squares indicate means ($n = 10$). (**B**) PIP_2 levels in the membrane fractions were measured by using PIP_2 ELISA kit. The levels of PIP_2 were decreased by a high-cholesterol diet ($n = 10$). * $p < 0.05$.

2.2. PLCβ1 Expression Increased in Rats on a High-Cholesterol Diet

The hydrolysis of PIP_2 mediated by PLC is the major catabolic pathway for PIP_2 [10]. Because we found that increased cholesterol levels led to increased PLCβ1 expression in the cultured cells [17], we examined whether high-cholesterol diet in rats affected expressions of PLCβ1 and PLCβ3. Levels of PLC expressions were measured from the cytosol and membrane fractions of the cerebral cortices using Western blot analysis. A typical Western blot for PLCβ1 and PLCβ3 is shown in Figure 2A. Full images of Western blot are shown in Supplementary Figures S1 and S2. All Western blot bands for PLCβ1 in membrane and cytosol fractions are shown in Supplementary Figure S3. We confirmed equal protein loading by β-tubulin levels, and the relative band densities of PLCβ1 and PLCβ3 compared with β-tubulin are shown in Figure 2B, C ($n = 10$). PLCβ1 expression slightly increased in membrane fractions from rats on the high-cholesterol diet although the effect was not statistically significant (Figure 2B; $p = 0.512$). The high-cholesterol diet did not affect the PLCβ1 expression in the cytosol fraction, and the PLCβ3 expression in both the cytosol and membrane fractions showed no differences between the two groups (Figure 2C).

2.3. PLCβ1 Expression Directly Correlated with PIP_2 Levels

We compared the PLCβ1 and PLCβ3 expression along with PIP_2 levels from individual rats. We hypothesized that if PLC expression directly affected PIP_2, these two would correlate. As shown in Figure 3A, we compared PLCβ1 expression (shown as relative to tubulin expression) with PIP_2 levels. There was a moderate correlation between them (RSQ, 0.297) from rat brains on the normal diet (open symbols, and a thin line). Interestingly, the PLCβ1 expression and PIP_2 levels were more tightly correlated in the rats on the high-cholesterol diet (closed symbols and a thick line; RSQ, 0.345). These results suggested that even in the normal diet rats, PIP_2 level in the brain correlated closely with PLCβ1 expression and that when cholesterol levels increased in high-cholesterol diet rats, the correlation and significance increased further. In the case of PLCβ3, however, there was no correlation between its expression levels and PIP_2 levels in either the normal (open symbols; Figure 3B) or the high-cholesterol (closed symbols; Figure 3B) diet rats. Consistent with this, we previously showed that increasing cholesterol levels in cultured cells decreased PIP_2 levels by increasing PLCβ1 expression, but not by increasing PLCβ3 expression [17].

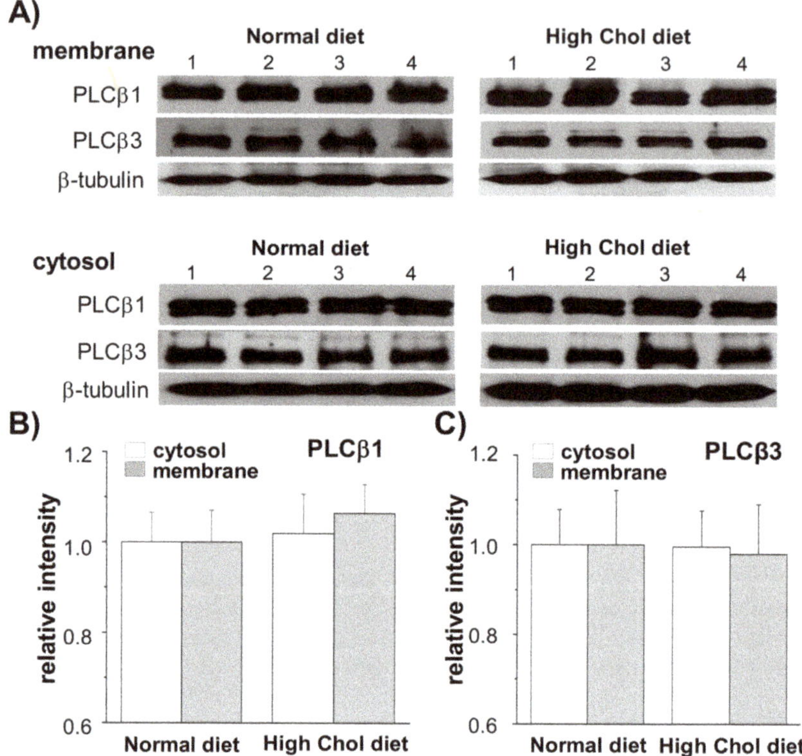

Figure 2. High-cholesterol diet slightly increased phospholipase C (PLCβ1) expression levels in the rat cerebral cortex. The expression levels of PLCβ1 and PLCβ3 were measured from the cytosol and membrane fractions of cerebral cortex using Western blot analysis. Representative Western blots are shown for PLCβ1 and PLCβ3. β-tubulin was used to confirm the amount of proteins loaded. (**A**) Expression levels of PLCβ1 and PLCβ3 were compared. (**B,C**) Bars indicate the levels of PLCβ1 (**B**) and PLCβ3 (**C**) obtained from densitometric analysis of Western bands in (**A**) (*n* = 10).

We next compared PIP_2 levels along with cholesterol levels. As shown in Figure 3C, there was no correlation between PIP_2 and cholesterol levels in either the normal (open symbols) or the high-cholesterol (closed symbols) diet rats, indicating that cholesterol level may not directly affect PIP_2. Together, these results may indicate that the increased cholesterol level in the brain slightly increased PLCβ1 expression, reducing levels of PIP_2. The effect of the high-cholesterol diet on the expression levels was specific for PLCβ1 given that we found no correlation between levels of PLCβ3 and PIP_2.

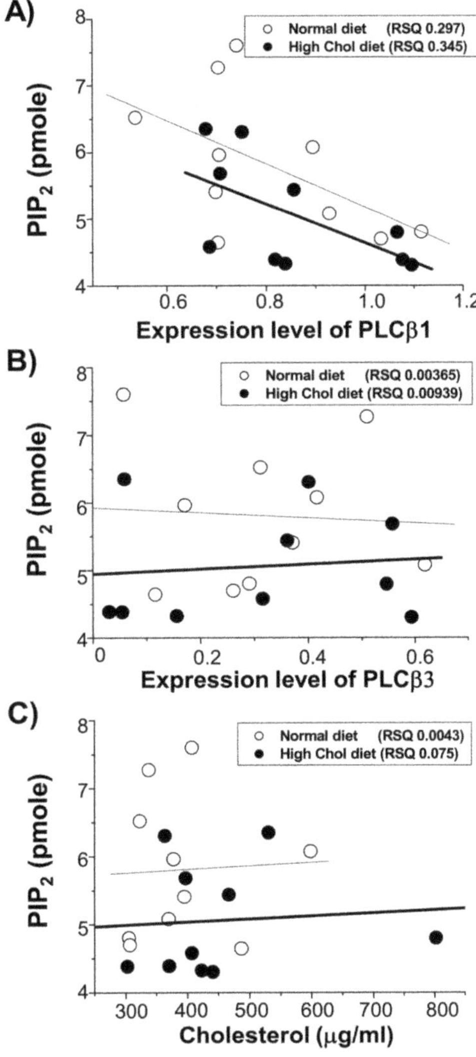

Figure 3. PIP$_2$ levels correlated with the PLCβ1 expression levels in the rat cerebral cortex. PIP$_2$ levels, PLCβ1 expression levels, PLCβ3 expression levels, and cholesterol levels from Figures 1 and 2 were re-plotted. Linear regressions were performed to obtain RSQ values from rats on normal diet (open symbols and a thin line), and from rats on a high-cholesterol diet (closed symbols and thick line). (**A**) PIP$_2$ levels correlated with the PLCβ1 expression levels. (**B**) There was no correlation between PIP$_2$ levels and PLCβ3 expression levels. (**C**) There was no correlation between PIP$_2$ levels and cholesterol levels.

3. Discussion

In this study, we demonstrated that a high-cholesterol diet significantly decreased PIP$_2$ levels and slightly increased PLCβ1 expression. The correlation between PIP$_2$ level and PLCβ1 expression further increased in high-cholesterol diet rats. In the high-cholesterol diet rats, cholesterol increased by 15% compared with the rats on normal diet, although the difference was not statistically significant, likely because of the varying cholesterol levels in each group. These varying levels may also explain why there was no significant effect of the high-cholesterol diet on the PLCβ1 expression. Importantly, however, there existed a correlation between the expression and PIP$_2$ levels even in the normal diet rats, and the correlation increased further in the high-cholesterol diet rats, indicating that the increased PLCβ1 expression by the high-cholesterol diet might have resulted in PIP$_2$ hydrolysis. These results suggest that there exist relationships among cholesterol, PIP$_2$, and PLCβ1 in the brain. We tested the effect of dietary cholesterol on levels of PIP$_2$ and PLCβ1 in healthy young male rat. However, it was shown that different physiological conditions such as gender and age could affect a dysregulation in brain cholesterol metabolism [22,23]. In addition, cholesterol concentration in serum significantly differed in male and female rats fed the cholesterol-containing diet [24]. Thus, further experiments will be needed to compare the effect of a high-cholesterol diet on PLCβ1 expression and PIP$_2$ levels by gender and age.

PIP$_2$ is a minor lipid component in the plasma membrane, localized to the cytoplasmic leaflet of the phospholipid bilayer [25]. It plays important roles in the attachment of the cytoskeleton to the plasma membrane, endocytosis, membrane trafficking, and enzyme activation [9]. A previous report suggests that there exist different pools of PIP$_2$ in the cell membrane to play its multiple roles [26]. PIP$_2$ levels are determined through their production by phosphoinositide kinases and through their breakdown by phosphatases and phosphoinositide-specific PLC. Our results showed that PIP$_2$ was closely related to PLCβ1 expression but there was no correlation with PLCβ3 expression, and these results are consistent with our previous report on cultured cells [17]. Taken together, we concluded that cholesterol enrichment specifically increased the expression of PLCβ1, resulting in the down-regulation of PIP$_2$. Lipid raft may serve as a platform for the specific regulation of PLCβ1 expression by cholesterol enrichment, since it is known as a microdomain enriched with cholesterol [27,28]. PIP$_2$ is shown to localize in lipid rafts [29–31]. Interestingly, PLCβ1 is also localized in lipid raft microdomains, when prepared from the synaptic plasma membrane fraction of rat brains [32]. However, further studies will be needed to clarify how cholesterol regulates PLCβ1PLCβ1 expression.

Cholesterol is essential for structural and physiological functions of neurons [4]. It has been suggested that changes in cholesterol homeostasis induce synaptic degeneration and the disruption of brain functions, contributing to neurodegenerative diseases [33]. Some studies showed that a high plasma cholesterol level increases the risk of developing PD [34], and the decrease of plasma cholesterol by statin might attenuate the deposition of α-synuclein in the brain [35]. The association of cholesterol with AD is one of the most studied topics [36]. Cholesterol levels are closely connected to the production of Aβ [37–39]. Also, cholesterol accumulates in senile plaques of AD patients and in transgenic APP(SW) mice [40]. Hypercholesterolemia has been shown to increase Aβ deposition and amyloid plaque formation in a transgenic mouse model [21]. Consistently, cholesterol synthesis inhibitors, statins, decrease Aβ production [41]. Since levels of PIP$_2$ are closely related to the production of Aβ [14], PIP$_2$ may serve as the important molecule that links cholesterol to the pathogenesis of neurodegenerative diseases such as AD.

4. Materials and Methods

4.1. High-Cholesterol Diet

Starting at 13 weeks of age, 10 male SD rats were placed on a high-cholesterol diet containing 5% cholesterol, 10% fat, and 2% sodium cholate for 6 weeks. A total of 10 animals were also placed on a normal diet containing 0.005% cholesterol and 10% fat for 6 weeks. The cerebral cortices were removed and immediately frozen and stored at −80 °C. All experiments on rats were carried out in accordance with the approved animal care and use guidelines of the Laboratory Animal Research Center in Sungkyunkwan University School of Medicine and all experimental protocols were approved by the Laboratory Animal Research Center in Sungkyunkwan University School of Medicine (#skkumed10-05, 10 January 2010).

4.2. Protein Extraction

The cerebral cortices were homogenized in Tris-buffered saline solution (20 mM Tris, 137 mM NaCl, pH 7.4) containing a protease inhibitor cocktail, and the extraction ratio (brain tissue: Tris-buffered saline) was 1:10 (*w/v*). Homogenate samples were sonicated for 1 min on ice and centrifuged at $1000\times g$ for 10 min at 4 °C to remove nuclei and debris. Supernatants were separated into membranes (pellet) and cytosols (supernatant) by centrifugation at $100,000\times g$ for 1 h at 4 °C. The pellet was lysed with Ripa buffer (25 mM Tris, 5 mM EDTA, 137 mM NaCl, 1% Triton X-100, 1% sodium deoxycholate, 0.1% SDS with a protease inhibitor cocktail, pH 7.4). The protein content was measured by Bradford assay (Bio-Rad Laboratories, Hercules, CA, USA).

4.3. Cholesterol Assay

We measured cholesterol using the Amplex Red cholesterol assay kit (Thermo Fisher Scientific, Halethorpe, CA, USA). Homogenate samples were centrifuged at $1000\times g$ for 10 min to remove nuclei and cell debris. Membranes were pelleted from the supernatants by centrifugation for 1 h at $100,000\times g$ at 4 °C and analyzed according to the supplier's instructions.

4.4. Western Blot Analysis

We resolved protein from each sample on SDS-PAGE and transferred the resolved protein to nitrocellulose membrane. Membranes were blocked with 5% nonfat milk powder in Tris-buffered saline/Tween 20 (TBST) for 1 h at room temperature, followed by incubation with anti-PLCβ1 (Santa Cruz Biotechnologies, Dallas, TX, USA), anti-PLCβ3 (Santa Cruz Biotechnologies), and anti-β-tubulin (Sigma-Aldrich, St. Louis, MO, USA) antibodies for overnight at 4 °C; the dilutions were 1:500 for the PLC isozymes and 1:4000 for β-tubulin. After being washed with TBST, the membranes were incubated with horseradish peroxidase-conjugated goat anti-rabbit IgG (Invitrogen, Waltham, MA, USA) for 1 h at room temperature. We visualized the peroxidase activity with enhanced chemiluminescence and quantified the detected signals using the Fujifilm LAS-3000 system with Multi Gauge software (Tokyo, Japan).

4.5. PIP$_2$ Assay

We measured the amount of PIP$_2$ extracted from the membrane fractions of the homogenate using a PIP$_2$ Mass ELISA kit (Echelon Biosciences Inc., Salt Lake City, CA, USA). We extracted the PIP$_2$ from the normal and high-cholesterol diet groups according to the supplier's instructions. We also estimated the cellular PIP$_2$ quantities by comparing the values from the standard curve, which showed linear relationships at concentrations ranging from 0.5 to 1000 pM.

4.6. Statistical Analysis

Data were expressed as mean ± SEM. We performed statistical comparisons between controls and treated experimental groups using one-way ANOVA and considered $p < 0.05$ statistically significant. Box plot graphs was used to show the distribution of the observed data variation, which displayed as minimum to maximum value together with distribution around the median and 25th and 75th percentile as edges.

Supplementary Materials: Supplementary materials can be found at http://www.mdpi.com/1422-0067/21/3/1161/s1.

Author Contributions: S.C. and Y.S.C. designed the study and wrote the manuscript. Y.S.C. performed all experiments. All authors have read and agreed to the published version of the manuscript.

Funding: This work was supported by the Basic Science Research Program through the National Research Foundation of Korea funded by the Ministry of Education, Science and Technology (2016R1D1A1A099) to S.C.

Conflicts of Interest: The authors declare no conflict of interest.

References

1. Zarroukeg, A.; Debbabi, M.; Bezine, M.; Karym, E.M.; Badreddine, A.; Rouaud, O.; Moreau, T.; Cherkaoui-Malki, M.; El Ayeb, M.; Nasser, B.; et al. Lipid Biomarkers in Alzheimer's Disease. *Curr. Alzheimer Res.* **2018**, *15*, 303–312. [CrossRef] [PubMed]
2. Mesa-Herrera, F.; Taoro-González, L.; Valdés-Baizabal, C.; Diaz, M.; Marín, R. Lipid and Lipid Raft Alteration in Aging and Neurodegenerative Diseases: A Window for the Development of New Biomarkers. *Int. J. Mol. Sci.* **2019**, *20*, 3810. [CrossRef] [PubMed]
3. Cerqueira, N.M.; Oliveira, E.F.; Gesto, D.S.; Santos-Martins, D.; Moreira, C.; Moorthy, H.N.; Ramos, M.J.; Fernandes, P.A. Cholesterol biosynthesis: A mechanistic overview. *Biochemistry* **2016**, *55*, 5483–5506. [CrossRef] [PubMed]
4. Pfrieger, F.W. Cholesterol homeostasis and function in neurons of the central nervous system. *Cell. Mol. Life Sci.* **2003**, *60*, 1158–1171. [CrossRef]
5. Van der Kant, R.; Zondervan, I.; Janssen, L.; Neefjes, J. Cholesterol-binding molecules MLN64 and ORP1L mark distinct late endosomes with transporters ABCA3 and NPC1. *J. Lipid Res.* **2013**, *54*, 2153–2165. [CrossRef] [PubMed]
6. Dietschy, J.M.; Turley, S.D. Thematic review series: Brain Lipids. Cholesterol metabolism in the central nervous system during early development and in the mature animal. *Lipid Res.* **2004**, *45*, 1375–1397. [CrossRef]
7. Arenas, F.; Garcia-Ruiz, C.; Fernandez-Checa, J.C. Intracellular Cholesterol Trafficking and Impact in Neurodegeneration. *Front. Mol. Neurosci.* **2017**, *10*, 382. [CrossRef] [PubMed]
8. Suh, B.C.; Hille, B. Regulation of ion channels by phosphatidylinositol 4,5-bisphosphate. *Curr. Opin. Neurobiol.* **2005**, *15*, 370–378. [CrossRef]
9. Di Paolo, G.; De Camilli, P. Phosphoinositides in cell regulation and membrane dynamics. *Nature* **2006**, *443*, 651–657. [CrossRef]
10. Rebecchi, M.J.; Pentyala, S.N. Structure, function, and control of phosphoinositide-specific phospholipase C. *Physiol. Rev.* **2000**, *80*, 1291–1335. [CrossRef]
11. Fitzjohn, S.M.; Collingridge, G.L. Calcium stores and synaptic plasticity. *Cell Calcium* **2002**, *32*, 405–411. [CrossRef] [PubMed]
12. West, A.E.; Chen, W.G.; Dalva, M.B.; Dolmetsch, R.E.; Kornhauser, J.M.; Shaywitz, A.J.; Takasu, M.A.; Tao, X.; Greenberg, M.E. Calcium regulation of neuronal gene expression. *Proc. Natl. Acad. Sci. USA* **2001**, *98*, 11024–11031. [CrossRef] [PubMed]
13. Trovò, L.; Ahmed, T.; Callaerts-Vegh, Z.; Buzzi, A.; Bagni, C.; Chuah, M.; Vandendriessche, T.; D'Hooge, R.; Balschun, D.; Dotti, C.G. Low hippocampal PI(4,5)P$_2$ contributes to reduced cognition in old mice as a result of loss of MARCKS. *Nat. Neurosci.* **2013**, *16*, 449–455. [CrossRef] [PubMed]
14. Landman, N.; Jeong, S.Y.; Shin, S.Y.; Voronov, S.V.; Serban, G.; Kang, M.S.; Park, M.K.; Di Paolo, G.; Chung, S.; Kim, T.-W. Presenilin mutations linked to familial Alzheimer's disease cause an imbalance in phosphatidylinositol 4,5-bisphosphate metabolism. *Proc. Natl. Acad. Sci. USA* **2006**, *103*, 9524–19529. [CrossRef]
15. Chun, Y.S.; Shin, S.; Kim, Y.; Cho, H.; Park, M.K.; Kim, T.-W.; Voronov, S.V.; Di Paolo, G.; Suh, B.C.; Chung, S. Cholesterol modulates ion channels via down-regulation of phosphatidylinositol 4,5-bisphosphate. *J. Neurochem.* **2010**, *112*, 1286–1294. [CrossRef]
16. Chun, Y.S.; Oh, H.G.; Park, M.K.; Cho, H.; Chung, S. Cholesterol regulates HERG K+ channel activation by increasing phospholipase C β1 expression. *Channels* **2013**, *7*, 1–13. [CrossRef]
17. Chun, Y.S.; Oh, H.G.; Park, M.K.; Kim, T.-W.; Chung, S. Increasing membrane cholesterol level increases the amyloidogenic peptide by enhancing the expression of phospholipase C. *J. Neurodegener. Dis.* **2013**, *2013*, 407903. [CrossRef]
18. Lu, J.; Wu, D.M.; Zheng, Z.H.; Zheng, Y.L.; Hu, B.; Zhang, Z.F. Troxerutin protects against high cholesterol-induced cognitive deficits in mice. *Brain* **2011**, *134*, 783–797. [CrossRef]

19. Ullrich, C.; Pirchl, M.; Humpel, C. Hypercholesterolemia in rats impairs the cholinergic system and leads to memory deficits. *Mol. Cell. Neurosci.* **2010**, *45*, 408–417. [CrossRef]

20. George, A.J.; Holsinger, R.M.; McLean, C.A.; Laughton, K.M.; Beyreuther, K.; Evin, G.; Masters, C.L.; Li, Q.X. APP Intracellular Domain Is Increased and Soluble Abeta is Reduced with Diet-Induced Hypercholesterolemia in a Transgenic Mouse Model of Alzheimer Disease. *Neurobiol. Dis.* **2004**, *16*, 124–132. [CrossRef]

21. Refolo, L.M.; Malester, B.; LaFrancois, J.; Bryant-Thomas, T.; Wang, R.; Tint, G.S.; Sambamurti, K.; Duff, K.; Pappolla, M.A. Hypercholesterolemia accelerates the Alzheimer's amyloid pathology in a transgenic mouse model. *Neurobiol. Dis.* **2000**, *7*, 321–331. [CrossRef] [PubMed]

22. Segatto, M.; Trapani, L.; Marino, M.; Pallottini, V. Age- and sex-related differences in extrahepatic low-density lipoprotein receptor. *J. Cell. Physiol.* **2011**, *226*, 2610–2616. [CrossRef] [PubMed]

23. Segatto, M.; Di Giovanni, A.; Marino, M.; Pallottini, V. Analysis of the Protein Network of Cholesterol Homeostasis in Different Brain Regions: An Age and Sex Dependent Perspective. *J. Cell. Physiol.* **2013**, *228*, 1561–1567. [CrossRef]

24. Marounek, M.; Volek, Z.; Skřivanová, E.; Czauderna, M. Gender-based differences in the effect of dietary cholesterol in rats. *Cent. Eur. J. Biol.* **2012**, *7*, 980–986. [CrossRef]

25. McLaughlin, S.; Wang, J.; Gambhir, A.; Murray, D. PIP(2) and proteins: Interactions, organization, and information flow. *Annu. Rev. Biophys. Biomol. Struct.* **2002**, *31*, 151–175. [CrossRef]

26. Janmey, P.A.; Lindberg, U. Cytoskeletal regulation: Rich in lipids. *Nat. Rev. Mol. Cell Biol.* **2004**, *5*, 658–666. [CrossRef] [PubMed]

27. Bodin, S.; Welch, M.D. Plasma membrane organization is essential for balancing competing pseudopod- and uropod-promoting signals during neutrophil polarization and migration. *Mol. Biol. Cell* **2005**, *16*, 5773–5783. [CrossRef]

28. Sánchez-Madrid, F.; Serrador, J.M. Bringing up the rear: Defining the roles of the uropod. *Nat. Rev. Mol. Cell Biol.* **2009**, *10*, 353–359. [CrossRef]

29. Liu, Y.; Casey, L.; Pike, L.J. Compartmentalization of phosphatidylinositol 4,5-bisphosphate in low-density membrane domains in the absence of caveolin. *Biochem. Biophys. Res. Commun.* **1998**, *245*, 684–690. [CrossRef]

30. Johnson, C.M.; Chichili, G.R.; Rodgers, W. Compartmentalization of phosphatidylinositol 4,5-bisphosphate signaling evidenced using targeted phosphatases. *J. Biol. Chem.* **2008**, *283*, 29920–29928. [CrossRef]

31. Johnson, C.M.; Rodgers, W. Spatial segregation of phosphatidylinositol 4,5-bisphosphate (PIP2) signaling in immune cell functions. Immunol. *Endocr. Metab. Agents Med. Chem.* **2008**, *8*, 349–357. [CrossRef] [PubMed]

32. Taguchi, K.; Kumanogoh, H.; Nakamura, S.; Maekawa, S. Localization of phospholipase Cβ1 on the detergent-resistant membrane microdomain prepared from the synaptic plasma membrane fraction of rat brain. *J. Neurosci. Res.* **2007**, *85*, 1364–1371. [CrossRef] [PubMed]

33. Koudinov, A.R.; Koudinova, N.V. Cholesterol homeostasis failure as a unifying cause of synaptic degeneration. *J. Neurol. Sci.* **2005**, *229*, 233–240. [CrossRef] [PubMed]

34. Hu, G.; Antikainen, R.; Jousilahti, P.; Kivipelto, M.; Tuomilehto, J. Total cholesterol and the risk of Parkinson disease. *Neurology* **2008**, *70*, 1972–1979. [CrossRef] [PubMed]

35. Roy, A.; Pahan, K. Prospects of statins in Parkinson disease. *Neuroscientist* **2011**, *17*, 244–255. [CrossRef]

36. Puglielli, L.; Tanzi, R.E.; Kovacs, D.M. Alzheimer's disease: The cholesterol connection. *Nat. Neurosci.* **2003**, *6*, 345–351. [CrossRef]

37. Notkola, I.L.; Sulkava, R.; Pekkanen, J.; Erkinjuntti, T.; Ehnholm, C.; Kivinen, P.; Tuomilehto, J.; Nissinen, A. Serum total cholesterol, apolipoprotein E epsilon 4 allele, and Alzheimer's disease. *Neuroepidemiology* **1998**, *17*, 14–20. [CrossRef]

38. Wolozin, B.; Kellman, W.; Ruosseau, P.; Celesia, G.G.; Siegel, G. Decreased prevalence of Alzheimer disease associated with 3-hydroxy-3-methyglutaryl coenzyme A reductase inhibitors. *Arch. Neurol.* **2000**, *57*, 1439–1443. [CrossRef]

39. Martins, I.J.; Berger, T.; Sharman, M.J.; Verdile, G.; Fuller, S.J.; Martins, R.N. Cholesterol metabolism and transport in the pathogenesis of Alzheimer's disease. *J. Neurochem.* **2009**, *111*, 1275–1308. [CrossRef]

40. Mori, T.; Paris, D.; Town, T.; Rojiani, A.M.; Sparks, D.L.; Delledonne, A.; Crawford, F.; Abdullah, L.I.; Humphrey, J.A.; Dickson, D.W.; et al. Cholesterol accumulates in senile plaques of Alzheimer disease patients and in transgenic APP(SW) mice. *J. Neuropathol. Exp. Neurol.* **2001**, *60*, 778–785. [CrossRef]
41. Jick, H.; Zornberg, G.L.; Jick, S.S.; Seshadri, S.; Drachman, D.A. Statins and the risk of dementia. *Lancet* **2000**, *356*, 1627–1631. [CrossRef]

 International Journal of
Molecular Sciences

Review

Modulation of DNA Damage Response by Sphingolipid Signaling: An Interplay that Shapes Cell Fate

Marina Francis [1], Alaa Abou Daher [1], Patrick Azzam [1], Manal Mroueh [1] and Youssef H. Zeidan [1,2,*]

[1] Department of Anatomy, Cell Biology and Physiology, Faculty of Medicine, American University of Beirut, Beirut 1107 2020, Lebanon; msf15@mail.aub.edu (M.F.); ara32@mail.aub.edu (A.A.D.); pa27@aub.edu.lb (P.A.); mym13@mail.aub.edu (M.M.)

[2] Department of Radiation Oncology, American University of Beirut Medical Center, Beirut 1107 2020, Lebanon

* Correspondence: yz09@aub.edu.lb; Tel.: +961-1-350000 (ext. 5091); Fax: +961-1-370795

Received: 16 April 2020; Accepted: 8 May 2020; Published: 24 June 2020

Abstract: Although once considered as structural components of eukaryotic biological membranes, research in the past few decades hints at a major role of bioactive sphingolipids in mediating an array of physiological processes including cell survival, proliferation, inflammation, senescence, and death. A large body of evidence points to a fundamental role for the sphingolipid metabolic pathway in modulating the DNA damage response (DDR). The interplay between these two elements of cell signaling determines cell fate when cells are exposed to metabolic stress or ionizing radiation among other genotoxic agents. In this review, we aim to dissect the mediators of the DDR and how these interact with the different sphingolipid metabolites to mount various cellular responses.

Keywords: DNA damage response; double strand breaks; ATM; ionizing radiation; metabolic stress; oxidative stress; p53; sphingolipids; nuclear sphingolipids

---

1. Introduction

An emerging body of literature indicates that sphingolipids and their metabolizing enzymes are involved in the modulation of the DNA damage response (DDR) [1]. The DNA damage induced by genotoxic stress (ionizing radiation (IR), ultraviolet (UV), chemotherapeutic agents, and metabolic stress) is often coupled to disturbances in the sphingolipids' homeostasis. Many bioactive sphingolipid metabolites including sphingomyelin (SM), ceramides (Cer), ceramide-1-phosphate (C1P), sphingosine and sphingosine-1-phosphate (S1P) have been proven to regulate important cellular processes such as cell growth, survival, senescence, inflammatory responses, and death [2–4]. Therefore, upregulation or downregulation of certain sphingolipids can shift cell fate from survival mode to cell death accordingly. Here, we revise the DDR and the sphingolipid pathway with particular attention paid to the various nuclear sphingolipid metabolites and enzymes. In addition, we shed the light on the interplay between these metabolites and DDR elements to shape cell fate.

2. Overview of the DNA Damage Response

Multiple exogenous (IR, UV, and chemotherapeutic agents) or endogenous (oxidative stress, metabolic stress, telomere attrition, and oncogenic mutations) stressors lead to the induction and accumulation of DNA damage [5]. For instance, IR has direct and indirect effects on cellular DNA. The targeted cells lose their ability to further undergo cellular proliferation and eventually die [6]. The ejected charged particles by incident photons can directly hit DNA molecules and disrupt their structure. This direct effect of radiation promotes cell death or induces carcinogenesis if damaged cells survive [7]. IR can also induce reactive oxygen species (ROS) production due to water radiolysis, with water being a major constituent of cells. ROS can attack vital macromolecules including nuclear DNA to induce biochemical damage, also known as indirect effect [8,9]. Most of the IR-induced damage is a consequence of this indirect effect [7]. Given that, the inherent cellular responses to genomic abrasions vary widely depending on the type and extent of inflicted DNA damage.

2.1. Types of DNA Damage

DNA damage is mainly constituted of double strands breaks (DSB), single strand breaks (SSB), and base damage (BD).

2.1.1. Double Strand Breaks

DSBs are formed when the chemical bonds of both strands of the DNA molecules are broken. Although the number of induced DSBs is relatively low (around 40 DSBs/Gy of IR), it is the most difficult form of DNA damage to fix. More than 50 min are required to repair 50% of the damage [10]. DSBs are also considered the most lethal forms of DNA lesions. Eventually, unrepaired DSBs result in unrepaired chromosome breaks and micronuclei formation [11]. A single residual DSB can lead to cell death or genomic rearrangements favoring cancer initiation [7,12]. According to Sharma et al., biologically relevant doses of hydrogen peroxide (H_2O_2) resulted in a significant increase in oxidative clustered DNA lesions and DSBs formation in DT40 cells at G1 phase [13]. However, some unrepairable DSBs are tolerated by the cell and might not contribute to cell death depending on their localization inside the nucleus and chromatin condensation events [14]. In this review, we mainly elaborate on the repair mechanisms of DSBs.

2.1.2. Single Strand Breaks

SSBs are caused by the cleavage of the phosphodiester bonds in only one strand of the DNA molecule. In fact, 1 Gy dose of IR can induce around 1000 SSBs, of which 50% get repaired within 10 to 20 min following radiation exposure [15,16]. Most ROS-induced DNA lesions are SSBs that can result in erroneous DNA replication or stalling of the replication fork. Some SSBs can eventually lead to DSB formation [17].

2.1.3. Base Damage

BD is a common type of DNA damage where a single dose of 1 Gy can induce around 3000 BDs. These chemical lesions mainly include oxidation, deamination, alkylation as well as hydrolysis that can cause serious genomic aberrations [12,17]. BDs can also be associated with sugar modifications and SSBs [12]. For instance, oxidative stress can induce different types of BDs [18]. ROS may attack any of the four nucleotide bases but most frequently guanine owing to its low oxidation potential. Consequently, the structure of these bases gets disrupted and their pairing properties become altered [19]. These oxidative BDs can cause alterations in gene expression besides mutations that either induce the activation of oncogenes or the inactivation of tumor suppressor genes [20].

2.2. Induction of the DNA Damage Response

In response to genomic injuries, cells develop a DNA damage response (DDR) to detect the lesions, indicate their presence, and launch their repair [1,21,22]. The complex signaling pathways involved in the DDR lead to cell cycle arrest and either DNA damage repair or cell death. The DDR serves to maintain the genomic integrity of the cells, which if compromised, would lead to severe disorders [1].

As mentioned earlier, DSBs are classified as the most lethal forms of DNA damage [23]. Since the non-homologous end-joining (NHEJ) pathway is the prevailing repair mechanism of DSBs in quiescent mammalian cells, it is thought to be the predominant element of DDR post-injury as most of the human body cells are in quiescence [24,25]. Furthermore, cells deficient in error prone-NHEJ develop elevated spontaneous chromosomal breaks which become partially repressed after the reduction of cellular oxygen tension [13]. NHEJ repair is based on the direct ligation of both ends of DSBs, mainly during the G1 phase [26]. However, different NHEJ sub-pathways are thought to co-exist, each of which with a distinct repair half-time [27]. On the other hand, homologous recombination (HR) repair pathway uses genetic information from the homologous chromosomes, i.e., from equivalent region found on the second undamaged DNA molecule [26]. This pathway accounts for the systematic repair of only around 15% of IR-induced DSBs [14,27]. Therefore, the lethal effect of IR can be well explained by the accumulation of unrepaired DSBs due to failure in NHEJ [27].

Both endogenous and exogenous stressors are associated with increased risk of SSBs, which are either generated directly or produced as DNA repair intermediates [28]. SSBs must be repaired completely before the initiation of DNA replication to avoid DSBs formation [29]. These SSBs are mainly repaired through base or nucleotide excision repair mechanisms (BER-NER) for a single or multiple and bulky base damage respectively [17]. Oxidative stress may lead to base damage besides SSBs and DSBs. These BDs are mainly repaired through BER pathway that is mediated by DNA *N*-glycosylases [30]. Although the occurrence of SSBs and BD is more frequent than DSBs, their repair mechanisms are more efficient and their contribution to cell lethality is lower.

2.2.1. Detection of Double Strands Breaks (DSBs) and Initiation of the DNA Damage Response (DDR)

The MRN complex, which consists of three subunits, namely Mre11, Nbs1, and Rad50, directly detects DNA damaged sites by binding to the DNA double stranded ends. This complex is responsible for signaling to the DDR upstream kinases including ataxia-telangiectasia mutated protein (ATM) and Rad3-related protein (ATR). As a consequence, ATM and ATR will be recruited to DSBs and RPA-coated SSBs respectively [31–33]. In turn, these kinases activate downstream effectors involved in the DDR signaling pathway in order to modulate the progression of the cell cycle and launch the repair [31]. For instance, ATM and/or ATR activate the tumor suppressor protein p53, which is known to be mutated in most cancers, in order to promote cell cycle arrest (p21 activation) or apoptosis (BAX-mediated caspase-3 cleavage) [34]. Moreover, ATM and ATR activate the checkpoint kinases (CHK1 and CHK2) which can either stimulate apoptosis and cell cycle arrest through p53 activation [35] or inhibit Cdc25 to arrest the cell cycle [35]. These checkpoint kinases can further activate BRCA2 to initiate a repair process [36]. Although the activation of ATM by DSBs is a well-established phenomenon, excessive SSBs generation during the repair mechanism can activate ATM independently of DSBs. The activated ATM delays cell cycle progression in a p53/p21-dependent pathway, thus allowing more time for SSBs repair before DNA replication. Consequently, ATM may also contribute to the prevention of DSBs formation [29].

2.2.2. Role of Ataxia-Telangiectasia Mutated Protein (ATM) in DSBs Repair

Members of the phosphatidylinositol-3 kinases family, including ATM, ATR and DNA-dependent protein kinase (DNAPK), are responsible for phosphorylation of the histone variant H2AX near DSBs [37,38]. In effect, γ-H2AX (phosphorylated H2AX) foci formation is a very early step in the DDR of mammalian cells. It plays an essential role in recruiting damage signaling factors to DSBs sites, which are essential for repair induction [37,38]. ATM is a key modulator of the DDR as it directs the activation of cell cycle checkpoints, DNA damage repair and the alteration of cellular metabolism following the induction of DSBs and oxidative stress [39]. Burma et al. demonstrated that ATM was the primary kinase responsible for the rapid phosphorylation of H2AX in response to DNA DSBs and that DNAPK, rather than ATR, took this job in the absence of ATM [40]. However, ATR is the kinase involved in the phosphorylation of H2AX in response to DNA SSBs and replicative stress [41,42]. Therefore, H2AX phosphorylation can be achieved via multiple pathways depending on the DNA damage type.

2.2.3. ATM Nuclear Shuttling

Strong evidence suggests that IR triggers the monomerization of cytoplasmic [14] and nuclear [43] ATM dimers through oxidation reactions resulting in p-ATM (phosphorylated ATM) monomers with increased kinase activity [14,44]. Guo et al. showed that oxidative stress could also trigger the monomerization and activation of ATM irrespective of DSBs induction [45]. Cytoplasmic ATM monomers then can bind to importins through their nuclear localization signals (NLS) and subsequently shuttle into the nucleus [46,47]. This phenomenon is proportionate to the delivered dose [14]. Once inside the nucleus, p-ATM is guided to DSBs sites by signals from the MRN complex. p-ATM phosphorylates the three subunits of the MRN complex as well as the histone variant H2AX resulting in the formation of nuclear foci. These events ensure DSBs recognition and initiation of the repair [47–49]. p-ATM nuclear shuttling and foci formation at DSBs is a very rapid process. The maximal number of nuclear p-ATM foci is usually reached within 10 min to 1 h post-injury [50]. However, certain cytoplasmic proteins like mutated huntingtin or tuberous sclerosis complex can bind activated ATM monomers and impede their shuttling towards the nucleus [49,51]. Therefore, DSBs recognition and repair may be influenced by the diffusion of p-ATM monomers as well as ATM re-dimerization or re-association with certain cytoplasmic proteins [14,47,49].

2.3. DNA Damage-Induced Cell Death and Senescence

The accumulation of DNA damage provokes multiple responses that lead to cellular senescence or death through various molecular mechanisms. The amount and the type of DNA damage significantly shapes cell fate [52]. With aging, the effectiveness of the DNA repair mechanisms appears to decline, resulting in the accumulation of DNA lesions in tissues [53]. In rodents and humans, altered phenotypes that are common in age-related pathologies, are caused by mutations of some genes involved in the machinery of DNA repair. These alterations include cardiovascular and metabolic disorders such as diabetes, which are linked to increased frequency of malignancies and decreased life expectancy [53–55].

Furthermore, cancer therapies involve the induction of unrepairable DNA damage to eradicate cancerous cells. Strong evidence suggests that their therapeutic outcome does not rely on apoptosis alone but rather involves various cell death mechanisms such as mitotic catastrophe, necrosis, and senescence. Cancer cell death does not occur instantaneously, and can take up to weeks from treatment cessation [56]. The guardian of the genome, p53, is a key player in these cellular responses [57]. During tumor progression, pro-apoptotic mechanisms are lost mainly due to impaired functional p53 in more than 50% of human malignant tumors [58–60]. The ATM/ATR-activated p53 signals cell cycle arrest and DNA damage repair, thus promoting cell survival. On the other hand, it can also stimulate the elimination of injured cells depending on the cell type and the extent of the DNA damage [61]. The most studied gene targets of p53 are mouse double minute 2 (MDM2), p21, p53 upregulated modulator of apoptosis (PUMA), and Bcl2 associated X (BAX). In normal cells, p53 expression is kept low due to the exhibition of regulatory feedback loops between p53 and MDM2 [62]. MDM2 has E3 ubiquitin ligase activity in addition to nuclear import and export signals. It has many targets including forkhead box O (FOXO) and p53 [63]. Therefore, it tags p53 by ubiquitin at lysine residues in the nucleus causing its nuclear export and subsequent proteasomal degradation [62]. Barak et al. showed that p53 could target MDM2 gene directly [64]; thus, the upregulation of p53 leads to the elevation of MDM2 expression which in turn downregulates p53 [65]. However, tumors may resist apoptosis by retaining functional p53 due to pro-apoptotic genes deactivation (BAX, Apaf1 ...) or anti-apoptotic genes stimulation (Bcl2, survivin ...) [66]. Hence, the regulation of p53 is crucial for determining cell fate post-genotoxic injury.

2.3.1. DNA Damage-Induced Senescence

Senescence is a condition of permanent cell cycle arrest in the G1 phase associated with alterations in gene expression and cellular morphology [52]. Senescent cells are flattened and enlarged cells with a granular cytoplasm that can preserve their metabolic activity and viability [57,67].

The accumulation of senescent stem cells, in particular, leads to impaired homeostasis and tissue regeneration in addition to metabolic dysfunction [68]. These accumulated senescent cells can also lead to p53-independent chronic inflammation in tissues through the release of pro-inflammatory chemokines and cytokines [68]. Moreover, senescence is clinically proven to occur in prostate cancers and desmoids tumors post-radiotherapy as a major p53-dependent mechanism for tumor regression [69–71]. Therefore, chronic DDR signaling and senescence hinder damage propagation to the next generation of cells [72].

2.3.2. DNA Damage-Induced Apoptosis

Cellular stress leads to the activation of apoptosis to eradicate aberrant cells when DNA damage repair becomes inefficient [52]. It remains unclear what influences the cell's decision, whether to launch senescence or apoptosis. Multiple studies demonstrated a cross-talk between senescence and apoptosis mainly at the level of p53 in response to genotoxic stress [73,74]. Activated p53 mounts the transcription of its target genes involved in apoptosis such as PUMA, p53 regulated apoptosis inducing protein 1 (p53AIP1), BAX and apoptotic protease activating factor 1 (Apaf1), to regulate the intrinsic pathway of apoptosis. Moreover, p53 activation can lead to the induction of the extrinsic apoptotic pathway, which is mediated by the upregulation of TRAIL (tumor necrosis factor-related apoptosis-inducing ligand) receptors, death receptors 4 and 5, and CD95 ligand and receptor [75]. Kim et al. identified ROR$_\alpha$, an orphan nuclear receptor and a direct p53 target gene, as a stabilizer for p53 and activator of p53 gene transcription. Consequently, ROR$_\alpha$ resulted in the activation of the p53 downstream effectors involved in apoptosis [76]. Caspase-2 also plays a role in linking DNA damage to the pathways of apoptosis. The upregulation of PIDD (p53-induced protein with a death domain) leads to the spontaneous cleavage and activation of caspase-2 in the nucleus [77] and consequently sensitizes cells to genotoxins-induced apoptosis [78]. For instance, there is an association between radiation-induced apoptosis and the activation of ATM/p53/BAX/cytochrome c/caspases pathway [79]. However, DNA damage-induced apoptosis might be p53-dependent or independent based on the damage and cell types [52].

2.3.3. DNA Damage-Induced Mitotic Catastrophe

Mitotic catastrophe designates cell death occurrence during or as a consequence of abnormal mitosis [80]. This results in aberrant segregation of chromosomes, followed by the generation of huge cells that hold abnormal nuclear morphologies, multiple nuclei [81–84], and/or multiple micronuclei [85]. It has been suggested that mitotic catastrophe occurs as a result of defective cell cycle checkpoints or aggregated DNA damage that triggers p53 mutation or inactivation [57,86]. p53-dependent centrosomes proliferation can also cause mitotic catastrophe after DNA damage induction and defective repair [87–91]. Delayed cell death induction by radiotherapy in solid tumors is mainly caused by the mitotic catastrophe [57] associated with p53/cytochrome c/caspases pathway [92]. Thus, mitotic catastrophe is considered an important cell death mechanism inflicted by DNA damage.

2.3.4. DNA Damage-Induced Necrosis

In contrast to apoptosis, necrosis is an uncontrolled cell death mechanism. Necrosis is characterized by various morphological changes, including increased cell volume, swelling of organelles, ruptured plasma membranes, and loss of intracellular components. It can be mediated by several catabolic and signal transduction pathways including TNFα/PARP/JNK/caspases [93,94]. Necrosis is a form of death that involves immunological activation and pro-inflammatory responses through the release of damage-associated molecular patterns like high mobility group (box1) and heat shock proteins (HSPs) [95,96]. As mitotic catastrophe is usually followed by necrosis, local inflammation tends to develop post-radiotherapy [97]. Given that, necrosis is one of the major contributors to cell death in response to genotoxic injuries.

3. Overview of Sphingolipids

Sphingolipids belong to a class of lipids characterized by a sphingosine backbone, which is an amino-alcohol compound of 18 carbon atoms. The sphingosine backbone is synthesized from non-sphingolipid precursors in the endoplasmic reticulum (ER). The diversity of sphingolipids arises from distinct variations in this basic structure [98].

Sphingolipids are no longer thought to be only major structural components of biological membranes but have recently proven to be, along with their active metabolites (sphingomyelin, ceramide, sphingosine, sphingosine-1-phosphate, ceramide-1-phosphate), key players in various human diseases [99,100]. These bioactive signaling molecules mediate various important biological processes such as cell growth, survival, senescence, inflammatory responses, and death by regulating their downstream targets [4,101]. For instance, Golgi-associated retrograde protein complex (GARP) is a mediator of retrograde vesicular transport from the endosome to Golgi, and thus important for the regulation of sphingolipids' recycling between the plasma membrane and endosomes. GARP deficiency ultimately leads to the development of progressive cerebello-cerebral atrophy type 2 (PCCA2) [102]. This severe neurodegenerative disease is attributed to dysfunctional lysosomes and dysregulated sphingolipid metabolism [102]. Niemann-Pick disease type C (NPC) is a lysosomal storage disease caused by mutations in lysosomal proteins NPC1 or 2. It involves the accumulation of sphingosine, sphingomyelin, glycosphingolipids, and cholesterol [103,104]. In Niemann–Pick disease, the accumulation of sphingomyelin is caused by acid sphingomyelinase deficiency [105]. Sphingosine contributes to lysosomal calcium release through two-pore channel 1 (TPC1) in normal fibroblasts. In contrast, cells derived from NPC patients exhibit reduced lysosomal calcium release due to accumulated sphingosine [104]. Moreover, sphingolipids are associated with various genetic (Gaucher disease) and non-genetic (diabetic nephropathy and focal segmental glomerulosclerosis) glomerular diseases. The dysregulation of sphingolipids in podocytes disrupts their proper functioning and subsequently compromises the glomerular filtration barrier [106]. Given that, sphingolipids altered metabolism mediates various inherited and non-inherited human diseases.

3.1. Sphingolipids Metabolic Pathway

Ceramide (Cer) is the central metabolite generated within the sphingolipid metabolism through three different pathways. (i) Ceramide de novo synthesis: Palmitoyl-CoA is condensed with serine by the action of serine palmitoyl transferase, followed by a set of reduction and acetylation reactions to generate ceramide; (ii) Sphingomyelin (SM) catabolism: SM is catabolized by sphingomyelinases to generate ceramide; (iii) Salvage pathway: *N*-acylation of fatty acids with a sphingosine backbone produces ceramide through the action of ceramide synthases [1]. The generated pro-apoptotic Cer can be phosphorylated by ceramide kinase into ceramide-1-phosphate (C1P) in trans-Golgi or plasma membranes. C1P plays an important role in inflammatory responses, cell survival and proliferation [98,99]. Afterwards, C1P can be dephosphorylated by C1P phosphatases or other unspecific lipid phosphate phosphatases (LPP family) [98,107]. Cer is also utilized to generate two major groups of complex glycosphingolipids. Glucosylceramide synthase generates glucosphingolipids by adding glucose as the first residue to Cer at C1 hydroxyl position, whereas galactosylceramide synthase generates galactosphingolipids by adding galactose to Cer [98]. Moreover, Cer can be catabolized by ceramidases into sphingosine which promotes cell cycle arrest and apoptosis. In its turn, sphingosine can be phosphorylated by sphingosine kinases into the pro-survival lipid sphingosine-1-phosphate (S1P) [101]. S1P can be dephosphorylated by S1P phosphatases [108,109] or unspecific LPP [110]. The generated sphingosine can be further used to produce Cer or S1P [109]. S1P lyase (SPL) is considered as the last enzyme in the sphingolipid catabolic pathway because it can irreversibly break down S1P into phosphoethanolamine and hexadecenal [111] (Figure 1).

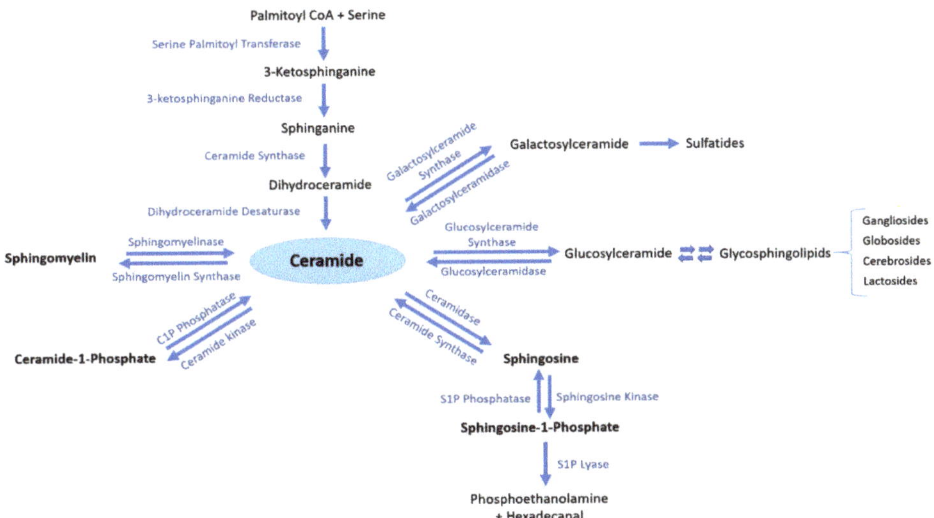

Figure 1. The sphingolipid metabolic pathway. Ceramide is the central metabolite generated in the sphingolipid metabolism by three distinct pathways. Ceramide de novo synthesis consists of Palmitoyl-CoA condensation with serine by the action of serine palmitoyl transferase, followed by a set of reduction and acetylation reactions to generate ceramide. Sphingomyelin (SM) catabolism generates ceramide through the action of sphingomyelinases. The salvage pathway involves N-acylation of fatty acids with a sphingosine backbone to produce ceramide by ceramide synthases. Ceramide can be further phosphorylated to ceramide-1-phosphate (C1P) by ceramide kinase, or converted into complex glycosphingolipids by glucosylceramide or galactosylceramide synthases. Ceramidase is responsible of catabolizing ceramide into sphingosine, which may be further metabolized by sphingosine kinases to generate sphingosine-1-phosphate (S1P).

3.2. Sphingolipids in the Nucleus

It took a long time to dismiss the hypothesis that the lipids found in the nuclear fraction result from contamination during extraction procedures. However, several years of dedicated research confirm that lipids are minor components of the nucleus (around 5% by weight) [112,113]. Under various physiological and pathological conditions, the composition, metabolism and behavior of these nuclear lipids are independent of the other lipids in cellular membranes and organelles [112,114]. It is of note that some exogenous stimuli influence only intranuclear signaling [115,116] while others can influence both the nuclear and cytoplasmic signaling [117,118]. In the nuclear fraction, lipids could be either polar or non-polar and consist of glycerophospholipids, plasmalogens, sphingolipids, gangliosides, cholesterol, arachidonic acid, and eicosanoids [112]. Their active metabolism is maintained by nuclear lipid enzymes which take responsibility of their anabolic and catabolic reactions [114]. Initially, the function of nuclear lipids was thought to be restricted to structural support maintenance of the nuclear membranes (nuclear envelope (NE)) as they contain the bulk of lipids [119]. However, besides the nuclear membranes, bioactive lipids were also identified in other subnuclear domains including the nuclear matrix [120], chromatin [121,122], and nucleolus [123]. Structurally, the NE consists of outer and inner nuclear membranes of distinct lipidomic profiles. The outer membrane is continuous with the ER, whereas the inner membrane is associated with the nuclear lamina and chromatin [124]. These nuclear membranes are separated by a perinuclear space and they are perforated by the nuclear pore complexes (NPC). The latter control nucleo-cytoplasmic communications mainly by regulating the bidirectional shuttling of ions, nucleotides, RNA and proteins [125]. Albi et al. demonstrated that the nuclear membranes' permeability and fluidity are heavily dependent on their lipid composition, in particular phosphatidylcholine (PC), sphingomyelin (SM) and cholesterol (CHO) [126]. Moreover, the inner membrane expresses a GM1 ganglioside-linked Na^+-Ca^{2+} exchanger that is responsible of mediating the transfer of nuclear Ca^{2+} to the perinuclear space as a cytoprotective mechanism [127]. Gangliosides are complex glycosphingolipids, which are mostly abundant in the central nervous system and involved in the regulation of nuclear calcium homeostasis [113]. The lipid content of the nuclear matrix is essential for conveying and maintaining its rigidity. It constitutes an anchor that organizes the chromatin and controls various important endonuclear events [128]. In fact, nuclear lipids are involved in multiple processes like DNA replication, transcription, splicing, and repair as well as Ca^{2+} homeostasis [129]. Therefore, stress-induced alterations in lipid metabolism can modulate cell growth, survival, differentiation, senescence and death [114].

Many studies have identified sphingolipids as important modulators of key nuclear processes. So far, various subnuclear compartments including the NE, nuclear matrix, nucleolus and chromatin have been described to host various sphingolipid species [113,123,130–136]. Although nuclear pores should permit nucleo-cytoplasmic exchange of sphingolipids, many of these nuclear metabolites are in dynamic state and undergo turnover. Therefore, the nuclear localization of sphingolipid metabolizing enzymes has been demonstrated. To this end, the utilization of several analytical, biochemical and microscopic techniques led to the identification and quantification of various nuclear sphingolipid species along with their metabolizing enzymes [137]. In fact, sphingomyelin (SM) is the dominant nuclear sphingolipid variant [119]. Through its metabolism, SM gives rise to ceramides, sphingosine, and S1P, which in turn give rise to other metabolites. Within the scope of this review, we describe and summarize the diverse functions of these nuclear sphingolipids based on their localization (Table 1).

Table 1. Nuclear sphingolipid metabolites and metabolizing enzymes. This table recapitulates the various nuclear sphingolipid metabolites and enzymes detected in the nuclear compartment with a brief description of their important nuclear functions. NE: nuclear envelop, dsRNA: double stranded RNA.

Nuclear Sphingolipid Metabolites	Nuclear Sphingolipid Producing Enzymes	Nuclear Sphingolipid Degrading or Converting Enzymes	Main Nuclear Functions
Sphingomyelin	Sphingomyelin synthase	Reverse sphingomyelin synthase Neutral sphingomyelinase	Maintenance of NE and nucleoplasm structure Regulation of NE permeability and Fluidity Stabilization of DNA and dsRNA
Ceramide	Ceramide synthase Ceramide desaturase Neutral sphingomyelinase	Ceramidase Ceramide kinase	Regulation of Cell cycle arrest, Senescence, and Apoptosis
Ceramide-1-phosphate	Ceramide kinase	C1P phosphatase	Regulation of cell growth and survival
Sphingosine	Ceramidase	Ceramide synthase Sphingosine kinase 2	Regulation of gene transcription and apoptosis
Sphingosine-1-phosphate	Sphingosine kinase 2	S1P lyase S1P phosphatase	Epigenetic modulation of gene transcription Regulation of cell cycle progression and apoptosis Stabilization of human telomerase

3.2.1. Nuclear Sphingomyelin and Metabolizing Enzymes

SM is the most abundant nuclear sphingolipid. It is primarily found in the nuclear envelope and to a lesser extent in the nuclear matrix and chromatin [113]. Besides SM, the nuclear membrane contains phosphatidylcholine (PC) and cholesterol (CHO). These are the important lipids that regulate the structure, function, and fluidity of nuclear membranes [138]. SM and CHO increase a membrane's rigidity, whereas PC increases its fluidity [125]. Therefore, a high SM-CHO/PC ratio will decrease the fluidity of the nuclear membrane and vice versa. As the fluidity of the nuclear membrane increases, the size of the nuclear pores changes allowing increased nucleo-cytoplasmic transport such as that of mRNA during cell proliferation [138]. Moreover, nuclear SM can either stabilize and/or destabilize the DNA molecules by influencing the helical to non-helical transition and vice versa [139]. The SM-DNA interaction is plausible due to the zwitterionic nature of SM. Its positively charged trimethylammonio group can bind to DNA anionic groups, whereas its negatively charged phosphate group repels the negatively charged phosphate of the DNA [112]. At low levels, the non-polar fatty acids of SM bind to the internal hydrophobic centers of helical DNA providing their stabilization. Conversely, increased SM concentration leads to space competition for the non-polar fatty acids. This change alters their binding to DNA and results in DNA molecule opening followed by rapid denaturation [112]. It has been proposed that SM also binds to and stabilizes double-stranded RNA in the nucleus and prevents its digestion by RNases [140,141]. In addition, SM is an essential component for the maintenance of the nuclear structure since it preferentially localizes within the peri-chromatin region. This is supported by the observation that sphingomyelinase microinjections into living cells' nuclei resulted in fast corrosion of the intranuclear architecture [142,143].

The metabolism of nuclear SM is independent from the Golgi complex and ER. Multiple factors such as stress, high fat diet, cell cycle and tissue regeneration can alter the levels of SM in the nucleus [124]. The enzymes responsible for SM synthesis and breakdown have been detected in nuclear fractions [124]. *De-novo* nuclear SM synthesis requires the production of Cer, followed by its conversion to SM by the action of SM synthase 1 or 2. The latter step requires phosphocholine derived from nuclear PC which is found in the NE or chromatin [144,145]. On the other hand, neutral sphingomyelinase, available in the nuclear envelope [130], nuclear matrix [120], and chromatin [146] of rat liver nuclei, metabolizes SM into pro-apoptotic ceramides. Reverse SM synthase was also detected in rat liver chromatin, which catalyzes the transfer of phosphocholine from SM into DAG, a mitogenic second messenger, forming PC [144]. Therefore, SM synthase and sphingomyelinase can modulate cell proliferation or death by regulating Cer to DAG ratio of chromatin.

3.2.2. Nuclear Ceramide, Ceramide-1-Phosphate and Metabolizing Enzymes

Cer is the central metabolite generated within the sphingolipid pathway. It serves as a precursor for complex sphingolipids production (SM and glycosphingolipids) and in turn can be metabolized to other bioactive species (sphingosine, C1P or S1P) [129]. After overexpression in HEK-293 cells, Cer synthases could be highly detected in the ER and NE [147–150]. Nuclear ceramidase activity was also reported in liver nuclear membranes, thus allowing further Cer metabolism [151]. Several studies showed that nuclear ceramides are key mediators of cell cycle arrest and apoptosis. Multiple exogenous stressors can alter the nuclear levels of Cer such as serum starvation, high-fat diet, bacterial infections, and apoptosis-inducing mediators (e.g., Fas ligand) [124,152]. For instance, Albi and colleagues reported that serum starvation was associated with nuclear Cer upregulation during the early phase of apoptosis. This was followed by extranuclear sphingomyelinases activation and cytoplasmic Cer accumulation during the late phase of apoptosis [153]. A high fat diet also resulted in increased nuclear ceramide levels by three-fold in rat liver nuclei along with the elevation of saturated fatty acid species (C:14, C:16, C:18) [154]. It remains unclear whether Cer nucleo-cytoplasmic shuttling is feasible via binding to Cer transport protein CERT and FAPP2 [155,156].

Cer can be phosphorylated into C1P by the action of ceramide kinase (CERK) previously reported in ER/Golgi organelles [157]. Then, C1P transfer protein (CPTP) transports C1P to the cytoplasmic membrane and other subcellular organelles including the nucleus [158]. Prior work detected nuclear import and export signals in the protein sequence of CERK [159]. It is plausible that nuclear ceramides may be further converted into C1P, however that remains to be fully established.

3.2.3. Nuclear Sphingosine, Sphingosine-1-Phosphate and Metabolizing Enzymes

Sphingosine levels, whether in whole cells or nuclear extracts, are much lower than Cer [133]. Nuclear ceramidases allow the hydrolysis of Cer into sphingosine which in turn can be converted into Cer by the action of Cer synthases [129,133]. Nuclear sphingosine is an important regulator of gene transcription. Sphingosine modulates the transcription of CYP17 and it is considered as a regulatory ligand for steroidogenic factor (SF-1) [160]. Under basal conditions, nuclear sphingosine binds to SF-1 with several co-repressors including Sin3A and histone deacetylase (HDAC). The stimulatory signals of the adrenocorticotropin hormone (ACTH) release sphingosine from bounded SF-1 through the activation of protein kinase A. Subsequently, the transcription of genes implicated in steroid hormone synthesis from cholesterol precursor will be initiated [161,162].

In addition, sphingosine levels can be modulated by the action of sphingosine kinases (SK) which phosphorylate sphingosine to sphingosine-1-phosphate (S1P). There are two isoforms of sphingosine kinases, SK1 and SK2 which differ by their subcellular localizations and functions. SK1 is mainly located in the cytoplasm due to its two functional nuclear export signals and regulates cell proliferation and growth. Conversely, SK2 is mainly located in the nucleus, due to the nuclear localizing signal at its N-terminus, and modulates apoptosis [163,164]. Both sphingosine kinases get altered after stimulation by growth and survival factors. They become subjected to post-translational modifications, translocations, protein-protein and lipid-protein interactions resulting in increased intracellular S1P levels [165]. Primally, nuclear SK activity was detected in the NE and nucleoplasm of Swiss 3T3 cells. This kinase activity got upregulated by the platelet derived growth factor and promoted cell cycle progression toward the S phase [118]. Therefore, S1P might be implicated in the regulation of cell cycle. In MCF-7 breast cancer cells, SK2 interacts with the histone variant H3 in chromatin and induces its acetylation. Thus, intranuclear S1P can exert epigenetic modulations of gene transcription. The nuclear S1P and dihydro-S1P can bind to the active sites of HDAC1 and 2 and consequently inhibit their activities [136]. Moreover, SK2 associates with HDAC at the promoter regions of p21 and c-*fos* genes resulting in histone acetylation, which favors their gene transcription and subsequent cell cycle arrest and apoptosis [136]. Recently, Selvam et al. suggested that S1P can bind and stabilize the human telomerase [166]. S1P can act as an intracellular messenger or an extracellular ligand for a family of five isoforms of G protein-coupled receptors (S1PR1-5) [167]. IHC and ICC techniques allowed the detection of all five isoforms of S1PR in both the cytoplasm and nucleus of healthy and cancerous human tissues of several organs [168]. All together, these studies suggest that S1P is a master regulator of cell proliferation and anti/pro-apoptotic processes.

To terminate S1P signaling at the ER, the cells opt either sphingosine phosphatases (SPP1-2) which dephosphorylate S1P into sphingosine [169] or S1P lyase (SPL) which terminally hydrolyzes S1P into ethanolamine phosphate and hexadecenal [111]. Schwiebs et al. reported that SPP-1 is expressed in the nuclear compartment of naïve dendritic cells and gets translocated to the cytoplasm upon inflammation [170]. Recently, Ebenezer et al. confirmed the nuclear localization of SPL in lung epithelial cells [171], as well as the crucial role of the generated nuclear hexadecenal in histone acetylation through interaction with HDAC1-2 [172]. The exact catabolic mechanisms of nuclear S1P remain to be fully elucidated in various tissues and cell lines.

4. Role of Sphingolipids in the DNA Damage Response

After reviewing the DDR and the various nuclear sphingolipid metabolites, we discuss how these two entities interplay in order to determine cell fate post-injury (Figure 2). Various chemotherapeutic drugs and DNA damaging agents target sphingolipid metabolizing enzymes. Strong evidence suggests that lipids are involved in DDR and determining cell fate [1]. Most of cancer treatments lead to Cer generation which is implicated in cell death response [173]. However, cancer cells tend to develop survival strategies like generating the pro-survival sphingolipid metabolite S1P after the phosphorylation of sphingosine generated by Cer hydrolysis [174]. Hence, the regulation of these metabolites production is of significant importance in determining the cells' fate in response to DNA damage [1].

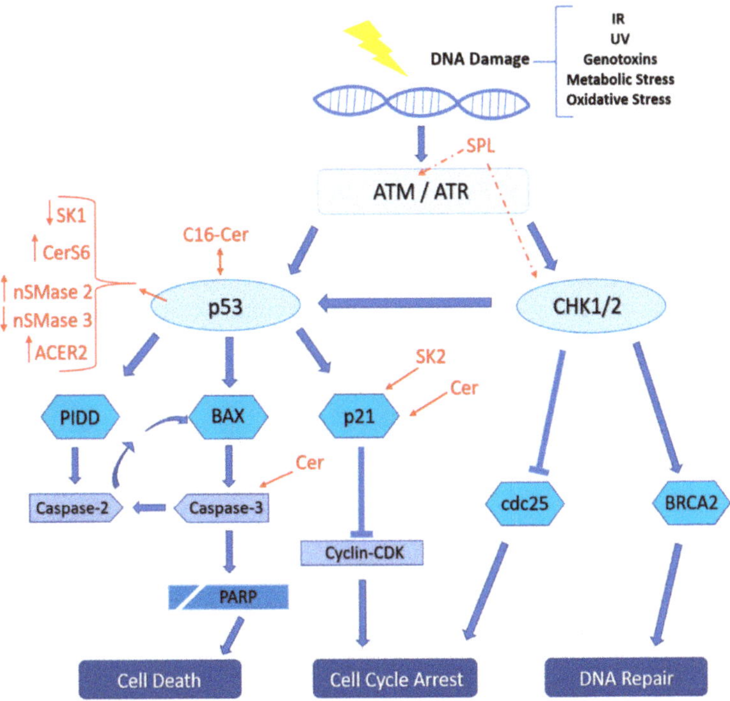

Figure 2. Interplay between DNA damage response (DDR) and sphingolipids to shape cell fate after DNA damage. After genotoxic injury, the DDR develops to sense the damage and amplify the transmitted signaling cascade. The transiently activated cell cycle arrest allows DNA repair. However, persisting unrepaired damage triggers cellular senescence or death to hinder damage propagation to the next generation of cells. The sphingolipid metabolism interacts mainly with p53 along with other elements of the DDR to determine the injured cell fate. ACER2: alkaline ceramidase 2, Cer: ceramide, CerS: ceramide synthase, nSMase: neutral sphingomyelinase, SK: sphingosine kinase, SPL: S1P lyase, double-headed arrow: interaction and activation, full arrow: upregulation and activation, and dashed arrow: downregulation.

Dbaibo et al. showed that p53 is involved in Cer-induced apoptosis. The accumulation of Cer in Molt4 lymphocyte leukemia cells post-irradiation and actinomycin D was p53-dependent since increased p53 preceded Cer up-regulation. Indeed, Cer accumulation can be impeded by p53 inhibition [175]. p53 is a downstream effector activated by ATM in DDR and numerous indications implicate its involvement in Cer accumulation. Prior work demonstrated that AT cells with a mutated ATM gene are resistant to IR-induced apoptosis. These cells maintained the first phase of Cer accumulation by acid sphingomyelinase but lost the second peak mediated by Cer synthase [176]. Thus, ATM mediates Cer synthase activation but not that of acid sphingomyelinase. Further studies point at the involvement of neutral sphingomyelinases 2 and 3 in DDR. ATM and p53 activate neutral sphingomyelinase 2 and down-regulate neutral sphingomyelinase 3 in order to induce apoptosis [177–179]. Furthermore, it has been shown that IR induces caspase-3 and PARP cleavage through Cer. Caspase-3 inhibition doesn't affect the levels of Cer whereas Cer depletion prevents the cleavage of caspase-3 and PARP [176,180]. These results suggest that Cer up-regulation is upstream of caspase-3 cleavage. Ceramides have been also implicated in cell cycle arrest during DDR. For instance, accumulation of Cer can arrest the cell cycle either at G0/G1 phase mediated by retinoblastoma protein (Rb) [181] or at G2 phase through the activation of p21 [182]. Others reported that human alkaline ceramidase 2 (ACER2) is a novel direct target gene of p53 that mediates the DDR [183]. Upregulation of ACER2 reduced the levels of Cer while accumulating sphingosine and S1P in H1299 cells. Extensive IR-induced DNA damage hyperactivated the p53-ACER2 pathway in HCT116 cells. This hyperactivation led to cell death because of a high pro-apoptotic sphingosine to pro-survival S1P ratio. In contrast, low levels of DNA damage moderately activated the p53-ACER2 pathway favoring cell cycle arrest and senescence. The pro-apoptotic and pro-senescence signals of sphingosine and Cer were balanced with the pro-survival and pro-proliferative signals of S1P [183]. In cancer cells, folate stress leads to the upregulation of C16-Cer coupled to a transient increase in Cer synthase 6 in a p53/PUMA-dependent manner [184]. Recent work proved a direct and highly specific interaction between p53 and C16-Cer. Under metabolic stress induced by serum or folate starvation, this interaction displaces MDM2 from p53 leading to the accumulation and activation of p53 [185]. However, the relationship between p53 and Cer remains elusive as many studies supported functional roles for Cer both upstream and downstream of p53.

The levels of the pro-survival bioactive lipid S1P are also involved in DDR and the determination of cell fate [1]. A link between p53 and SK1, which catalyzes S1P production, was established after treating Molt-4 leukemia cells with multiple chemotherapeutic agents and γ-rays. The DDR leads to a decrease in the protein but not mRNA levels of SK1 coupled to p53 up-regulation [186]. Conversely, SK2 was shown to exacerbate the apoptotic response and its overexpression was associated with the induction of apoptosis. In addition, SK2 was proven as critical for p21 expression which was needed for cell cycle arrest in a p53-independent route [164,187]. These results could be potentially explained by the fact that mainly SK1 is cytoplasmic and SK2 is nuclear. Hence, increased SK2 expression and intranuclear production of S1P can inactivate HDACs and promote the transcription of genes involved in cell cycle arrest and apoptosis. On the other hand, SPP1 knockdown had a protective role against DNA damage and cell death induced by daunorubicin in MCF7 cells [188]. Furthermore, augmenting SPL activity through its overexpression or by treatment with etoposide favored apoptosis through caspase-3, annexin-V, PARP and nuclear condensation [189]. Similarly, SPL modulates DNA repair, G2 cell cycle arrest, and apoptosis post-irradiation. SPL upregulation exacerbates stress-induced accumulation of Cer through acid sphingomyelinase. Interestingly, overexpression of SPL results in decreased ATM and checkpoint kinase 1 protein levels when compared to control cells [190]. Taken together, these findings reveal the importance of sphingolipids in mediating the DNA damage response.

Our group previously showed that the specific expression of sphingomyelin phosphodiesterase acid-like 3b (SMPDL3b) in podocytes to be a modulator of radiation stress signaling [191]. Strong evidence suggests that SMPDL3b is critical for preserving podocytes proper functioning [192]. Novel work pinpointed that SMPDL3b regulates C1P to Cer levels in podocytes by interacting with CERK and C1P [193,194]. Furthermore, IR results in a time-dependent drop in the protein levels of SMPDL3b, downregulation of sphingosine and S1P, upregulation of various pro-apoptotic Cer species, and loss of podocytes' filopodia [191]. Conversely, SMPDL3b overexpression confers radioresistance by partially reversing these changes and enhancing DNA damage repair evidenced by reduced γ-H2AX nuclear foci in comparison to wild-type podocytes post IR [191]. It remains of great interest to elucidate the proper mechanism by which SMPDL3b modulates the DDR.

In a nutshell, all non-surgical cancer therapies aim to eradicate tumor cells while sparing normal tissues through complex cell signaling pathways. Research over the past few decades has confirmed that the stress induced by these therapies involves the accumulation of ceramide. However, any dysregulation in this process, due to either decreased generation or increased metabolism of ceramide, confers resistance against these therapies [195]. From this perspective, emerging therapeutic and clinical interventions are under investigation to maximize the positive outcomes of these therapies by a combinatorial approach. For instance, as recombinant human acid sphingomyelinase (rhASM) was previously evaluated in patients with Niemann–Pick disease, the idea of its administration in cancer therapies flourishes. rhASM might be used to induce pro-apoptotic ceramide levels beyond the tolerance of cells. This treatment is more likely to affect tumors than normal tissues [196]. Moreover, a recent study reported that gentamicin, a commonly used anti-microbial drug, can potentially play a role in cancer therapies. The administration of gentamicin highly upregulated acid sphingomyelinase and induced apoptosis in human gastric cancer cells [197]. As cancer cells can develop survival strategies like generating the pro-survival S1P, inhibitors for both SK1 and SK2 were developed. However, sphingosine kinase inhibitors exhibited some downstream off-target effects such as inhibiting ERK and Akt pathways [198]. Hence, further studies should address the development of more specific sphingosine kinase targets for possible clinical trials. Interestingly, the total plasma levels of ceramide, measured in early days after the combined treatment of radio-chemotherapy, can predict tumor responses in patients with liver and lung metastases of colorectal cancer. It allows the identification of patients with high risks of metastases [199]. Therefore, successful discoveries of sphingolipid therapeutic targets or sphingolipid biomarkers of tumor response will potentially enhance the outcomes of standard of care therapies.

5. Conclusions

Up to now, a large body of evidence supports the interplay between the DDR and the sphingolipid metabolic pathway in determining cell fate after exposure to genotoxins or metabolic stress. These findings elicit sphingolipids as master modulators of the overall mechanism by which cells respond to genotoxic injuries. Further studies are needed to clarify some debatable theories and to further understand the complex interactions between the various sphingolipid metabolites and the DDR mediators. For instance, although SK2 was reported to mediate cell cycle arrest and apoptosis [164,187], downregulation of SK2 but not SK1 was effective in the suppression of tumor cell proliferation and migration [200]. Despite the numerous studies that emphasized important roles of sphingolipids in cancer and metabolic diseases, a very limited number of sphingolipid-targeting drugs entered clinical trials. Complete understanding of these bioactive metabolites and enzymes remains to be fully elucidated. The first-in-class clinical inhibitor of SK2, ABC29464, successfully achieved phase I clinical trial and proceeded to phase II as an anti-cancer drug [201]. Therefore, new effective sphingolipid therapeutic targets are expected to develop as a result of the numerous emerging studies in the field. These potential discoveries are predicted to advance the standard of care therapies by overcoming tumor resistance and developing new effective diagnostic and prognostic sphingolipidomic-based tests.

Author Contributions: Conceptualization, Y.H.Z. and M.F.; writing—original draft preparation, M.F.; writing—review and editing, M.F., P.A., A.A.D., M.M. and Y.H.Z.; project administration, Y.H.Z.; funding acquisition, Y.H.Z. All authors have read and agreed to the published version of the manuscript.

Acknowledgments: Youssef H. Zeidan is supported by NIH grant 1R01CA227493 and AUB MPP grant.

Conflicts of Interest: The authors declare no conflict of interest.

Abbreviations

Apaf1	Apoptotic protease activating factor 1
ATM	Ataxia telangiectasia mutated
ATR	Ataxia telangiectasia and Rad-3 related
BD	Base damage
Cer	Ceramide
CERK	Ceramide kinase
CHO	Cholesterol
C1P	Ceramide-1-phosphate
DDR	DNA damage response
DNAPK	DNA-dependent protein kinase
DSB	Double strand break
Gy	Gray (absorbed dose)
HR	Homologous recombination
IR	Ionizing radiation

MDM2	Mouse double minute 2
NHEJ	Non-homologous end joining
PC	Phosphatidylcholine
PUMA	p53 upregulated modulator of apoptosis
p53AIP1	p53 regulated apoptosis inducing protein 1
ROS	Reactive oxygen species
SK	Sphingosine kinase
SM	Sphingomyelin
SMPDL3b	Sphingomyelin phosphodiesterase acid like-3b
S1P	Sphingosine-1-phosphate
SPP	S1P phosphatase
SSB	Single strand break
TRAIL	TNF-related apoptosis-inducing ligand

References

1. Carroll, B.; Donaldson, J.C.; Obeid, L. Sphingolipids in the DNA damage response. *Adv. Biol. Regul.* **2015**, *58*, 38–52. [CrossRef] [PubMed]
2. Presa, N.; Gomez-Larrauri, A.; Dominguez-Herrera, A.; Trueba, M.; Gomez-Muñoz, A. Novel signaling aspects of ceramide 1-phosphate. *Biochim. Biophys. Acta (BBA) Mol. Cell Biol. Lipids* **2020**, *1865*, 158630. [CrossRef] [PubMed]
3. Taniguchi, M.; Okazaki, T. The role of sphingomyelin and sphingomyelin synthases in cell death, proliferation and migration—From cell and animal models to human disorders. *Biochim. Biophys. Acta (BBA) Mol. Cell Biol. Lipids* **2014**, *1841*, 692–703. [CrossRef] [PubMed]
4. Hannun, Y.A.; Obeid, L.M. Sphingolipids and their metabolism in physiology and disease. *Nat. Rev. Mol. Cell Biol.* **2018**, *19*, 175. [CrossRef] [PubMed]
5. López-Otín, C.; Blasco, M.A.; Partridge, L.; Serrano, M.; Kroemer, G. The hallmarks of aging. *Cell* **2013**, *153*, 1194–1217. [CrossRef]
6. Jackson, S.P.; Bartek, J. The DNA-damage response in human biology and disease. *Nature* **2009**, *461*, 1071. [CrossRef]
7. Saha, G.B. *Physics and Radiobiology of Nuclear Medicine*; Springer Science & Business Media: Berlin, Germany, 2012.
8. Hall, E.J.; Giaccia, A.J. *Radiobiology for the Radiologist*; Lippincott Williams & Wilkins: Philadelphia, PA, USA, 2006; Volume 6.
9. Azzam, E.I.; Jay-Gerin, J.-P.; Pain, D. Ionizing radiation-induced metabolic oxidative stress and prolonged cell injury. *Cancer Lett.* **2012**, *327*, 48–60. [CrossRef]
10. Negritto, C. Repairing double-strand DNA breaks. *Nat. Educ.* **2010**, *3*, 26.
11. Cornforth, M.N.; Bedford, J.S. A quantitative comparison of potentially lethal damage repair and the rejoining of interphase chromosome breaks in low passage normal human fibroblasts. *Radiat. Res.* **1987**, *111*, 385–405. [CrossRef]
12. Bauer, N.C.; Corbett, A.H.; Doetsch, P.W. The current state of eukaryotic DNA base damage and repair. *Nucleic Acids Res.* **2015**, *43*, 10083–10101. [CrossRef]
13. Sharma, V.; Collins, L.B.; Chen, T.-h.; Herr, N.; Takeda, S.; Sun, W.; Swenberg, J.A.; Nakamura, J. Oxidative stress at low levels can induce clustered DNA lesions leading to nhej mediated mutations. *Oncotarget* **2016**, *7*, 25377. [CrossRef] [PubMed]
14. Bodgi, L.; Foray, N. The nucleo-shuttling of the atm protein as a basis for a novel theory of radiation response: Resolution of the linear-quadratic model. *Int. J. Radiat. Biol.* **2016**, *92*, 117–131. [CrossRef]
15. Isaksson, M.; Raaf, C.L. *Environmental Radioactivity and Emergency Preparedness*; CRC Press: Boca Raton, FL, USA, 2017.
16. Thompson, L.H. Recognition, signaling, and repair of DNA double-strand breaks produced by ionizing radiation in mammalian cells: The molecular choreography. *Mutat. Res./Rev. Mutat. Res.* **2012**, *751*, 158–246. [CrossRef]
17. Lindahl, T. Instability and decay of the primary structure of DNA. *Nature* **1993**, *362*, 709–715. [CrossRef]

18. Cadet, J.; Wagner, J.R. DNA base damage by reactive oxygen species, oxidizing agents, and uv radiation. *Cold Spring Harb. Perspect. Biol.* **2013**, *5*, a012559. [CrossRef] [PubMed]

19. Bokhari, B.; Sharma, S. Stress marks on the genome: Use or lose? *Int. J. Mol. Sci.* **2019**, *20*, 364. [CrossRef] [PubMed]

20. Klaunig, J.E.; Kamendulis, L.M.; Hocevar, B.A. Oxidative stress and oxidative damage in carcinogenesis. *Toxicol. Pathol.* **2010**, *38*, 96–109. [CrossRef]

21. Harper, J.W.; Elledge, S.J. The DNA damage response: Ten years after. *Mol. Cell* **2007**, *28*, 739–745. [CrossRef]

22. Rouse, J.; Jackson, S.P. Interfaces between the detection, signaling, and repair of DNA damage. *Science* **2002**, *297*, 547–551. [CrossRef]

23. Jeggo, P.; Löbrich, M. DNA double-strand breaks: Their cellular and clinical impact? *Oncogene* **2007**, *26*, 7717. [CrossRef]

24. Woodbine, L.; Gennery, A.R.; Jeggo, P.A. The clinical impact of deficiency in DNA non-homologous end-joining. *DNA Repair* **2014**, *16*, 84–96. [CrossRef] [PubMed]

25. Riballo, E.; Kühne, M.; Rief, N.; Doherty, A.; Smith, G.C.; Recio, M.a.-J.; Reis, C.; Dahm, K.; Fricke, A.; Krempler, A. A pathway of double-strand break rejoining dependent upon atm, artemis, and proteins locating to γ-h2ax foci. *Mol. Cell* **2004**, *16*, 715–724. [CrossRef] [PubMed]

26. Sonoda, E.; Hochegger, H.; Saberi, A.; Taniguchi, Y.; Takeda, S. Differential usage of non-homologous end-joining and homologous recombination in double strand break repair. *DNA Repair* **2006**, *5*, 1021–1029. [CrossRef] [PubMed]

27. Beucher, A.; Birraux, J.; Tchouandong, L.; Barton, O.; Shibata, A.; Conrad, S.; Goodarzi, A.A.; Krempler, A.; Jeggo, P.A.; Löbrich, M. Atm and artemis promote homologous recombination of radiation-induced DNA double-strand breaks in g2. *EMBO J.* **2009**, *28*, 3413–3427. [CrossRef]

28. Abbotts, R.; Wilson III, D.M. Coordination of DNA single strand break repair. *Free Radic. Biol. Med.* **2017**, *107*, 228–244. [CrossRef]

29. Khoronenkova, S.V.; Dianov, G.L. Atm prevents dsb formation by coordinating ssb repair and cell cycle progression. *Proc. Natl. Acad. Sci. USA* **2015**, *112*, 3997–4002. [CrossRef]

30. Fortini, P.; Pascucci, B.; Parlanti, E.; D'errico, M.; Simonelli, V.; Dogliotti, E. The base excision repair: Mechanisms and its relevance for cancer susceptibility. *Biochimie* **2003**, *85*, 1053–1071. [CrossRef]

31. Czornak, K.; Chughtai, S.; Chrzanowska, K.H. Mystery of DNA repair: The role of the mrn complex and atm kinase in DNA damage repair. *J. Appl. Genet.* **2008**, *49*, 383–396. [CrossRef]

32. Cimprich, K.A.; Cortez, D. Atr: An essential regulator of genome integrity. *Nat. Rev. Mol. Cell Biol.* **2008**, *9*, 616–627. [CrossRef]

33. Shiloh, Y. Atm and related protein kinases: Safeguarding genome integrity. *Nat. Rev. Cancer* **2003**, *3*, 155–168. [CrossRef]

34. Kastan, M.B.; Onyekwere, O.; Sidransky, D.; Vogelstein, B.; Craig, R.W. Participation of p53 protein in the cellular response to DNA damage. *Cancer Res.* **1991**, *51*, 6304–6311. [CrossRef]

35. Goto, H.; Izawa, I.; Li, P.; Inagaki, M. Novel regulation of checkpoint kinase 1: Is checkpoint kinase 1 a good candidate for anti-cancer therapy? *Cancer Sci.* **2012**, *103*, 1195–1200. [CrossRef]

36. Dasika, G.K.; Lin, S.-C.J.; Zhao, S.; Sung, P.; Tomkinson, A.; Lee, E.Y.P. DNA damage-induced cell cycle checkpoints and DNA strand break repair in development and tumorigenesis. *Oncogene* **1999**, *18*, 7883–7899. [CrossRef] [PubMed]

37. Paull, T.T.; Rogakou, E.P.; Yamazaki, V.; Kirchgessner, C.U.; Gellert, M.; Bonner, W.M. A critical role for histone h2ax in recruitment of repair factors to nuclear foci after DNA damage. *Curr. Biol.* **2000**, *10*, 886–895. [CrossRef]

38. Rappold, I.; Iwabuchi, K.; Date, T.; Chen, J. Tumor suppressor p53 binding protein 1 (53bp1) is involved in DNA damage–signaling pathways. *J. Cell Biol.* **2001**, *153*, 613–620. [CrossRef] [PubMed]

39. Paull, T.T. Mechanisms of atm activation. *Annu. Rev. Biochem.* **2015**, *84*, 711–738. [CrossRef] [PubMed]

40. Burma, S.; Chen, B.P.; Murphy, M.; Kurimasa, A.; Chen, D.J. Atm phosphorylates histone h2ax in response to DNA double-strand breaks. *J. Biol. Chem.* **2001**, *276*, 42462–42467. [CrossRef]

41. Ward, I.M.; Chen, J. Histone h2ax is phosphorylated in an atr-dependent manner in response to replicational stress. *J. Biol. Chem.* **2001**, *276*, 47759–47762. [CrossRef]

42. Ward, I.M.; Minn, K.; Chen, J. Uv-induced ataxia-telangiectasia-mutated and rad3-related (atr) activation requires replication stress. *J. Biol. Chem.* **2004**, *279*, 9677–9680. [CrossRef]

43. Bensimon, A.; Schmidt, A.; Ziv, Y.; Elkon, R.; Wang, S.-Y.; Chen, D.J.; Aebersold, R.; Shiloh, Y. Atm-dependent and-independent dynamics of the nuclear phosphoproteome after DNA damage. *Sci. Signal.* **2010**, *3*, rs3. [CrossRef]

44. Bakkenist, C.J.; Kastan, M.B. DNA damage activates atm through intermolecular autophosphorylation and dimer dissociation. *Nature* **2003**, *421*, 499. [CrossRef] [PubMed]

45. Guo, Z.; Kozlov, S.; Lavin, M.F.; Person, M.D.; Paull, T.T. Atm activation by oxidative stress. *Science* **2010**, *330*, 517–521. [CrossRef] [PubMed]

46. Canman, C.E.; Lim, D.-S.; Cimprich, K.A.; Taya, Y.; Tamai, K.; Sakaguchi, K.; Appella, E.; Kastan, M.B.; Siliciano, J.D. Activation of the atm kinase by ionizing radiation and phosphorylation of p53. *Science* **1998**, *281*, 1677–1679. [CrossRef] [PubMed]

47. Bodgi, L.; Granzotto, A.; Devic, C.; Vogin, G.; Lesne, A.; Bottollier-Depois, J.-F.; Victor, J.-M.; Maalouf, M.; Fares, G.; Foray, N. A single formula to describe radiation-induced protein relocalization: Towards a mathematical definition of individual radiosensitivity. *J. Theor. Biol.* **2013**, *333*, 135–145. [CrossRef]

48. Ouenzar, F.; Hendzel, M.J.; Weinfeld, M. Shuttling towards a predictive assay for radiotherapy. *Transl. Cancer Res.* **2016**, *5*, S742–S746. [CrossRef]

49. Ferlazzo, M.L.; Sonzogni, L.; Granzotto, A.; Bodgi, L.; Lartin, O.; Devic, C.; Vogin, G.; Pereira, S.; Foray, N. Mutations of the huntington's disease protein impact on the atm-dependent signaling and repair pathways of the radiation-induced DNA double-strand breaks: Corrective effect of statins and bisphosphonates. *Mol. Neurobiol.* **2014**, *49*, 1200–1211. [CrossRef]

50. Pereira, S.; Bodgi, L.; Duclos, M.; Canet, A.; Ferlazzo, M.L.; Devic, C.; Granzotto, A.; Deneuve, S.; Vogin, G.; Foray, N. Fast and binary assay for predicting radiosensitivity based on the theory of atm nucleo-shuttling: Development, validation, and performance. *Int. J. Radiat. Oncol. Biol. Phys.* **2018**, *100*, 353–360. [CrossRef]

51. Ferlazzo, M.L.; Bach-Tobdji, M.K.E.; Djerad, A.; Sonzogni, L.; Devic, C.; Granzotto, A.; Bodgi, L.; Bachelet, J.-T.; Djefal-Kerrar, A.; Hennequin, C. Radiobiological characterization of tuberous sclerosis: A delay in the nucleo-shuttling of atm may be responsible for radiosensitivity. *Mol. Neurobiol.* **2018**, *55*, 4973–4983. [CrossRef]

52. Surova, O.; Zhivotovsky, B. Various modes of cell death induced by DNA damage. *Oncogene* **2013**, *32*, 3789–3797. [CrossRef]

53. Lombard, D.B.; Chua, K.F.; Mostoslavsky, R.; Franco, S.; Gostissa, M.; Alt, F.W. DNA repair, genome stability, and aging. *Cell* **2005**, *120*, 497–512. [CrossRef] [PubMed]

54. Hasty, P.; Campisi, J.; Hoeijmakers, J.; Van Steeg, H.; Vijg, J. Aging and genome maintenance: Lessons from the mouse? *Science* **2003**, *299*, 1355–1359. [CrossRef]

55. Tchkonia, T.; Zhu, Y.; Van Deursen, J.; Campisi, J.; Kirkland, J.L. Cellular senescence and the senescent secretory phenotype: Therapeutic opportunities. *J. Clin. Investig.* **2013**, *123*, 966–972. [CrossRef]

56. Baskar, R.; Lee, K.A.; Yeo, R.; Yeoh, K.-W. Cancer and radiation therapy: Current advances and future directions. *Int. J. Med Sci.* **2012**, *9*, 193. [CrossRef] [PubMed]

57. Eriksson, D.; Stigbrand, T. Radiation-induced cell death mechanisms. *Tumor Biol.* **2010**, *31*, 363–372. [CrossRef] [PubMed]

58. Hollstein, M.; Sidransky, D.; Vogelstein, B.; Harris, C.C. P53 mutations in human cancers. *Science* **1991**, *253*, 49–53. [CrossRef] [PubMed]

59. Soussi, T.; Béroud, C. Assessing tp53 status in human tumours to evaluate clinical outcome. *Nat. Rev. Cancer* **2001**, *1*, 233. [CrossRef] [PubMed]

60. Soussi, T.; Lozano, G. P53 mutation heterogeneity in cancer. *Biochem. Biophys. Res. Commun.* **2005**, *331*, 834–842. [CrossRef]

61. Helton, E.S.; Chen, X. P53 modulation of the DNA damage response. *J. Cell. Biochem.* **2007**, *100*, 883–896. [CrossRef]

62. Wang, Z.; Li, B. Mdm2 links genotoxic stress and metabolism to p53. *Protein Cell* **2010**, *1*, 1063–1072. [CrossRef]

63. Fu, W.; Ma, Q.; Chen, L.; Li, P.; Zhang, M.; Ramamoorthy, S.; Nawaz, Z.; Shimojima, T.; Wang, H.; Yang, Y. Mdm2 acts downstream of p53 as an e3 ligase to promote foxo ubiquitination and degradation. *J. Biol. Chem.* **2009**, *284*, 13987–14000. [CrossRef]

64. Barak, Y.; Gottlieb, E.; Juven-Gershon, T.; Oren, M. Regulation of mdm2 expression by p53: Alternative promoters produce transcripts with nonidentical translation potential. *Genes Dev.* **1994**, *8*, 1739–1749. [CrossRef] [PubMed]

65. Iwakuma, T.; Lozano, G. Mdm2, an introduction. *Mol. Cancer Res.* **2003**, *1*, 993–1000. [PubMed]

66. Igney, F.H.; Krammer, P.H. Death and anti-death: Tumour resistance to apoptosis. *Nat. Rev. Cancer* **2002**, *2*, 277. [CrossRef] [PubMed]

67. Gewirtz, D.A.; Holt, S.E.; Elmore, L.W. Accelerated senescence: An emerging role in tumor cell response to chemotherapy and radiation. *Biochem. Pharmacol.* **2008**, *76*, 947–957. [CrossRef]

68. Rodier, F.; Coppé, J.-P.; Patil, C.K.; Hoeijmakers, W.A.; Muñoz, D.P.; Raza, S.R.; Freund, A.; Campeau, E.; Davalos, A.R.; Campisi, J. Persistent DNA damage signalling triggers senescence-associated inflammatory cytokine secretion. *Nat. Cell Biol.* **2009**, *11*, 973–979. [CrossRef]

69. Batalni, J.P.; Belloir, C.; Mazabraud, A.; Pilleron, J.P.; Cartigny, A.; Jaulerry, C.; Ghossein, N.A. Desmoid tumors in adults: The role of radiotherapy in their management. *Am. J. Surg.* **1988**, *155*, 754–760. [CrossRef]

70. Cox, J.D.; Kline, R.W. Do prostatic biopsies 12 months or more after external irradiation for adenocarcinoma, stage iii, predict long-term survival? *Int. J. Radiat. Oncol. Biol. Phys.* **1983**, *9*, 299–303. [CrossRef]

71. Shay, J.W.; Roninson, I.B. Hallmarks of senescence in carcinogenesis and cancer therapy. *Oncogene* **2004**, *23*, 2919. [CrossRef]

72. Suzuki, K.; Mori, I.; Nakayama, Y.; Miyakoda, M.; Kodama, S.; Watanabe, M. Radiation-induced senescence-like growth arrest requires tp53 function but not telomere shortening. *Radiat. Res.* **2001**, *155*, 248–253. [CrossRef]

73. Ninomiya, Y.; Cui, X.; Yasuda, T.; Wang, B.; Yu, D.; Sekine-Suzuki, E.; Nenoi, M. Arsenite induces premature senescence via p53/p21 pathway as a result of DNA damage in human malignant glioblastoma cells. *BMB Rep.* **2014**, *47*, 575. [CrossRef]

74. Pawlowska, E.; Szczepanska, J.; Szatkowska, M.; Blasiak, J. An interplay between senescence, apoptosis and autophagy in glioblastoma multiforme—Role in pathogenesis and therapeutic perspective. *Int. J. Mol. Sci.* **2018**, *19*, 889. [CrossRef] [PubMed]

75. Hock, A.K.; Vousden, K.H. Tumor suppression by p53: Fall of the triumvirate? *Cell* **2012**, *149*, 1183–1185. [CrossRef] [PubMed]

76. Kim, H.; Lee, J.M.; Lee, G.; Bhin, J.; Oh, S.K.; Kim, K.; Pyo, K.E.; Lee, J.S.; Yim, H.Y.; Kim, K.I. DNA damage-induced rorα is crucial for p53 stabilization and increased apoptosis. *Mol. Cell* **2011**, *44*, 797–810. [CrossRef]

77. Baliga, B.C.; Colussi, P.A.; Read, S.H.; Dias, M.M.; Jans, D.A.; Kumar, S. Role of prodomain in importin-mediated nuclear localization and activation of caspase-2. *J. Biol. Chem.* **2003**, *278*, 4899–4905. [CrossRef] [PubMed]

78. Tinel, A.; Tschopp, J. The piddosome, a protein complex implicated in activation of caspase-2 in response to genotoxic stress. *Science* **2004**, *304*, 843–846. [CrossRef]

79. Schmitt, C.A. Senescence, apoptosis and therapy—Cutting the lifelines of cancer. *Nat. Rev. Cancer* **2003**, *3*, 286. [CrossRef]

80. Galluzzi, L.; Maiuri, M.; Vitale, I.; Zischka, H.; Castedo, M.; Zitvogel, L.; Kroemer, G. Cell Death Modalities: Classification and Pathophysiological Implications. *Cell Death Differ.* **2007**, *14*, 1237–1243. [CrossRef]

81. Eriksson, D.; Joniani, H.M.; Sheikholvaezin, A.; Löfroth, P.-O.; Johansson, L.; Åhlström, K.R.; Stigbrand, T. Combined low dose radio-and radioimmunotherapy of experimental hela hep 2 tumours. *Eur. J. Nucl. Med. Mol. Imaging* **2003**, *30*, 895–906. [CrossRef]

82. Eriksson, D.; Löfroth, P.-O.; Johansson, L.; Riklund, K.Å.; Stigbrand, T. Cell cycle disturbances and mitotic catastrophes in hela hep2 cells following 2.5 to 10 gy of ionizing radiation. *Clin. Cancer Res.* **2007**, *13*, 5501s–5508s. [CrossRef]

83. Castedo, M.; Kroemer, G. Mitotic catastrophe: A special case of apoptosis. *J. Soc. Biol.* **2004**, *198*, 97–103. [CrossRef]

84. Erenpreisa, J.; Kalejs, M.; Ianzini, F.; Kosmacek, E.A.; Mackey, M.; Emzinsh, D.; Cragg, M.S.; Ivanov, A.; Illidge, T.M. Segregation of genomes in polyploid tumour cells following mitotic catastrophe. *Cell Biol. Int.* **2005**, *29*, 1005–1011. [CrossRef]

85. Roninson, I.B.; Broude, E.V.; Chang, B.-D. If not apoptosis, then what? Treatment-induced senescence and mitotic catastrophe in tumor cells. *Drug Resist. Updates* **2001**, *4*, 303–313. [CrossRef] [PubMed]

86. Ianzini, F.; Bertoldo, A.; Kosmacek, E.A.; Phillips, S.L.; Mackey, M.A. Lack of p53 function promotes radiation-induced mitotic catastrophe in mouse embryonic fibroblast cells. *Cancer Cell Int.* **2006**, *6*, 11. [CrossRef] [PubMed]

87. Bourke, E.; Dodson, H.; Merdes, A.; Cuffe, L.; Zachos, G.; Walker, M.; Gillespie, D.; Morrison, C.G. DNA damage induces chk1-dependent centrosome amplification. *EMBO Rep.* **2007**, *8*, 603–609. [CrossRef] [PubMed]

88. Dodson, H.; Wheatley, S.P.; Morrison, C.G. Involvement of centrosome amplification in radiation-induced mitotic catastrophe. *Cell Cycle* **2007**, *6*, 364–370. [CrossRef]

89. Kawamura, K.; Fujikawa-Yamamoto, K.; Ozaki, M.; Iwabuchi, K.; Nakashima, H.; Domiki, C.; Morita, N.; Inoue, M.; Tokunaga, K.; Shiba, N. Centrosome hyperamplification and chromosomal damage after exposure to radiation. *Oncology* **2004**, *67*, 460–470. [CrossRef] [PubMed]

90. Kawamura, K.; Morita, N.; Domiki, C.; Fujikawa-Yamamoto, K.; Hashimoto, M.; Iwabuchi, K.; Suzuki, K. Induction of centrosome amplification in p53 sirna-treated human fibroblast cells by radiation exposure. *Cancer Sci.* **2006**, *97*, 252–258. [CrossRef]

91. Hanashiro, K.; Kanai, M.; Geng, Y.; Sicinski, P.; Fukasawa, K. Roles of cyclins a and e in induction of centrosome amplification in p53-compromised cells. *Oncogene* **2008**, *27*, 5288. [CrossRef]

92. Vakifahmetoglu, H.; Olsson, M.; Zhivotovsky, B. Death through a tragedy: Mitotic catastrophe. *Cell Death Differ.* **2008**, *15*, 1153. [CrossRef]

93. Kroemer, G.; Galluzzi, L.; Vandenabeele, P.; Abrams, J.; Alnemri, E.S.; Baehrecke, E.; Blagosklonny, M.; El-Deiry, W.; Golstein, P.; Green, D. Classification of cell death: Recommendations of the nomenclature committee on cell death 2009. *Cell Death Differ.* **2009**, *16*, 3. [CrossRef]

94. Brandsma, D.; Stalpers, L.; Taal, W.; Sminia, P.; van den Bent, M.J. Clinical features, mechanisms, and management of pseudoprogression in malignant gliomas. *Lancet Oncol.* **2008**, *9*, 453–461. [CrossRef]

95. Krysko, O.; Aaes, T.L.; Bachert, C.; Vandenabeele, P.; Krysko, D. Many faces of damps in cancer therapy. *Cell Death Dis.* **2013**, *4*, e631. [CrossRef]

96. Proskuryakov, S.Y.; Konoplyannikov, A.G.; Gabai, V.L. Necrosis: A specific form of programmed cell death? *Exp. Cell Res.* **2003**, *283*, 1–16. [CrossRef]

97. Cohen–Jonathan, E.; Bernhard, E.J.; McKenna, W.G. How does radiation kill cells? *Curr. Opin. Chem. Biol.* **1999**, *3*, 77–83. [CrossRef]

98. Gault, C.R.; Obeid, L.M.; Hannun, Y.A. An overview of sphingolipid metabolism: From synthesis to breakdown. In *Sphingolipids as Signaling and Regulatory Molecules*; Springer: Berlin, Germany, 2010; pp. 1–23.

99. Abou Daher, A.; El Jalkh, T.; Eid, A.; Fornoni, A.; Marples, B.; Zeidan, Y. Translational aspects of sphingolipid metabolism in renal disorders. *Int. J. Mol. Sci.* **2017**, *18*, 2528. [CrossRef] [PubMed]

100. Zeidan, Y.H.; Hannun, Y.A. Translational aspects of sphingolipid metabolism. *Trends Mol. Med.* **2007**, *13*, 327–336. [CrossRef]

101. Hannun, Y.A.; Obeid, L.M. Principles of bioactive lipid signalling: Lessons from sphingolipids. *Nat. Rev. Mol. Cell Biol.* **2008**, *9*, 139. [CrossRef]

102. Fröhlich, F.; Petit, C.; Kory, N.; Christiano, R.; Hannibal-Bach, H.-K.; Graham, M.; Liu, X.; Ejsing, C.S.; Farese, R.V., Jr.; Walther, T.C. The garp complex is required for cellular sphingolipid homeostasis. *Elife* **2015**, *4*, e08712. [CrossRef]

103. te Vruchte, D.; Lloyd-Evans, E.; Veldman, R.J.; Neville, D.C.; Dwek, R.A.; Platt, F.M.; van Blitterswijk, W.J.; Sillence, D.J. Accumulation of glycosphingolipids in niemann-pick c disease disrupts endosomal transport. *J. Biol. Chem.* **2004**, *279*, 26167–26175. [CrossRef]

104. Höglinger, D.; Haberkant, P.; Aguilera-Romero, A.; Riezman, H.; Porter, F.D.; Platt, F.M.; Galione, A.; Schultz, C. Intracellular sphingosine releases calcium from lysosomes. *Elife* **2015**, *4*, e10616. [CrossRef]

105. Schuchman, E. The pathogenesis and treatment of acid sphingomyelinase-deficient niemann–pick disease. *J. Inherit. Metab. Dis.* **2007**, *30*, 654. [CrossRef] [PubMed]

106. Merscher, S.; Fornoni, A. Podocyte pathology and nephropathy–sphingolipids in glomerular diseases. *Front. Endocrinol.* **2014**, *5*, 127. [CrossRef] [PubMed]

107. Boath, A.; Graf, C.; Lidome, E.; Ullrich, T.; Nussbaumer, P.; Bornancin, F. Regulation and traffic of ceramide 1-phosphate produced by ceramide kinase comparative analysis to glucosylceramide and sphingomyelin. *J. Biol. Chem.* **2008**, *283*, 8517–8526. [CrossRef]

108. Ogawa, C.; Kihara, A.; Gokoh, M.; Igarashi, Y. Identification and characterization of a novel human sphingosine-1-phosphate phosphohydrolase, hspp2. *J. Biol. Chem.* **2003**, *278*, 1268–1272. [CrossRef] [PubMed]

109. Mandala, S.M.; Thornton, R.; Galve-Roperh, I.; Poulton, S.; Peterson, C.; Olivera, A.; Bergstrom, J.; Kurtz, M.B.; Spiegel, S. Molecular cloning and characterization of a lipid phosphohydrolase that degrades sphingosine-1-phosphate and induces cell death. *Proc. Natl. Acad. Sci. USA* **2000**, *97*, 7859–7864. [CrossRef]

110. Pyne, S.; Long, J.; Ktistakis, N.; Pyne, N. Lipid phosphate phosphatases and lipid phosphate signalling. *Biochem. Soc. Trans.* **2005**, *33*, 1370–1374. [CrossRef] [PubMed]

111. Ikeda, M.; Kihara, A.; Igarashi, Y. Sphingosine-1-phosphate lyase spl is an endoplasmic reticulum-resident, integral membrane protein with the pyridoxal 5′-phosphate binding domain exposed to the cytosol. *Biochem. Biophys. Res. Commun.* **2004**, *325*, 338–343. [CrossRef] [PubMed]

112. Albi, E. Role of intranuclear lipids in health and disease. *Clin. Lipidol.* **2011**, *6*, 59–69. [CrossRef]

113. Ledeen, R.W.; Wu, G. Thematic review series: Sphingolipids. Nuclear sphingolipids: Metabolism and signaling. *J. Lipid Res.* **2008**, *49*, 1176–1186. [CrossRef]

114. Bernardini, I.; Bartoccini, E.; Viola Magni, M. Nuclear lipids and cell fate. *Dyn. Cell Biol.* **2017**, *1*, 42–47.

115. Divecha, N.; Banfic, H.; Irvine, R.F. Inositides and the nucleus and inositides in the nucleus. *Cell* **1993**, *74*, 405–407. [CrossRef]

116. Cocco, L.; Martelli, A.M.; Gilmour, R.S.; Rhee, S.G.; Manzoli, F.A. Nuclear phospholipase c and signaling. *Biochim. Biophys. Acta* **2001**, *1530*, 1. [CrossRef]

117. Maraldi, N.; Cocco, L.; Capitani, S.; Mazzotti, G.; Barnabei, O.; Manzoli, F. Lipid-dependent nuclear signalling: Morphological and functional features. *Adv. Enzym. Regul.* **1994**, *34*, 129–143. [CrossRef]

118. Kleuser, B.; Maceyka, M.; Milstien, S.; Spiegel, S. Stimulation of nuclear sphingosine kinase activity by platelet-derived growth factor. *FEBS Lett.* **2001**, *503*, 85–90. [CrossRef]

119. Ledeen, R.W.; Wu, G. Sphingolipids of the nucleus and their role in nuclear signaling. *Biochim. Biophys. Acta (BBA) Mol. Cell Biol. Lipids* **2006**, *1761*, 588–598. [CrossRef]

120. Neitcheva, T.; Peeva, D. Phospholipid composition, phospholipase a2 and sphingomyelinase activities in rat liver nuclear membrane and matrix. *Int. J. Biochem. Cell Biol.* **1995**, *27*, 995–1001. [CrossRef]

121. Pliss, A.; Kuzmin, A.N.; Kachynski, A.V.; Prasad, P.N. Nonlinear optical imaging and raman microspectrometry of the cell nucleus throughout the cell cycle. *Biophys. J.* **2010**, *99*, 3483–3491. [CrossRef]

122. Albi, E.; Mersel, M.; Leray, C.; Tomassoni, M.; Viola-Magni, M. Rat liver chromatin phospholipids. *Lipids* **1994**, *29*, 715–719. [CrossRef]

123. Cave, C.F.; Gahan, P. A cytochemical and autoradiographic investigation of nucleolar phospholipids. *Caryologia* **1970**, *23*, 303–312. [CrossRef]

124. Fu, P.; Ebenezer, D.L.; Ha, A.W.; Suryadevara, V.; Harijith, A.; Natarajan, V. Nuclear lipid mediators: Role of nuclear sphingolipids and sphingosine-1-phosphate signaling in epigenetic regulation of inflammation and gene expression. *J. Cell. Biochem.* **2018**, *119*, 6337–6353. [CrossRef]

125. Tomassoni, M.-L.; Amori, D.; Magni, M.V. Changes of nuclear membrane lipid composition affect rna nucleocytoplasmic transport. *Biochem. Biophys. Res. Commun.* **1999**, *258*, 476–481. [CrossRef] [PubMed]

126. Albi, E.; Tomassoni, M.L.; Viola-Magni, M. Effect of lipid composition on rat liver nuclear membrane fluidity. *Cell Biochem. Funct. Cell. Biochem. Modul. Act. Agents Dis.* **1997**, *15*, 181–190. [CrossRef]

127. Xie, X.; Wu, G.; Lu, Z.H.; Ledeen, R.W. Potentiation of a sodium–calcium exchanger in the nuclear envelope by nuclear gm1 ganglioside. *J. Neurochem.* **2002**, *81*, 1185–1195. [CrossRef]

128. Albi, E.; Magni, M.V. Chromatin-associated sphingomyelin: Metabolism in relation to cell function. *Cell Biochem. Funct. Cell. Biochem. Modul. Act. Agents Dis.* **2003**, *21*, 211–215. [CrossRef]

129. Lucki, N.C.; Sewer, M.B. Nuclear sphingolipid metabolism. *Annu. Rev. Physiol.* **2012**, *74*, 131–151. [CrossRef] [PubMed]

130. Alessenko, A.; Chatterjee, S. Neutral sphingomyelinase: Localization in rat liver nuclei and involvement in regeneration/proliferation. *Mol. Cell. Biochem.* **1995**, *143*, 169–174. [CrossRef] [PubMed]

131. Wu, G.; Lu, Z.H.; Ledeen, R.W. Gm1 ganglioside in the nuclear membrane modulates nuclear calcium homeostasis during neurite outgrowth. *J. Neurochem.* **1995**, *65*, 1419–1422. [CrossRef] [PubMed]

132. Micheli, M.; Albi, E.; Leray, C.; Magni, M.V. Nuclear sphingomyelin protects rna from rnase action. *FEBS Lett.* **1998**, *431*, 443–447. [CrossRef]

133. Tsugane, K.; Tamiya-Koizumi, K.; Nagino, M.; Nimura, Y.; Yoshida, S. A possible role of nuclear ceramide and sphingosine in hepatocyte apoptosis in rat liver. *J. Hepatol.* **1999**, *31*, 8–17. [CrossRef]

134. Rossi, G.; Magni, M.V.; Albi, E. Sphingomyelin-cholesterol and double stranded rna relationship in the intranuclear complex. *Arch. Biochem. Biophys.* **2007**, *459*, 27–32. [CrossRef]

135. Albi, E.; Cataldi, S.; Rossi, G.; Magni, M.V.; Toller, M.; Casani, S.; Perrella, G. The nuclear ceramide/diacylglycerol balance depends on the physiological state of thyroid cells and changes during uv-c radiation-induced apoptosis. *Arch. Biochem. Biophys.* **2008**, *478*, 52–58. [CrossRef]

136. Hait, N.C.; Allegood, J.; Maceyka, M.; Strub, G.M.; Harikumar, K.B.; Singh, S.K.; Luo, C.; Marmorstein, R.; Kordula, T.; Milstien, S. Regulation of histone acetylation in the nucleus by sphingosine-1-phosphate. *Science* **2009**, *325*, 1254–1257. [CrossRef] [PubMed]

137. Gupta, S.; Maurya, M.R.; Merrill, A.H., Jr.; Glass, C.K.; Subramaniam, S. Integration of lipidomics and transcriptomics data towards a systems biology model of sphingolipid metabolism. *BMC Syst. Biol.* **2011**, *5*, 26. [CrossRef]

138. Albi, E.; Viola Magni, M. Sphingomyelin: A small-big molecule in the nucleus. *Recent Res. Dev. Biophys. Biochem.* **2006**, *37*, 211–227.

139. Exton, J. Signaling through phosphatidylcholine breakdown. *J. Biol. Chem.* **1990**, *265*, 1–4.

140. Reszka, A.A.; Halasy-Nagy, J.; Rodan, G.A. Nitrogen-bisphosphonates block retinoblastoma phosphorylation and cell growth by inhibiting the cholesterol biosynthetic pathway in a keratinocyte model for esophageal irritation. *Mol. Pharmacol.* **2001**, *59*, 193–202. [CrossRef] [PubMed]

141. Novello, F.; Muchmore, J.; Bonora, B.; Capitani, S.; Manzoli, F. Effect of phospholipids on the activity of DNA polymerase i from e. Coli. *Ital. J. Biochem.* **1975**, *24*, 325–334.

142. Martelli, A.M.; Follo, M.Y.; Evangelisti, C.; Fala, F.; Fiume, R.; Billi, A.M.; Cocco, L. Nuclear inositol lipid metabolism: More than just second messenger generation? *J. Cell. Biochem.* **2005**, *96*, 285–292. [CrossRef]

143. Scassellati, C.; Albi, E.; Cmarko, D.; Tiberi, C.; Cmarkova, J.; Bouchet-Marquis, C.; Verschure, P.J.; Van Driel, R.; Magni, M.V.; Fakan, S. Intranuclear sphingomyelin is associated with transcriptionally active chromatin and plays a role in nuclear integrity. *Biol. Cell* **2010**, *102*, 361–375. [CrossRef]

144. Albi, E.; Lazzarini, R.; Magni, M.V. Reverse sphingomyelin-synthase in rat liver chromatin. *FEBS Lett.* **2003**, *549*, 152–156. [CrossRef]

145. Albi, E.; Magni, M.V. Sphingomyelin synthase in rat liver nuclear membrane and chromatin. *FEBS Lett.* **1999**, *460*, 369–372. [CrossRef]

146. Albi, E.; Magni, M.V. Chromatin neutral sphingomyelinase and its role in hepatic regeneration. *Biochem. Biophys. Res. Commun.* **1997**, *236*, 29–33. [CrossRef]

147. Venkataraman, K.; Riebeling, C.; Bodennec, J.; Riezman, H.; Allegood, J.C.; Sullards, M.C.; Merrill, A.H.; Futerman, A.H. Upstream of growth and differentiation factor 1 (uog1), a mammalian homolog of the yeast longevity assurance gene 1 (lag1), regulatesn-stearoyl-sphinganine (c18-(dihydro) ceramide) synthesis in a fumonisin b1-independent manner in mammalian cells. *J. Biol. Chem.* **2002**, *277*, 35642–35649. [CrossRef] [PubMed]

148. Riebeling, C.; Allegood, J.C.; Wang, E.; Merrill, A.H.; Futerman, A.H. Two mammalian longevity assurance gene (lag1) family members, trh1 and trh4, regulate dihydroceramide synthesis using different fatty acyl-coa donors. *J. Biol. Chem.* **2003**, *278*, 43452–43459. [CrossRef]

149. Mizutani, Y.; Kihara, A.; Igarashi, Y. Mammalian lass6 and its related family members regulate synthesis of specific ceramides. *Biochem. J.* **2005**, *390*, 263–271. [CrossRef]

150. Min, J.; Mesika, A.; Sivaguru, M.; Van Veldhoven, P.P.; Alexander, H.; Futerman, A.H.; Alexander, S. (dihydro) ceramide synthase 1–regulated sensitivity to cisplatin is associated with the activation of p38 mitogen-activated protein kinase and is abrogated by sphingosine kinase 1. *Mol. Cancer Res.* **2007**, *5*, 801–812. [CrossRef] [PubMed]

151. Shiraishi, T.; Imai, S.; Uda, Y. The presence of ceramidase activity in liver nuclear membrane. *Biol. Pharm. Bull.* **2003**, *26*, 775–779. [CrossRef]

152. Watanabe, M.; Kitano, T.; Kondo, T.; Yabu, T.; Taguchi, Y.; Tashima, M.; Umehara, H.; Domae, N.; Uchiyama, T.; Okazaki, T. Increase of nuclear ceramide through caspase-3-dependent regulation of the "sphingomyelin cycle" in fas-induced apoptosis. *Cancer Res.* **2004**, *64*, 1000–1007. [CrossRef] [PubMed]

153. Albi, E.; Cataldi, S.; Bartoccini, E.; Magni, M.V.; Marini, F.; Mazzoni, F.; Rainaldi, G.; Evangelista, M.; Garcia-Gil, M. Nuclear sphingomyelin pathway in serum deprivation-induced apoptosis of embryonic hippocampal cells. *J. Cell. Physiol.* **2006**, *206*, 189–195. [CrossRef]

154. Chocian, G.; Chabowski, A.; Żendzian-Piotrowska, M.; Harasim, E.; Łukaszuk, B.; Górski, J. High fat diet induces ceramide and sphingomyelin formation in rat's liver nuclei. *Mol. Cell. Biochem.* **2010**, *340*, 125–131. [CrossRef]

155. Schroeder, F.; Petrescu, A.D.; Huang, H.; Atshaves, B.P.; McIntosh, A.L.; Martin, G.G.; Hostetler, H.A.; Vespa, A.; Landrock, D.; Landrock, K.K. Role of fatty acid binding proteins and long chain fatty acids in modulating nuclear receptors and gene transcription. *Lipids* **2008**, *43*, 1–17. [CrossRef] [PubMed]

156. Yamaji, T.; Kumagai, K.; Tomishige, N.; Hanada, K. Two sphingolipid transfer proteins, cert and fapp2: Their roles in sphingolipid metabolism. *Iubmb Life* **2008**, *60*, 511–518. [CrossRef]

157. Sugiura, M.; Kono, K.; Liu, H.; Shimizugawa, T.; Minekura, H.; Spiegel, S.; Kohama, T. Ceramide kinase, a novel lipid kinase molecular cloning and functional characterization. *J. Biol. Chem.* **2002**, *277*, 23294–23300. [CrossRef] [PubMed]

158. Simanshu, D.K.; Kamlekar, R.K.; Wijesinghe, D.S.; Zou, X.; Zhai, X.; Mishra, S.K.; Molotkovsky, J.G.; Malinina, L.; Hinchcliffe, E.H.; Chalfant, C.E. Non-vesicular trafficking by a ceramide-1-phosphate transfer protein regulates eicosanoids. *Nature* **2013**, *500*, 463–467. [CrossRef] [PubMed]

159. Rovina, P.; Schanzer, A.; Graf, C.; Mechtcheriakova, D.; Jaritz, M.; Bornancin, F. Subcellular localization of ceramide kinase and ceramide kinase-like protein requires interplay of their pleckstrin homology domain-containing n-terminal regions together with c-terminal domains. *Biochim. Biophys. Acta (BBA) Mol. Cell Biol. Lipids* **2009**, *1791*, 1023–1030. [CrossRef] [PubMed]

160. Urs, A.N.; Dammer, E.; Kelly, S.; Wang, E.; Merrill, A.H., Jr.; Sewer, M.B. Steroidogenic factor-1 is a sphingolipid binding protein. *Mol. Cell. Endocrinol.* **2007**, *265*, 174–178. [CrossRef] [PubMed]

161. Urs, A.N.; Dammer, E.; Sewer, M.B. Sphingosine regulates the transcription of cyp17 by binding to steroidogenic factor-1. *Endocrinology* **2006**, *147*, 5249–5258. [CrossRef]

162. Sewer, M.B.; Waterman, M.R. Acth modulation of transcription factors responsible for steroid hydroxylase gene expression in the adrenal cortex. *Microsc. Res. Tech.* **2003**, *61*, 300–307. [CrossRef] [PubMed]

163. Spiegel, S.; Milstien, S. Functions of the multifaceted family of sphingosine kinases and some close relatives. *J. Biol. Chem.* **2007**, *282*, 2125–2129. [CrossRef] [PubMed]

164. Maceyka, M.; Sankala, H.; Hait, N.C.; Le Stunff, H.; Liu, H.; Toman, R.; Collier, C.; Zhang, M.; Satin, L.S.; Merrill, A.H. Sphk1 and sphk2, sphingosine kinase isoenzymes with opposing functions in sphingolipid metabolism. *J. Biol. Chem.* **2005**, *280*, 37118–37129. [CrossRef] [PubMed]

165. Alemany, R.; van Koppen, C.J.; Danneberg, K.; Ter Braak, M.; Zu Heringdorf, D.M. Regulation and functional roles of sphingosine kinases. *Naunyn-Schmiedeberg's Arch. Pharmacol.* **2007**, *374*, 413–428. [CrossRef] [PubMed]

166. Selvam, S.P.; De Palma, R.M.; Oaks, J.J.; Oleinik, N.; Peterson, Y.K.; Stahelin, R.V.; Skordalakes, E.; Ponnusamy, S.; Garrett-Mayer, E.; Smith, C.D. Binding of the sphingolipid s1p to htert stabilizes telomerase at the nuclear periphery by allosterically mimicking protein phosphorylation. *Sci. Signal.* **2015**, *8*, ra58. [CrossRef] [PubMed]

167. Stunff, H.L.; Milstien, S.; Spiegel, S. Generation and metabolism of bioactive sphingosine-1-phosphate. *J. Cell. Biochem.* **2004**, *92*, 882–899. [CrossRef] [PubMed]

168. Wang, C.; Mao, J.; Redfield, S.; Mo, Y.; Lage, J.M.; Zhou, X. Systemic distribution, subcellular localization and differential expression of sphingosine-1-phosphate receptors in benign and malignant human tissues. *Exp. Mol. Pathol.* **2014**, *97*, 259–265. [CrossRef] [PubMed]

169. Lépine, S.; Allegood, J.; Park, M.; Dent, P.; Milstien, S.; Spiegel, S. Sphingosine-1-phosphate phosphohydrolase-1 regulates er stress-induced autophagy. *Cell Death Differ.* **2011**, *18*, 350–361. [CrossRef] [PubMed]

170. Schwiebs, A.; Thomas, D.; Kleuser, B.; Pfeilschifter, J.M.; Radeke, H.H. Nuclear translocation of sgpp-1 and decrease of sgpl-1 activity contribute to sphingolipid rheostat regulation of inflammatory dendritic cells. *Mediat. Inflamm.* **2017**, *2017*, 5187368. [CrossRef]

171. Ebenezer, D.; Fu, P.; Berdyshev, E.; Natarajan, V. Nuclear s1p lyase regulates histone acetylation in pseudomonas aeruginosa-induced lung inflammation. *FASEB J.* **2015**, *29*, 863.26.

172. Ebenezer, D.L.; Fu, P.; Mangio, L.A.; Berdyshev, E.; Schumacher, F.; Kleuser, B.; Van Veldhoven, P.P.; Natarajan, V. Δ-2 hexadecenal generated from s1p by nuclear s1p lyase is a regulator of hdac1/2 activity and histone acetylation in lung epithelial cells. *FASEB J.* **2019**, *33*, 489.3.

173. Reynolds, C.P.; Maurer, B.J.; Kolesnick, R.N. Ceramide synthesis and metabolism as a target for cancer therapy. *Cancer Lett.* **2004**, *206*, 169–180. [CrossRef]

174. Gault, C.R.; Obeid, L.M. Still benched on its way to the bedside: Sphingosine kinase 1 as an emerging target in cancer chemotherapy. *Crit. Rev. Biochem. Mol. Biol.* **2011**, *46*, 342–351. [CrossRef]

175. Dbaibo, G.S.; Pushkareva, M.Y.; Rachid, R.A.; Alter, N.; Smyth, M.J.; Obeid, L.M.; Hannun, Y.A. P53-dependent ceramide response to genotoxic stress. *J. Clin. Investig.* **1998**, *102*, 329–339. [CrossRef] [PubMed]

176. Vit, J.-P.; Rosselli, F. Role of the ceramide-signaling pathways in ionizing radiation-induced apoptosis. *Oncogene* **2003**, *22*, 8645. [CrossRef] [PubMed]

177. Sawada, M.; Nakashima, S.; Kiyono, T.; Nakagawa, M.; Yamada, J.; Yamakawa, H.; Banno, Y.; Shinoda, J.; Nishimura, Y.; Nozawa, Y. P53 regulates ceramide formation by neutral sphingomyelinase through reactive oxygen species in human glioma cells. *Oncogene* **2001**, *20*, 1368. [CrossRef] [PubMed]

178. Corcoran, C.A.; He, Q.; Ponnusamy, S.; Ogretmen, B.; Huang, Y.; Sheikh, M.S. Neutral sphingomyelinase-3 is a DNA damage and nongenotoxic stress-regulated gene that is deregulated in human malignancies. *Mol. Cancer Res.* **2008**, *6*, 795–807. [CrossRef] [PubMed]

179. Jaffrézou, J.-P.; Bruno, A.P.; Moisand, A.; Levade, T.; Laurent, G. Activation of a nuclear sphingomyelinase in radiation-induced apoptosis. *FASEB J.* **2001**, *15*, 123–133. [CrossRef]

180. Ravid, T.; Tsaba, A.; Gee, P.; Rasooly, R.; Medina, E.A.; Goldkorn, T. Ceramide accumulation precedes caspase-3 activation during apoptosis of a549 human lung adenocarcinoma cells. *Am. J. Physiol. Lung Cell. Mol. Physiol.* **2003**, *284*, L1082–L1092. [CrossRef]

181. Dbaibo, G.S.; Pushkareva, M.Y.; Jayadev, S.; Schwarz, J.K.; Horowitz, J.M.; Obeid, L.M.; Hannun, Y.A. Retinoblastoma gene product as a downstream target for a ceramide-dependent pathway of growth arrest. *Proc. Natl. Acad. Sci. USA* **1995**, *92*, 1347–1351. [CrossRef]

182. Phillips, D.; Hunt, J.; Moneypenny, C.; Maclean, K.; McKenzie, P.; Harris, L.; Houghton, J. Ceramide-induced g 2 arrest in rhabdomyosarcoma (rms) cells requires p21 cip1/waf1 induction and is prevented by mdm2 overexpression. *Cell Death Differ.* **2007**, *14*, 1780. [CrossRef]

183. Xu, R.; Garcia-Barros, M.; Wen, S.; Li, F.; Lin, C.-L.; Hannun, Y.A.; Obeid, L.M.; Mao, C. Tumor suppressor p53 links ceramide metabolism to DNA damage response through alkaline ceramidase 2. *Cell Death Differ.* **2017**. [CrossRef]

184. Hoeferlin, L.A.; Fekry, B.; Ogretmen, B.; Krupenko, S.A.; Krupenko, N.I. Folate stress induces apoptosis via p53-dependent de novo ceramide synthesis and up-regulation of ceramide synthase 6. *J. Biol. Chem.* **2013**, *288*, 12880–12890. [CrossRef]

185. Fekry, B.; Jeffries, K.A.; Esmaeilniakooshkghazi, A.; Szulc, Z.M.; Knagge, K.J.; Kirchner, D.R.; Horita, D.A.; Krupenko, S.A.; Krupenko, N.I. C 16-ceramide is a natural regulatory ligand of p53 in cellular stress response. *Nat. Commun.* **2018**, *9*, 1–12. [CrossRef] [PubMed]

186. Taha, T.A.; Osta, W.; Kozhaya, L.; Bielawski, J.; Johnson, K.R.; Gillanders, W.E.; Dbaibo, G.S.; Hannun, Y.A.; Obeid, L.M. Down-regulation of sphingosine kinase-1 by DNA damage dependence on proteases and p53. *J. Biol. Chem.* **2004**, *279*, 20546–20554. [CrossRef] [PubMed]

187. Sankala, H.M.; Hait, N.C.; Paugh, S.W.; Shida, D.; Lépine, S.; Elmore, L.W.; Dent, P.; Milstien, S.; Spiegel, S. Involvement of sphingosine kinase 2 in p53-independent induction of p21 by the chemotherapeutic drug doxorubicin. *Cancer Res.* **2007**, *67*, 10466–10474. [CrossRef] [PubMed]

188. Johnson, K.R.; Johnson, K.Y.; Becker, K.P.; Bielawski, J.; Mao, C.; Obeid, L.M. Role of human sphingosine-1-phosphate phosphatase 1 in the regulation of intra-and extracellular sphingosine-1-phosphate levels and cell viability. *J. Biol. Chem.* **2003**, *278*, 34541–34547. [CrossRef]

189. Oskouian, B.; Sooriyakumaran, P.; Borowsky, A.D.; Crans, A.; Dillard-Telm, L.; Tam, Y.Y.; Bandhuvula, P.; Saba, J.D. Sphingosine-1-phosphate lyase potentiates apoptosis via p53-and p38-dependent pathways and is down-regulated in colon cancer. *Proc. Natl. Acad. Sci. USA* **2006**, *103*, 17384–17389. [CrossRef] [PubMed]

190. Kumar, A.; Oskouian, B.; Fyrst, H.; Zhang, M.; Paris, F.; Saba, J. S1p lyase regulates DNA damage responses through a novel sphingolipid feedback mechanism. *Cell Death Dis.* **2011**, *2*, e119. [CrossRef]

191. Ahmad, A.; Mitrofanova, A.; Bielawski, J.; Yang, Y.; Marples, B.; Fornoni, A.; Zeidan, Y.H. Sphingomyelinase-like phosphodiesterase 3b mediates radiation-induced damage of renal podocytes. *FASEB J.* **2016**, *31*, 771–780. [CrossRef] [PubMed]

192. Fornoni, A.; Sageshima, J.; Wei, C.; Merscher-Gomez, S.; Aguillon-Prada, R.; Jauregui, A.N.; Li, J.; Mattiazzi, A.; Ciancio, G.; Chen, L. Rituximab targets podocytes in recurrent focal segmental glomerulosclerosis. *Sci. Transl. Med.* **2011**, *3*, ra46–ra85. [CrossRef]

193. Mitrofanova, A.; Mallela, S.; Ducasa, G.; Yoo, T.; Rosenfeld-Gur, E.; Zelnik, I.; Molina, J.; Santos, J.V.; Ge, M.; Sloan, A. Smpdl3b modulates insulin receptor signaling in diabetic kidney disease. *Nat. Commun.* **2019**, *10*, 2692. [CrossRef]

194. Mallela, S.K.; Mitrofanova, A.; Merscher, S.; Fornoni, A. Regulation of the amount of ceramide-1-phosphate synthesized in differentiated human podocytes. *Biochim. Biophys. Acta (BBA) Mol. Cell Biol. Lipids* **2019**, *1864*, 158517. [CrossRef]

195. Beckham, T.H.; Cheng, J.C.; Marrison, S.T.; Norris, J.S.; Liu, X. Interdiction of sphingolipid metabolism to improve standard cancer therapies. In *Advances in Cancer Research*; Elsevier: Amsterdam, The Netherlands, 2013; Volume 117, pp. 1–36.

196. Savić, R.; Schuchman, E.H. Use of acid sphingomyelinase for cancer therapy. In *Advances in Cancer Research*; Elsevier: Amsterdam, The Netherlands, 2013; Volume 117, pp. 91–115.

197. Albi, E.; Cataldi, S.; Ceccarini, M.R.; Conte, C.; Ferri, I.; Fettucciari, K.; Patria, F.F.; Beccari, T.; Codini, M. Gentamicin targets acid sphingomyelinase in cancer: The case of the human gastric cancer nci-n87 cells. *Int. J. Mol. Sci.* **2019**, *20*, 4375. [CrossRef]

198. Cao, M.; Ji, C.; Zhou, Y.; Huang, W.; Ni, W.; Tong, X.; Wei, J.-F. Sphingosine kinase inhibitors: A patent review. *Int. J. Mol. Med.* **2018**, *41*, 2450–2460. [CrossRef] [PubMed]

199. Dubois, N.; Rio, E.; Ripoche, N.; Ferchaud-Roucher, V.; Gaugler, M.-H.; Campion, L.; Krempf, M.; Carrie, C.; Mahé, M.; Mirabel, X. Plasma ceramide, a real-time predictive marker of pulmonary and hepatic metastases response to stereotactic body radiation therapy combined with irinotecan. *Radiother. Oncol.* **2016**, *119*, 229–235. [CrossRef] [PubMed]

200. Gao, P.; Smith, C.D. Ablation of sphingosine kinase-2 inhibits tumor cell proliferation and migration. *Mol. Cancer Res.* **2011**, *9*, 1509–1519. [CrossRef] [PubMed]

201. Britten, C.D.; Garrett-Mayer, E.; Chin, S.H.; Shirai, K.; Ogretmen, B.; Bentz, T.A.; Brisendine, A.; Anderton, K.; Cusack, S.L.; Maines, L.W. A phase i study of abc294640, a first-in-class sphingosine kinase-2 inhibitor, in patients with advanced solid tumors. *Clin. Cancer Res.* **2017**, *23*, 4642–4650. [CrossRef]

International Journal of
Molecular Sciences

Article

Acid and Neutral Sphingomyelinase Behavior in Radiation-Induced Liver Pyroptosis and in the Protective/Preventive Role of rMnSOD

Samuela Cataldi [1], Antonella Borrelli [2], Maria Rachele Ceccarini [1], Irina Nakashidze [1], Michela Codini [1], Oleg Belov [3], Alexander Ivanov [3], Eugene Krasavin [3], Ivana Ferri [4], Carmela Conte [1], Federica Filomena Patria [1], Tommaso Beccari [1], Aldo Mancini [5], Francesco Curcio [6], Francesco Saverio Ambesi-Impiombato [6] and Elisabetta Albi [1,*]

[1] Department of Pharmaceutical Sciences, University of Perugia, 06126 Perugia, Italy; samuelacataldi@libero.it (S.C.); chele@hotmail.it (M.R.C.); irinanakashidze@yahoo.com (I.N.); michela.codini@unipg.it (M.C.); carmela.conte@unipg.it (C.C.); patriafederica@gmail.com (F.F.P.); tommaso.beccari@unipg.it (T.B.)
[2] MolecularBiology and Viral Oncology Unit, Istituto Nazionale Tumori IRCCS "Fondazione G. Pascale", 80131 Napoli, Italy; a.borrelli@istitutotumori.na.it
[3] Laboratory of Radiation Biology, Joint Institute for Nuclear Research, 141980 Dubna, Russia; dem@jinr.ru (O.B.); a1931192@mail.ru (A.I.); krasavin@jinr.ru (E.K.)
[4] Division of Pathological Anatomy and Histology, Department of Experimental Medicine, School of Medicine and Surgery, University of Perugia, 06126 Perugia, Italy; ivanaferri@gmail.com
[5] Laedhexa Biotechnologies Inc., San Francisco, CA 94130, USA; aldo_mancini@tiscali.it
[6] Dipartimento di Area Medica, University of Udine, 33100 Udine, Italy; francesco.curcio@uniud.it (F.C.); saverio.ambesi@uniud.it (F.S.A.-I.)
* Correspondence: elisabetta.albi@unipg.it; Tel./Fax: +39-07-5585-7906

Received: 30 March 2020; Accepted: 4 May 2020; Published: 6 May 2020

Abstract: Sphingomyelins (SMs) are a class of relevant bioactive molecules that act as key modulators of different cellular processes, such as growth arrest, exosome formation, and the inflammatory response influenced by many environmental conditions, leading to pyroptosis, a form of programmed cell death due to Caspase-1 involvement. To study liver pyroptosis and hepatic SM metabolism via both lysosomal acid SMase (aSMase) and endoplasmic reticulum/nucleus neutral SMase (nSMase) during the exposure of mice to radiation and to ascertain if this process can be modulated by protective molecules, we used an experimental design (previously used by us) to evaluate the effects of both ionizing radiation and a specific protective molecule (rMnSOD) in the brain in collaboration with the Joint Institute for Nuclear Research, Dubna (Russia). As shown by the Caspase-1 immunostaining of the liver sections, the radiation resulted in the loss of the normal cell structure alongside a progressive and dose-dependent increase of the labelling, treatment, and pretreatment with rMnSOD, which had a significant protective effect on the livers. SM metabolic analyses, performed on aSMase and nSMase gene expression, as well as protein content and activity, proved that rMnSOD was able to significantly reduce radiation-induced damage by playing both a protective role via aSMase and a preventive role via nSMase.

Keywords: acid sphingomyelinase; neutral sphingomyelinase; radiation; SOD; liver

---

1. Introduction

Sphingomyelins (SMs) are a class of bioactive lipid molecules that act as key modulators of different pathophysiologic processes, including cell growth, cell death, autophagy, stress, inflammatory responses, and cancer [1]. Sphingomyelinases (SMase) are a family of key enzymes in SM metabolism that generate the ceramide and phosphorylcholine headgroups. From a cellular perspective, the existence of the isoenzyme's multiplicity has functional reasons. Our improved understanding of these isoenzymes has provided information on the different roles of SMs [2]. SMases are named based on their optimal pH activity as an acid, neutral, or alkaline SMase, with different locations and functions inside the cells [3]. Alkaline SMase (Alk-SMase) shares no structural similarities with the other two SMases; it belongs to the ecto-nucleotide pyrophosphatase phosphodiesterase (NPP) family and is located in the mucosal membrane of the intestinal tract [4]. Acid SMase (aSMase) and neutral SMases (nSMases) have organelle-specific activities and distinct regulatory mechanisms. Thus, the aSMase isoform is located in the lysosome and is involved in apoptosis signaling [5]. Moreover, four nSMase isoforms are located in the inner and outer leaflet of the plasma membrane, endoplasmic reticulum, mitochondria, and nucleus [6]. Enzyme localization influences the biologically relevant activation mechanisms. Moreover, the concentration of the SM and ceramide is organelle dependent. nSMase isoforms have been identified on the basis of four genes that were cloned or purified: nSMase1 (gene name = SMPD2), nSMase2 (SMPD3), nSMase3 (SMPD4), and MA-nSMase (mitochondrial-associated nSMase) (SMPD5) [6]. nSMase1 is located in the reticulum endoplasmic/Golgi apparatus [7], as well as in the cell nucleus [8]. It is activated in response to stress by inducing apoptosis [9] and also suppresses hepatocellular carcinoma [10]. nSMase2 is specific to the inner leaflet of the plasma membrane and is involved in many cell responses, such as cell growth arrest, exosome formation, and inflammatory response [11]. nSMase3 is located in the endoplasmic reticulum and is involved in TNF-α mediated signaling, tumorigenesis [6], and cellular stress response [12]. MA-nSMase, which has sequence homology with nSMase2 and zebrafish mitochondrial N-SMase [13], was identified only in 2010 [14].

It has been demonstrated that SM metabolism is finely modulated in the liver [15], an active metabolic organ influenced by many environmental conditions, including radiation [16]. Irradiation induces DNA repair activities after DNA damage in hepatocytes (and inflammatory reactions in other cell types [16]) by leading to pyroptosis, a form of programmed cell death due to Caspase1 involvement [17], which is characterized by membrane rupture, pore formation, and the release of pro-inflammatory cytokines [18]. Therefore, pyroptosis is important as a final event of radiation-induced damage.

The effect of ionizing radiation on liver SM metabolism has not yet been clarified. It was previously shown that in thyroid cells, ionizing radiation exposure induces proapoptotic signals via ceramide production from the SMs [19]. Moreover, proton beams move quiescent thyroid cells towards a proapoptotic state and proliferating thyroid cells towards an initial apoptotic state by altering the nuclear SM-metabolism [20]. In the same experimental model, ultraviolet radiation enriched the ceramide pool due to the ability of both aSMase and nSMase to induce apoptosis [21].

Thus, the apoptotic process requires the action of both aSMase and nSMase (especially nSMase1 [9]), but whether there is cooperation between the two enzymes and whether they behave differently in relation to the same apoptotic stimulus has not yet been investigated. To study what happens to the metabolism of hepatic SM via both the lysosomal and endoplasmic reticulum/nucleus SMase during irradiation, and to ascertain if this process can be modulated by protective molecules, we used the same experimental model previously applied to evaluate the effects of both ionizing radiation and a specific protective molecule in the brain [22]. This research originated from a collaborative project among Italian research groups and the Joint Institute for Nuclear Research, Dubna (Russia), in which mice were exposed to a set of minor γ radiation, neutrons, and a spectrum of neutrons, simulating the radiation levels that cosmonauts are exposed to during deep-space long-term missions. In the brain, radiation was shown to deconstruct neurofilaments in a dose-dependent manner with an increase of the nSMase3 gene and protein expression. Human recombinant manganese superoxide dismutase (rMnSOD), which has a protective and preventive role on brain damage, strongly increased nSMase expression and activity [22]. Since ionizing radiation induces oxidative stress, and rMnSOD has specific antioxidant and anti-free radical activity, in the previous study, we analyzed the behavior of the nSMase3 that is stimulated in the cellular stress response [12]. Here, we analyzed the behaviors of two enzymes that are involved in apoptosis, aSMase and nSMase1, which are capable of degrading MS in two different cellular districts. We performed our study on the liver, which is an organ that actively reacts to radiation though parenchymal cells or hepatocytes in the G0 phase of the cell cycle, thereby regulating the metabolism of many factors, including lipids [16]. Thus, considering our extensive experience with the role of SM metabolism in liver proliferation and apoptosis [23,24], we evaluated aSMase and nSMase in relation to radiation-induced pyroptosis and their response to rMnSOD treatment. The present paper reports the results of an observational study of a novel experiment that could be useful to the scientific community as the basis for future work.

2. Results

2.1. Ionizing Radiation Effects on the Liver and the Role of rMnSOD

The microscopy analysis, performed on histological microsections of the control livers (CTR, rMnSOD untreated, and un-irradiated mice) subjected to Caspase-1 immunostaining, showed normal cells with a very low percentage of labelling (Figure 1a,b). Radiation induced a loss of the normal cell structure alongside a progressive and dose-dependent increase in labelling. The images provide evidence of a significant increase in irregular cellular shapes and membrane ruptures compared to the CTR sample (Figure 1a). The quantification of Caspase-1 showed an increase of 2.8, 3.9, and 5.1 times in the labelling (0.25 Gy, 0.5 Gy, and 1.0 Gy, respectively) compared to the CTR samples. The treatment with rMnSOD alone did not induce significant variation with respect to the CTR but reduced the effect of radiation when administered preventively (see Materials and Methods). The labelling increased by 2.0, 2.5, and 4.1 times with 0.25 Gy, 0.5 Gy, and 1.0 Gy, respectively, compared to the CTR samples. Pretreatment with rMnSOD for preventive purposes had an even greater effect, as labelling increased by only 3.4 times among the mice receiving 1.0 Gy of radiation (Figure 1a,b).

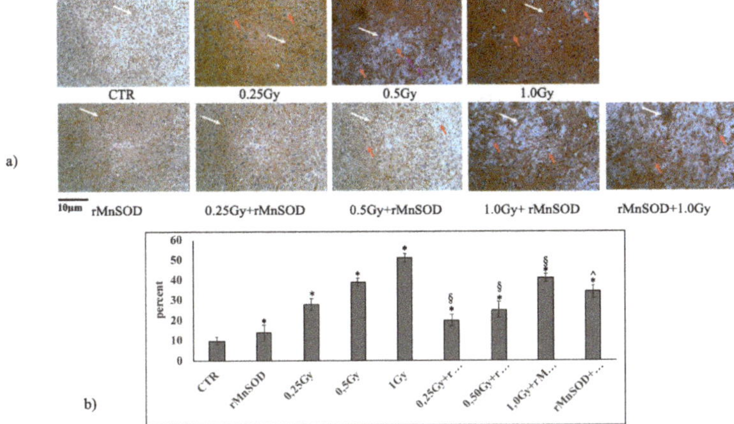

Figure 1. Mouse liver after irradiation with or without protective or preventive rMnSOD treatment (a) representative liver histology by Caspase-1 immunohistochemical staining. CTR, control mice; rMnSOD, mice treated with human recombinant manganese superoxide dismutase; 0.25 Gy, 0.5 Gy, and 1.0 Gy, mice exposed to increasing radiation doses; 0.25 Gy + rMnSOD, 0.5 Gy + rMnSOD, and 1.0 Gy + rMnSOD, mice exposed to increasing radiation doses and treated with rMnSOD (protective role of rMnSOD); rMnSOD+1.0Gy, mice pretreated with rMnSOD and exposed to 1.0 Gy radiation (preventive role of rMnSOD). Images are representative of 3 similar images from each group of mice (20× magnification). White arrows indicate positive caspase labelling; red arrows indicate cell membrane ruptures. (b) Quantification of Caspase-1 staining was performed using the ImageFocus software. Positive staining is indicated as low (+), medium (++), or high (+++). Only high positive staining was considered and was measured as a percentage of the total area. Data represent the mean + S.D. of three livers for each group. Significance, * $p < 0.05$ with respect to the CTR, § $p < 0.05$ with respect to the irradiated samples, ˆ $p < 0.05$ with respect to 1.0 Gy + rMnSOD.

2.2. Changes of Sphingomyelin Metabolism

Our previous studies indicated that radiation targets SMase in the thyroid [20,21] and brain [22]. As there are two SMases involved in the apoptotic process (lysosomal aSMase and endoplasmic reticulum/nucleus nSMase1), we defined their behavior in the liver, where radiation upregulated Caspase-1, thereby triggering pyroptosis. We first measured SMPD1 (coding for aSMase) and SMPD2 (coding for nSMase1) gene expression in livers from a) CTR mice, b) rMnSOD treated mice, and un-irradiated mice; c) 0.25 Gy, 0.5 Gy, and 1.0 Gy irradiated mice and mice untreated with rMnSOD; d) 0.25 Gy, 0.5 Gy, and 1.0 Gy irradiated and rMnSOD treated mice; and e) mice pretreated with rMnSOD and irradiated with 1.0 Gy radiation (Figure 2). The results show that SMPD1 was overexpressed by 2.23 + 0.34, 7.05 + 0.42, and 14.1 + 1.47 times with 0.25 Gy, 0.5 Gy, and 1.0 Gy radiation, respectively. The gene expression of SMPD1 did not vary when treated with rMnSOD alone. Treatment with rMnSOD limited the effects of radiation among the irradiated mice and reduced the effects of 0.25 Gy by 19.3%, that of 0.5 Gy by 62%, and that of 1.0 Gy by 75%. The use of rMnSOD as a method of damage prevention was less effective. Notably, the effect of 1.0 Gy radiation was reduced by 44%. These results suggest that rMnSOD plays a limited role in controlling SMPD1 expression when it is used as a preventive molecule for radiation-induced damage, while also being an effective protective molecule.

We then tested the expression of the SMPD2 gene coding for nSMase1. Its variations under radiation treatment, with or without rMnSOD, were very low (Figure 2).

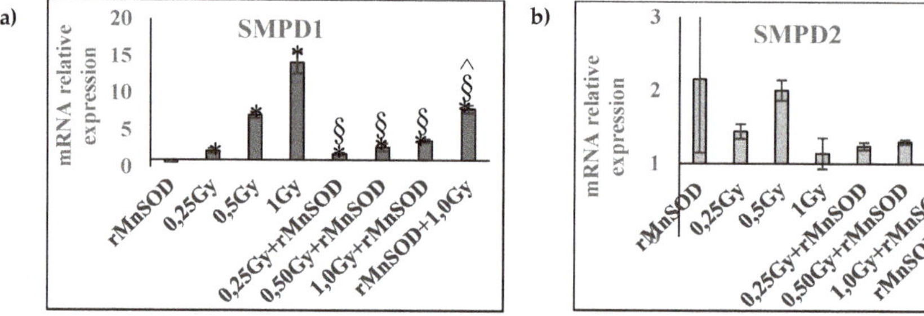

Figure 2. Effect of radiation and rMnSOD on SMPD1 and SMPD2 gene expression in the liver. SMPD1 and SMPD2 gene expression evaluated by RTqPCR as reported in the "Materials and Methods" section. Liver from mice treated with increasing doses of radiation with or without rMnSOS. (**a**) SMPD1 (**b**) SMPD2. Data are expressed as the mean + SD of three liver samples, each carried out in triplicate. Significance: (a) * $p < 0.05$ versus the control sample (CTR); (b) § $p < 0.05$ rMnSOD treated and irradiated samples versus the irradiated samples; (c)^ $p < 0.05$ pretreated and 1.0 Gy irradiated sample versus 1.0 Gy irradiated and rMnSOD treated samples. CTR, control mice; rMnSOD, mice treated with human recombinant manganese superoxide dismutase; 0.25 Gy, 0.5 Gy, and 1.0 Gy, mice exposed to increasing radiation doses; 0.25 Gy + rMnSOD, 0.5 Gy + rMnSOD, and 1.0 Gy + rMnSOD, mice exposed to increasing radiation doses and treated with rMnSOD (protective role of rMnSOD); rMnSOD + 1.0 Gy, mice pretreated with rMnSOD and exposed to 1.0 Gy radiation (preventive role of rMnSOD).

To date, the changes of both aSMase and nSMase1 proteins induced by increasing radiation doses and/or rMnSOD have not been analyzed. Thus, we determined if the changes caused by radiation at the genetic level were consistent with protein variation. Using aSMase and nSMase1 specific antibodies, we were able to measure the level of proteins relative to the CTR samples (Figure 3a). The results related to aSMase, normalized for β-tubulin, showed that the enzyme was reduced by 18%, 52%, and 34% with 0.25 Gy, 0.5 Gy, and 1.0 Gy, respectively (Figure 3b). The reduction of protein levels despite increased gene overexpression strongly suggests an increased degradation of the enzyme. Treatment with rMnSOD alone caused a significant reduction in protein compared to the CTR, even when gene expression did not change, possibly because rMnSOD slowed the synthesis of the enzyme due to its proapoptotic role. This effect remained evident with 0.25 Gy and 0.5 Gy of radiation, but very high radiation (1.0 Gy) strongly limited the action of the rMnSOD for both protective and preventive purposes (Figure 3a,b). Therefore, a high radiation dose would hinder the action of rMnSOD.

Conversely, nSMase1 content did not change with irradiation (Figure 3a,b). Surprisingly, rMnSOD alone strongly reduced the nSMase1 form with an apparent 48 kDa molecular weight while inducing the formation of a band with an apparent molecular weight of approximately 28 kDa. Radiation hindered the strong protein reduction obtained via rMnSOD with milder action at a 0.25 Gy dose and a much more intense effect at 0.5 Gy and 1.0 Gy doses (Figure 3a,b).

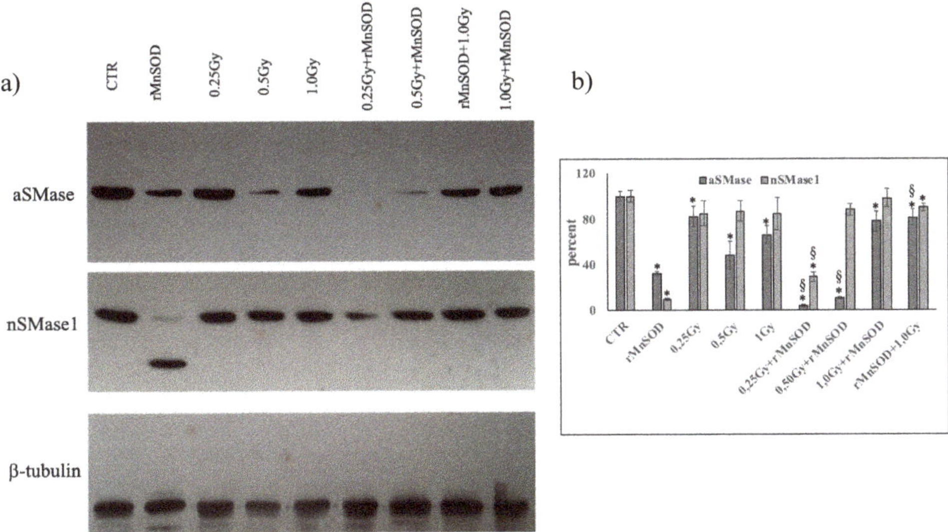

Figure 3. Effect of radiation and rMnSOD on the aSMase and nSMase1 protein level in the liver. The aSMase and nSMase1 level evaluated by Western Blotting, as reported in the "Materials and Methods" section. Liver from the mice treated with increasing doses of radiation with or without rMnSOS. (**a**) western blotting panel; (**b**) densitometric analysis performed using the ImageFocus software. Data are expressed as a percentage with respect to the control sample and represent the mean + SD of three liver samples, each carried out in triplicate. Significance: (a) * $p < 0.05$ versus the control sample (CTR); (b) § $p < 0.05$ rMnSOD treated and irradiated samples versus irradiated samples; (c) ^ $p < 0.05$ pretreated and 1.0 Gy irradiated samples versus the 1.0 Gy irradiated and rMnSOD treated samples. CTR, control mice; rMnSOD, mice treated with human recombinant manganese superoxide dismutase; 0.25 Gy, 0.5 Gy, and 1.0 Gy, mice exposed to increasing radiation doses; 0.25 Gy + rMnSOD, 0.5 Gy + rMnSOD, and 1.0 Gy + rMnSOD, mice exposed to increasing radiation doses and treated with rMnSOD (protective role of rMnSOD); rMnSOD + 1.0 Gy, mice pretreated with rMnSOD and exposed to 1.0 Gy radiation (preventive role of rMnSOD).

To investigate whether ionizing radiation could affect SMase activity, we employed a specific assay kit, as reported in the Materials and Methods. This is a powerful analytical technique that measures the enzymatic activity of SMases, distinguishing aSMase from nSMase by their pH values. We observed that aSMase activity did not change significantly with radiation in both the absence and presence of rMnSOD in the CTR samples (Figure 4). Interestingly, the activity of the nSMase pool was very high in the CTR and was inhibited by radiation. rMnSOD alone increased the activity by 1.65 times in comparison with CTR, but its ability to stimulate this activity was reduced by radiation in a dose-dependent manner. Pretreatment with rMnSOD was able to strongly increase nSMase activity, with values 2× higher than those of CTR.

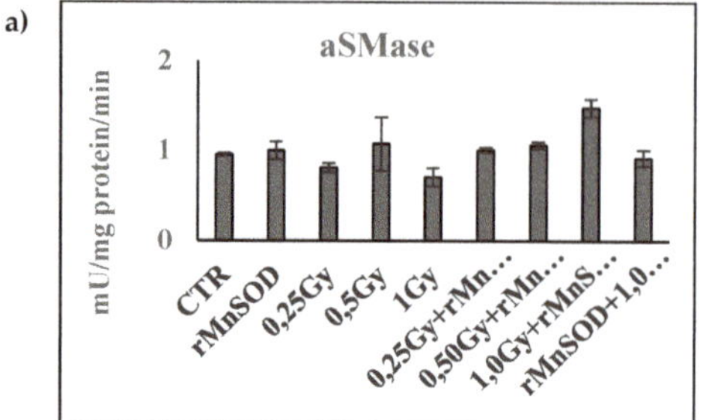
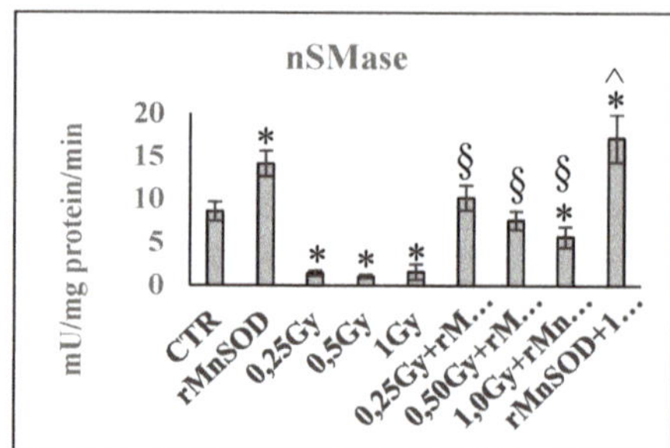

Figure 4. Effect of radiation and rMnSOD on aSMase and nSMase activity in the liver. aSMase and nSMase activity evaluated using an Amplex Red Sphingomyelinase assay kit, as reported in the "Materials and Methods" section. Livers from the mice treated with increasing doses of radiation with or without rMnSOS. (**a**) aSMase; (**b**) nSMase. Data are expressed as mU/mg protein/min and represent the mean+SD of three liver samples, each carried out in triplicate. Significance: (a) * $p < 0.05$ versus the control sample (CTR); (b) § $p < 0.05$ rMnSOD treated and irradiated samples versus the irradiated samples; (c) ˆ $p < 0.05$ pretreated and 1.0 Gy irradiated sample versus the 1.0 Gy irradiated and rMnSOD treated sample. CTR, control mice; rMnSOD, mice treated with human recombinant manganese superoxide dismutase; 0.25 Gy, 0.5 Gy, and 1.0 Gy, mice exposed to increasing radiation doses; 0.25 Gy + rMnSOD, 0.5 Gy + rMnSOD, and 1.0 Gy + rMnSOD, mice exposed to increasing radiation doses and treated with rMnSOD (protective role of rMnSOD); rMnSOD + 1.0 Gy, mice pretreated with rMnSOD and exposed to 1.0 Gy radiation (preventive role of rMnSOD).

3. Discussion

Caspase-1 is recognized as a molecule involved in the canonical signaling pathway that induces pyroptosis [25,26], a specific type of programmed cell death [27] characterized by membrane rupture and pore formation. Since pyroptosis occurs in different liver diseases [28] and is stimulated by radiation [29], we studied the damage induced in the liver with ionizing radiation treatment in mice, observing histological microsections of liver tissue fixed and stained with anti-Caspase1 antibodies. The specific experimental model used for the present study permitted us to simultaneously investigate the effect of ionizing radiation and the protective/preventive effect of rMnSOD against radiation-induced liver damage. This was a unique opportunity under a collaborative project among Italian research groups and the Joint Institute for Nuclear Research, Dubna (Russia) [22]. Our results provide evidence that both prominent signs of pyroptosis (i.e., the loss of cell membrane integrity and the overexpression of Caspase-1) are induced by radiation treatment in a dose-dependent manner. These effects were limited by rMnSOD administered as a protective agent. Therefore, rMnSOD reduced radiation-induced pyroptosis. If the rMnSOD was administered before irradiation as a preventive agent, its effect was less marked. Thus, the data presented here reveal that rMnSOD played an important role in protection from radiation-induced damage. In our previous study, conducted on the same experimental model, we clearly demonstrated that rMnSOD is able to limit radiation-induced damage to the brain by protecting the brain from the destructuring of neurofilaments with the involvement of nSMase [22]. These results induced us to investigate the behavior of aSMase and nSMase in the liver in association with the expression of Caspase-1.

Interestingly, SMPD1 (the gene coding for aSMase) was overexpressed following radiation treatment in a dose dependent manner. Considering the important effect of aSMase in apoptosis [30], we expected an upregulation of the enzyme in the liver featuring the characteristics of pyroptosis. Surprisingly, the content of the aSMase protein was lower than that of the control sample, and the enzyme activity remained at low control values. We hypothesized that radiation induced SMPD1 gene overexpression would lead to the synthesis of aSMase, which, in turn, would stimulate the synthesis of Caspase-1. The latter, triggering cell damage, might be responsible for lysosomal rupture with a consequent loss of the aSMase protein. A time-dependent study could have clarified this phenomenon, but such a study was not compatible with the irradiation windows available to us in Dubna when the above results were collected. In support of our data however, recent research has reported the role of aSMase in activating Caspase-1 via the inflammasome, which mediates radiation-induced pyroptosis [31] and the consequent rupture of the lysosome [32]. In this scenario, the use of rMnSOD as a molecule administered alone showed a strong reduction of the aSMasi protein, even though its gene expression did not change. We hypothesize that rMnSOD slowed the synthesis of the enzyme because of its proapoptotic role. Thus, rMnSOD was able to reduce the synthesis of aSMase in the presence of low and medium radiation doses but not at a higher dosage. However, the values of the protein content at all radiation doses always remained lower than those of the control samples in accordance with the mechanism hypothesized above. In case of significant aSMase protein damage due to pyroptosis, rMnSOD might protect the aSMase protein, thereby reducing cellular and tissue damage. Indeed, despite the reduced SMPD1 overexpression following 1.0 Gy irradiation + rMnSOD and rMnSOD + 1.0 Gy compared to the 1.0 Gy irradiated sample, aSMase protein expression was not significantly higher in the 1.0 Gy irradiation + rMnSOD group but instead significantly higher in the rMnSOD + 1.0 Gy group, thereby indicating reduced aSMase damage

The relevance of nSMase isoforms in cellular physiopathology, underscored by the role of nSMase1 in apoptosis, led us to investigate the behavior of this isoenzyme in ionizing radiation-induced pyrotosis. Our results demonstrate that nSMase1 gene and protein expression did not change with radiation. rMnSOD alone strongly reduced nSMase1 with its apparent molecular weight present in all other samples (48 kDa). Unexpectedly, a much lower molecular weight band (28 kDa) appeared in our Western Blotting. Interestingly, a specific isoform of nSMase1 (isoform) with an apparent molecular weight of 28 kDa, inhibited by reduced glutathione, was recently reported [33]. In the presence of radiation, the reduction of nSMase1 by rMnSOD was attenuated, indicating that MnSOD was unable to reduce the synthesis of nSMase1 at medium / high radiation doses. As reported in the results, the method used for nSMase activity was not specific for the nSMase1 isoform, but the results indicated the activity of the nSMase pool. We thus carried out experiments to evaluate whether nSMase activity might change in liver pyroptosis, as in previously reported midbrain radiation-induced damage [22]. Interestingly, radiation inhibited nSMase activity more than the control samples, while rMnSOD alone increased strongly, but this effect was reduced by radiation in a dose-dependent manner when used as a protective agent. Conversely, the use of rMnSOD as a preventive agent strongly increased nSMase activity. Since nSMase1 mainly resides in the nuclear matrix, and to a lesser extent in the endoplasmic reticulum [8], rMnSOD was likely able to act strongly on the plasma membrane nSMase but unable to act significantly at a nuclear level.

These results contrast with previous studies showing the stimulation of nSMase with ionizing radiation (γ rays) [19], protons [20], and ultraviolet radiation [21]. The explanation for this result is twofold. 1) First, there is the issue of tissue-specificity. The experiments in previous studies were carried out on normal and cancer thyroid cells, while our data focus on the liver. 2) Also important are the dose and type of radiation. Sautin et al. [19] used gamma radiation at a dose of 2–5 Gy, while we used a set of minor γ radiation and neutrons simulating ionizing radiation during space flight at doses of 0.25 Gy, 0.5 Gy, and 1.0 Gy. There are no data in the literature that compare the amount of ROS production in response to the different types of radiation in the liver. Therefore, it is very difficult to establish if the ROS–SMase relationship is also dependent on tissue specificity and / or radiation dose.

The results of the work show a lack of correlation between gene expression, protein expression, and the activities of both the aSMase and nSMase enzymes. We believe this is due to the much more complex in vivo experimental model than the in vitro system, especially for lipid metabolism. Indeed, the in vivo metabolism of lipids, including sphingolipids, is very rapid and influenced by hormones and various metabolic interactions with the production of molecules that can act as stimulators or inhibitors in response to stress conditions, induced by different factors such as radiation or drug treatments. The actions of these molecules can take place at the gene level or during the long process of protein synthesis; conversely, these actions can directly influence the activity of an enzyme. In this specific experiment, we do not know the exact mechanisms that were activated. We have images of the results and can only make assumptions relative to the literature data.

In conclusion, we currently have no reliable evidence on the direct effect of radiation and rMnSOD on SMases but we have data indicating their variations after different treatments during liver pyroptosis. The possibility of an indirect effect with the involvement of other signal molecules cannot be excluded.

4. Materials and Methods

4.1. Chemicals

The rMnSOD protein was provided by the Molecular Biology and Viral Oncology Unit, Department of Experimental Oncology, "Istituto Nazionale Tumori Fondazione G. Pascale"—IRCCS, Naples, Italy [34]. Anti-aSMase, anti-nSMase1, anti-Caspase1, and anti-β-tubulin were obtained from Abcam (Cambridge, UK). Horseradish peroxidase-conjugated goat anti-rabbit secondary antibodies were obtained from Santa Cruz. The TaqMan SNP Genotyping Assay and Reverse Transcription kit were obtained from Applied Biosystems (Foster City, CA, USA). The RNAqueous®-4PCR kit was obtained from Ambion Inc. (Austin, Texas, USA). The SDS-PAGE molecular weight standards were purchased from Bio-Rad Laboratories (Hercules, CA, USA). Chemiluminescence kits were purchased from Amersham (Rainham, Essex, UK).

4.2. Experimental Model

The experimental model was the same previously reported [22]. Animals: 54 female mice weighting approximately 25–30 g were obtained from the Laboratory Animal Nursery of the Russian Academy of Sciences (Pushchino, Russia). Mice were adapted to the vivarium at the "Joint Institute for Nuclear Research (JINR)" in Dubna over a period of 10 days. Then, they were divided into 9 cages with 6 mice each, each receiving standard briquetted fodder and water ad libitum. All procedures were performed according to the Russian Guidelines for the Care and Use of Experimental Animals and Bioethics Instructions (Order of the USSR Ministry of Health No. 755 12.08.1987). The mice were divided into groups and numbered with progressive numbers: a) numbers 1,3,4, and 5 were treated with daily subcutaneous injections of sterile PBS solution for 7 days from the day of irradiation; b) mice 2, 6, 7, and 8 were treated with daily subcutaneous injections of rMnSOD in sterile PBS for 7 days from the day of irradiation; c) 9 received a total of 10 injections; they were pretreated with rMnSOD in sterile PBS for 3 days prior to irradiation and then for 7 days from the day of irradiation. All animals were irradiated at the JINR; they were exposed to a set of minor γ radiation and neutrons from a Phasatron with high Relative Biological Effectiveness (RBE) and a spectrum of neutrons to simulate space flight exposure. Animals in groups 3 and 6 were exposed to a dose of 0.25 Gy, those in groups 4 and 7 were exposed to a dose of 0.50 Gy, and those in groups 5 and 8 were exposed to a dose of 1.00 Gy. Mice in groups 1 (mock-treated with PBS) and 2 (rMnSOD-treated) were not exposed to radiation and were considered a biological control. At the end of the experiment, all mice were beheaded and had their livers immediately frozen.

4.3. Immunohistochemical Analysis

Three livers from each group were fixed in 4% neutral phosphate-buffered formaldehyde solution for 24 h and dropped in a specific orientation into paraffin. Immunohistochemical analyses were performed as previously reported [35] using the anti-Caspase-1 antibody. A bond Dewax solution was used for removal of paraffin from tissue sections before rehydration and immunostaining was performed in the Bond automated system (Leica Biosystems Newcastle Ltd, UK). The observations were performed using inverted microscopy EUROMEX FE 2935 (ED Amhem, The Netherlands) equipped with a CMEX 5000 camera system (20× magnification). The analysis of labelling was performed using the ImageFocus software.

4.4. Reverse Transcription Quantitative PCR (RTqPCR)

Total RNA was extracted from the livers using an RNAqueous-4PCR kit. Its integrity was evaluated, and the cDNA was synthesized as previously reported [35]. RTqPCR was performed using a TaqMan®Gene Expression Master Mix and a 7500 RT-PCR instrument (Applied Biosystems), targeting genes of SM phosphodiesterase 1 (SMPD1, Hs03679347_g1) and SM phosphodiesterase 2 (SMPD2, Hs04187047_g1) genes. The mRNA expression levels were then normalized to those of the glyceraldehyde-3-phosphate dehydrogenase (GAPDH, Hs99999905_m1) housekeeping gene (Thermo Fisher Scientific, MA, USA). The relative mRNA expression levels were calculated as $2-\Delta\Delta Ct$ and compared to the results of the treated samples with the control and/or with those of the untreated ones [35].

4.5. Western Blotting

Protein concentrations were analyzed and electrophoresis was performed as previously reported [36]. Proteins were transferred onto a 0.45 µm cellulose nitrate strip membrane (Sartorius Stedim Biotech S.A.) in a transfer buffer for 1 h at 100 V at 4 °C. Membranes were blocked with 5% (w/v) non-fat dry milk in PBS, pH7.5, for 1 h at room temperature. The blot was incubated overnight at 4 °C with the specific antibodies, anti-aSMase and anti nSMase1 (1:1000), and then treated with horseradish peroxidase-conjugated goat anti-rabbit secondary antibodies (1:5000). A Super Signal West Pico Chemiluminescent Substrate (ThermoFisher Scientific) was used to detect the chemiluminescent (ECL) HRP substrate. The apparent molecular weights of the proteins were calculated in reference to the migration rate of the molecular size standards. The area density of the bands was evaluated by densitometric scanning and analyzed using Scion Image.

4.6. aSMase and nSMasi Activity Assay

The aSMase and nSMase activity was assayed according to Conte et al. [37]. Liver homogenates were suspended in 0.1% NP-40 detergent in PBS, sonicated for 30 s on ice at 20 watt, kept on ice for 30 min, and centrifuged at 16,000× g for 10 min. The supernatants were then used for the aSMase and nSMase assay. The enzyme activity was assayed in 60 µg proteins/10 µL Tris-MgCl2, pH 5.0, for aSMase and 7.4 for nSMase using an Amplex Red Sphingomyelinase assay kit (Invitrogen, Monza, Italy), according to the manufacturer's instructions. The fluorescence was measured with a FLUOstar Optima fluorimeter (BMG Labtech, Germany) using a filter set with a 360 nm excitation and 460 nm emissions.

4.7. Statistical Analysis

Data were expressed as the means ± SD of three livers, and their significance was checked by an ANOVA test. Significance: (a) * $p < 0.05$ versus the control sample (CTR); (b) § $p < 0.05$ rMnSOD treated and irradiated samples versus irradiated samples; (c) $\hat{}$ $p < 0.05$ the pretreated and 1.0 Gy irradiated sample versus the 1.0 Gy irradiated and rMnSOD treated sample.

Author Contributions: The conception and design of the study (F.S.A.-I., E.A., A.B., A.M.), experiments (F.S.A.-I., A.B., O.B., S.C., M.R.C., I.N., I.F., F.F.P., A.I., E.K.) acquisition of data (M.C., C.C.), analysis of data (S.C., F.C., B.F.), interpretation of data (E.A., F.C.), drafting the article or revising it critically for important intellectual content (E.A., T.B., A.M.) , founding acquisition (F.C.), final approval of the version to be submitted (E.A., F.C., F.S.A.-I.). All authors have read and agreed to the published version of the manuscript.

Funding: This study was supported by a grant from the Italian Space Agency, Project RASC, contract number 2015-008-R.0.

Conflicts of Interest: The authors declare no conflict of interest.

Abbreviations

aSMase: acid sphingomyelinase; nSMase1, neutral sphingomyelinase1; rMnSOD, recombinant manganese-containing superoxide dismutase; ROS, reactive oxygen species; RTqPCR, reverse transcription quantitative PCR; SM, sphingomyelin; SMPD, SM phosphodiesterase; SOD, superoxide dismutase.

References

1. Slotte, J.P. Biological functions of sphingomyelins. *Prog. Lipid Res.* **2013**, *52*, 424–437. [CrossRef]
2. Bartke, N.; Hannun, Y.A. Bioactive sphingolipids: metabolism and function. *J. Lipid Res.* **2009**, *50*, S91–S96. [CrossRef]
3. Goñi, F.M.; Alonso, A. Sphingomyelinases: enzymology and membrane activity. *FEBS Lett.* **2002**, *531*, 38–46. [CrossRef]
4. Zhang, P.; Chen, Y.; Zhang, T.; Zhu, J.; Zhao, L.; Li, J.; Wang, G.; Li, Y.; Xu, S.; Nilsson, A.; et al. Deficiency of alkaline SMase enhances dextran sulfate sodium-induced colitis in mice with upregulation of autotaxin. *J. Lipid Res.* **2018**, *59*, 1841–1850. [CrossRef]
5. Jenkins, R.W.; Canals, D.; Hannun, Y.A. Roles and regulation of secretory and lysosomal acid sphingomyelinase. *Cell. Signal.* **2009**, *21*, 836–846. [CrossRef]
6. Airola, M.V.; Hannun, Y.A. Sphingolipid metabolism and neutral sphingomyelinases. *Handb. Exp. Pharmacol.* **2013**, *215*, 57–76. [CrossRef]
7. Tomiuk, S.; Hofmann, K.; Nix, M.; Zumbansen, M.; Stoffel, W. Cloned mammalian neutral sphingomyelinase: Functions in sphingolipid signaling? *Proc. Natl. Acad. Sci. USA* **1998**, *95*, 3638–3643. [CrossRef] [PubMed]
8. Mizutani, Y.; Tamiya-Koizumi, K.; Nakamura, N.; Kobayashi, M.; Hirabayashi, Y.; Yoshida, S. Nuclear localization of neutral sphingomyelinase 1: biochemical and immunocytochemical analyses. *J. Cell Sci.* **2001**, *114*, 3727–3736. [PubMed]
9. Yabu, T.; Shiba, H.; Shibasaki, Y.; Nakanishi, T.; Imamura, S.; Touhata, K.; Yamashita, M. Stress-induced ceramide generation and apoptosis via the phosphorylation and activation of nSMase1 by JNK signaling. *Cell Death Differ.* **2014**, *22*, 258–273. [CrossRef] [PubMed]
10. Lin, M.; Liao, W.; Dong, M.; Zhu, R.; Xiao, J.; Sun, T.; Chen, Z.; Wu, B.; Jin, J. Exosomal neutral sphingomyelinase 1 suppresses hepatocellular carcinoma via decreasing the ratio of sphingomyelin/ceramide. *FEBS J.* **2018**, *285*, 3835–3848. [CrossRef] [PubMed]
11. Milhas, D.; Clarke, C.J.; Idkowiak-Baldys, J.; Canals, D.; Hannun, Y.A. Anterograde and retrograde transport of neutral sphingomyelinase-2 between the Golgi and the plasma membrane. *Biochim. Biophys. Acta Bioenerg.* **2010**, *1801*, 1361–1374. [CrossRef]
12. Corcoran, C.A.; He, Q.; Ponnusamy, S.; Ogretmen, B.; Huang, Y.; Sheikh, M.S. Neutral sphingomyelinase-3 is a DNA damage and nongenotoxic stress-regulated gene that is deregulated in human malignancies. *Mol. Cancer Res.* **2008**, *6*, 795–807. [CrossRef] [PubMed]
13. Yabu, T.; Shimuzu, A.; Yamashita, M. A Novel Mitochondrial Sphingomyelinase in Zebrafish Cells. *J. Biol. Chem.* **2009**, *284*, 20349–20363. [CrossRef]
14. Wu, B.X.; Rajagopalan, V.; Roddy, P.L.; Clarke, C.J.; Hannun, Y.A. Identification and Characterization of Murine Mitochondria-associated Neutral Sphingomyelinase (MA-nSMase), the Mammalian Sphingomyelin Phosphodiesterase 5. *J. Biol. Chem.* **2010**, *285*, 17993–18002. [CrossRef] [PubMed]
15. Albi, E.; Magni, M.V. Chromatin-associated sphingomyelin: metabolism in relation to cell function. *Cell Biochem. Funct.* **2003**, *21*, 211–215. [CrossRef] [PubMed]
16. Nakajima, T.; Ninomiya, Y.; Nenoi, M. Radiation-Induced Reactions in The Liver — Modulation of Radiation Effects by Lifestyle-Related Factors. *Int. J. Mol. Sci.* **2018**, *19*, 3855. [CrossRef]
17. Fink, S.L.; Cookson, B.T. Caspase-1-dependent pore formation during pyroptosis leads to osmotic lysis of infected host macrophages. *Cell. Microbiol.* **2006**, *8*, 1812–1825. [CrossRef]
18. Fink, S.L.; Cookson, B.T. Apoptosis, Pyroptosis, and Necrosis: Mechanistic Description of Dead and Dying Eukaryotic Cells. *Infect. Immun.* **2005**, *73*, 1907–1916. [CrossRef]
19. Sautin, Y.; Takamura, N.; Shklyaev, S.; Nagayama, Y.; Ohtsuru, A.; Namba, H.; Yamashita, S. Ceramide-Induced Apoptosis of Human Thyroid Cancer Cells Resistant to Apoptosis by Irradiation. *Thyroid.* **2000**, *10*, 733–740. [CrossRef]
20. Albi, E.; Perrella, G.; Lazzarini, A.; Cataldi, S.; Lazzarini, R.; Floridi, A.; Ambesi-Impiombato, F.S.; Curcio, F. Critical Role for the Protons in FRTL-5 Thyroid Cells: Nuclear Sphingomyelinase Induced-Damage. *Int. J. Mol. Sci.* **2014**, *15*, 11555–11565. [CrossRef]
21. Albi, E.; Cataldi, S.; Rossi, G.; Magni, M.V.; Toller, M.; Casani, S.; Perrella, G. The nuclear ceramide/diacylglycerol balance depends on the physiological state of thyroid cells and changes during UV-C radiation-induced apoptosis. *Arch. Biochem. Biophys.* **2008**, *478*, 52–58. [CrossRef] [PubMed]

22. Cataldi, S.; Borrelli, A.; Ceccarini, M.R.; Nakashidze, I.; Codini, M.; Belov, O.; Ivanov, A.; Krasavin, E.; Ferri, I.; Conte, C.; et al. Neutral Sphingomyelinase Modulation in the Protective/Preventive Role of rMnSOD from Radiation-Induced Damage in the Brain. *Int. J. Mol. Sci.* **2019**, *20*, 5431. [CrossRef] [PubMed]

23. Albi, E.; Pieroni, S.; Magni, M.V.; Sartori, C. Chromatin sphingomyelin changes in cell proliferation and/or apoptosis induced by ciprofibrate. *J. Cell. Physiol.* **2003**, *196*, 354–361. [CrossRef] [PubMed]

24. Albi, E.; Lazzarini, A.; Lazzarini, R.; Floridi, A.; Damaskopoulou, E.; Curcio, F.; Cataldi, S. Nuclear Lipid Microdomain as Place of Interaction between Sphingomyelin and DNA during Liver Regeneration. *Int. J. Mol. Sci.* **2013**, *14*, 6529–6541. [CrossRef] [PubMed]

25. Broz, P.; Dixit, V.M. Inflammasomes: mechanism of assembly, regulation and signalling. *Nat. Rev. Immunol.* **2016**, *16*, 407–420. [CrossRef]

26. Xiang, H.; Zhu, F.; Xu, Z.; Xiong, J. Role of Inflammasomes in Kidney Diseases via Both Canonical and Non-canonical Pathways. *Front. Cell Dev. Biol.* **2020**, *8*, 106. [CrossRef]

27. Lamkanfi, M. Emerging inflammasome effector mechanisms. *Nat. Rev. Immunol.* **2011**, *11*, 213–220. [CrossRef]

28. Wu, J.; Lin, S.; Wan, B.; Velani, B.; Zhu, Y. Pyroptosis in Liver Disease: New Insights into Disease Mechanisms. *Aging Dis.* **2019**, *10*, 1094–1108. [CrossRef]

29. Liu, Y.-G.; Chen, J.-K.; Zhang, Z.-T.; Ma, X.-J.; Chen, Y.-C.; Du, X.-M.; Liu, H.; Zong, Y.; Lu, G.-C. NLRP3 inflammasome activation mediates radiation-induced pyroptosis in bone marrow-derived macrophages. *Cell Death Dis.* **2017**, *8*, e2579. [CrossRef]

30. Haimovitz-Friedman, A.; Kan, C.C.; Ehleiter, D.; Persaud, R.S.; McLoughlin, M.; Fuks, Z.; Kolesnick, R.N. Ionizing radiation acts on cellular membranes to generate ceramide and initiate apoptosis. *J. Exp. Med.* **1994**, *180*, 525–535. [CrossRef]

31. Li, C.; Guo, S.; Pang, W.; Zhao, Z. Crosstalk Between Acid Sphingomyelinase and Inflammasome Signaling and Their Emerging Roles in Tissue Injury and Fibrosis. Front. *Cell Dev. Biol.* **2020**, *7*, 378. [CrossRef]

32. Wang, Y.; Tang, M. Dysfunction of various organelles provokes multiple cell death after quantum dot exposure. *Int. J. Nanomed.* **2018**, *13*, 2729–2742. [CrossRef] [PubMed]

33. Jung, S.Y.; Suh, J.H.; Park, H.J.; Jung, K.-M.; Kim, M.Y.; Na, D.S.; Kim, D.K. Identification of Multiple Forms of Membrane-Associated Neutral Sphingomyelinase in Bovine Brain. *J. Neurochem.* **2002**, *75*, 1004–1014. [CrossRef] [PubMed]

34. Borrelli, A.; Schiattarella, A.; Mancini, A.; Morrica, B.; Cerciello, V.; Mormile, M.; D'Alesio, V.; Bottalico, L.; Morelli, F.; D'Armiento, M.; et al. A recombinant MnSOD is radioprotective for normal cells and radiosensitizing for tumor cells. *Free Radic. Biol. Med.* **2009**, *46*, 110–116. [CrossRef] [PubMed]

35. Albi, E.; Cataldi, S.; Ferri, I.; Sidoni, A.; Traina, G.; Fettucciari, K.; Ambesi-Impiombato, F.S.; Lazzarini, A.; Curcio, F.; Ceccarini, M.R.; et al. VDR independent induction of acid-sphingomyelinase by 1,23(OH)2 D3 in gastric cancer cells: Impact on apoptosis and cell morphology. *Biochim.* **2018**, *146*, 35–42. [CrossRef] [PubMed]

36. Cataldi, S.; Codini, M.; Hunot, S.; Légeron, F.-P.; Ferri, I.; Siccu, P.; Sidoni, A.; Ambesi-Impiombato, F.S.; Beccari, T.; Curcio, F.; et al. e-Cadherin in 1-Methyl-4-phenyl-1,2,3,6-tetrahydropyridine-Induced Parkinson Disease. *Mediat. Inflamm.* **2016**, *2016*, 1–7. [CrossRef]

37. Conte, C.; Arcuri, C.; Cataldi, S.; Mecca, C.; Codini, M.; Ceccarini, M.R.; Patria, F.F.; Beccari, T.; Albi, E. Niemann-Pick Type A Disease: Behavior of Neutral Sphingomyelinase and Vitamin D Receptor. *Int. J. Mol. Sci.* **2019**, *20*, 2365. [CrossRef]

International Journal of
Molecular Sciences

Article

Neutral Sphingomyelinase Modulation in the Protective/Preventive Role of rMnSOD from Radiation-Induced Damage in the Brain

Samuela Cataldi [1], Antonella Borrelli [2], Maria Rachele Ceccarini [1], Irina Nakashidze [1], Michela Codini [1], Oleg Belov [3], Alexander Ivanov [3], Eugene Krasavin [3], Ivana Ferri [4], Carmela Conte [1], Federica Filomena Patria [1], Giovanna Traina [1], Tommaso Beccari [1], Aldo Mancini [5], Francesco Curcio [6], Francesco Saverio Ambesi-Impiombato [6] and Elisabetta Albi [1,*]

[1] Department of Pharmaceutical Sciences, University of Perugia, 06126 Perugia, Italy; samuelacataldi@libero.it (S.C.); chele@hotmail.it (M.R.C.); irinanakashidze@yahoo.com (I.N.); michela.codini@unipg.it (M.C.); carmela.conte@unipg.it (C.C.); patriafederica@gmail.com (F.F.P.); giovanna.traina@unipg.it (G.T.); tommaso.beccari@unipg.it (T.B.)
[2] Molecular Biology and Viral Oncology Unit, Istituto Nazionale Tumori IRCCS "Fondazione G. Pascale", 80131 Napoli, Italy; a.borrelli@istitutotumori.na.it
[3] Laboratory of Radiation Biology, Joint Institute for Nuclear Research, 141980 Dubna, Russia; dem@jinr.ru (O.B.); a1931192@mail.ru (A.I.); krasavin@jinr.ru (E.K.)
[4] Division of Pathological Anatomy and Histology, Department of Experimental Medicine, School of Medicine and Surgery, University of Perugia, 06126 Perugia, Italy; ivanaferri@gmail.com
[5] Laedhexa Biotechnologies Inc., San Francisco, CA QB3@953, USA; aldo_mancini@tiscali.it
[6] Dipartimento di Area Medica, University of Udine, 33100 Udine, Italy; francesco.curcio@uniud.it (F.C.); ambesis@me.com (F.S.A.-I.)
* Correspondence: elisabetta.albi@unipg.it; Tel.: +39-075-585-7906

Received: 9 October 2019; Accepted: 29 October 2019; Published: 31 October 2019

Abstract: Studies on the relationship between reactive oxygen species (ROS)/manganese superoxide dismutase (MnSOD) and sphingomyelinase (SMase) are controversial. It has been demonstrated that SMase increases the intracellular ROS level and induces gene expression for MnSOD protein. On the other hand, some authors showed that ROS modulate the activation of SMase. The human recombinant manganese superoxide dismutase (rMnSOD) exerting a radioprotective effect on normal cells, qualifies as a possible pharmaceutical tool to prevent and/or cure damages derived from accidental exposure to ionizing radiation. This study aimed to identify neutral SMase (nSMase) as novel molecule connecting rMnSOD to its radiation protective effects. We used a new, and to this date, unique, experimental model to assess the effect of both radiation and rMnSOD in the brain of mice, within a collaborative project among Italian research groups and the Joint Institute for Nuclear Research, Dubna (Russia). Mice were exposed to a set of minor γ radiation and neutrons and a spectrum of neutrons, simulating the radiation levels to which cosmonauts will be exposed during deep-space, long-term missions. Groups of mice were treated or not-treated (controls) with daily subcutaneous injections of rMnSOD during a period of 10 days. An additional group of mice was also pretreated with rMnSOD for three days before irradiation, as a model for preventive measures. We demonstrate that rMnSOD significantly protects the midbrain cells from radiation-induced damage, inducing a strong upregulation of nSMase gene and protein expression. Pretreatment with rMnSOD before irradiation protects the brain with a value of very high nSMase activity, indicating that high levels of activity might be sufficient to exert the rMnSOD preventive role. In conclusion, the protective effect of rMnSOD from radiation-induced brain damage may require nSMase enzyme.

Keywords: neutral sphingomyelinase; radiation; sphingomyelin metabolism; pathology; cell signaling; brain

---

1. Introduction

Evidence for the involvement of sphingomyelin (SM) in radiation-induced apoptosis relies on studies focused on specific enzymes as sphingomyelinase (SMase) [1–3]. SMase cleaves SM, generating ceramide and choline phosphate. The ceramide pathway, in turn, is responsible for the generation of various lipid mediators for cell signaling. Studies on the relationship between reactive oxygen species (ROS) and SMase are controversial. It has been demonstrated that SMase increases the intracellular ROS level in several experimental models [4,5]. Accordingly, ceramides increase ROS level [6,7]. Neutral SMase (nSMase)-protein kinase Cζ(PKCζ)-NADPH oxidase is essential for ROS production [8]. On the other hand, ceramides derived from SMase activity are involved in ROS production [9] and the activation of ceramide-p47phox-ROS signaling cascade is essential for apoptosis [10]. Moreover, both SMase and ceramide induce MnSOD gene expression [11]. Although numerous studies indicate a stimulatory effect of SMase and ceramide in ROS production, some authors suggest the positive role of ROS in the ceramide generation [12]. In line with this, previous published results showed that p53-induced ROS modulate the activation of nSMase [13]. In addition, by inhibiting ROS production nSMase stimulation and ceramide generation are suppressed [14].

The human recombinant manganese superoxide dismutase (rMnSOD) has specific antioxidant and anti-free radical activity as the native superoxide dismutase (SOD) [15]. MnSOD enzyme has been proposed to be useful in the prevention and treatment of damage caused by physical agents, such as ionizing radiations [16]. More rMnSOD has been identified as a possible pharmaceutical tool to prevent and/or cure the accidental damage derived from exposure to ionizing radiation [15]. Thus, rMnSOD has the invaluable advantage, over the native enzyme, of being able to easily enter into the cells and tissues thanks to the persistence in the recombinant mature protein of its leader peptide. Consequently, treatment side effects should be significantly reduced in comparison with traditional treatments [17,18].

To address a comprehensive analysis of nSMase involvement in the mechanism of rMnSOD protection or prevention from radiation damage in the brain, we performed experiments within a collaborative project among Italian research groups and the Joint Institute for Nuclear Research, Dubna (Russia). Mice, exposed to a set of minor γ radiation and neutrons and a spectrum of neutrons, simulating the radiation levels to which cosmonauts are exposed during deep-space long-term missions, were injected with rMnSOD either at the time (protection) or before (prevention) irradiation. The study aimed to identify a possible target molecule of the rMnSOD administered in order to reduce radiation-induced damage. Thus, the nSMase, known to be involved in the production of ROS and in the response to MnSOS, has been studied in our experimental model.

2. Results

2.1. Protective and Preventive Effect of rMnSOD on Radiation-Induced Structural Changes in Midbrain Tissue

Before evaluating the effect of rMnSOD in mice brain, we confirmed its capacity to freely diffuse through the blood-brain barrier, locating itself within brain tissues. This was accomplished by immunohistochemical analysis (Figure 1).

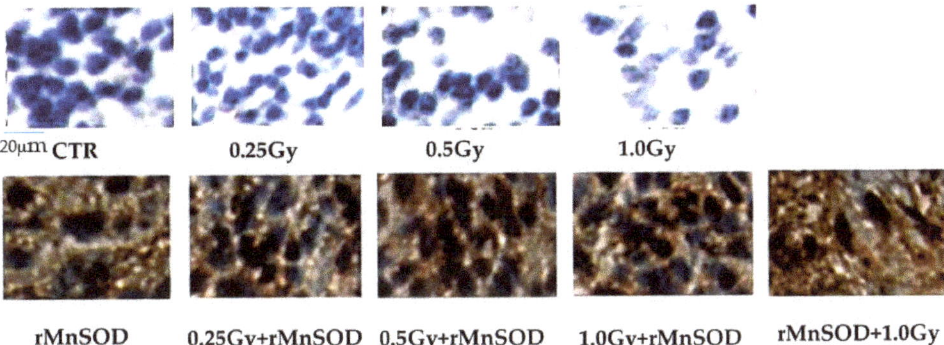

Figure 1. Localization of rMnSOD in brain tissue. Immunohistochemical analysis was performed by using specific antibody. The immunostaining was evident only in brain samples from rMnSOD-treated mice.

We also found that the brains of not-irradiated control mice, with or without rMnSOD treatment, had comparable cell numbers. Only medium and high radiation doses induced a loss of the cells with progressive increase of intercellular spaces (Figure 2a). As expected, in these conditions we observed a robust protective effect with rMnSOD (Figure 2b). We thus investigated the possibility that rMnSOD might play a protective role within the neurofilament structure. To date, 200–220 kDa heavy neurofilament (NF200) is considered the specific marker of large myelinated A-βfiber neurons [19]. The analysis of NF200by immunohistochemistry showed that the irradiation caused an accumulation of the labeling in rounded areas with loss of the characteristic length and thickness of neurofilaments (Figure 2c). Such effects were not evident in mice treated or pretreated with rMnSOD; in those samples, normal heavy neurofilaments were present (Figure 2c). Morphological evaluation is however only qualitative, due to the loss of the normal neuritic structure.

2.2. nSMase Is Required for the Protective and Preventive Effect of rMnSOD

We have previously reported that space radiation stimulated cellular and nuclear SMase in mice thyroids after their long stay in the International Space Station [20]. Our results indicated that radiation increased nSMase gene expression (Figure 3b) in comparison with control samples (Figure 3a). rMnSOD alone had a similar effect (Figure 3a). In comparison with rMnSOD alone considered as control, the nSMase gene expression strongly increased with medium- and high radiation exposure (Figure 3b). The effect was attenuated by rMnSOD pre-treatment (Figure 3b). Treatment and pretreatment with rMnSOD were responsible for the nSMase gene expression increase in irradiated samples (Figure 3b compared withFigure 3a). To analyze the extent of the nSMase response to radiation and rMnSOD treatment, we studied the protein expression. We found that the content of nSMase was low in the control sample and increased with irradiation (Figure 3c). Thus, the presence of rMnSOD resulted in overexpression of nSMase protein, responding to radiation in a dose-dependent mode. The densitometry analysis, performed with Scion Image program by using the corresponding beta-tubulin as control, showed that the enzyme increases about 100–130% over controls with radiation (Figure 3e) and 65% with rMnSOD alone (Figure 3e). By using rMnSOD alone as control, the enzyme increase in response to radiation in a dose-dependent mode in the presence of rMnSOD, was confirmed (Figure 3e).

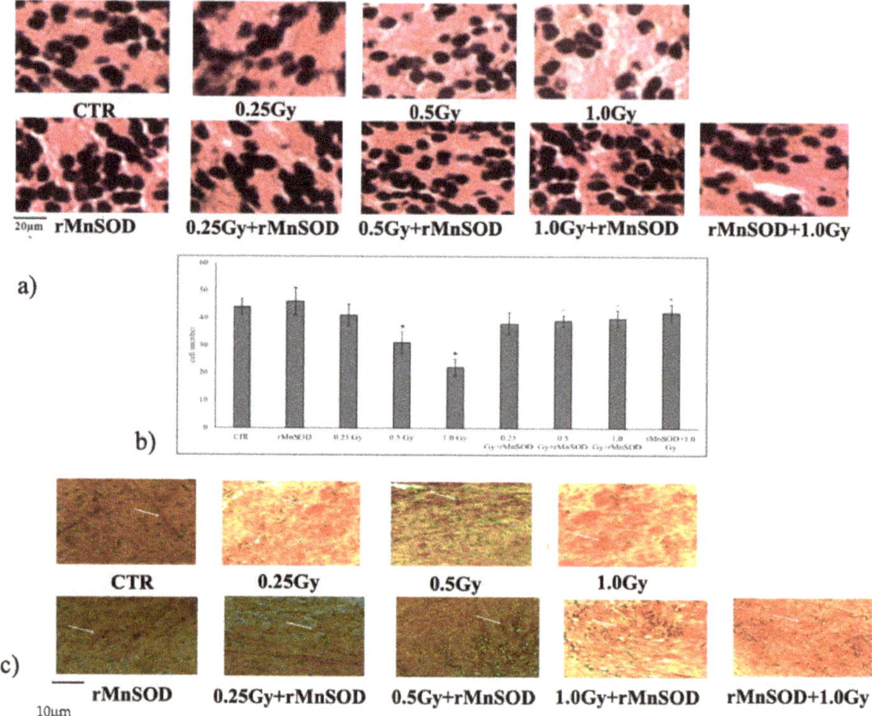

Figure 2. Midbrain nuclei and heavy neurofilament 200 kDa (NF200). (**a**) Hematoxylin-eosin-stained midbrain sections from normal mice exposed to 0.25, 0.5 and 1.0Gy radiation doses, treated in the presence or absence of rMnSOD. rMnSOD + 1.0Gy: mice pretreated with rMnSOD and exposed to 1.0 Gy, then treated with rMnSOD for 3 days. Shown images are representative of similar findings in the nuclear regions of 3 midbrains from each group of mice (40× magnification); (**b**) cell numbers were counted as described in Results. * $p < 0.05$ irradiated samples vs. not-irradiated control samples (CRT), ^ $p < 0.05$ irradiated and rMnSOD-treated samples vs. corresponding irradiated samples, ° $p < 0.05$ rMnSOD-pretreated and 1.0 Gy irradiated sample vs. 1.0 Gy irradiated sample; (**c**) NF200 immunohistochemical staining. Normal mice were exposed to 0.25, 0.5 and 1.0 Gy radiation doses, with or without rMnSOD administration. rMnSOD + 1.0Gy: mice pretreated with rMnSOD for 3 days before 1.0 Gy radiation. Images are representative of similar images showing heavy neurofilaments in 3 midbrains from each group of mice (20× magnification). Arrows indicate normal neurofilaments (CTR), accumulation of labeling in round areas with reduction of length and thickness in neurofilaments (irradiated samples), and the presence of normal neurofilaments in rMnSOD-treated samples, thus demonstrating its radioprotective effect.

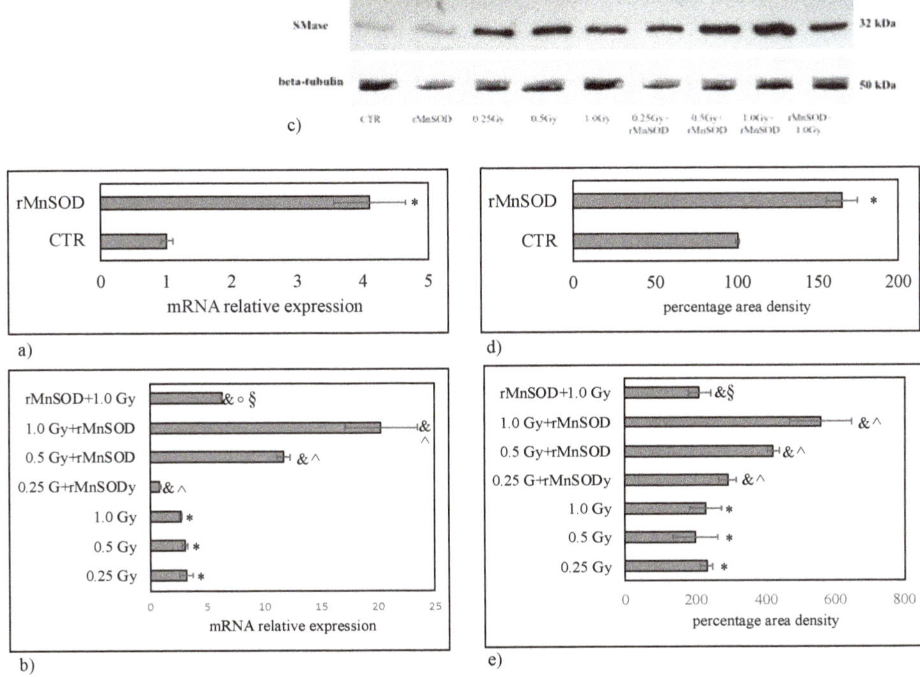

Figure 3. Effect of rMnSOD on nSMase gene and protein expression. Samples from mice treated with 0.25, 0.5 and 1.0 Gy radiation doses, in the presence or absence of rMnSOD. rMnSOD + 1.0 Gy: mice pretreated with rMnSOD, exposed to a 1.0 Gy radiation dose, then treated with rMnSOD for 3 days. Left panels (**a**,**b**): RTqPCR analyses. GAPDH was used as housekeeping control gene. Data are expressed as mean ± SD of mRNA expression (folds increase). Top panel (**c**): immunoblotting analysis of nSMase; tubulin was used as control. Right panels (**d**,**e**): immunoblotting densitometric analysis, normalized with beta-tubulin. Data are expressed as the mean ± SD of 3 independent experiments, each carried out in triplicate. Significance: (**a**,**d**) not-irradiated rMnSOD-treated samples compared to not-irradiated and not-rMnSOD-treated samples (CRT) (* $p < 0.05$); (**b**,**e**): irradiated not-rMnSOD-treated samples and irradiated rMnSOD-treated and pretreated samples. Irradiated samples compared to CTR samples (* $p < 0.05$), irradiated and rMnSOD-treated and pretreated samples vs. rMnSOD-treated samples (& $p < 0.05$), irradiated and rMnSOD-treated samples vs. their correspondent irradiated samples (^ $p < 0.05$), rMnSOD-pretreated and 1.0 Gy irradiated sample vs. 1.0 Gy irradiated sample (° $p < 0.05$), rMnSOD-pretreated and 1.0 Gy irradiated sample vs. 1.0 Gy irradiated and rMnSOD-treated sample (§ $p < 0.05$).

To investigate the biological role of nSMase, we also assayed the enzyme activity. Only high radiation was able to further stimulate the nSMase activity (Figure 4b) with value similar to those obtained with rMnSOD alone (Figure 4a). The treatment with rMnSOD in irradiated mice strongly increased the nSMase activity in radiation dose-dependent mode (Figure 4b). Surprisingly, the pretreatment of mice with rMnSOD for 3 days before exposure to high radiation doses increased the enzyme activity even further (Figure 4b).

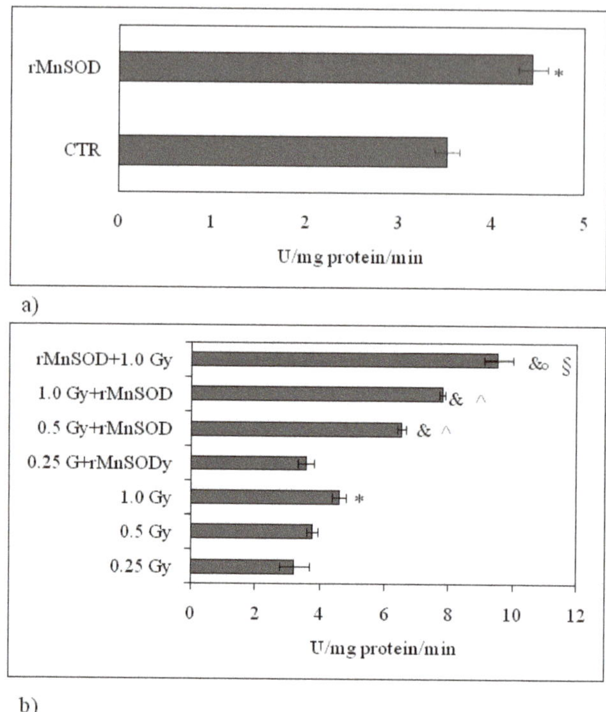

Figure 4. Activity of nSMase. Significance: (**a**) not-irradiated rMnSOD-treated samples compared to not-irradiated and not-rMnSOD-treated samples (CRT) (* $p < 0.05$); (**b**) irradiated not-rMnSOD-treated samples and irradiated rMnSOD-treated and pretreated samples. Significance: irradiated samples compared to CTR samples (* $p < 0.05$), irradiated and rMnSOD-treated and pretreated samples vs. rMnSOD-treated samples (& $p < 0.05$), irradiated and rMnSOD-treated samples vs. their correspondent irradiated samples (^ $p < 0.05$), rMnSOD pretreated and 1.0 Gy irradiated sample vs. 1.0 Gy irradiated sample (° $p < 0.05$), rMnSOD pretreated and 1.0 Gy irradiated sample vs. 1.0 Gy irradiated and rMnSOD-treated sample (§ $p < 0.05$).

3. Discussion

The aim of this study was to elucidate the rMnSOD-dependent nSMase changes and their role in the protective-preventive effect from ionizing radiation in the brain. Mice were exposed to a set of minor γ radiation and neutrons and a spectrum of neutrons, simulating the radiation levels to which cosmonauts will be exposed during deep-space, long-term missions. At the moment we do not know whether γ radiation only was unable to induce nSMase activation. We show that rMnSOD particles are capable to cross the blood-brain barrier and localize in the midbrain. Here, rMnSOD has the ability to limit radiation-induced damage, such as the loss of neuron number and the alteration of neurofilaments. The effect of radiation on brain damage has previously been reported by other authors [21,22]. Although many studies elucidated the radioprotective role of rMnSOD through antioxidant mechanisms in mitochondria [15–18], no data exist about its effect on nSMase. We show that rMnSOD stimulates nSMase gene and protein expression and enzymatic activity. The study was articulated with the aim of analyzing the response of nSMase in samples in which rMnSOD has a protective effect (non-pretreated samples) and in samples in which rMnSOD has a preventive effect (pretreated samples). In samples for protective effect study, rMnSOD increases strongly nSMase gene and protein expression in irradiated samples in a dose-dependent manner. Small variations of nSMase enzyme activities are obtained only with high radiation doses. This observation is relevant, as nSMase plays a crucial role in the regulation of ROS generation and in the maintenance of homeostasis between proapoptotic and prosurvival signals [1,4]. In line with this, we show that highest radiation levels induce significant nSMase responses only in the presence of rMnSOD. Although the present results indicate that nSMase may be required for the action of rMnSOD, we cannot prove the direct action of rMnSOD on this enzyme. To our knowledge, this is the first study indicating SMase as a potential effector of rMnSOD during its protection action. In samples for preventive effect study, nSMase gene and protein expression is reduced in comparison with the samples in which the protective effect of rMnSOD is studied but the enzyme activity is much higher. At this moment we can only speculate why nSMase levels are reduced with pretreatment, a possible explanation being that an extended treatment time could stimulate different metabolic patterns, leading to inhibition of nSMase gene expression. Merging already published data with our findings, it is possible to suggest that in brain tissues exposed to radiation, significant amounts of ROS are generated, stimulating nSMase production to significantly increase MnSOD levels. This may in turn stimulate nSMase. At this point, ROS levels may be reduced, with the end result of limiting the damage. In the above described series of events, nSMase may play predominant roles.

4. Materials and Methods

4.1. Chemicals

The rMnSOD radioprotective protein, discovered and obtained in the recombinant form by A.Mancini, was provided by the Molecular Biology and Viral Oncology Unit, Department of Experimental Oncology, Istituto Nazionale Tumori Fondazione G. Pascale—IRCCS, Naples, Italy [15]. Sterile 2 µg aliquots of rMnSOD in 0.5 mL of sterile saline phosphate buffer (PBS) were prepared and stored at −80 °C. Anti-nSMase, and anti-βtubulin were from Abcam (Cambridge, UK). Horseradish peroxidase-conjugated goat anti-rabbit secondary antibodies were from Santa Cruz. Anti-rMnSOD was from InBios International (Washington, WA, USA). TaqMan SNP Genotyping Assay and Reverse Transcription kit were purchased from Applied Biosystems (Foster City, CA, USA). RNAqueous®-4PCR kit was from Ambion Inc. (Austin, TX, USA). SDS-PAGE molecular weight standards were purchased from Bio-Rad Laboratories (Hercules, CA, USA). Chemiluminescence kits were purchased fromAmersham (Rainham, Essex, UK).

4.2. Experimental Design and Animal Care

Animals: Fifty-fourfemale ICR mice weighting approximately 25–30 g were obtained from the Laboratory Animal Nursery of the Russian Academy of Sciences (Pushchino, Russia). After transportation, the animals were adapted to the vivarium at the Joint Institute for Nuclear Research (JINR) in Dubna during a period of 10 days. The animals were divided into 9 cages, 6 mice each, receiving the standard briquetted fodder and water ad libitum. All procedures were performed according to the Russian Guidelines for the Care and Use of Experimental Animals and Bioethics Instructions (Order of the USSR Ministry of Health No. 755 12.08.1987) accepted in the vivarium of the Institute of Biomedical Problems, part of the above-mentioned JINR.

Animal treatments: Mice in groups number 1,3,4 and 5 were treated with daily subcutaneous injections of sterile PBS solution for 7 days from the day of irradiation. Animals in groups 2,6,7,8 and 9 were treated with daily subcutaneous injections of rMnSOD in sterile PBS. In particular, mice in groups 2,6,7 and 8 received 7 injections of rMnSOD while animals in group 9 received a total of 10 injections, being pretreated with rMnSOD for 3 additional days prior to irradiation. All animals were irradiated at the JINR, being exposed to a set of minor γ radiation and neutrons of a Phasatron with high Relative Biological Effectiveness (RBE) and a spectrum of neutrons, to simulate space flight exposure. Animals in groups 3&6, 4&7 and 5&8 were exposed to doses of 0.25, 0.50 and 1.00 Gy (respectively). Mice in groups 1 (mock-treated with PBS) and 2 (rMnSOD-treated) were not exposed to radiation and considered a biological control. At the end of the experiment, all mice were beheaded and brains immediately frozen.

4.3. Morphological and Immunohistochemistry Analysis

Three brains from each group were fixed in 4% neutral phosphate-buffered formaldehyde solution for 24 h and dropped with specific orientation in paraffin. Morphological and immunohistochemical analyses were performed as previously reported [19].

4.4. Reverse Transcription Quantitative PCR (RTqPCR)

Total RNA was extracted from mice brain using RNAqueous-4PCR kit (Ambion Inc., Austin, TX, USA), as previously reported [23]. Before cDNA synthesis, the integrity of RNA was evaluated by electrophoresis in TAE 1.2% agarose gel prepared in our lab. cDNA was synthesized using 1 µg total RNA for all samples by High-Capacity cDNA Reverse Transcription kit (Applied Biosystems, Foster City, CA, USA) under the following conditions: 50 °C for 2 min, 95 °C for 10 min, 95 °C for 15 s and 60 °C for 1 min, for a total of 40 cycles [23]. RTqPCR was performed using TaqMan®Gene Expression Master Mix and 7500 RT-PCR instrument (Applied Biosystems), SM phosphodiesterase 4 (SMPD4, Hs04187047_g1) genes. mRNA expression levels were then normalized to the glyceraldehyde-3-phosphate dehydrogenase (GAPDH, Hs99999905_m1) housekeeping gene (Thermo Fisher Scientific, Austin, MA, USA). mRNA relative expression levels were calculated as $2^{-\Delta\Delta Ct}$, comparing the results of the treated samples with those of the untreated ones [23].

4.5. Protein Concentration and Western Blotting

Protein concentrations were determined according to the Bradford method, as previously reported [23]. Forty-µgproteins were submitted to 12% SDS (sodium dodecyl sulfate) polyacrylamide gel electrophoresis at 200 V for 60 min [23]. Briefly, proteins were transferred onto 0.45 µm cellulose nitrate strips membrane (Sartorius Stedim Biotech S.A., Aubagne, France) in transfer buffer for 1 h at 100 V at 4 °C. Membranes were blocked with 5% (*w/v*) non-fat dry milk in PBS, pH 7.5 for 1 h at room temperature. The blot was incubated overnight at 4 °C with specific antibodies (1:1000) and then treated with horseradish peroxidase-conjugated goat anti-rabbit secondary antibodies (1:5000). Super Signal West Pico Chemiluminescent Substrate (Thermo Fisher Scientific) was used to detect chemiluminescent (ECL) HRP substrate. The apparent molecular weight of proteins was calculated referringto the migration rate of molecular size standards. The area density of the bands was evaluated by densitometry scanning and analyzing themwith Scion Image.

4.6. nSMaseActivity Assay

nSMase activity was assayed according to Ceccarini et al. [24]. Brain homogenates were suspended in 0.1% NP-40 detergent in PBS, sonicated for 30 s once at 20 watts, then kept on ice for 30 min and centrifuged at 16,000× *g* for 10 min. Supernatants were used for nSMase assay. Sixty µg/10µL proteins were incubated with 10 µL HMU-PC substrate for 10 min at 37 °C. The reaction was stopped by adding 200 µL stop buffer [21]. The fluorescence of 6-hexadecanoyl-4-methylumbelliferone (HMU) was measured with FLUOstar Optima fluorimeter (BMG Labtech, Ortenberg, Germany), using the filter set of 4-methylumbelliferone (MU), 360 nm excitation, and 460 nm emission. The fluorimeter was calibrated with MU in stop buffer.

4.7. Statistical Analysis

Data were expressed as means ± SD and their significance was checked by ANOVA test. Significance: (a) * $p < 0.05$ irradiated samples versus not-irradiated control sample (CTR); (b) & $p < 0.05$ irradiated and rMnSOD-treated samples versus rMnSOD-treated sample (rMnSOD); (c) ^ $p < 0.05$ irradiated and rMnSOD-treated samples versus respective irradiated samples;

(d) ° $p < 0.05$ rMnSOD-pretreated and 1.00 Gy irradiated sample versus 1.00 Gy irradiated sample; (e) § $p < 0.05$ rMnSOD-pretreated and 1.00 Gy irradiated sample versus 1.0 Gy irradiated and rMnSOD-treated sample.

Author Contributions: The conception and design of the study (F.S.A.-I., E.A., A.B., A.M.), experiments (F.S.A.-I., A.B., O.B., A.B., S.C., M.R.C., I.N., I.F., F.F.P., A.I., E.K.) acquisition of data (M.C., C.C.), analysis of data (S.C., F.C., G.T., I.F.), interpretation of data (F.C.), drafting the article or revising it critically for important intellectual content (E.A., T.B., A.M.), founding acquisition (F.C.), final approval of the version to be submitted (E.A., F.S.A.-I.).

Funding: This study was supported by the grant from theItalian Space Agency "Project RASC, number of contract: 2015-008-R.0".

Conflicts of Interest: The authors declare no conflict of interest.

References

1. Haimovitz-Friedman, A.; Kan, C.C.; Ehleiter, D.; Persaud, R.S.; McLoughlin, M.; Fuks, Z.; Kolesnick, R.N. Ionizingradiationacts on cellular membranes to generate ceramide and initiate apoptosis. *J. Exp. Med.* **1994**, *180*, 525–535. [CrossRef] [PubMed]

2. Chmura, S.J.; Mauceri, H.J.; Advani, S.; Heimann, R.; Beckett, M.A.; Nodzenski, E.; Quintans, J.; Kufe, D.W.; Weichselbaum, R.R. Decreasing the apoptotic threshold of tumor cells through protein kinase C inhibition and sphingomyelinaseactivation increases tumor killing by ionizingradiation. *Cancer Res.* **1997**, *57*, 4340–4347. [PubMed]

3. Albi, E.; Cataldi, S.; Lazzarini, A.; Codini, M.; Beccari, T.; Ambesi-Impiombato, F.S.; Curcio, F. Radiation and thyroid cancer. *Int. J. Mol. Sci.* **2017**, *18*, E911. [CrossRef] [PubMed]

4. Won, J.S.; Singh, I. Sphingolipid signaling and redox regulation. *Free Radic Biol. Med.* **2016**, *40*, 1875–1888.

5. Kitatani, K.; Akiba, S.; Sato, T. Ceramide-induced enhancement of secretory phospholipase A2 expression via generation of reactive oxygen species in tumor necrosis factor-alpha-stimulated mesangial cells. *Cell Signal.* **2004**, *16*, 967–974. [CrossRef] [PubMed]

6. Li, H.; Junk, P.; Huwiler, A.; Burkhardt, C.; Wallerath, T.; Pfeilschifter, J.; Förstermann, U. Dual effect of ceramide on human endothelial cells: Induction of oxidative stress and transcriptional upregulation of endothelial nitric oxide synthase. *Circulation* **2002**, *106*, 2250–2256. [CrossRef]

7. Li, X.; Becker, K.A.; Zhang, Y. Ceramide in redox signaling and cardiovascular diseases. *Cell. Physiol. Biochem.* **2010**, *26*, 41–48. [CrossRef]

8. Frazziano, G.; Moreno, L.; Moral-Sanz, J.; Menendez, C.; Escolano, L.; Gonzalez, C.; Villamor, E.; Alvarez-Sala, J.L.; Cogolludo, A.L.; Perez-Vizcaino, F. Neutral sphingomyelinase, NADPH oxidase and reactive oxygen species. Role in acute hypoxic pulmonary vasoconstriction. *J. Cell. Physiol.* **2011**, *226*, 2633–2640. [CrossRef]

9. JazvinšćakJembrek, M.; Hof, P.R.; Šimić, G. Ceramides in Alzheimer's disease: Key mediators of neuronal apoptosis induced by oxidative stress and Aβ accumulation. *Oxid. Med. Cell. Longev.* **2015**, *346783*.

10. Chen, Y.Y.; Hsu, M.J.; Sheu, J.R.; Lee, L.W.; Hsieh, C.Y. Andrographolide, a Novel NF-κB Inhibitor, Induces Vascular Smooth Muscle Cell Apoptosis via a Ceramide-p47phox-ROSSignaling Cascade. *Evid. Based. Complement. Alternat. Med.* **2013**, *2013*. [CrossRef]

11. Pahan, K.; Dobashi, K.; Ghosh, B.; Singh, I. Induction of the manganese superoxide dismutase gene by sphingomyelinase and ceramide. *J. Neurochem.* **1999**, *73*, 513–520. [CrossRef] [PubMed]

12. He, X.; Schuchman, E.H. Ceramide and ischemia/reperfusion injury. *J. Lipids.* **2018**, *2018*. [CrossRef] [PubMed]

13. Dumitru, C.A.; Zhang, Y.; Li, X.; Gulbins, E. Ceramide: A novel player in reactive oxygen species-induced signaling? *Antioxid. Redox. Signal.* **2007**, *9*, 1535–1540. [CrossRef] [PubMed]

14. Bezombes, C.; de Thonel, A.; Apostolou, A.; Louat, T.; Jaffrézou, J.P.; Laurent, G.; Quillet-Mary, A. Overexpression of protein kinase Czeta confers protection against antileukemic drugs by inhibiting the redox-dependentsphingomyelinaseactivation. *Mol. Pharmacol.* **2002**, *62*, 1446–1455. [CrossRef]

15. Borrelli, A.; Schiattarella, A.; Mancini, R.; Morrica, B.; Cerciello, V.; Mormile, M.; D'Alesio, V.; Bottalico, L.; Morelli, F.; D'Armiento, M.; et al. A recombinant MnSOD is radioprotective for normal cells and radiosensitizing for tumor cells. *Free Radic. Biol. Med.* **2009**, *46*, 110–116. [CrossRef]

Int. J. Mol. Sci. **2019**, *20*, 5431

16. Epperly, M.W.; Gretton, J.E.; Sikora, C.A.; Jefferson, M.; Bernarding, M.; Nie, S.; Greenberger, J.S. Mitochondrial localization of superoxide dismutase is required for decreasing radiation cellular damage. *Radiat. Res.* **2003**, *160*, 568–578. [CrossRef]

17. Borrelli, A.; Schiattarella, A.; Bonelli, P.; Tuccillo, F.M.; Buonaguro, F.M.; Mancini, A. The functional role of MnSOD as a biomarker of human diseases and therapeutic potential of a new isoform of a human recombinant MnSOD. *Biomed. Res. Int.* **2014**, *2014*. [CrossRef]

18. Pica, A.; Di Santi, A.; D'Angelo, V.; Iannotta, A.; Ramaglia, M.; Di Martino, M.; Pollio, M.L.; Schiattarella, A.; Borrelli, A.; Mancini, A.; et al. Effect ofrMnSODon survival signaling in pediatric high risk T-cell acute lymphoblastic leukaemia. *J. Cell Physiol.* **2015**, *230*, 1086–1093. [CrossRef]

19. Cataldi, S.; Codini, M.; Hunot, S.; Légeron, F.P.; Ferri, I.; Siccu, P.; Sidoni, A.; Ambesi-Impiombato, F.S.; Beccari, T.; Curcio, F.; et al. e-Cadherin in 1-Methyl-4-phenyl-1,2,3,6-tetrahydropyridine-Induced Parkinson Disease. *Mediators Inflamm.* **2016**, *2016*. [CrossRef]

20. Albi, E.; Ambesi-Impiombato, S.; Villani, M.; DePol, I.; Spelat, R.; Lazzarini, R.; Perrella, G. Thyroid cell growth: Sphingomyelin metabolism as non-invasive marker for cell damage acquired during spaceflight. *Astrobiology* **2018**, *10*, 811–820. [CrossRef]

21. Le, O.; Palacio, L.; Bernier, G.; Batinic-Haberle, I.; Hickson, G.; Beauséjour, C. INK4a/ARF Expression Impairs Neurogenesis in theBrainof Irradiated Mice. *Stem. Cell Rep.* **2018**, *10*, 1721–1733. [CrossRef] [PubMed]

22. Constanzo, J.; Dumont, M.; Lebel, R.; Tremblay, L.; Whittingstall, K.; Masson-Côté, L.; Geha, S.; Sarret, P.; Lepage, M.; Paquette, B.; et al. Diffusion MRI monitoring of specific structures in the irradiated ratbrain. *Magn. Reson. Med.* **2018**, *80*, 1614–1625. [CrossRef] [PubMed]

23. Albi, E.; Cataldi, S.; Ferri, I.; Sidoni, A.; Traina, G.; Fettucciari, K.; Ambesi-Impiombato, F.S.; Lazzarini, A.; Curcio, F.; Ceccarini, M.R.; et al. VDR independent induction of acid-sphingomyelinase by 1,23(OH)$_2$ D3 in gastric cancer cells: Impact on apoptosis and cell morphology. *Biochimie* **2018**, *146*, 35–42. [CrossRef] [PubMed]

24. Ceccarini, M.R.; Codini, M.; Cataldi, S.; Vannini, S.; Lazzarini, A.; Floridi, A.; Moretti, M.; Villarini, M.; Fioretti, B.; Beccari, T.; et al. Acid sphingomyelinaseas target of LyciumChinense: Promising new action for cell health. *Lipids. Health. Dis.* **2016**, *15*, 183. [CrossRef] [PubMed]

International Journal of
Molecular Sciences

Review

Lipids and Lipid Mediators Associated with the Risk and Pathology of Ischemic Stroke

Anna Kloska [1], **Marcelina Malinowska** [1], **Magdalena Gabig-Cimińska** [1,2,*] and **Joanna Jakóbkiewicz-Banecka** [1,*]

[1] Department of Medical Biology and Genetics, Faculty of Biology, University of Gdańsk, Wita Stwosza 59, 80-308 Gdańsk, Poland; anna.kloska@ug.edu.pl (A.K.); marcelina.malinowska@ug.edu.pl (M.M.)
[2] Laboratory of Molecular Biology, Institute of Biochemistry and Biophysics, Polish Academy of Sciences, Kładki 24, 80-822 Gdańsk, Poland
*   Correspondence: magdalena.gabig-ciminska@ug.edu.pl (M.G.-C.); joanna.jakobkiewicz-banecka@biol.ug.edu.pl (J.J.-B.); Tel.: +48-585-236-046 (M.G.-C.); +48-585-236-043 (J.J.-B.)

Received: 30 April 2020; Accepted: 19 May 2020; Published: 20 May 2020

Abstract: Stroke is a severe neurological disorder in humans that results from an interruption of the blood supply to the brain. Worldwide, stoke affects over 100 million people each year and is the second largest contributor to disability. Dyslipidemia is a modifiable risk factor for stroke that is associated with an increased risk of the disease. Traditional and non-traditional lipid measures are proposed as biomarkers for the better detection of subclinical disease. In the central nervous system, lipids and lipid mediators are essential to sustain the normal brain tissue structure and function. Pathways leading to post-stroke brain deterioration include the metabolism of polyunsaturated fatty acids. A variety of lipid mediators are generated from fatty acids and these molecules may have either neuroprotective or neurodegenerative effects on the post-stroke brain tissue; therefore, they largely contribute to the outcome and recovery from stroke. In this review, we provide an overview of serum lipids associated with the risk of ischemic stroke. We also discuss the role of lipid mediators, with particular emphasis on eicosanoids, in the pathology of ischemic stroke. Finally, we summarize the latest research on potential targets in lipid metabolic pathways for ischemic stroke treatment and on the development of new stroke risk biomarkers for use in clinical practice.

Keywords: eicosanoids; cholesterol; ischemic stroke; ischemia; lipoproteins; polyunsaturated fatty acids

1. Introduction

Stroke is a severe neurological disorder in humans that results from an interruption of the blood supply to the brain caused by a vascular occlusion (ischemic stroke, IS) or a blood vessel rupture or leakage (hemorrhagic stroke, HS). Within seconds, the insufficient blood supply leads to a strong oxygen-glucose deprivation (OGD) of the brain tissue, which initiates a cascade of pathophysiological response consequently leading to neuronal death and severe neurological deterioration. According to the Global Burden of Disease study, in 2017, stroke affected 104.2 million people worldwide; of them, 82.4 million were affected by ischemic stroke. Stroke is the second largest contributor to disability-adjusted life years in the world after ischemic heart disease, resulting in up to 50% of survivors being chronically disabled, and is the second most common cause of death [1,2]. The age-standardized global rate of new strokes reached 150.5 per 100,000 people in 2017 [3].

Primary and secondary prevention of stroke is crucial, considering that over 80% of the global stroke burden is attributable to a few risk factors that can be improved significantly. Many risk factors for stroke have been documented, including hypertension, current smoking, diabetes, abdominal obesity, poor diet, inactivity, excessive alcohol consumption, cardiac causes and stress/depression [4]. Dyslipidemia is a modifiable risk factor for stroke and is associated with a 1.8- to 2.6-times relative risk of stroke.

In the central nervous system, lipids and lipid mediators are essential to sustain the normal structure and function of brain tissue. Pathways leading to post-stroke brain deterioration include the metabolism of polyunsaturated fatty acids (PUFAs). The rapid and extensive release of PUFAs from cell membranes starts in the brain tissue immediately after the onset of ischemia. The lipids released are utilized in either enzymatic or non-enzymatic reactions, generating diverse classes of short-lived, lipid mediators, e.g., eicosanoids. These molecules have either neuroprotective or neurodegenerative effects on the post-stroke brain tissue; thus, they largely contribute to the outcome and recovery from stroke. Because the brain infarct area consists of two zones—the ischemic core, which is generally considered unsalvageable because of the greatest damage, and the surrounding area called the penumbra, where the blood flow during stroke is only partially reduced—the brain cells located in the penumbra may be rescued from degeneration by timely intervention. Numerous PUFA-derived lipid mediators contribute to the brain injury occurring after the ischemic stroke.

In this review, we provide an overview of serum lipids associated with the risk of ischemic stroke. We also discuss the role of lipid mediators, with particular emphasis on eicosanoids, in the pathology of ischemic stroke. Finally, we summarize the latest research on potential targets in lipid metabolic pathways for ischemic stroke treatment and on the development of new stroke risk biomarkers for use in clinical practice.

2. Association of Serum Lipids with the Risk of Stroke

Dyslipidemia is conventionally considered to play an important role in the pathogenesis of stroke, primarily ischemic stroke. Traditional lipid parameters, represented by increased concentrations of total cholesterol (TC), triglycerides (TGs), low-density lipoprotein cholesterol (LDL-C), and decreased high-density lipoprotein cholesterol (HDL-C), have been identified as risk factors and predictors of cardiovascular disease, including stroke [5–7]. The assessment of lipid ratios such as TC/HDL-C, TG/HDL-C, and LDL-C/HDL-C is recognized as a better predictor of vascular risk compared to traditional lipid parameters [8]. In addition to standard lipid components testing, the analysis of composition, particle size, and density of lipids and lipoproteins [e.g., lipoprotein(a), [Lp(a)]], have been proposed as biomarkers for the better detection of subclinical diseases [9].

Lipid profile components and the risk of ischemic stroke are discussed in subsequent paragraphs and summarized in Figure 1.

2.1. Cholesterol and the Risk of Stroke

As one modifiable risk factor, TC has been shown by many studies to be associated with the risk of stroke [10]. Although the relationship between lipid levels and coronary heart disease (CHD) is well established, the results of observational studies investigating the relationship between lipid profile and stroke are less conclusive. Some data from Asian and American studies indicate that TC is not identified as being associated or shows only weak relationships with various stroke subtypes [11,12]. Until the beginning of the 21st century, research on this topic yielded inconsistent results [13–15] and led to conflicting views on the importance of circulating cholesterol in IS [5,16]. Part of the controversy is related to the methodological shortcomings of previous studies. Many did not investigate the final point of interest, namely the ischemic stroke incident, but focused rather on endpoints such as fatal stroke or a combination of IS and HS, which are clearly different pathophysiological units. Some studies lack data on cholesterol subfraction levels or rely on lipid profile measurements only after a stroke has occurred. Additionally, few studies are large enough to examine the dose–response pattern in each sex group, which is another important aspect of the nature of this disease [11].

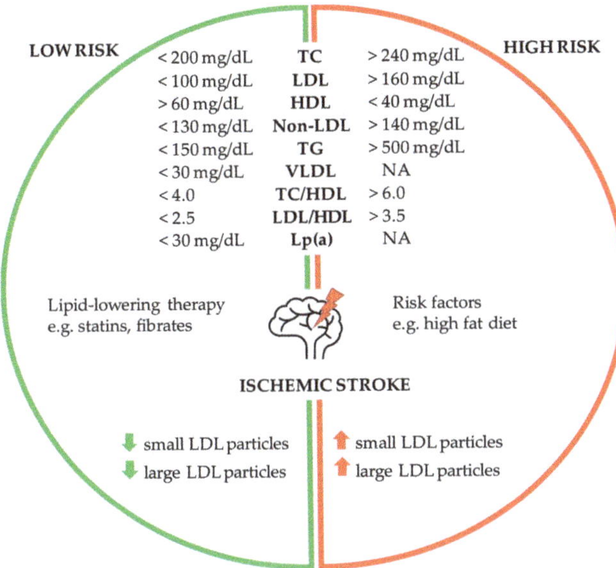

Figure 1. Lipid profile components and the risk of ischemic stroke. The reference values of the specified parameters may vary slightly according to different diagnostic recommendations. The arrow down indicates a low number of particles; the arrow up indicates a high number of particles. NA—not analyzed.

In later years, a number of studies appeared that took into account the previous suggestions of scientists. Some reports have shown a positive association between TC and IS [17], whereas an inverse relationship between TC and HS was found in others [17,18]. In a meta-analysis, primarily in Europe and North America, low-density lipoprotein cholesterol is associated with ischemic stroke but not with hemorrhagic stroke [19]. In addition, elevated low-density lipoprotein cholesterol is shown to be one of the most important risk factors for coronary artery disease and stroke, while high-density lipoprotein cholesterol is protective [20–22]. A study by Tirschwell et al. shows that elevated cholesterol and reduced HDL-C levels are associated with an increased risk of IS [6]. The effect of hyperlipidemia may be different depending on the ischemic stroke subtype. A significant and positive relationship was found in the case of atherothrombotic infarction, but a negative association was observed in the case of cardioembolic infarction [23].

Too low values of lipid components may also be responsible for an increased risk of stroke. The risk of HS is higher when TC is lower than 120 mg/dL. LDL-C and TC seem to be associated with hemorrhagic stroke. In contrast, the risk of ischemic and hemorrhagic stroke may be higher when HDL-C is lower than 50 mg/dL [24].

2.2. Hypertriglyceridemia and Ischemic Stroke

Few reports to date have shown the role of triglycerides in acute stroke and recovery after stroke. The studies that have evaluated this association report diverse associations [7,25,26]. According to a study published in 2012 in the *Stroke* journal, the strongest predictor of stroke risk in women is TC, the most-overlooked lipid in the cholesterol profile. The authors found that people with the highest TC levels are 56% more likely to have ischemic stroke than those with the lowest levels [27]. Triglyceride levels and their role in ischemic stroke have already been the subject of intensive research [28,29]. Large-scale epidemiological studies also detected a relationship between high non-fasting triglyceride levels and the risk of ischemic stroke [30,31], while results in smaller cohorts remain inconclusive [29]. Weir et al. suggested the opposite results and showed that low TG level, not low TC concentration, independently predicts poor outcome after acute stroke [32]. The mechanisms by which triglycerides affect ischemic stroke are still unexplained [33]. Triglyceride-rich lipoproteins and high TC levels can have a direct atherogenic effect and appear to be an indicator of atherosclerotic and prothrombotic changes. Elevated TC levels are associated with abnormalities in the coagulation cascade and fibrinolysis that is associated with ischemic stroke [34]. The effects of triglycerides are likely to be multifactorial.

2.3. Non-Traditional Lipid Profiles as Stroke Predictors

Human lipid profiles are currently actively studied, including not only traditional but also non-traditional lipid profiles, primarily as independent predictors of cardiovascular disease (CVD) [35,36]. Some studies show that low-density lipoprotein cholesterol, non-high-density lipoprotein cholesterol (non-HDL-C) and the TC/HDL-C ratio are significant predictors of CVD [37,38]. The relationship between LDL-C/HDL-C lipid profiles is found to be a more useful indicator of CVD risk than single, isolated lipid values [39]. In addition, some reports show that non-HDL-C is a better indicator of the development of CVD than LDL-C [36]. Lipid profiles as the main indicator of stroke prevention are still subject to significant uncertainty, in contrast to the clear results for CVD. However, several studies attempted to assess the role of traditional and non-traditional lipid indices in predicting stroke risk (Figure 1). Among the various non-traditional lipid variables, elevated baseline TC/HDL-C ratio and TC/HDL-C ratio increases future vascular risk after stroke, but only elevated TC/HDL-C ratio is associated with stroke recurrence risk [8]. Other studies estimated that the recurrence of cerebral ischemia increases with age and the increased composition of non-traditional lipid variables values: TC/HDL-C and LDL-C/HDL-C [40]. Zheng et al. showed that LDL-C, non-HDL-C and LDL-C/HDL-C are associated with the future all stroke status, and TC, LDL-C, non-HDL-C, TC/HDL-C and LDL-C/HDL-C are associated with the future ischemic stroke state [41]. Other studies show a positive relationship between cholesterol levels and the risk of IS in men, while an inverse trend between TC and the risk of hemorrhagic stroke is observed in the female group. A positive relationship was found between TC/HDL-C ratio and risk of ischemic stroke in both sexes; however, these links are not as clear after adjusting for body mass index, blood pressure and history of diabetes [17]. Other authors also studied the impact of lipid profiles separately in men and women. TC/HDL-C ratio is mainly associated with ischemic stroke and total stroke in men, while TG is more important in predicting ischemic and total stroke in women. The authors of this study suggest that these two lipid indexes have the most important prognostic value for identifying high risk participants predisposed to stroke for each sex separately, and there may be potential goals for stroke prevention [42]. Some researchers recommend that the level of non-traditional lipid profiles should be considered in the daily treatment of ischemic stroke for the first prevention in clinical practice [8].

2.4. Number, Size and Composition of Lipoprotein Particle

Standard measurements of circulating lipids lack the ability to distinguish between the size, density or concentration and composition of lipoproteins that may be important in assessing CVD risk [43]. Therefore, in addition to the standard lipid components tested, several biomarkers of lipids and lipoproteins are proposed as potential risk factors for the better detection of subclinical diseases [44]. Conventional HDL-C measures include the sum of cholesterol carried in HDL particles, but ignore their composition, particle size, and subclass concentration. Additionally, TG may show divergent relationships with vascular disease when transported in different lipoprotein molecules [9]. Above all, the interest focuses on lipoprotein parameters such as the number and size of LDL and HDL particles, and the number of intermediate-density lipoprotein (IDL) particles and lipoprotein(a) [Lp(a)]. One of the most commonly used methods for measuring the size and concentration of lipoprotein particles is nuclear magnetic resonance (NMR) spectroscopy. This technique simultaneously determines the average size (in nanometers) and concentration (in mol/L) of lipoprotein particles [45]. Holmes et al. assessed the relationship between metabolic markers and the risk of three cardiovascular diseases, including ischemic stroke. The study shows that the subclasses of lipoproteins and their lipid components are associated with stroke risk. Cholesterol and triglycerides in apolipoprotein B-containing lipoproteins (very low-density lipoprotein [VLDL], intermediate-density lipoprotein [IDL] and low-density lipoproteins [LDL]) are positively associated with the risk of stroke. In contrast, cholesterol in large and medium high-density lipoprotein particles inversely associates with the ischemic stroke risk, while triglycerides in HDL particles positively associate with the risk of this disease. VLDL particle concentrations are at least as strongly associated with ischemic stroke as LDL particles. In addition, TGs are more consistently associated with IS across the entire lipoprotein subfraction spectrum than cholesterol [9].

2.5. Elevated Lipoprotein(a) and the Risk of Ischemic Stroke

Lipoprotein(a) is an LDL particle with an added apolipoprotein(a). The link between Lp(a) and stroke has been questioned by some researchers. Several studies show no association between Lp(a) levels and the risk of stroke. Hachinski et al. did not notice a significant difference in Lp(a) levels among the patient and control groups [46]. Similarly, Glader et al. found no association between baseline plasma Lp(a) values and future ischemic cerebral infarction [47]. In a prospective study in Finland, no relationship was found between the initial Lp(a) plasma levels and the future risk of total (all types of stroke) or thromboembolic stroke among those participating in the study [48]. More recent meta-analyses summarize the existing evidence for Lp(a) and stroke from both control and prospective studies and show a significant and independent association of elevated Lp(a) with an increased risk of ischemic stroke [49–51]. The largest meta-analysis carried out by the Emerging Risk Factor Collaboration analyzed the data from 13 prospective studies and confirmed that elevated Lp(a) is an independent risk factor for ischemic stroke [49]. As many previously published reports investigating the risk of stroke in patients with elevated Lp(a) levels did not show the underlying cause of the stroke event; differences in the etiology of stroke between cohorts may explain some of the inconsistent results reported in the literature. Data suggest that Lp(a) primarily increases the risk of large-artery atherosclerosis stroke [52]. According to the guidelines of the European Society of Atherosclerosis, it is recommended to measure Lp(a) in patients with a medium or high risk of cardiovascular disease, considering levels lower than 50 mg/dL [53]. Studies with larger cohorts are needed to see whether higher cut-off values than conventional and/or interactions with other risk factors are necessary to establish the role of Lp(a) as a risk factor for ischemic stroke. Therefore, a more detailed study of the relationship between individual subclasses of lipoprotein particles and lipid-associated characteristics with the risk of CVD and stroke subtypes may be important and informative.

2.6. Polyunsaturated Fatty Acids and Risk of Ischemic Stroke

The n-3 polyunsaturated fatty acids (n-3 PUFAs) are a class of essential unsaturated fatty acids necessary for proper biological activity and function in living organisms. The n-3 PUFAs are poorly synthesized in the human body and have to be orally supplemented. Fish, such as mackerel, salmon, tuna, sardines, herring, and halibut, are a major source of n-3 PUFAs in the human diet, and they contain docosahexaenoic acid (DHA), docosapentaenoic acid (DPA), and eicosapentaenoic acid (EPA) [54]. The n-3 PUFAs have potent anti-inflammatory activity, reduce platelet aggregation, stabilize atherosclerotic plaques, and reduce major cardiovascular risk factors, including hypertension and hyperlipidemia [55]. They can act as an antioxidant in reducing cerebral lipid peroxides and play a role in regulating oxidative stress by increasing the oxidative burden and improving antioxidant defense capabilities [56]. In addition, n-3 PUFAs trigger other responses, such as neurogenesis and revascularization in stroke, which could be used in the development of acute-phase ischemic stroke therapy [54]. Moreover, because ischemic stroke is a heterogeneous disorder with different pathophysiological pathways and separate etiological subtypes, various mechanisms of PUFA action should be considered. DHA plays a greater role in reducing the risk of atherothrombotic stroke by reducing endothelial dysfunction and atherosclerosis [57], while EPA and DPA have a greater impact on the risk of cardioembolic stroke because of their effect on clotting and atrial fibrillation [58]. The second key component of a heart-healthy dietary pattern is a high content of n-6 PUFAs, obtained mainly from vegetable oil, nuts, and seeds. Dietary n-6 PUFAs primarily include linoleic acid (LA) and arachidonic acid (AA). The n-6 PUFAs have not generally been associated with stroke risk [59,60].

The Japan EPA Lipid Intervention Study (JELIS) has shown that treatment with highly purified EPA and low-dose statins significantly reduces the incidence of coronary artery diseases and stroke compared to statin therapy alone, without altering the reduction in low-density lipoprotein cholesterol levels [61]. Nishizaki et al. reported that ratios of serum n-3 PUFAs to n-6 PUFAs, such as EPA/AA and DHA/AA ratios, could be useful markers to determine the incidence of coronary events, peripheral artery diseases, and early neurological deterioration after acute ischemic stroke [62]. Thies et al. examined the effect of fish oil administration on plaque regression and found that giving fish oil to patients resulted not only in plaque regression but also an increases in EPA and DHA within the plaque and a decrease in macrophage counts [63]. In addition, Ajami et al. reported that DHA+EPA provided neuroprotection against ischemic brain injury by increasing the levels of antiapoptotic proteins, such as Bcl-2 and Bcl-xL, thereby suppressing the inflammatory response [64].

Attention has also been focused on assessing the relationship between PUFA levels and early neurological deterioration (END) in acute-phase ischemic stroke [65]. END occurs in approximately one-third of patients in the acute phase of ischemic stroke and is associated with neurological and functional decline. It also strongly correlates with poor functional outcome and usually leads to a significant increase in mortality rate [66]. According to a recommended definition, END occurs when there is an increase in the total National Institutes of Health Stroke Scale score of ≥ 2 points within 72 h 1–3 times a day after admission [67]. Suda et al. revealed that END is negatively associated with the EPA/AA, DHA/AA, and EPA+DHA/AA ratios. The study shows that a low serum n-3 PUFA/n-6 PUFA ratio might be an indication of possible END in patients with acute ischemic stroke, as demonstrated in the population of Japanese stroke patients [65].

2.7. Fatty Acids and Cardioembolic Stroke

Elevated fatty acid (FA) levels are associated with several risk factors for atherosclerosis, including abdominal obesity [68], arterial hypertension [69], and insulin resistance [70], as well as coronary artery disease (CAD) [71], arrhythmia [72], and atrial fibrillation [73]. Because an association with both atherosclerosis and arrhythmia has been reported, FAs may correlate with ischemic stroke. However, the effect of FAs on ischemic stroke is poorly understood. Because ischemic stroke is a heterogeneous disorder with a variety of pathophysiological pathways, including atherothrombosis and cardioembolism, the etiological subtypes of ischemic stroke should be analyzed separately [74]. Almost all long-term correlation studies of FAs and stroke risk estimate the level of FAs obtained with food on the basis of self-reported questionnaires. Unfortunately, this is unclear for individual FAs that are not well separated by dietary questionnaire data. For a more detailed assessment, Saber et al. measured the levels of circulating n-3 PUFA phospholipids and examined their association with ischemic stroke incidence, including atherosclerotic and cardioembolic stroke subtypes [75]. Patients evaluated for phospholipid levels were recruited for three separate prospective cohort studies in the US: the Cardiovascular Health Study (CHS), Nurses' Health Study (NHS), and Health Professionals Follow-Up Study (HPFS) [75]. The authors of these studies show that, among ischemic stroke subtypes, DHA is inversely associated with atherothrombotic stroke and DPA is associated with cardioembolic stroke. These relationships remain significant after including demographic, lifestyle, and vascular risk factors. By comparison, EPA is not associated with total ischemic, atherothrombotic, or cardioembolic stroke. The authors confirmed the hypothesis that individual FAs in serum have various associations with ischemic, atherothrombotic, and cardioembolic stroke [75]. Earlier studies have shown that elevated FA levels are associated with cardioembolic (CE) stroke, but this association was not seen in non-CE stroke. Atrial fibrillation may potentially act as an intermediary between FAs and stroke caused by cardioembolism [73]. Another study also found that an elevated FA concentration may serve as a marker of stroke caused by cardioembolism. In addition, the assessment of FA concentration can predict the stroke recurrence following a CE stroke [76]. In patients with acute stroke, significantly elevated FA levels are observed in groups with a higher risk of cardioembolism. These results suggest that enhanced thrombogenicity may be the main mechanism explaining elevated FA levels in patients with cardioembolic stroke [77].

2.8. The Role of Fish-Derived Fatty Acids in Stroke

For many years, intensive research has been carried out to assess the biological effects of consuming fish-derived PUFAs [78]. These studies confirm the view that n-3 PUFAs can affect several cellular processes known to be important in the development of cardiovascular disease, stroke, and protective effects [79,80]. Most studies have used doses of fish oils exceeding what is usually found in the diet. Surprisingly, significant vascular benefits are observed even with modest fish consumption [81]. Long-term studies have shown that increased intake of n-3 PUFAs, in particular EPA and DHA, can have a beneficial effect on serum lipids [82], platelet aggregation [83], and bleeding time [84] and, thus, may lead to a reduced risk of atherosclerosis and thrombotic complications [63]. Prolonged fish consumption leads to increased incorporation of n-3 PUFAs into plasma lipids, erythrocytes, and platelets [85]. After supplementation with fish oil, elevated EPA and DHA content in plasma lipids, platelets, and erythrocyte membranes are observed with a simultaneous decrease in AA content [86]. In order to compare the effects of n-3 PUFA consumption, the following four fish-derived sources of n-3 PUFA were used: three rich sources (raw fatty fish [smoked salmon], cooked fatty fish [salmon fillet], or fish oil [cod liver oil]) and one poor source (fish low in n-3 PUFA [cod fillet]). Therefore, the following blood parameters were assessed: blood lipid composition and functional properties of blood cells, as measured by the potential of lipopolysaccharide (LPS) to generate activation products in whole blood [87]. Elvevoll et al. did not notice any significant differences between the effects of eating cooked fish compared to raw fish (smoked salmon). They found that the intake of fatty fish is more effective in increasing EPA and DHA than supplementing with fish oil and is more likely to have a beneficial effects on HDL cholesterol and whole blood activation reactions [87]. However, another group of researchers showed that the action of DHA-rich oil (without EPA) had a comparable hypotriglyceridemic effect as a fish diet and fish oil supplementation. Moreover, a fish diet and fish oil supplementation increased the proportion of n-3 PUFA in plasma lipids, platelets, and erythrocyte membranes [88].

2.9. Lipid-Lowering Therapy for Prevention of Ischemic Stroke

Over the past two decades, compelling evidence from clinical trials revealed the importance of low-density lipoprotein cholesterol-lowering therapies in reducing cardiovascular and stroke morbidity and mortality [89,90]. At present, there is a more than four-fold increase in the use of cholesterol-lowering agents in our population compared to in 2000. There is strong evidence for the role of statins in stroke prevention and association with an approximately their 20% risk reduction [91], in particular a decreased risk of ischemic stroke [92]. Moreover, more aggressive statin treatment improves the long-term functional outcome of patients discharged after an acute ischemic stroke more than less aggressive treatment [93]. One of the most commonly used statins to lower high cholesterol is atorvastatin. It is a common statin used to investigate the efficacy of statin therapy, especially high-intensity statin therapy in patients with ischemic stroke. Preliminary studies have shown that in patients with recent stroke or transient ischemic attack and without coronary heart disease, 80 mg atorvastatin daily reduces the overall incidence of stroke and cardiovascular events, despite a slight increase in the incidence of hemorrhagic stroke [94]. Many studies have confirmed the effect of atorvastatin at lowering lipids and decreasing the number and frequency of vascular incidents, as well as its clinical efficacy at reducing the burden of disease after stroke or transient ischemic attack [95,96]. Atorvastatin improves endothelial function, enhances the stability of atherosclerotic plaque, and inhibits inflammatory and thrombogenic responses in arterial walls [97]. In the latest

Stroke Prevention by Aggressive Reduction in Cholesterol Levels (SPARCLs) cohort trial, atorvastatin was compared with a placebo in patients with recent stroke or transient ischemic attack. Atorvastatin was found to reduce the first occurrence of stroke and the first occurrence of composite vascular events [98].

During recent years, new drug classes proved their efficacy and safety in lowering cholesterol and preventing cardiovascular incidents in randomized controlled studies. PCSK9 (proprotein convertase subtilisin/kexin type 9) inhibitors increase the number of available LDL receptors on the surface of hepatocytes, which leads to a higher cleavage of LDL-C from the circulation. The mechanism of action of PCSK9 inhibitors involves halting the metabolic breakdown of LDL-R, resulting in increased LDL-C clearance. The two currently available antibodies (Alirocumab and Evolocumab) against PCSK9 are fully human IgG subtypes that bind with an approximate 1:1 stoichiometry to circulating PCSK9 and prohibit its binding to the LDL-R. The mean percentage change in LDL-C levels decreases by 50% or more from baseline in patients receiving PCSK9 inhibitors [99]. Several non-antibody therapies have also been developed to inhibit PCSK9 function. Gene silencing or editing technologies were used, such as antisense oligonucleotides [100], small interfering RNAs [101], small-molecule inhibitors [102], mimetic peptides [103], adnectins [104], and vaccinations [105].

According to the American Heart Association and American Stroke Association guidelines for stroke prevention, which recognize triglyceride as a risk factor for stroke, fibric acid derivatives can be considered in patients with hypertriglyceridemia [106]. Fibric acid derivatives (e.g., gemfibrozil, fenofibrate, and bezafibrate) lower triglyceride levels and increase HDL cholesterol. The Veterans Administration HDL Intervention Trial of men with low HDL-C and associated coronary artery disease showed that gemfibrozil reduces the risk of all strokes, mainly ischemic strokes [107].

The issue of elevated Lp(a) level therapy is slightly different. Lp(a) levels are essentially unresponsive to traditional lipid-lowering drugs such as statins or fibrates. Some lipid-lowering agents that are not specific for Lp(a) reduce Lp(a) levels (e.g., niacin, PCSK9 inhibitors, and CETP inhibitors) [108]; however, to date, no randomized controlled trial has demonstrated that the lowering of Lp(a) leads to decreased risk of cardiovascular disease. Recent findings show that antisense oligonucleotides can be used to inactivate genes involved in the pathogenesis of vascular diseases. To lower Lp(a), synthetic oligonucleotides have been developed. In clinical trials, Mipomersen (Genzyme/ISIS Pharmaceuticals), an antisense oligonucleotide targeted to apolipoprotein B, reduced Lp(a) by 21% to 39% [109]. AKCEA-APO(a)-LRx (Akcea Therapeutics/Ionis Pharmaceuticals) is the latest antisense oligonucleotide targeted to apolipoprotein(a). In a phase II trial, the specific antisense oligonucleotide resulted in a dose-dependent reduction of 66% to 92% in blood circulating Lp(a) and is expected to enter into a phase III study in 2020.

3. Lipids of the Brain during Ischemic Stroke

The brain has the second highest lipid content among the organs of the human body, accounting for about 50% of its dry weight [110]. Lipids that are essential for the central nervous system are classified into five major subcategories: fatty acids, triglycerides, phospholipids, sterol lipids, and sphingolipids. They serve as the structural components of biological membranes, act as messengers in cellular signaling pathways and contribute to the energy supply [111]. For example, about 20% of the brain total energy requirements comes from the oxidation of fatty acids, which takes place in astrocytes [112].

Brain tissue is characterized by cellular heterogeneity; thus, the fatty acid composition varies between different cell types. In the ischemic brain, the levels of different lipids change compared to the control conditions [113]. It is even possible to distinguish the ischemic core area from the penumbra according to the post-stroke lipid profiles [114]. Generally, brain tissue is characterized by a high proportion of polyunsaturated fatty acids: arachidonic acid (AA; omega-6; C20:4ω6), eicosapentaenoic acid (EPA; omega-3; C20:5ω3), and docosahexaenoic acid (DHA; omega-3; C22:6ω3). Arachidonic acid and docosahexaenoic acid make up ~20% of fatty acids in the mammalian brain [115].

All of these lipids are located in cellular membranes and are essential for the normal structure and function of the central nervous system [110]. However, cerebral ischemia initiates a cascade of events that stimulates the release of free fatty acids from the membrane. In particular, there is a rapid accumulation of arachidonic acid, docosahexaenoic acid, diacylglycerol, and platelet-activating factor (PAF; 1-Oalkyl-2-acyl-sn-3-phosphocholine).

The released arachidonic acid is highly reactive and prone to downstream enzymatic reactions that generate different classes of eicosanoids in the brain area affected by the ischemia (Figure 2). These molecules play a major role in cerebral vasoconstriction, edema, neurotoxicity and neuroprotection that occur after ischemia and reperfusion in the brain following stroke episodes. The cytosolic phospholipase A_2 alpha (cPLA$_{2\alpha}$), an enzyme responsible for arachidonic acid release from membrane phospholipids, plays a key role in post-ischemic brain pathology. The down-regulation of cytosolic phospholipase A_2 alpha expression [116] or blockage of its activity with specific antibodies [117] attenuates ischemic brain damage in the mouse model of cerebral ischemia-reperfusion injury.

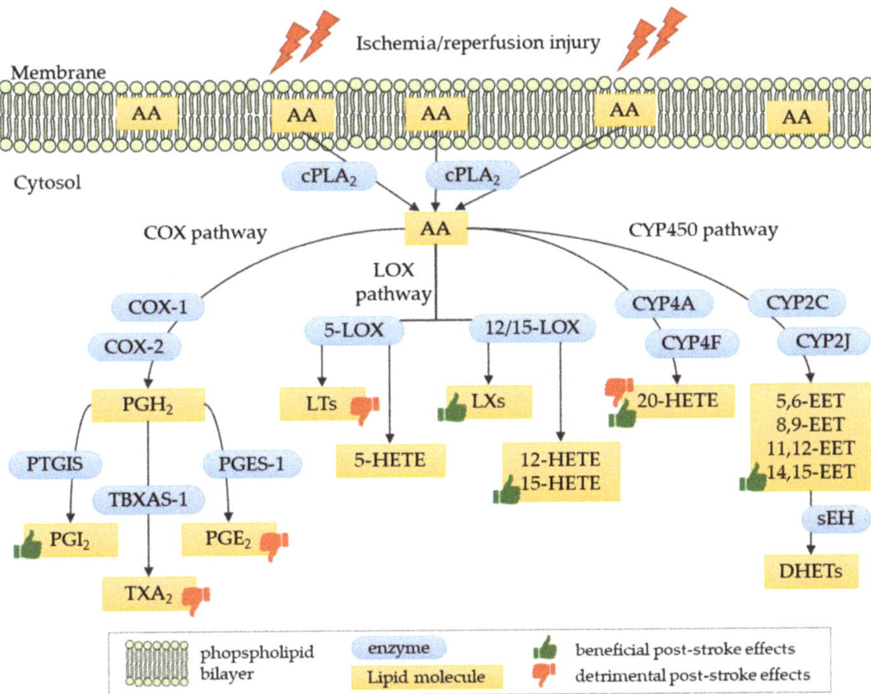

Figure 2. Eicosanoid biosynthesis pathways. Enzymes involved in biosynthetic pathways are denoted as: cPLA$_2$, cytosolic phospholipase A_2; COX-1, cyclooxygenase 1; COX-2, cyclooxygenase 2; PGIS, prostaglandin-I synthase; TBXAS-1, thromboxane-A synthase; PGES-1, prostaglandin E synthase; 12/15-LOX, 12/15-lipoxygenase; 5-LOX, 5-lipoxygenase; CYP4A, cytochrome P450 4A subfamily; CYP4F, cytochrome P450 4F subfamily; CYP2C, cytochrome P450 2C subfamily; CYP2J, cytochrome P450 2J subfamily; sEH, soluble epoxide hydrolase. Lipid molecules generated in biosynthesis are denoted as: AA, arachidonic acid; PGH$_2$, prostaglandin H$_2$; PGI$_2$, prostaglandin I$_2$, prostacyclin; TXA$_2$, thromboxane A$_2$; PGE$_2$, prostaglandin E$_2$; LTs, leukotrienes; 5-HETE, 5-hydroxyeicosatetraenoic acid; LXs, lipoxins; 12-HETE, 12-hydroxyeicosatetraenoic acid; 15-HETE, 15-hydroxyeicosatetraenoic acid; 20-HETE, 20-hydroxyeicosatetraenoic acid; 5,6-EET, 5,6-epoxyeicosatrienoic acid; 8,9-EET, 8,9-epoxyeicosatrienoic acid; 11,12-EET, 11,12-epoxyeicosatrienoic acid; 14,15-EET, 14,15-epoxyeicosatrienoic acid; DHETs, dihydroxyeicosatrienoic acids. Thumb symbol denotes beneficial or detrimental effects on the post-stroke brain of eicosanoids discussed in this review.

The DHA released during the ischemic event is converted enzymatically into lipid messengers. One of them is the neuroprotectin D1 (NPD1), a potent cell-protective, pro-survival and anti-inflammatory lipid mediator in experimental stroke models [118,119]. This experimental post-stroke DHA treatment improves the neurobehavioral recovery and reduces brain infarct volume. This research shows that the additional DHA enhances the synthesis of NPD1 in the penumbra, revealing a highly desirable neuroprotective effect [118].

The excessive accumulation of platelet-activating factor in the ischemic brain contributes to excitotoxicity, Ca^{2+} uptake, and mitochondrial dysfunction, leading to neuronal death. However, the disruption or inhibition of PAF-receptor has a neuroprotective effect [120,121].

4. Eicosanoids in Ischemic Stroke

Eicosanoids belong to oxylipins, a family of oxidized forms of polyunsaturated fatty acids. They are produced by enzymatic reactions catalyzed by three different enzymes—cyclooxygenases, lipoxygenases and cytochrome P450 monooxygenases. This diverse group of compounds includes prostanoids (i.e., prostaglandins, prostacyclin, and thromboxanes), leukotrienes, lipoxins, hydroxyeicosatetraenoic acids (HETEs), epoxyeicosatrienoic acids (EETs), resolvins, and eoxins. In general, eicosanoids are biologically active compounds involved in the regulation of many physiological functions, but they are also related to pathological processes, including ischemic stroke.

4.1. Biosynthesis and Physiological Role of Eicosanoids

Eicosanoids are synthesized from arachidonic acid, a precursor, 20-carbon polyunsaturated fatty acid released from cell membranes by phospholipase A_2 (PLA_2). Free arachidonic acid, as a highly reactive compound is prone to further oxidation and is readily converted into eicosanoids in their downstream enzymatic pathways (Figure 2). Cyclooxygenases (COX-1 and COX-2) convert the free arachidonic acid into the unstable prostaglandin H_2 (PGH_2), which is next converted by downstream prostanoid synthases into prostaglandins D_2, E_2, F_2, and I_2 (PGD_2, PGE_2, PGF_2 and PGI_2, respectively) or thromboxane A_2 (TXA_2) [122]. Arachidonic acid is also converted by lipoxygenases (LOX enzymes) to produce leukotrienes, lipoxins and HETEs, while cytochrome P450 converts the arachidonic acid into 20-HETE and EETs molecules.

Eicosanoids are known for their various and often contradictory roles in the human body. On the one hand, they regulate many physiological functions in the cardiovascular, gastrointestinal, urogenital, and nervous system. They play critical roles in immunity, acting as pro- and anti-inflammatory agents, regulating fever, and triggering platelet aggregation, blood clotting, muscle contraction, vasoconstriction, and vasodilation. On the other hand, eicosanoids play an important role in cardiovascular diseases and stroke, renal diseases, rheumatoid arthritis, Alzheimer's disease, cancer, and even infectious diseases [123,124].

4.2. The Role of Eicosanoids in Ischemic Stroke Pathology

The role of eicosanoids in stroke is reported to be both beneficial and detrimental. The levels of many eicosanoids are altered following the ischemic insult and often stay elevated during the recovery phase, thus playing an important role in either brain tissue injury or protection. The balance between neuroprotection and neuronal cell death caused by the post-stroke disruption of eicosanoid homeostasis is detrimental for the size of the disease pathology and the extent of stroke patient recovery. In the following sections, principal findings of the current investigations on the role of different classes of eicosanoids in ischemic stroke pathology as well as possible targets for therapeutic interventions will be reviewed.

4.2.1. Prostanoids in Ischemic Stroke

Prostanoids group arachidonic-acid derivatives which are generated by cyclooxygenase enzymes. These include prostaglandins PGE_2, PGD_2, $PGF_{2\alpha}$, PGI_2, and thromboxane A_2 (Figure 2). Two distinct isoforms of cyclooxygenase are involved in the prostanoid biosynthetic process: COX-1, which is found in the kidney, stomach and platelets, and COX-2, located in macrophages, leukocytes and fibroblasts. In the central nervous system, COX-1 is constitutively expressed in neurons, astrocytes, and microglial cells, while COX-2 is up-regulated under pathological conditions [125].

Special interest in the pathogenesis of ischemic stroke concerns the role of two prostanoids: prostacyclin, also known as prostaglandin I_2 (PGI_2), and thromboxane A_2 (TXA_2). Prostacyclin is a potent vasodilator and inhibitor of platelet aggregation. In contrast, thromboxane A_2 is a strong vasoconstrictor and inductor of platelet aggregation [126]. The over-production of TXA_2 is one of the key factors causing thrombosis, stroke, and heart disease. PGI_2 is the primary arachidonic acid metabolite in vascular walls; due to its contrasting biological activity to TXA_2, PGI_2 represents the most potent endogenous vascular protector, acting as an inhibitor of platelet aggregation and a strong vasodilator on vascular beds [127]. PGI_2 acts mostly as an immune-inhibitory molecule through multiple cell types such as dendritic cells, macrophages, and T-cells. The decreased biosynthesis of anti-thrombotic PGI_2 alongside with the excessive production of pro-inflammatory PGE_2 and pro-thrombotic TXA_2 increases the risk of stroke and heart attack in gene-knockout models [128]. Another study using the transgenic mouse model revealed that the redirection of arachidonic acid conversion toward favoring PGI_2 production over TXA_2 and PGE_2 clearly contributes to resistance to induced ischemic stroke [128]. This experimental attempt uses genetic engineering to link the COX-1 enzyme, which generates the unstable prostaglandin H_2 (PGH_2) from arachidonic acid, with the PGIS enzyme that generates PGI_2 from PGH_2 intermediates. Since PGH_2 serves as a substrate for the production of all of the downstream prostanoids (including PGI_2, PGE_2) and thromboxane A_2, this approach enables the regulation of prostanoid biosynthesis in favor of PGI_2 rather than TXA_2 and PGE_2 in cells by channeling the PGH_2 intermediate to the PGIS biosynthetic pathway [128].

The post-stroke accumulation of PGE_2 is also accompanied by a marked induction of the prostanoid EP2 receptor expression in the ischemic hemisphere, particularly in neurons of the penumbra [129]. The EP2 signaling pathway contributes to ischemic injury but its inhibition efficiently protects against the inflammatory neurodegeneration observed in post-ischemic brain tissue. In addition, the other EP1 prostanoid receptor is also up-regulated following ischemic stroke and its expression is not only detected in neurons but also in the endothelial cells [130]. The PGE_2 action, mediated by the EP1 receptor, leads to the disruption of the blood–brain barrier which, in turn, significantly contributes to progressive neuronal death in the stroke penumbral region. Additionally, in this case, inhibition of the EP1 receptor reduces the infarct volume, blood–brain barrier disruption and permeability, neutrophil infiltration and hemorrhagic transformation in the animal stroke model.

4.2.2. Leukotrienes in Ischemic Stroke

Leukotrienes (LTs) are synthesized form arachidonic acid by the 5-lipoxygenase (5-LOX) enzyme; this process also requires the presence of 5-LOX-activating protein (FLAP). The pathway generates two groups of leukotrienes: dihydroxy acid leukotriene B_4 (LTB_4) and cysteinyl-leukotrienes (i.e., LTC_4, LTD_4, and LTE_4) (Figure 2). Leukotrienes are thought to be involved in atherosclerosis and plaque formation, a process leading to cardio- and cerebrovascular pathology that eventually leads to occlusion resulting in stroke. However, as leukotriene levels increase after ischemia, they are also considered the factors responsible for the post-stroke pathology.

Numerous studies pinpoint leukotriene B_4 (LTB_4) as the factor responsible for pathological events occurring during the post-ischemic state. Both the expression of 5-lipoxygenase and the levels of leukotrienes increase following focal cerebral ischemia in rats and persist for days after reperfusion in the ischemic core and the boundary zones [131]. The increased and sustained plasma level of leukotriene B_4 is associated with the poor functional recovery of patients with acute middle cerebral artery infarction [132]. However, studies using animal or cell culture models consistently show that the regulation of 5-lipoxygenase expression results in a neuroprotective effect in cases of ischemia [133–135]. An evident post-ischemic brain injury is also mediated by the cysteinyl-leukotrienes but is associated with neuronal damage and astrogliosis [136].

4.2.3. Lipoxin A_4 in Ischemic Stroke

Lipoxins (LXs) are a class of eicosanoids that exert anti-inflammatory and pro-resolving activities, generally reducing tissue injury and chronic inflammation. The synthesis of LXs from arachidonic acid involves three major lipoxygenases: 5-LOX, 15-LOX, and 12-LOX (Figure 2). Additionally, LXs may be synthesized in an aspirin-triggered pathway [137].

One representative of this group of molecules, lipoxin A_4 (LXA_4), has been shown to be very effective in neuroprotection after brain ischemia-reperfusion injury. The experimental treatment with LXA_4 effectively reduces infarct volume and brain edema, improving the neurological outcome, as shown for rats after middle cerebral artery occlusion [138,139]. The authors of the studies suggest that the effect obtained with the treatment may be generated by the inhibition of 5-lipoxygenase translocation triggered by LXA_4 and the resulting reduced biosynthesis of pro-inflammatory leukotrienes [138].

The protective effect of lipoxin A_4 also applies to astrocytes. The treatment of these cells with LXA_4 protects against cell damage and attenuates the production of reactive oxygen species under oxygen-glucose deprivation and re-oxygenation conditions [140]. Up-regulation of the nuclear factor erythroid 2-related factor 2 (Nrf2) signaling pathway links lipoxin A_4 with reduction of oxidative stress that most likely underlies its protective effect.

4.2.4. Hydroxyeicosatetraenoic Acids in Ischemic Stroke

Hydroxyeicosatetraenoic acids (HETEs) are formed from arachidonic acid in three different metabolic pathways [141]. The predominant pathway involves lipoxygenases (5-LOX, 12-LOX, and 15-LOX enzymes) that form a variety of HETE molecules, including 5-HETE, 12-HETE, and 15-HETE (Figure 2). Small amounts of 11-HETE and 15-HETE can be generated by the cyclooxygenases COX-1 and COX-2. Some HETEs, with the 20-HETE molecule at the forefront, are generated by cytochrome P450 hydroxylases. The analysis of HETEs composition and quantity in rat brains exposed to ischemia reveals that the amount of HETE molecules tends to increase with time following the occlusion [142]. After 72 h, almost all lipoxygenase-generated HETEs are significantly increased in the affected brain tissue.

Animal studies show that hypoxia up-regulates the expression of 15-lipoxygenase in the brain artery endothelium after stroke and leads to the increased production of 15-HETE [143]. 15-HETE promotes angiogenesis and neuronal recovery that protects against ischemic brain infarction and improves neurological function in a mouse model of focal ischemia. It also stimulates the proliferation and migration of brain microvascular endothelial cells with the PI3K/Akt signaling pathway involved. Another study shows that 15-HETE boosts the angiogenesis in mouse ischemic brain by up-regulation of the vascular endothelial growth factor (VEGF) [144].

Contradictory results are obtained on the role of the 20-hydroxyeicosatetraenoic acid (20-HETE) in stroke pathology as it is either described as detrimental or beneficial. 20-HETE is produced by cytochrome P450 members, mainly CYP4A and CYP4F. The role of 20-HETE in ischemic brain pathology depends on whether it is determined in the early or late post-stroke phase. On the one hand, elevated 20-HETE levels that are detected in the plasma of ischemic stroke patients associate with greater lesion size and neurological impairment, reduced cognitive functions and poorer functional independence in daily living [145]. Some authors propose that this particular eicosanoid may serve as a valuable predictor of neurological deterioration prognosis in acute minor ischemic stroke as their findings indicate the association of high 20-HETE plasma levels and neurological deterioration and poorer prognosis in patients [146]. The increased production of 20-HETE is also observed in animal ischemia-reperfusion models [147,148]. For example, the expression of 20-HETE synthase increases in the plasma and brain of animals after experimental cerebral ischemia and clearly contributes to oxidative stress and endothelial dysfunction [148].

On the other hand, the increase in CYP4A expression and 20-HETE production following oxygen-glucose deprivation noted in astrocytes promotes angiogenesis via the induction of endothelial cell proliferation, tube formation and migration in a later stage of stroke [149]. The specific cross-talk discovered between astrocytes and endothelial cells is mediated by 20-HETE and involves HIF-1α/VEGF and JNK signaling pathways. The beneficial effects of 20-HETE on recovery from stroke are confirmed—the inhibition of CYP4A, an enzyme involved in 20-HETE synthesis, resulted in diminished peri-infarct angiogenesis and worsened neurological deficits in mice. The authors highlight that, although 20-HETE induces neuronal and vascular injury in the early post-stroke phase, in the later post-stroke phase, this mediator is necessary for neurovascular repair and remodeling to obtain functional recovery after stroke.

4.2.5. Epoxyeicosatrienoic Acids in Ischemic Stroke

The synthesis of epoxyeicosatrienoic acids (EETs) from arachidonic acid is catalyzed by cytochrome P450 epoxygenases from the CYP2C and CYP2J families. Four biologically active EETs belong to this group of eicosanoids: 5,6-EET, 8,9-EET, 11,12-EET, and 14,15-EET (Figure 2). These molecules are metabolized to less bioactive dihydroxyeicosatrienoic acids (DHETs) by the soluble epoxide hydrolase (sEH).

EETs exert strong neuroprotective effects in the case of cerebral ischemia, especially with regard to components of the neurovascular unit. While oxygen-glucose deprivation decreases the viability of cerebral smooth muscle cells, different types of EETs prevent these events during in vitro experiments [150]. 14,15-EET shows the strongest anti-apoptotic effect under these conditions [150,151]. Two pathways are linked to the protective effects of EETs: PI3K/Akt pathway and ATP-sensitive potassium channels contribute to the EETs-protective effects on cerebral microvascular smooth muscle cells [151], while the JNK/c-Jun and mTOR signaling pathway contributes to the anti-apoptotic effect of 14,15-EET under conditions of oxygen-glucose deprivation [150].

4.3. Eicosanoids as the Target for Ischemic Stroke Treatment

The emphasis on the role of eicosanoids in the pathology of ischemic stroke has opened up new perspectives for the development of treatments based on modifications within the biosynthetic pathways of these lipid mediators.

The modulation of prostanoid biosynthesis with cyclooxygenase inhibitors has been proposed as one potential treatment strategy. To date, aspirin is the only antiplatelet agent that is used effectively in the early treatment of acute ischemic stroke. Aspirin irreversibly inhibits COX activity in platelets and prevents the conversion of AA to thromboxane A_2. The decline in the risk of mortality and morbidity when aspirin is initiated within 48 h of acute ischemic stroke is small though significant [152]. Targeting prostanoid biosynthesis with COX-2 inhibitors, such as nonsteroidal anti-inflammatory drugs (NSAIDs), the classical pain medications, or the novel COX-2 selective inhibitors (e.g., celecoxib or rofecoxib), has been shown to reduce edema, neuroinflammation, and infarct size in rodent stroke models [153]. Although administration of these drugs reveals a neuroprotective effect in stroke models, adverse effects, including an increased risk of stroke, occur following long-term usage. A recent increase in the number of heart attacks related the use of COX-2 inhibitors is attributed to their ability to reduce PGI_2 and increase TXA_2 biosynthesis in vascular walls and platelets [154]. A disrupted production balance in favor of TXA_2 over PGI_2, the most important vascular protector, is shown to be responsible, at least in part, for the pro-thrombotic and pro-atherogenic effects [155]. A clinical trial of a small group of ischemic stroke patients confirmed that treatment with intravenous prostacyclin infusions results in clinical improvement, with regression of hemiplegia or hemiparesis, the disappearance of aphasia, and clearing of consciousness within a few hours after administration [156]. Vasodilation of cerebral microvessels is the effect that may account for the benefit of PGI_2 on post-ischemic brain tissue. Interestingly, statins, apart from their lipid-lowering activity, have also been shown to be effective in reducing TXA_2 levels in ischemic stroke patients [157]; thus, extensive studies should investigate the role of statin use as a post-stroke treatment.

The lowering of leukotriene B_4 synthesis as an ischemic stroke treatment has also been intensively investigated. For example, the experimental inhibition of 5-lipoxygenase with zileuton, an anti-asthmatic drug, attenuates inflammation in the ischemic zone, reduces brain damage, neuronal apoptosis, infarct volume and even improves neurological deficits [133–135]. Similar neuroprotection against the injury caused by ischemia-reperfusion is also mediated by the regulation of 5-lipoxygenase expression by microRNA [158]. The disruption or inhibition of the FLAP to block leukotriene synthesis is as efficient in reducing brain edema and neuroinflammation [159]. Antagonists of cysteinyl-leukotriene receptor 1, pranlukast [136], or receptor 2, HAMI 3379 [160], protect against cerebral ischemic injury, ameliorate neuron loss, inhibit astrocyte proliferation, decrease cytokines release, microglial activation and neutrophil accumulation in ischemic regions.

The level of 20-HETE can be effectively modulated by selective inhibitors, such as TS-011 or HET0016, or by the 20-HETE antagonist, 20-hydroxyeicosa-6(Z),15(Z)-dienoic acid. Administration of these agents results in reduced infarct volume, improved microcirculation in the infarct area, and better post-ischemic neurological outcomes in animal stroke models [147,161–163]. However, as the role of 20-HETE is reported to be both detrimental and beneficial for stroke pathology, the appropriate modulation of the level of this eicosanoid may be challenging in diverse post-stroke stages.

Increasing lipoxin A_4 levels seem to be a promising treatment approach; however, physiologically, LXA_4 is rapidly inactivated. Thus, synthetic analogues could be more promising agents for that kind of stroke therapy. For example, the BML-111 analog of LXA_4 reduces the infarct volume and improves sensorimotor function in the early post-ischemic phase in rats, but does not affect behavioral deficits in the long-term [164]. However, the administration of an LXA_4 analog was efficient in reducing levels of pro-inflammatory cytokines and chemokines and increasing the anti-inflammatory cell populations in the post-ischemic brain. Another lipoxin A_4 analog, LXA_4 methyl ester (LXA_4ME), was shown to reduce the brain injury by ameliorating the blood–brain barrier dysfunction in a rat model of transient or permanent cerebral ischemic injury [165].

Reduction of the infarct volume, apoptosis in the ischemic area and the amelioration of neurological deficits can also be achieved by blocking of the downstream EET metabolism by inhibitors of sEH [166–168]. Both cell culture and animal studies prove that sEH inhibitors mediate cerebral protection by increasing the levels of EETs. For example, astrocytes treated with sEH inhibitors increase the concentration of EETs after the ischemia-like event and, as a result, release higher levels of protective neurotrophic factors, such as vascular endothelial growth factor (VEGF), which prevent neuronal cell death [167]. Another study showed that 14,15-EET itself or the inhibition of sEH attenuates neuronal apoptosis, astrogliosis and microglia activation, reduces inflammatory responses and promotes angiogenesis in the rat brain after occlusion of the middle cerebral artery [169].

5. Conclusions

Stroke is a devastating brain injury with tremendous consequences for human health. Understanding the pathophysiology of ischemic stroke is critical for reducing the burden of the disease or developing therapies. Lipids and lipid mediators, such as eicosanoids, largely contribute to the pathophysiology of ischemic stroke and possible mechanisms of their involvement, discussed in this review, are summarized on Figure 3.

The prevalence of stroke continuously increases because the elderly population is growing faster than the populations of other ages. Stroke is one of the main causes leading to disability and reduced quality of life worldwide; thus, the development of preventive and therapeutic interventions is urgently needed. Approved treatments for ischemic stroke are limited to aspirin, recombinant tissue plasminogen activator (rtPA) and mechanical thrombectomy. Each year, we observe the progress in stroke research, but the findings are not always successfully translated into the clinic.

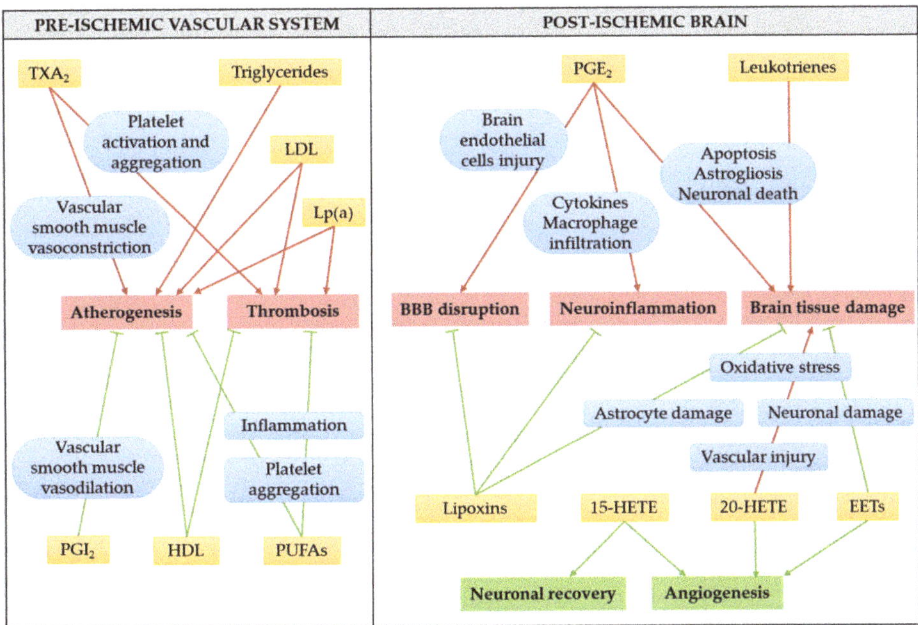

Figure 3. Possible mechanisms of involvement of serum lipids and eicosanoids in the pathophysiology of ischemic stroke. Arrows indicate stimulation, while bar-headed lines indicate inhibition. Green color stands for beneficial, while red for detrimental effects. Abbreviations used: 15-HETE, 15-hydroxyeicosatetraenoic acid; 20-HETE, 20-hydroxyeicosatetraenoic acid; EETs, epoxyeicosatrienoic acids; HDL, high-density lipoprotein; LDL, low-density lipoprotein; Lp(a), lipoprotein(a); PGE$_2$, prostaglandin E$_2$; PGI$_2$, prostaglandin I$_2$, prostacyclin; PUFAs, polyunsaturated fatty acids; TXA$_2$, thromboxane A$_2$.

An effective stroke treatment must focus on rescue of the penumbra zone from ischemic injury to preserve the viability and vitality of as many cells in this zone as possible. Studies on cell culture and animal stroke models have indicated many potential targets for therapeutic interventions within the biosynthetic pathways of lipids and lipid mediators, especially eicosanoids, derived from polyunsaturated fatty acids (summarized in Table 1). Specific enzyme inhibitors, expression modulators or receptor antagonists are able to redirect the metabolite flow in a way that beneficial rather than detrimental effects in the post-stoke brain tissue can be obtained. Neuroprotective, anti-inflammatory and pro-angiogenic effects of these experimental treatments attenuate ischemic brain damage, reduce the infarct area, boost the microcirculation and improve neurological deficits and recovery; however, a lot of research is necessary to determine the molecular mechanisms, efficacy and safety of that kind of intervention.

A number of serum lipids are associated with the risk of ischemic stroke. In clinical practice, the most common laboratory tests determine only the traditional lipid profile. However, research findings show that, for example, lipid ratios are better predictors of vascular risk than a single isolated lipid value. Conventional serum lipid measurements often ignore additional parameters like composition, particle size or the concentration of lipid subclass. Studies show that non-traditional lipids may also be good predictors of stroke risk. More evidence is needed to fully unravel the relationship between lipids, type of lipids, lipid profile (traditional and non-traditional) and stroke. However, it is worth considering including non-traditional lipid measures in daily clinical practice to better predict the risk of stroke in a patient or monitor the recovery of a patient who had already had a stroke episode.

Table 1. Therapeutic interventions targeting lipids and lipid mediators in the prevention and treatment of ischemic stroke.

Therapeutic Intervention [1]	Effect	Application [2]	Reference
Lipid-lowering therapies for stroke prevention			
Lowering cholesterol with statins	Anti-atherogenic: LDL cholesterol decrease	Clinical use	[91,92,97]
Lowering cholesterol with PCSK9 inhibitors, siRNAs, mimetic peptides, adnectins, vaccinations	Anti-atherogenic: increased LDL-C clearance from circulation	Experimental	[99–105]
Reduction of hypertriglyceridemia with fibric acid and derivatives	Anti-atherogenic: triglyceride decrease; HDL cholesterol increase	Clinical use	[106,107]
Lowering Lp(a) levels with niacin, PCSK9 or CETP inhibitors, or antisense oligonucleotides	Anti-atherogenic: reduced Lp(a) in circulation	Experimental	[108,109]
Eicosanoid-targeted therapies for post-ischemic brain tissue rescue			
Lowering prostaglandins level with COX inhibitors	Anti-atherogenic: anti-platelet effect of aspirin Neuroprotection: anti-inflammatory actions	Clinical use Experimental	[152,153,170]
Increasing prostacyclin levels with intravenous infusions	Vasodilation of cerebral microvessels	Experimental	[156]
Lowering TXA$_2$ levels with statins	Anti-thrombogenic: TXA$_2$ level decrease, anti-platelet activity	Experimental	[157]
Decreasing leukotriene B$_4$ synthesis with 5-LOX inhibitors, microRNAs, FLAP inhibitors, or CysLT receptors antagonists	Neuroprotection: anti-inflammatory and anti-apoptotic actions, reduced neuronal loss	Experimental	[133–136,158–160]
Increasing lipoxin A$_4$ levels with LXA$_4$ or LXA$_4$ME analogues	Neuroprotection: anti-inflammatory action, blood–brain barrier rescue	Experimental	[164,165]
Lowering 20-HETE levels with inhibitors or antagonists	Neuroprotection: improved microcirculation	Experimental	[147,161–163]
Increasing EETs levels with sEH inhibitors	Neuroprotection: anti-apoptotic, anti-inflammatory, pro-angiogenic, astrogliosis-preventive actions	Experimental	[166–169]

[1] Abbreviations: 20-HETE, 20-hydroxyeicosatetraenoic acid; 5-LOX, 5-lipoxygenase; CETP, cholesterylester transfer protein; COX, cyclooxygenase; CysLT, cysteinyl leukotriene; EETs, epoxyeicosatrienoic acids; FLAP, 5-LOX-activating protein; Lp(a), lipoprotein(a); LXA$_4$, lipoxin A$_4$; LXA$_4$ME, lipoxin A$_4$ methyl ester; PCSK9, proprotein convertase subtilisin/kexin type-9; sEH, soluble epoxide hydrolase; TXA$_2$, thromboxane A$_2$. [2] Clinical use: therapeutic interventions used as a standard prevention or treatment for stroke patients. Experimental: therapeutic interventions tested in preclinical or clinical studies, but currently not translated into the clinic.

Author Contributions: Writing—original draft preparation, A.K., J.J.-B., M.M., M.G.-C. All authors have read and agreed to the published version of the manuscript.

Funding: No funding supported the preparation of this manuscript.

Acknowledgments: This work was supported by the Institute of Biochemistry and Biophysics of the Polish Academy of Sciences (task grant no. T-32.1).

Conflicts of Interest: The authors declare no conflict of interest.

Abbreviations

AA	arachidonic acid
CAD	coronary artery disease
CE	cardioembolic
CETP	cholesterylester transfer protein
COX	cyclooxygenase
cPLA$_2$	cytosolic phospholipase A$_2$
CVD	cardiovascular disease
CYP	cytochrome P450
CysLT	cysteinyl leukotriene
DHA	docosahexaenoic acid
DPA	docosapentaenoic acid
EET	epoxyeicosatrienoic acid
END	early neurological deterioration
EPA	eicosapentaenoic acid
FA	fatty acid
FLAP	5-LOX-activating protein
HDL	high-density lipoprotein
HDL-C	high-density lipoprotein cholesterol
HETE	hydroxyeicosatetraenoic acid
HS	hemorrhagic stroke
IDL	intermediate-density lipoprotein
IS	ischemic stroke
LA	linoleic acid
LDL	low-density lipoprotein
LDL-C	low-density lipoprotein cholesterol
LOX	lipoxygenase
Lp(a)	lipoprotein(a)
LT	leukotriene
LX	lipoxin
LXA$_4$ME	lipoxin A$_4$ methyl ester
NMR	nuclear magnetic resonance
non-HDL-C	non-high-density lipoprotein cholesterol
NPD1	neuroprotectin D1
OGD	oxygen-glucose deprivation
PAF	platelet-activating factor
PCSK9	proprotein convertase subtilisin/kexin type-9
PG	prostaglandin
PGE$_2$	prostaglandin E$_2$
PGI$_2$	prostaglandin I$_2$, prostacyclin
PUFA	polyunsaturated fatty acid
sEH	soluble epoxide hydrolase
TC	total cholesterol
TG	triglyceride
TXA$_2$	thromboxane A$_2$
VEGF	vascular endothelial growth factor
VLDL	very low-density lipoprotein

References

1. James, S.L.; Abate, D.; Hassen Abate, K.; Abay, S.M.; Abbafati, C.; Abbasi, N.; Abbastabar, H.; Abd-Allah, F.; Abdela, J.; Abdelalim, A.; et al. Global, regional, and national incidence, prevalence, and years lived with disability for 354 diseases and injuries for 195 countries and territories, 1990–2017: A systematic analysis for the Global Burden of Disease Study 2017. *Lancet* **2018**, *392*, 1789–1858. [CrossRef]

2. World Health Organization. *Global Health Estimates 2016: Disease burden by Cause, Age, Sex, by Country and by Region, 2000–2016*; WHO: Geneva, Switzerland, 2018.

3. Avan, A.; Digaleh, H.; Di Napoli, M.; Stranges, S.; Behrouz, R.; Shojaeianbabaei, G.; Amiri, A.; Tabrizi, R.; Mokhber, N.; Spence, J.D.; et al. Socioeconomic status and stroke incidence, prevalence, mortality, and worldwide burden: An ecological analysis from the Global Burden of Disease Study 2017. *BMC Med.* **2019**, *17*, 191. [CrossRef] [PubMed]

4. O'Donnell, M.J.; Xavier, D.; Liu, L.; Zhang, H.; Chin, S.L.; Rao-Melacini, P.; Rangarajan, S.; Islam, S.; Pais, P.; McQueen, M.J.; et al. Risk factors for ischaemic and intracerebral haemorrhagic stroke in 22 countries (the INTERSTROKE study): A case-control study. *Lancet* **2010**, *376*, 112–123. [CrossRef]

5. Gorelick, P.B.; Mazzone, T. Plasma Lipids and Stroke. *Eur. J. Cardiovasc. Risk* **1999**, *6*, 217–221. [CrossRef]

6. Tirschwell, D.L.; Smith, N.L.; Heckbert, S.R.; Lemaitre, R.N.; Longstreth, W.T.; Psaty, B.M. Association of cholesterol with stroke risk varies in stroke subtypes and patient subgroups. *Neurology* **2004**, *63*, 1868–1875. [CrossRef]

7. Jain, M.; Jain, A.; Yerragondu, N.; Brown, R.D.; Rabinstein, A.; Jahromi, B.S.; Vaidyanathan, L.; Blyth, B.; Ganti Stead, L. The Triglyceride Paradox in Stroke Survivors: A Prospective Study. *Corp. Neurosci. J.* **2013**, *2013*. [CrossRef]

8. Park, J.H.; Lee, J.; Ovbiagele, B. Nontraditional serum lipid variables and recurrent stroke risk. *Stroke* **2014**, *45*, 3269–3274. [CrossRef]

9. Holmes, M.V.; Millwood, I.Y.; Kartsonaki, C.; Hill, M.R.; Bennett, D.A.; Boxall, R.; Guo, Y.; Xu, X.; Bian, Z.; Hu, R.; et al. Lipids, Lipoproteins, and Metabolites and Risk of Myocardial Infarction and Stroke. *J. Am. Coll. Cardiol.* **2018**, *71*, 620–632. [CrossRef]

10. Goldstein, L.B.; Bushnell, C.D.; Adams, R.J.; Appel, L.J.; Braun, L.T.; Chaturvedi, S.; Creager, M.A.; Culebras, A.; Eckel, R.H.; Hart, R.G.; et al. Guidelines for the primary prevention of stroke: A Guideline for Healthcare Professionals from the American Heart Association/American Stroke Association. *Stroke* **2011**, *42*, 517–584. [CrossRef]

11. Shahar, E.; Chambless, L.E.; Rosamond, W.D.; Boland, L.L.; Ballantyne, C.M.; McGovern, P.G.; Sharrett, A.R. Plasma lipid profile and incident ischemic stroke: The Atherosclerosis Risk in Communities (ARIC) Study. *Stroke* **2003**, *34*, 623–631. [CrossRef]

12. Suh, I.; Jee, S.H.; Kim, H.C.; Nam, C.M.; Kim, I.S.; Appel, L.J. Low serum cholesterol and haemorrhagic stroke in men: Korea Medical Insurance Corporation Study. *Lancet* **2001**, *357*, 922–925. [CrossRef]

13. Hart, C.L.; Hole, D.J.; Smith, G.D. Risk factors and 20-year stroke mortality in men and women in the Renfrew/Paisley study in Scotland. *Stroke* **1999**, *30*, 1999–2007. [CrossRef] [PubMed]

14. Tanizaki, Y.; Kiyohara, Y.; Kato, I.; Iwamoto, H.; Nakayama, K.; Shinohara, N.; Arima, H.; Tanaka, K.; Ibayashi, S.; Fujishima, M. Incidence and risk factors for subtypes of cerebral infarction in a general population: The Hisayama study. *Stroke* **2000**, *31*, 2616–2622. [CrossRef] [PubMed]

15. Sacco, R.L.; Benson, R.T.; Kargman, D.E.; Boden-Albala, B.; Tuck, C.; Lin, I.F.; Cheng, J.F.; Paik, M.C.; Shea, S.; Berglund, L. High-density lipoprotein cholesterol and ischemic stroke in the elderly the northern manhattan stroke study. *J. Am. Med. Assoc.* **2001**, *285*, 2729–2735. [CrossRef]

16. Spence, J.D. Statins for prevention of stroke. *Lancet* **1998**, *352*, 909. [CrossRef]

17. Zhang, Y.; Tuomilehto, J.; Jousilahti, P.; Wang, Y.; Antikainen, R.; Hu, G. Total and high-density lipoprotein cholesterol and stroke risk. *Stroke* **2012**, *43*, 1768–1774. [CrossRef]

18. Wang, X.; Dong, Y.; Qi, X.; Huang, C.; Hou, L. Cholesterol levels and risk of hemorrhagic stroke: A systematic review and meta-analysis. *Stroke* **2013**, *44*, 1833–1839. [CrossRef]

19. Di Angelantonio, E.; Sarwar, N.; Perry, P.; Kaptoge, S.; Ray, K.K.; Thompson, A.; Wood, A.M.; Lewington, S.; Sattar, N.; Packard, C.J.; et al. Major lipids, apolipoproteins, and risk of vascular disease. *JAMA J. Am. Med. Assoc.* **2009**, *302*, 1993–2000.

20. Ali, K.M.; Wonnerth, A.; Huber, K.; Wojta, J. Cardiovascular disease risk reduction by raising HDL cholesterol—Current therapies and future opportunities. *Br. J. Pharmacol.* **2012**, *167*, 1177–1194. [CrossRef]

21. Ballantyne, C.M.; Herd, J.A.; Ferlic, L.L.; Dunn, J.K.; Farmer, J.A.; Jones, P.H.; Schein, J.R.; Gotto, A.M. Influence of low HDL on progression of coronary artery disease and response to fluvastatin therapy. *Circulation* **1999**, *99*, 736–743. [CrossRef]

22. Santos-Gallego, C.G.; Badimón, J.J. High-Density Lipoprotein and Cardiovascular Risk Reduction: Promises and Realities. *Rev. Española Cardiol.* **2012**, *65*, 305–308. [CrossRef] [PubMed]

23. Imamura, T.; Doi, Y.; Arima, H.; Yonemoto, K.; Hata, J.; Kubo, M.; Tanizaki, Y.; Ibayashi, S.; Iida, M.; Kiyohara, Y. LDL cholesterol and the development of stroke subtypes and coronary heart disease in a general Japanese population the Hisayama study. *Stroke* **2009**, *40*, 382–388. [CrossRef] [PubMed]

24. Gu, X.; Li, Y.; Chen, S.; Yang, X.; Liu, F.; Li, Y.; Li, J.; Cao, J.; Liu, X.; Chen, J.; et al. Association of Lipids With Ischemic and Hemorrhagic Stroke: A Prospective Cohort Study Among 267 500 Chinese. *Stroke* **2019**, *50*, 3376–3384. [CrossRef] [PubMed]

25. Dziedzic, T.; Slowik, A.; Gryz, E.A.; Szczudlik, A. Lower serum triglyceride level is associated with increased stroke severity. *Stroke* **2004**, *35*, e151–e152. [CrossRef]

26. Simundic, A.M.; Nikolac, N.; Topic, E.; Basic-Kes, V.; Demarin, V. Are serum lipids measured on stroke admission prognostic? *Clin. Chem. Lab. Med.* **2008**, *46*, 1163–1167. [CrossRef]

27. Berger, J.S.; McGinn, A.P.; Howard, B.V.; Kuller, L.; Manson, J.E.; Otvos, J.; Curb, J.D.; Eaton, C.B.; Kaplan, R.C.; Lynch, J.K.; et al. Lipid and lipoprotein biomarkers and the risk of ischemic stroke in postmenopausal women. *Stroke* **2012**, *43*, 958–966. [CrossRef]

28. Ebinger, M.; Sievers, C.; Klotsche, J.; Schneider, H.J.; Leonards, C.O.; Pieper, L.; Wittchen, H.U.; Stalla, G.K.; Endres, M. Triglycerides and stroke risk prediction: Lessons from a prospective cohort study in German primary care patients. *Front. Neurol.* **2010**. [CrossRef]

29. Leonards, C.; Ebinger, M.; Batluk, J.; Malzahn, U.; Heuschmann, P.; Endres, M. The role of fasting versus non-fasting triglycerides in ischemic stroke: A systematic review. *Front. Neurol.* **2010**, *1*, 133. [CrossRef]

30. Bansal, S.; Buring, J.E.; Rifai, N.; Mora, S.; Sacks, F.M.; Ridker, P.M. Fasting compared with nonfasting triglycerides and risk of cardiovascular events in women. *J. Am. Med. Assoc.* **2007**, *298*, 309–316. [CrossRef]

31. Freiberg, J.J.; Tybjærg-Hansen, A.; Jensen, J.S.; Nordestgaard, B.G. Nonfasting triglycerides and risk of ischemic stroke in the general population. *JAMA J. Am. Med. Assoc.* **2008**, *300*, 2142–2152. [CrossRef]

32. Weir, C.J.; Sattar, N.; Walters, M.R.; Lees, K.R. Low triglyceride, not low cholesterol concentration, independently predicts poor outcome following acute stroke. *Cerebrovasc. Dis.* **2003**, *16*, 76–82. [CrossRef] [PubMed]

33. Chapman, M.J.; Ginsberg, H.N.; Amarenco, P.; Andreotti, F.; Borén, J.; Catapano, A.L.; Descamps, O.S.; Fisher, E.; Kovanen, P.T.; Kuivenhoven, J.A.; et al. Triglyceride-rich lipoproteins and high-density lipoprotein cholesterol in patients at high risk of cardiovascular disease: Evidence and guidance for management. *Eur. Heart J.* **2011**, *32*, 1345–1361. [CrossRef] [PubMed]

34. Grundy, S.M. Hypertriglyceridemia, atherogenic dyslipidemia, and the metabolic syndrome. *Am. J. Cardiol.* **1998**, *81*, 18B–25B. [CrossRef]

35. Arsenault, B.J.; Rana, J.S.; Stroes, E.S.G.; Després, J.P.; Shah, P.K.; Kastelein, J.J.P.; Wareham, N.J.; Boekholdt, S.M.; Khaw, K.T. Beyond Low-Density Lipoprotein Cholesterol. Respective Contributions of Non-High-Density Lipoprotein Cholesterol Levels, Triglycerides, and the Total Cholesterol/High-Density Lipoprotein Cholesterol Ratio to Coronary Heart Disease Risk in Apparently Healt. *J. Am. Coll. Cardiol.* **2009**, *55*, 35–41. [CrossRef]

36. Ridker, P.M.; Rifai, N.; Cook, N.R.; Bradwin, G.; Buring, J.E. Non-HDL cholesterol, apolipoproteins A-I and B 100, standard lipid measures, lipid ratios, and CRP as risk factors for cardiovascular disease in women. *J. Am. Med. Assoc.* **2005**, *294*, 326–333. [CrossRef]

37. Mathews, S.C.; Mallidi, J.; Kulkarni, K.; Toth, P.P.; Jones, S.R. Achieving Secondary Prevention Low-Density Lipoprotein Particle Concentration Goals Using Lipoprotein Cholesterol-Based Data. *PLoS ONE* **2012**, *7*, e33692. [CrossRef]

38. Elshazly, M.B.; Quispe, R.; Michos, E.D.; Sniderman, A.D.; Toth, P.P.; Banach, M.; Kulkarni, K.R.; Coresh, J.; Blumenthal, R.S.; Jones, S.R.; et al. Patient-level discordance in population percentiles of the total cholesterol to high-density lipoprotein cholesterol ratio in comparison with low-density lipoprotein cholesterol and non-high-density lipoprotein cholesterol: The very large database of lipi. *Circulation* **2015**, *132*, 667–676. [CrossRef]

39. Katakami, N.; Kaneto, H.; Osonoi, T.; Saitou, M.; Takahara, M.; Sakamoto, F.; Yamamoto, K.; Yasuda, T.; Matsuoka, T.A.; Matsuhisa, M.; et al. Usefulness of lipoprotein ratios in assessing carotid atherosclerosis in Japanese type 2 diabetic patients. *Atherosclerosis* **2011**, *214*, 442–447. [CrossRef]

40. De la Riva, P.; Zubikarai, M.; Sarasqueta, C.; Tainta, M.; Muñoz-Lopetegui, A.; Andrés-Marín, N.; González, F.; Díez, N.; de Arce, A.; Bergareche, A.; et al. Nontraditional Lipid Variables Predict Recurrent Brain Ischemia in Embolic Stroke of Undetermined Source. *J. Stroke Cerebrovasc. Dis.* **2017**, *26*, 1670–1677. [CrossRef]

41. Zheng, J.; Sun, Z.; Zhang, X.; Li, Z.; Guo, X.; Xie, Y.; Sun, Y.; Zheng, L. Non-traditional lipid profiles associated with ischemic stroke not hemorrhagic stroke in hypertensive patients: Results from an 8.4 years follow-up study. *Lipids Health Dis.* **2019**, *18*, 9. [CrossRef]

42. Guo, X.; Li, Z.; Sun, G.; Guo, L.; Zheng, L.; Yu, S.; Yang, H.; Pan, G.; Zhang, Y.; Sun, Y. Comparison of four nontraditional lipid profiles in relation to ischemic stroke among hypertensive Chinese population. *Int. J. Cardiol.* **2015**, *201*, 123–125. [CrossRef] [PubMed]

43. Würtz, P.; Havulinna, A.S.; Soininen, P.; Tynkkynen, T.; Prieto-Merino, D.; Tillin, T.; Ghorbani, A.; Artati, A.; Wang, Q.; Tiainen, M.; et al. Metabolite profiling and cardiovascular event risk: A prospective study of 3 population-based cohorts. *Circulation* **2015**, *131*, 774–785. [CrossRef] [PubMed]

44. Björnheden, T.; Babyi, A.; Bondjers, G.; Wiklund, O. Accumulation of lipoprotein fractions and subfractions in the arterial wall, determined in an in vitro perfusion system. *Atherosclerosis* **1996**, *123*, 43–56. [CrossRef]

45. Jeyarajah, E.J.; Cromwell, W.C.; Otvos, J.D. Lipoprotein Particle Analysis by Nuclear Magnetic Resonance Spectroscopy. *Clin. Lab. Med.* **2006**, *26*, 847–870. [CrossRef] [PubMed]

46. Hachinski, V.; Graffagnino, C.; Beaudry, M.; Bernier, G.; Buck, C.; Donner, A.; Spence, J.D.; Doig, G.; Wolfe, B.M.J. Lipids and stroke: A paradox resolved. *Arch. Neurol.* **1996**, *53*, 303–308. [CrossRef] [PubMed]

47. Glader, C.A.; Stegmayr, B.; Boman, J.; Stenlund, H.; Weinehall, L.; Hallmans, G.; Dahlén, G.H. Chlamydia pneumoniae antibodies and high lipoprotein(a) levels do not predict ischemic cerebral infarctions. Results from a nested case-control study in Northern Sweden. *Stroke* **1999**, *30*, 2013–2018. [CrossRef] [PubMed]

48. Alfthan, G.; Pekkanen, J.; Jauhiainen, M.; Pitkäniemi, J.; Karvonen, M.; Tuomilehto, J.; Salonen, J.T.; Ehnholm, C. Relation of serum homocysteine and lipoprotein(a) concentrations to atherosclerotic disease in a prospective Finnish population based study. *Atherosclerosis* **1994**, *106*, 9–19. [CrossRef]

49. Tipping, R.W.; Ford, C.E.; Simpson, L.M.; Walldius, G.; Jungner, I.; Folsom, A.R.; Chambless, L.; Panagiotakos, D.; Pitsavos, C.; Chrysohoou, C.; et al. Lipoprotein(a) concentration and the risk of coronary heart disease, stroke, and nonvascular mortality. *JAMA J. Am. Med. Assoc.* **2009**, *302*, 412–423.

50. Nave, A.H.; Lange, K.S.; Leonards, C.O.; Siegerink, B.; Doehner, W.; Landmesser, U.; Steinhagen-Thiessen, E.; Endres, M.; Ebinger, M. Lipoprotein (a) as a risk factor for ischemic stroke: A meta-analysis. *Atherosclerosis* **2015**, *242*, 496–503. [CrossRef]

51. Journal, B.; Lan, Y.; Zhao, X.; Wang, X.; Song, X.; Chen, J.; Zhang, Y.; Zhang, Z. Lipoprotein(a) as a Risk Factor for Predicting Coronary Artery Disease Events: A Meta-analysis. *Biomark J.* **2018**, *4*, 17.

52. Yaghi, S.; Elkind, M.S.V. Lipids and Cerebrovascular Disease. *Stroke* **2015**, *46*, 3322–3328. [CrossRef] [PubMed]

53. Nordestgaard, B.G.; Chapman, M.J.; Ray, K.; Boré, J.; Andreotti, F.; Watts, G.F.; Ginsberg, H.; Amarenco, P.; Catapano, A.; Descamps, O.S.; et al. Lipoprotein(a) as a cardiovascular risk factor: Current status. *Eur. Heart J.* **2010**, *31*, 2844–2853. [CrossRef] [PubMed]

54. Bu, J.; Dou, Y.; Tian, X.; Wang, Z.; Chen, G. The Role of Omega-3 Polyunsaturated Fatty Acids in Stroke. *Oxid. Med. Cell. Lomgev.* **2016**. [CrossRef] [PubMed]

55. Mozaffarian, D.; Wu, J.H.Y. Omega-3 fatty acids and cardiovascular disease: Effects on risk factors, molecular pathways, and clinical events. *J. Am. Coll. Cardiol.* **2011**, *58*, 2047–2067. [CrossRef]

56. Rebiger, L.; Lenzen, S.; Mehmeti, I. Susceptibility of brown adipocytes to pro-inflammatory cytokine toxicity and reactive oxygen species. *Biosci. Rep.* **2016**, *36*, e00306. [CrossRef]

57. Geleijnse, J.M.; Giltay, E.J.; Grobbee, D.E.; Donders, A.R.T.; Kok, F.J. Blood pressure response to fish oil supplementation: Metaregression analysis of randomized trials. *J. Hypertens.* **2002**, *20*, 1493–1499. [CrossRef]

58. Mozaffarian, D.; Wu, J.H.Y. (n-3) Fatty Acids and Cardiovascular Health: Are Effects of EPA and DHA Shared or Complementary? *J. Nutr.* **2012**, *142*, 614S–625S. [CrossRef]

59. Wallström, P.; Sonestedt, E.; Hlebowicz, J.; Ericson, U.; Drake, I.; Persson, M.; Gullberg, B.; Hedblad, B.; Wirfält, E. Dietary fiber and saturated fat intake associations with cardiovascular disease differ by sex in the Malmö diet and cancer cohort: A prospective study. *PLoS ONE* **2012**, *7*, e31637. [CrossRef]

60. Yaemsiri, S.; Sen, S.; Tinker, L.F.; Robinson, W.R.; Evans, R.W.; Rosamond, W.; Wasserthiel-Smoller, S.; He, K. Serum fatty acids and incidence of ischemic stroke among postmenopausal women. *Stroke* **2013**, *44*, 2710–2717. [CrossRef]

61. Tanaka, K.; Ishikawa, Y.; Yokoyama, M.; Origasa, H.; Matsuzaki, M.; Saito, Y.; Matsuzawa, Y.; Sasaki, J.; Oikawa, S.; Hishida, H.; et al. Reduction in the recurrence of stroke by eicosapentaenoic acid for hypercholesterolemic patients: Subanalysis of the JELIS trial. *Stroke* **2008**, *39*, 2052–2058. [CrossRef]

62. Nishizaki, Y.; Shimada, K.; Tani, S.; Ogawa, T.; Ando, J.; Takahashi, M.; Yamamoto, M.; Shinozaki, T.; Miyauchi, K.; Nagao, K.; et al. Significance of imbalance in the ratio of serum n-3 to n-6 polyunsaturated fatty acids in patients with acute coronary syndrome. *Am. J. Cardiol.* **2014**, *113*, 441–445. [CrossRef] [PubMed]

63. Thies, F.; Garry, J.M.C.; Yaqoob, P.; Rerkasem, K.; Williams, J.; Shearman, C.P.; Gallagher, P.J.; Calder, P.C.; Grimble, R.F. Association of n-3 polyunsaturated fatty acids with stability of atherosclerotic plaques: A randomised controlled trial. *Lancet* **2003**, *361*, 477–485. [CrossRef]

64. Ajami, M.; Eghtesadi, S.; Razaz, J.M.; Kalantari, N.; Habibey, R.; Nilforoushzadeh, M.A.; Zarrindast, M.; Pazoki-Toroudi, H. Expression of Bcl-2 and Bax after hippocampal ischemia in DHA + EPA treated rats. *Neurol. Sci.* **2011**, *32*, 811–818. [CrossRef] [PubMed]

65. Suda, S.; Katsumata, T.; Okubo, S.; Kanamaru, T.; Suzuki, K.; Watanabe, Y.; Katsura, K.I.; Katayama, Y. Low serum n-3 polyunsaturated fatty acid/n-6 polyunsaturated fatty acid ratio predicts neurological deterioration in japanese patients with acute ischemic stroke. *Cerebrovasc. Dis.* **2013**, *36*, 388–393. [CrossRef]

66. Alawneh, J.A.; Moustafa, R.R.; Baron, J.C. Hemodynamic factors and perfusion abnormalities in early neurological deterioration. *Stroke* **2009**, *40*, e443–e450. [CrossRef]

67. Huang, Z.X.; Wang, Q.Z.; Dai, Y.Y.; Lu, H.K.; Liang, X.Y.; Hu, H.; Liu, X.T. Early neurological deterioration in acute ischemic stroke: A propensity score analysis. *J. Chinese Med. Assoc.* **2018**, *81*, 865–870. [CrossRef]

68. Poirier, P.; Giles, T.D.; Bray, G.A.; Hong, Y.; Stern, J.S.; Pi-Sunyer, F.X.; Eckel, R.H. Obesity and cardiovascular disease: Pathophysiology, evaluation, and effect of weight loss: An update of the 1997 American Heart Association Scientific Statement on obesity and heart disease from the Obesity Committee of the Council on Nutrition, Physical. *Circulation* **2006**, *113*, 898–918. [CrossRef]

69. Fagot-Campagna, A.; Balkau, B.; Simon, D.; Warnet, J.-M.; Claude, J.-R.; Ducimetiere, P.; Eschwege, E. High free fatty acid concentration: An independent risk factor for hypertension in the Paris Prospective Study. *Int. J. Epidemiol.* **1998**, *27*, 808–813. [CrossRef]

70. Bays, H.; Mandarino, L.; DeFronzo, R.A. Role of the Adipocyte, Free Fatty Acids, and Ectopic Fat in Pathogenesis of Type 2 Diabetes Mellitus: Peroxisomal Proliferator-Activated Receptor Agonists Provide a Rational Therapeutic Approach. *J. Clin. Endocrinol. Metab.* **2004**, *89*, 463–478. [CrossRef]

71. O'Donoghue, M.; De Lemos, J.A.; Morrow, D.A.; Murphy, S.A.; Buros, J.L.; Cannon, C.P.; Sabatine, M.S. Prognostic utility of heart-type fatty acid binding protein in patients with acute coronary syndromes. *Circulation* **2006**, *114*, 550–557. [CrossRef]

72. Cocco, G.; Chu, D. Drug points: Rimonabant may induce atrial fibrillation. *BMJ* **2009**, *339*, 296.

73. Khawaja, O.; Bartz, T.M.; Ix, J.H.; Heckbert, S.R.; Kizer, J.R.; Zieman, S.J.; Mukamal, K.J.; Tracy, R.P.; Siscovick, D.S.; Djoussé, L. Plasma free fatty acids and risk of atrial fibrillation (from the Cardiovascular Health Study). *Am. J. Cardiol.* **2012**, *110*, 212–216. [CrossRef] [PubMed]

74. Iso, H.; Sato, S.; Umemura, U.; Kudo, M.; Koike, K.; Kitamura, A.; Imano, H.; Okamura, T.; Naito, Y.; Shimamoto, T. Linoleic acid, other fatty acids, and the risk of stroke. *Stroke* **2002**, *33*, 2086–2093. [CrossRef] [PubMed]

75. Saber, H.; Yakoob, M.Y.; Shi, P.; Longstreth, W.T.; Lemaitre, R.N.; Siscovick, D.; Rexrode, K.M.; Willett, W.C.; Mozaffarian, D. Omega-3 fatty acids and incident ischemic stroke and its atherothrombotic and cardioembolic subtypes in 3 US cohorts. *Stroke* **2017**, *48*, 2678–2685. [CrossRef]

76. Choi, J.Y.; Kim, J.S.; Kim, J.H.; Oh, K.; Koh, S.B.; Seo, W.K. High free fatty acid level is associated with recurrent stroke in cardioembolic stroke patients. *Neurology* **2014**, *82*, 1142–1148. [CrossRef]

77. Seo, W.K.; Jung, J.M.; Kim, J.H.; Koh, S.B.; Bang, O.Y.; Oh, K. Free fatty acid is associated with thrombogenicity in cardioembolic stroke. *Cerebrovasc. Dis.* **2017**, *44*, 160–168. [CrossRef]

78. Erkkilä, A.T.; Lehto, S.; Pyörälä, K.; Uusitupa, M.I.J. n-3 Fatty acids and 5-y risks of death and cardiovascular disease events in patients with coronary artery disease. *Am. J. Clin. Nutr.* **2003**, *78*, 65–71. [CrossRef]

79. Watanabe, Y.; Tatsuno, I. Omega-3 polyunsaturated fatty acids for cardiovascular diseases: Present, past and future. *Expert Rev. Clin. Pharmacol.* **2017**, *10*, 865–873. [CrossRef]

80. Owen, A.J.; Magliano, D.J.; O'Dea, K.; Barr, E.L.M.; Shaw, J.E. Polyunsaturated fatty acid intake and risk of cardiovascular mortality in a low fish-consuming population: A prospective cohort analysis. *Eur. J. Nutr.* **2016**, *55*, 1605–1613. [CrossRef]

81. Wennberg, M.; Jansson, J.-H.; Norberg, M.; Skerfving, S.; Strömberg, U.; Wiklund, P.-G.; Bergdahl, I.A. Fish consumption and risk of stroke: A second prospective case-control study from northern Sweden. *Nutr. J.* **2016**, *15*, 98. [CrossRef]

82. Mensink, R.P. *Effects of Saturated Fatty Acids on Serum Lipids and Lipoproteins: A Systematic Review and Regression Analysis*; World Health Organization: Geneva, Switzerland, 2016; Available online: https://apps.who.int/iris/bitstream/handle/10665/246104/9789241565349-eng.pdf (accessed on 11 May 2020).

83. McEwen, B.J.; Morel-Kopp, M.C.; Chen, W.; Tofler, G.H.; Ward, C.M. Effects of omega-3 polyunsaturated fatty acids on platelet function in healthy subjects and subjects with cardiovascular disease. *Semin. Thromb. Hemost.* **2013**, *39*, 25–32. [CrossRef] [PubMed]

84. Cohen, M.G.; Rossi, J.S.; Garbarino, J.; Bowling, R.; Motsinger-Reif, A.A.; Schuler, C.; Dupont, A.G.; Gabriel, D. Insights into the inhibition of platelet activation by omega-3 polyunsaturated fatty acids: Beyond aspirin and clopidogrel. *Thromb. Res.* **2011**, *128*, 335–340. [CrossRef] [PubMed]

85. Browning, L.M.; Walker, C.G.; Mander, A.P.; West, A.L.; Madden, J.; Gambell, J.M.; Young, S.; Wang, L.; Jebb, S.A.; Calder, P.C. Incorporation of eicosapentaenoic and docosahexaenoic acids into lipid pools when given as supplements providing doses equivalent to typical intakes of oily fish. *Am. J. Clin. Nutr.* **2012**, *96*, 748–758. [CrossRef] [PubMed]

86. Ghasemi Fard, S.; Wang, F.; Sinclair, A.J.; Elliott, G.; Turchini, G.M. How does high DHA fish oil affect health? A systematic review of evidence. *Crit. Rev. Food Sci. Nutr.* **2019**, *59*, 1684–1727. [CrossRef] [PubMed]

87. Elvevoll, E.O.; Barstad, H.; Breimo, E.S.; Brox, J.; Eilertsen, K.E.; Lund, T.; Olsen, J.O.; Østerud, B. Enhanced incorporation of n-3 fatty acids from fish compared with fish oils. *Lipids* **2006**, *41*, 1109–1114. [CrossRef] [PubMed]

88. Vidgren, H.M.; Ågren, J.J.; Schwab, U.; Rissanen, T.; Hänninen, O.; Uusitupa, M.I.J. Incorporation of n-3 fatty acids into plasma lipid fractions, and erythrocyte membranes and platelets during dietary supplementation with fish, fish oil, and docosahexaenoic acid-rich oil among healthy young men. *Lipids* **1997**, *32*, 697–705. [CrossRef] [PubMed]

89. Saposnik, G.; Fonarow, G.C.; Pan, W.; Liang, L.; Hernandez, A.F.; Schwamm, L.H.; Smith, E.E. Guideline-directed low-density lipoprotein management in high-risk patients with ischemic stroke: Findings from get with the guidelines-stroke 2003 to 2012. *Stroke* **2014**, *45*, 3343–3351. [CrossRef]

90. Amarenco, P.; Lavallée, P.; Touboul, P.J. Stroke prevention, blood cholesterol, and statins. *Lancet Neurol.* **2004**, *3*, 271–278. [CrossRef]

91. Wafa, H.A.; Wolfe, C.D.A.; Rudd, A.; Wang, Y. Long-term trends in incidence and risk factors for ischaemic stroke subtypes: Prospective population study of the South London Stroke Register. *PLoS Med.* **2018**, *15*, e1002669. [CrossRef]

92. Tramacere, I.; Boncoraglio, G.B.; Banzi, R.; Del Giovane, C.; Kwag, K.H.; Squizzato, A.; Moja, L. Comparison of statins for secondary prevention in patients with ischemic stroke or transient ischemic attack: A systematic review and network meta-analysis. *BMC Med.* **2019**, *17*, 67. [CrossRef]

93. Tziomalos, K.; Giampatzis, V.; Bouziana, S.D.; Spanou, M.; Kostaki, S.; Papadopoulou, M.; Angelopoulou, S.M.; Konstantara, F.; Savopoulos, C.; Hatzitolios, A.I. Comparative effects of more versus less aggressive treatment with statins on the long-term outcome of patients with acute ischemic stroke. *Atherosclerosis* **2015**, *243*, 65–70. [CrossRef] [PubMed]

94. Amarenco, P.; Bogousslavsky, J.; Callahan, A.; Goldstein, L.B.; Hennerici, M.; Rudolph, A.E.; Sillesen, H.; Simunovic, L.; Szarek, M.; Welch, K.M.A.; et al. High-dose atorvastatin after stroke or transient ischemic attack. *N. Engl. J. Med.* **2006**, *355*, 549–559. [CrossRef] [PubMed]

95. Muscari, A.; Puddu, G.M.; Santoro, N.; Serafini, C.; Cenni, A.; Rossi, V.; Zoli, M. The atorvastatin during ischemic stroke study: A pilot randomized controlled trial. *Clin. Neuropharmacol.* **2011**, *34*, 141–147. [CrossRef] [PubMed]

96. Tuttolomondo, A.; Di Raimondo, D.; Pecoraro, R.; Maida, C.; Arnao, V.; Corte, V.D.; Simonetta, I.; Corpora, F.; Di Bona, D.; Maugeri, R.; et al. Early high-dosage atorvastatin treatment improved serum immune-inflammatory markers and functional outcome in acute ischemic strokes classified as large artery atherosclerotic stroke: A randomized trial. *Medicine* **2016**, *95*, e3186. [CrossRef]

97. Yu, Y.; Zhu, C.; Liu, C.; Gao, Y. Clinical Study Effect of Prior Atorvastatin Treatment on the Frequency of Hospital Acquired Pneumonia and Evolution of Biomarkers in Patients with Acute Ischemic Stroke: A Multicenter Prospective Study. *BioMed Res. Int.* **2017**, *2017*, 5642704. [CrossRef]

98. Szarek, M.; Amarenco, P.; Callahan, A.; DeMicco, D.; Fayyad, R.; Goldstein, L.B.; Laskey, R.; Sillesen, H.; Welch, K.M. Atorvastatin Reduces First and Subsequent Vascular Events Across Vascular Territories: The SPARCL Trial. *J. Am. Coll. Cardiol.* **2020**, *75*, 2110–2118. [CrossRef]

99. Sabatine, M.S.; Giugliano, R.P.; Keech, A.C.; Honarpour, N.; Wiviott, S.D.; Murphy, S.A.; Kuder, J.F.; Wang, H.; Liu, T.; Wasserman, S.M.; et al. Evolocumab and clinical outcomes in patients with cardiovascular disease. *N. Engl. J. Med.* **2017**, *376*, 1713–1722. [CrossRef]

100. Van Poelgeest, E.P.; Hodges, M.R.; Moerland, M.; Tessier, Y.; Levin, A.A.; Persson, R.; Lindholm, M.W.; Dumong Erichsen, K.; Ørum, H.; Cohen, A.F.; et al. Antisense-mediated reduction of proprotein convertase subtilisin/kexin type 9 (PCSK9): A first-in-human randomized, placebo-controlled trial. *Br. J. Clin. Pharmacol.* **2015**, *80*, 1350–1361. [CrossRef]

101. Ray, K.K.; Landmesser, U.; Leiter, L.A.; Kallend, D.; Dufour, R.; Karakas, M.; Hall, T.; Troquay, R.P.T.; Turner, T.; Visseren, F.L.J.; et al. Inclisiran in patients at high cardiovascular risk with elevated LDL cholesterol. *N. Engl. J. Med.* **2017**, *376*, 1430–1440. [CrossRef]

102. Lintner, N.G.; McClure, K.F.; Petersen, D.; Londregan, A.T.; Piotrowski, D.W.; Wei, L.; Xiao, J.; Bolt, M.; Loria, P.M.; Maguire, B.; et al. Selective stalling of human translation through small-molecule engagement of the ribosome nascent chain. *PLoS Biol.* **2017**, *15*, e2001882. [CrossRef]

103. Alghamdi, R.H.; O'Reilly, P.; Lu, C.; Gomes, J.; Lagace, T.A.; Basak, A. LDL-R promoting activity of peptides derived from human PCSK9 catalytic domain (153–421): Design, synthesis and biochemical evaluation. *Eur. J. Med. Chem.* **2015**, *92*, 890–907. [CrossRef] [PubMed]

104. Mitchell, T.; Chao, G.; Sitkoff, D.; Lo, F.; Monshizadegan, H.; Meyers, D.; Low, S.; Russo, K.; DiBella, R.; Denhez, F.; et al. Pharmacologic profile of the adnectin BMS-962476, a small protein biologic alternative to PCSK9 antibodies for low-density lipoprotein lowering. *J. Pharmacol. Exp. Ther.* **2014**, *350*, 412–424. [CrossRef] [PubMed]

105. Landlinger, C.; Pouwer, M.G.; Juno, C.; van der Hoorn, J.W.A.; Pieterman, E.J.; Jukema, J.W.; Staffler, G.; Princen, H.M.G.; Galabova, G. The AT04A vaccine against proprotein convertase subtilisin/kexin type 9 reduces total cholesterol, vascular inflammation, and atherosclerosis in APOE*3Leiden.CETP mice. *Eur. Heart J.* **2017**, *38*, 2499–2507. [CrossRef] [PubMed]

106. Meschia, J.F.; Bushnell, C.; Boden-Albala, B.; Braun, L.T.; Bravata, D.M.; Chaturvedi, S.; Creager, M.A.; Eckel, R.H.; Elkind, M.S.; Fornage, M.; et al. Guidelines for the primary prevention of stroke: A statement for healthcare professionals from the American Heart Association/American Stroke Association. *Stroke* **2014**, *45*, 3754–3832. [CrossRef] [PubMed]

107. Bloomfield Rubins, H.; Davenport, J.; Babikian, V.; Brass, L.M.; Collins, D.; Wexler, L.; Wagner, S.; Papademetriou, V.; Rutan, G.; Robins, S.J. Reduction in Stroke With Gemfibrozil in Men With Coronary Heart Disease and Low HDL Cholesterol. *Circulation* **2001**, *103*, 2828–2833. [CrossRef] [PubMed]

108. Van Capelleveen, J.C.; Van Der Valk, F.M.; Stroes, E.S.G. Current therapies for lowering lipoprotein (a). *J. Lipid Res.* **2016**, *57*, 1612–1618. [CrossRef]

109. Santos, R.D.; Raal, F.J.; Catapano, A.L.; Witztum, J.L.; Steinhagen-Thiessen, E.; Tsimikas, S. Mipomersen, an Antisense Oligonucleotide to Apolipoprotein B-100, Reduces Lipoprotein(a) in Various Populations with Hypercholesterolemia: Results of 4 Phase III Trials. *Arterioscler. Thromb. Vasc. Biol.* **2015**, *35*, 689–699. [CrossRef]

110. Hamilton, J.A.; Hillard, C.J.; Spector, A.A.; Watkins, P.A. Brain uptake and utilization of fatty acids, lipids and lipoproteins: Application to neurological disorders. *J. Mol. Neurosci.* **2007**, *33*, 2–11. [CrossRef]

111. Tracey, T.J.; Steyn, F.J.; Wolvetang, E.J.; Ngo, S.T. Neuronal Lipid Metabolism: Multiple Pathways Driving Functional Outcomes in Health and Disease. *Front. Mol. Neurosci.* **2018**, *11*, 10. [CrossRef]

112. Ebert, D.; Haller, R.G.; Walton, M.E. Energy contribution of octanoate to intact rat brain metabolism measured by 13C nuclear magnetic resonance spectroscopy. *J. Neurosci.* **2003**, *23*, 5928–5935. [CrossRef]

113. Hankin, J.A.; Farias, S.E.; Barkley, R.M.; Heidenreich, K.; Frey, L.C.; Hamazaki, K.; Kim, H.-Y.; Murphy, R.C. MALDI mass spectrometric imaging of lipids in rat brain injury models. *J. Am. Soc. Mass Spectrom.* **2011**, *22*, 1014–1021. [CrossRef] [PubMed]

114. Mulder, I.A.; Ogrinc Potocnik, N.; Broos, L.A.M.; Prop, A.; Wermer, M.J.H.; Heeren, R.M.A.; van den Maagdenberg, A.M.J.M. Distinguishing core from penumbra by lipid profiles using Mass Spectrometry Imaging in a transgenic mouse model of ischemic stroke. *Sci. Rep.* **2019**, *9*, 1090. [CrossRef] [PubMed]

115. Rapoport, S.I. Arachidonic acid and the brain. *J. Nutr.* **2008**, *138*, 2515–2520. [CrossRef] [PubMed]

116. Wu, H.; Liu, H.; Zuo, F.; Zhang, L. Adenoviruses-mediated RNA interference targeting cytosolic phospholipase A2alpha attenuates focal ischemic brain damage in mice. *Mol. Med. Rep.* **2018**, *17*, 5601–5610.

117. Liu, H.; Zuo, F.; Wu, H. Blockage of cytosolic phospholipase A2 alpha by monoclonal antibody attenuates focal ischemic brain damage in mice. *Biosci. Trends* **2017**, *11*, 439–449. [CrossRef]

118. Belayev, L.; Khoutorova, L.; Atkins, K.D.; Eady, T.N.; Hong, S.; Lu, Y.; Obenaus, A.; Bazan, N.G. Docosahexaenoic Acid therapy of experimental ischemic stroke. *Transl. Stroke Res.* **2011**, *2*, 33–41. [CrossRef]

119. Eady, T.N.; Belayev, L.; Khoutorova, L.; Atkins, K.D.; Zhang, C.; Bazan, N.G. Docosahexaenoic acid signaling modulates cell survival in experimental ischemic stroke penumbra and initiates long-term repair in young and aged rats. *PLoS ONE* **2012**, *7*, e46151. [CrossRef]

120. De Brito Toscano, E.C.; Silva, B.C.; Victoria, E.C.G.; de Souza Cardoso, A.C.; de Miranda, A.S.; Sugimoto, M.A.; Sousa, L.P.; de Carvalho, B.A.; Kangussu, L.M.; da Silva, D.G.; et al. Platelet-activating factor receptor (PAFR) plays a crucial role in experimental global cerebral ischemia and reperfusion. *Brain Res. Bull.* **2016**, *124*, 55–61. [CrossRef]

121. Belayev, L.; Eady, T.N.; Khoutorova, L.; Atkins, K.D.; Obenaus, A.; Cordoba, M.; Vaquero, J.J.; Alvarez-Builla, J.; Bazan, N.G. Superior Neuroprotective Efficacy of LAU-0901, a Novel Platelet-Activating Factor Antagonist, in Experimental Stroke. *Transl. Stroke Res.* **2012**, *3*, 154–163. [CrossRef]

122. Smyth, E.M.; Grosser, T.; Wang, M.; Yu, Y.; FitzGerald, G.A. Prostanoids in health and disease. *J. Lipid Res.* **2009**, *50*, S423–S428. [CrossRef]

123. Huang, H.; Al-Shabrawey, M.; Wang, M.-H. Cyclooxygenase- and cytochrome P450-derived eicosanoids in stroke. *Prostaglandins Other Lipid Mediat.* **2016**, *122*, 45–53. [CrossRef] [PubMed]

124. *The Role of Bioactive Lipids in Cancer, Inflammation and Related Diseases*, 1st ed.; Honn, K.V., Zeldin, D.C., Eds.; Springer International Publishing: Cham, Switzerland, 2019; ISBN 978-3-030-21636-8.

125. Nandakishore, R.; Yalavarthi, P.R.; Kiran, Y.R.; Rajapranathi, M. Selective cyclooxygenase inhibitors: Current status. *Curr. Drug Discov. Technol.* **2014**, *11*, 127–132. [CrossRef] [PubMed]

126. Trettin, A.; Bohmer, A.; Suchy, M.-T.; Probst, I.; Staerk, U.; Stichtenoth, D.O.; Frolich, J.C.; Tsikas, D. Effects of paracetamol on NOS, COX, and CYP activity and on oxidative stress in healthy male subjects, rat hepatocytes, and recombinant NOS. *Oxid. Med. Cell. Longev.* **2014**, *2014*, 212576. [CrossRef] [PubMed]

127. Ruan, K.-H.; Deng, H.; So, S.-P. Engineering of a protein with cyclooxygenase and prostacyclin synthase activities that converts arachidonic acid to prostacyclin. *Biochemistry* **2006**, *45*, 14003–14011. [CrossRef] [PubMed]

128. Ling, Q.-L.; Mohite, A.J.; Murdoch, E.; Akasaka, H.; Li, Q.-Y.; So, S.-P.; Ruan, K.-H. Creating a mouse model resistant to induced ischemic stroke and cardiovascular damage. *Sci. Rep.* **2018**, *8*, 1653. [CrossRef] [PubMed]

129. Liu, Q.; Liang, X.; Wang, Q.; Wilson, E.N.; Lam, R.; Wang, J.; Kong, W.; Tsai, C.; Pan, T.; Larkin, P.B.; et al. PGE2 signaling via the neuronal EP2 receptor increases injury in a model of cerebral ischemia. *Proc. Natl. Acad. Sci. USA* **2019**, *116*, 10019–10024. [CrossRef] [PubMed]

130. Frankowski, J.C.; DeMars, K.M.; Ahmad, A.S.; Hawkins, K.E.; Yang, C.; Leclerc, J.L.; Dore, S.; Candelario-Jalil, E. Detrimental role of the EP1 prostanoid receptor in blood-brain barrier damage following experimental ischemic stroke. *Sci. Rep.* **2015**, *5*, 17956. [CrossRef]

131. Zhou, Y.; Wei, E.-Q.; Fang, S.-H.; Chu, L.-S.; Wang, M.-L.; Zhang, W.-P.; Yu, G.-L.; Ye, Y.-L.; Lin, S.-C.; Chen, Z. Spatio-temporal properties of 5-lipoxygenase expression and activation in the brain after focal cerebral ischemia in rats. *Life Sci.* **2006**, *79*, 1645–1656. [CrossRef]

132. Chan, S.J.; Ng, M.P.E.; Zhao, H.; Ng, G.J.L.; De Foo, C.; Wong, P.T.-H.; Seet, R.C.S. Early and Sustained Increases in Leukotriene B4 Levels Are Associated with Poor Clinical Outcome in Ischemic Stroke Patients. *Neurotherapeutics* **2020**, *17*, 282–293. [CrossRef]

133. Shi, S.; Yang, W.; Tu, X.; Wang, C.; Chen, C.; Chen, Y. 5-Lipoxygenase inhibitor zileuton inhibits neuronal apoptosis following focal cerebral ischemia. *Inflammation* **2013**, *36*, 1209–1217. [CrossRef]

134. Silva, B.C.; de Miranda, A.S.; Rodrigues, F.G.; Silveira, A.L.M.; Resende, G.H.d.s.; Moraes, M.F.D.; de Oliveira, A.C.P.; Parreiras, P.M.; Barcelos, L.d.S.; Teixeira, M.M.; et al. The 5-lipoxygenase (5-LOX) Inhibitor Zileuton Reduces Inflammation and Infarct Size with Improvement in Neurological Outcome Following Cerebral Ischemia. *Curr. Neurovasc. Res.* **2015**, *12*, 398–403. [CrossRef] [PubMed]

135. Tu, X.; Yang, W.; Wang, C.; Shi, S.; Zhang, Y.; Chen, C.; Yang, Y.; Jin, C.; Wen, S. Zileuton reduces inflammatory reaction and brain damage following permanent cerebral ischemia in rats. *Inflammation* **2010**, *33*, 344–352. [CrossRef] [PubMed]

136. Fang, S.H.; Wei, E.Q.; Zhou, Y.; Wang, M.L.; Zhang, W.P.; Yu, G.L.; Chu, L.S.; Chen, Z. Increased expression of cysteinyl leukotriene receptor-1 in the brain mediates neuronal damage and astrogliosis after focal cerebral ischemia in rats. *Neuroscience* **2006**, *140*, 969–979. [CrossRef] [PubMed]

137. Chandrasekharan, J.A.; Sharma-Walia, N. Lipoxins: Nature's way to resolve inflammation. *J. Inflamm. Res.* **2015**, *8*, 181–192. [PubMed]

138. Wu, L.; Miao, S.; Zou, L.-B.; Wu, P.; Hao, H.; Tang, K.; Zeng, P.; Xiong, J.; Li, H.-H.; Wu, Q.; et al. Lipoxin A4 inhibits 5-lipoxygenase translocation and leukotrienes biosynthesis to exert a neuroprotective effect in cerebral ischemia/reperfusion injury. *J. Mol. Neurosci.* **2012**, *48*, 185–200. [CrossRef] [PubMed]

139. Wu, L.; Liu, Z.J.; Miao, S.; Zou, L.B.; Cai, L.; Wu, P.; Ye, D.Y.; Wu, Q.; Li, H.H. Lipoxin A4 ameliorates cerebral ischaemia/reperfusion injury through upregulation of nuclear factor erythroid 2-related factor 2. *Neurol. Res.* **2013**, *35*, 968–975. [CrossRef]

140. Wu, L.; Li, H.-H.; Wu, Q.; Miao, S.; Liu, Z.-J.; Wu, P.; Ye, D.-Y. Lipoxin A4 Activates Nrf2 Pathway and Ameliorates Cell Damage in Cultured Cortical Astrocytes Exposed to Oxygen-Glucose Deprivation/Reperfusion Insults. *J. Mol. Neurosci.* **2015**, *56*, 848–857. [CrossRef]

141. Powell, W.S.; Rokach, J. Biosynthesis, biological effects, and receptors of hydroxyeicosatetraenoic acids (HETEs) and oxoeicosatetraenoic acids (oxo-ETEs) derived from arachidonic acid. *Biochim. Biophys. Acta* **2015**, *1851*, 340–355. [CrossRef]

142. Usui, M.; Asano, T.; Takakura, K. Identification and quantitative analysis of hydroxy-eicosatetraenoic acids in rat brains exposed to regional ischemia. *Stroke* **1987**, *18*, 490–494. [CrossRef]

143. Wang, D.; Liu, Y.; Chen, L.; Li, P.; Qu, Y.; Zhu, Y.; Zhu, Y. Key role of 15-LO/15-HETE in angiogenesis and functional recovery in later stages of post-stroke mice. *Sci. Rep.* **2017**, *7*, 46698. [CrossRef]

144. Chen, L.; Zhu, Y.-M.; Li, Y.-N.; Li, P.-Y.; Wang, D.; Liu, Y.; Qu, Y.-Y.; Zhu, D.-L.; Zhu, Y.-L. The 15-LO-1/15-HETE system promotes angiogenesis by upregulating VEGF in ischemic brains. *Neurol. Res.* **2017**, *39*, 795–802. [CrossRef] [PubMed]

145. Ward, N.C.; Croft, K.D.; Blacker, D.; Hankey, G.J.; Barden, A.; Mori, T.A.; Puddey, I.B.; Beer, C.D. Cytochrome P450 metabolites of arachidonic acid are elevated in stroke patients compared with healthy controls. *Clin. Sci.* **2011**, *121*, 501–507. [CrossRef] [PubMed]

146. Yi, X.; Han, Z.; Zhou, Q.; Lin, J.; Liu, P. 20-Hydroxyeicosatetraenoic Acid as a Predictor of Neurological Deterioration in Acute Minor Ischemic Stroke. *Stroke* **2016**, *47*, 3045–3047. [CrossRef] [PubMed]

147. Tanaka, Y.; Omura, T.; Fukasawa, M.; Horiuchi, N.; Miyata, N.; Minagawa, T.; Yoshida, S.; Nakaike, S. Continuous inhibition of 20-HETE synthesis by TS-011 improves neurological and functional outcomes after transient focal cerebral ischemia in rats. *Neurosci. Res.* **2007**, *59*, 475–480. [CrossRef] [PubMed]

148. Dunn, K.M.; Renic, M.; Flasch, A.K.; Harder, D.R.; Falck, J.; Roman, R.J. Elevated production of 20-HETE in the cerebral vasculature contributes to severity of ischemic stroke and oxidative stress in spontaneously hypertensive rats. *Am. J. Physiol. Heart Circ. Physiol.* **2008**, *295*, H2455–H2465. [CrossRef] [PubMed]

149. Liu, Y.; Li, Y.; Zhan, M.; Liu, Y.; Li, Z.; Li, J.; Cheng, G.; Teng, G.; Lu, L. Astrocytic cytochrome P450 4A/20-hydroxyeicosatetraenoic acid contributes to angiogenesis in the experimental ischemic stroke. *Brain Res.* **2019**, *1708*, 160–170. [CrossRef]

150. Qu, Y.; Liu, Y.; Zhu, Y.; Chen, L.; Sun, W.; Zhu, Y. Epoxyeicosatrienoic Acid Inhibits the Apoptosis of Cerebral Microvascular Smooth Muscle Cells by Oxygen Glucose Deprivation via Targeting the JNK/c-Jun and mTOR Signaling Pathways. *Mol. Cells* **2017**, *40*, 837–846.

151. Qu, Y.-Y.; Yuan, M.-Y.; Liu, Y.; Xiao, X.-J.; Zhu, Y.-L. The protective effect of epoxyeicosatrienoic acids on cerebral ischemia/reperfusion injury is associated with PI3K/Akt pathway and ATP-sensitive potassium channels. *Neurochem. Res.* **2015**, *40*, 1–14. [CrossRef]

152. Bansal, S.; Sangha, K.S.; Khatri, P. Drug treatment of acute ischemic stroke. *Am. J. Cardiovasc. Drugs* **2013**, *13*, 57–69. [CrossRef]

153. Ahmad, M.; Zhang, Y.; Liu, H.; Rose, M.E.; Graham, S.H. Prolonged opportunity for neuroprotection in experimental stroke with selective blockade of cyclooxygenase-2 activity. *Brain Res.* **2009**, *1279*, 168–173. [CrossRef]

154. Pirlamarla, P.; Bond, R.M. FDA labeling of NSAIDs: Review of nonsteroidal anti-inflammatory drugs in cardiovascular disease. *Trends Cardiovasc. Med.* **2016**, *26*, 675–680. [CrossRef] [PubMed]

155. Stitham, J.; Midgett, C.; Martin, K.A.; Hwa, J. Prostacyclin: An inflammatory paradox. *Front. Pharmacol.* **2011**, *2*, 24. [CrossRef] [PubMed]

156. Gryglewski, R.J.; Nowak, S.; Kostka-Trabka, E.; Kusmiderski, J.; Dembinska-Kiec, A.; Bieron, K.; Basista, M.; Blaszczyk, B. Treatment of ischaemic stroke with prostacyclin. *Stroke* **1983**, *14*, 197–202. [CrossRef] [PubMed]

157. Zhao, J.; Zhang, X.; Dong, L.; Wen, Y.; Cui, L. The Many Roles of Statins in Ischemic Stroke. *Curr. Neuropharmacol.* **2014**, *12*, 564–574. [CrossRef] [PubMed]

158. Chen, Z.; Yang, J.; Zhong, J.; Luo, Y.; Du, W.; Hu, C.; Xia, H.; Li, Y.; Zhang, J.; Li, M.; et al. MicroRNA-193b-3p alleviates focal cerebral ischemia and reperfusion-induced injury in rats by inhibiting 5-lipoxygenase expression. *Exp. Neurol.* **2020**, *327*, 113223. [CrossRef] [PubMed]

159. Corser-Jensen, C.E.; Goodell, D.J.; Freund, R.K.; Serbedzija, P.; Murphy, R.C.; Farias, S.E.; Dell'Acqua, M.L.; Frey, L.C.; Serkova, N.; Heidenreich, K.A. Blocking leukotriene synthesis attenuates the pathophysiology of traumatic brain injury and associated cognitive deficits. *Exp. Neurol.* **2014**, *256*, 7–16. [CrossRef]

160. Shi, Q.J.; Wang, H.; Liu, Z.X.; Fang, S.H.; Song, X.M.; Lu, Y.B.; Zhang, W.P.; Sa, X.Y.; Ying, H.Z.; Wei, E.Q. HAMI 3379, a CysLT2R antagonist, dose- and time-dependently attenuates brain injury and inhibits microglial inflammation after focal cerebral ischemia in rats. *Neuroscience* **2015**, *291*, 53–69. [CrossRef]

161. Marumo, T.; Eto, K.; Wake, H.; Omura, T.; Nabekura, J. The inhibitor of 20-HETE synthesis, TS-011, improves cerebral microcirculatory autoregulation impaired by middle cerebral artery occlusion in mice. *Br. J. Pharmacol.* **2010**, *161*, 1391–1402. [CrossRef]

162. Renic, M.; Klaus, J.A.; Omura, T.; Kawashima, N.; Onishi, M.; Miyata, N.; Koehler, R.C.; Harder, D.R.; Roman, R.J. Effect of 20-HETE inhibition on infarct volume and cerebral blood flow after transient middle cerebral artery occlusion. *J. Cereb. Blood Flow Metab.* **2009**, *29*, 629–639. [CrossRef]

163. Liu, Y.; Wang, D.; Wang, H.; Qu, Y.; Xiao, X.; Zhu, Y. The protective effect of HET0016 on brain edema and blood-brain barrier dysfunction after cerebral ischemia/reperfusion. *Brain Res.* **2014**, *1544*, 45–53. [CrossRef]

164. Hawkins, K.E.; DeMars, K.M.; Alexander, J.C.; de Leon, L.G.; Pacheco, S.C.; Graves, C.; Yang, C.; McCrea, A.O.; Frankowski, J.C.; Garrett, T.J.; et al. Targeting resolution of neuroinflammation after ischemic stroke with a lipoxin A4 analog: Protective mechanisms and long-term effects on neurological recovery. *Brain Behav.* **2017**, *7*, e00688. [CrossRef]

165. Wu, Y.; Wang, Y.-P.; Guo, P.; Ye, X.-H.; Wang, J.; Yuan, S.-Y.; Yao, S.-L.; Shang, Y. A lipoxin A4 analog ameliorates blood-brain barrier dysfunction and reduces MMP-9 expression in a rat model of focal cerebral ischemia-reperfusion injury. *J. Mol. Neurosci.* **2012**, *46*, 483–491. [CrossRef] [PubMed]

166. Wang, S.-B.; Pang, X.-B.; Zhao, Y.; Wang, Y.-H.; Zhang, L.; Yang, X.-Y.; Fang, L.-H.; Du, G.-H. Protection of salvianolic acid A on rat brain from ischemic damage via soluble epoxide hydrolase inhibition. *J. Asian Nat. Prod. Res.* **2012**, *14*, 1084–1092. [CrossRef] [PubMed]

167. Zhang, Y.; Hong, G.; Lee, K.S.S.; Hammock, B.D.; Gebremedhin, D.; Harder, D.R.; Koehler, R.C.; Sapirstein, A. Inhibition of soluble epoxide hydrolase augments astrocyte release of vascular endothelial growth factor and neuronal recovery after oxygen-glucose deprivation. *J. Neurochem.* **2017**, *140*, 814–825. [CrossRef] [PubMed]

168. Li, R.; Xu, X.; Chen, C.; Yu, X.; Edin, M.L.; Degraff, L.M.; Lee, C.R.; Zeldin, D.C.; Wang, D.W. Cytochrome P450 2J2 is protective against global cerebral ischemia in transgenic mice. *Prostaglandins Other Lipid Mediat.* **2012**, *99*, 68–78. [CrossRef]

169. Liu, Y.; Wan, Y.; Fang, Y.; Yao, E.; Xu, S.; Ning, Q.; Zhang, G.; Wang, W.; Huang, X.; Xie, M. Epoxyeicosanoid Signaling Provides Multi-target Protective Effects on Neurovascular Unit in Rats After Focal Ischemia. *J. Mol. Neurosci.* **2016**, *58*, 254–265. [CrossRef]

170. Hackam, D.G.; Spence, J.D. Antiplatelet therapy in ischemic stroke and transient ischemic attack: An overview of major trials and meta-analyses. *Stroke* **2019**, *50*, 773–778. [CrossRef]

International Journal of
Molecular Sciences

Review

Lipid Mediators Regulate Pulmonary Fibrosis: Potential Mechanisms and Signaling Pathways

Vidyani Suryadevara [1], Ramaswamy Ramchandran [2], David W. Kamp [3,4] and Viswanathan Natarajan [2,5,*]

[1]   Department of Pathology & Laboratory Medicine, Indiana University School of Medicine, Indianapolis, IN 46202, USA; visurya@iu.edu
[2]   Departments of Pharmacology & Regenerative Medicine, University of Illinois, Chicago, IL 60612, USA; ramchan@uic.edu
[3]   Department of Medicine, Division of Pulmonary & Critical Care Medicine, Jesse Brown VA Medical Center, Chicago, IL 60612, USA; d-kamp@northwestern.edu
[4]   Department of Medicine, Northwestern University Feinberg School of Medicine, Chicago, IL 60611, USA
[5]   Department of Medicine, University of Illinois, Chicago, IL 60612, USA
*   Correspondence: visnatar@uic.edu; Tel.: +1-312-355-5896

Received: 15 May 2020; Accepted: 12 June 2020; Published: 15 June 2020

Abstract: Idiopathic pulmonary fibrosis (IPF) is a progressive lung disease of unknown etiology characterized by distorted distal lung architecture, inflammation, and fibrosis. The molecular mechanisms involved in the pathophysiology of IPF are incompletely defined. Several lung cell types including alveolar epithelial cells, fibroblasts, monocyte-derived macrophages, and endothelial cells have been implicated in the development and progression of fibrosis. Regardless of the cell types involved, changes in gene expression, disrupted glycolysis, and mitochondrial oxidation, dysregulated protein folding, and altered phospholipid and sphingolipid metabolism result in activation of myofibroblast, deposition of extracellular matrix proteins, remodeling of lung architecture and fibrosis. Lipid mediators derived from phospholipids, sphingolipids, and polyunsaturated fatty acids play an important role in the pathogenesis of pulmonary fibrosis and have been described to exhibit pro- and anti-fibrotic effects in IPF and in preclinical animal models of lung fibrosis. This review describes the current understanding of the role and signaling pathways of prostanoids, lysophospholipids, and sphingolipids and their metabolizing enzymes in the development of lung fibrosis. Further, several of the lipid mediators and enzymes involved in their metabolism are therapeutic targets for drug development to treat IPF.

Keywords: pulmonary fibrosis; lipid mediators; sphingolipids; sphingosine-1-phosphate; sphingosine kinase 1; prostaglandins; lysophosphatidic acid; autotaxin; G-protein coupled receptors; lysocardiolipin acyltransferase; phospholipase D; oxidized phospholipids

---

1. Introduction

Lipids are the principal constituents of cell membranes and play an essential role in several physiological and pathophysiological processes by mediating intracellular and extracellular cues. Phospholipids and sphingolipids, which are the structural components of the membranes, regulate cell shape, ion transport, intra- and inter-cellular communication and signaling. Many of the lipid-derived mediators are short-lived second messengers, and regulate cellular functions as migration, proliferation, apoptosis, redox balance, and cytoskeletal organization. Alteration or aberration in the generation of lipids mediators has been shown to regulate the physiology and pathophysiology of several disorders including, but not limited to, cancer, brain injury, cardiovascular diseases, kidney diseases, and pulmonary complications [1–6]. This review will specifically highlight the involvement of lipid mediators in pulmonary fibrosis (PF) and provide some mechanistic insights into the regulation of idiopathic pulmonary fibrosis (IPF) pathology by various lipid metabolites in animal models that mimic IPF.

IPF is a progressive fibrotic disease of the lung of unknown etiology that occurs in older adults, diagnosed as usual interstitial pneumonia with a clinicopathologic criteria [7], wherein the lung tissue becomes thickened from scarring [8–10]. The compromised architecture leads to disturbed gas exchange, decreased lung compliance, and respiratory failure and death [11]. Recurrent injury to the lung epithelium triggers pro-inflammatory and pro-fibrotic signaling involving the alveolar epithelial cells (AECs), alveolar macrophages (AM), fibroblasts, and endothelial cells contributing to the fibrotic foci and progression of IPF [12–15]. This pathogenesis is characterized by epithelial cell apoptosis, epithelial-to-mesenchymal cell-transition (EMT), endothelial-mesenchymal transition (EndMT), activation of fibroblasts which differentiates into contractile myofibroblasts leading to deposition of the extracellular matrix and scar tissue formation [12].

TGF-β is a critical cytokine that drives development of fibrosis. In mammals, three major isoforms of TGF-β have been identified, namely TGF-β1, -2, and -3 [16], and TGF-β1 is the predominant isoform expressed in lungs of IPF patients and preclinical models [17]. Activation of TGF-β1 binding to TGF-βRII via the SMAD2/3/4-dependent pathway leads to the fibrogenic program with extracellular matrix synthesis. TGFβ1-mediation of PF is recognized additionally via SMAD-independent non-canonical pathways [18]. Some of these known regulators include JNK kinase, MAPKinase, PI3K/Akt, and Rho kinase pathways, and their inhibitors are being targeted for clinical interventions in IPF [19]. Inhibition of bleomycin-induced PF and extracellular matrix (ECM) deposition was demonstrated by imatinib, a tyrosine kinase inhibitor, suggesting a crucial role for cAbl kinase [20]. The inhibition of the lectin, galecin-3, presumably derived from alveolar macrophages has also been shown to diminish bleomycin and TGFβ-induced fibrosis in a SMAD-independent manner [21]. Currently, IPF has only two drugs, Nintedanib and Pirfenidone, approved by the Food and Drug Administration for treatment [12]. Unfortunately, these drugs do not cure the disease, but only aid in slowing the progression of the disease. Thus, there is a crucial requirement for identifying new targets and signaling pathways that underlie the mechanisms behind IPF [22]. This review is specifically focused on the role of lipid-derived mediators and their signaling pathways modulating pulmonary fibrosis, in humans, and preclinical models. It is beyond the scope of this review to touch on all aspects of the disease since several recent reviews do justice in this context and will also sway the subject away from lipid signaling [9,10,12].

2. Plasma Lipid Profile in IPF Patients

Aberrations in phospholipids and sphingolipids metabolism have been identified as potential contributors to the pathophysiology of IPF. A recent lipidomics study revealed that several lipids were found to be altered in plasma of IPF patients. Several glycerophospholipids that were screened in this study were found to be lowered in IPF patients. 30 out of the 159 glycerolipids were distinct between control and IPF patients [23]. These altered lipid profiles in plasma can be exploited as potential biomarkers for IPF.

2.1. Fatty Acids and Fatty Acid Elongation in Pulmonary Fibrosis

Biosynthesis of palmitic acid (C16:0) and other long-chain saturated and unsaturated fatty acids is central to the generation of triglycerides, phospholipids, and sphingolipids with different fatty acid molecular species that dictate their function and metabolic fate. Cellular levels of free fatty acids are very low in tissues; however, elevated levels of palmitic acid were detected in the lungs of patients with IPF compared with control subjects [24]. Palmitic acid-rich high fat diet-induced epithelial cell death and a prolonged pro-apoptotic endoplasmic reticulum (ER) stress response after bleomycin-induced lung fibrosis. Palmitic acid accumulation in IPF lungs could be due to a defect in long-chain fatty acid family member 6 (Elovl6) enzyme that converts palmitoyl CoA to stearoyl CoA or a defect in stearoyl CoA desaturase that converts palmitic acid to palmitoleic acid (C16:1 n-7) or stearic acid to oleic acid (C18:1 n-9). The expression of *Elovl6* was decreased in lungs of patients with IPF and lungs of bleomycin-treated mice [25]. *Elovl6* depletion in LA-4 epithelial cells increased the cellular levels of palmitoleic acid (C16:1 n-7) and decreased stearic acid (C18:0) whereas the levels of linoleic acid (C18:2 n-6) and arachidonic acid (C20:4 n-6) were unchanged. Further, depletion of *Elovl6* with siRNA or exogenous addition of palmitic acid to LA-4 epithelial cells increased reactive oxygen species (ROS) and apoptosis, which was inhibited by oleic or linoleic acid, while ROS generation was elevated in *Elovl6$^{-/-}$* mice [25]. Similarly, deficiency of stearoyl CoA desaturase-1, which catalyzes conversion of saturated to monounsaturated fatty acid, induced ER stress, and experimental PF in mice [26]. These findings suggest that lipotoxicity due to accumulation of saturated fatty acids may have a detrimental role in the development of lung fibrosis in IPF and in animal models of lung fibrosis by inducing ER stress and apoptosis in AECs.

2.2. Nitrated Fatty Acids in Pulmonary Fibrosis

Nitrated fatty acids (NFAs) are produced by non-enzymatic reactions between nitric oxide (NO), unsaturated fatty acids such as oleic acid, and linoleic acid to generate 10-nitro-oleic acid (OA-NO2), and 12-nitrolinoleic acid (LNO2), respectively [27–30]. OA-N2 and LNO2 are the most abundant NFAs in human plasma [27], and both are physiological activators of the nuclear hormone receptor peroxisome-activated receptor γ (PPARγ), which exhibits tissue-protective and wound healing properties [31]. PPARγ agonists have been shown to exhibit antifibrotic activity in vitro [32], and in bleomycin-induced PF [33]. NFAs upregulated PPARγ and blocked TGF-β-induced fibroblast differentiation in vitro and administration of OA-NO2, in mice, post-bleomycin challenge ameliorated and reversed bleomycin-induced PF, suggesting therapeutic potential of NFAs in resolving PF [34].

2.3. Prostaglandins and Leukotrienes in Pulmonary Fibrosis

Prostaglandins and leukotrienes are lipid autacoids derived by the action of cyclooxygenases (COXs) 1 and 2 and lipoxygenases, respectively, that oxygenate and cyclize arachidonic acid released from membrane phospholipids such as phosphatidylcholine by the action of phospholipase (PL) A$_2$. The oxygenated arachidonic acid intermediate thus generated leads to production of prostaglandin (PG) E2 (PGE2), prostacyclin (PGI2), prostaglandin D2 (PGD2), prostaglandin F2α (PGF2α), thromboxane A2 and other eicosanoids catalyzed by a specific PG synthase [35]. Cytosolic PLA$_2$ has been shown to play a key role in generating proinflammatory eicosanoids, and depletion of cytosolic PLA$_2$ attenuated bleomycin-induced PF and inflammation by reducing production of leukotrienes and thromboxanes [36]. Since cytosolic PLA$_2$ is the rate-limiting enzyme in eicosanoid biosynthesis, there has been an ongoing push to develop specific inhibitors for cytosolic PLA$_2$ by researchers and Pharma companies. One such potent and selective inhibitor of cytosolic PLA$_2$, AK106-001616, reduced the disease score of bleomycin-induced lung fibrosis in rats [37], but it is unclear if this inhibitor has been clinically tested for its efficacy against IPF.

2.4. PGD2 in Pulmonary Fibrosis

Prostaglandin D2 (PGD2) is synthesized by hematopoietic PGD synthase (H-PGDS) in hematopoietic linage cells including mast cells and Th2 lymphocytes. Depletion of H-PGDS was found to accelerate bleomycin-induced PF and increase vascular permeability [38]. This was found to be predominantly mediated by inflammation, as seen by increased expression of H-PGDS in the neutrophils and monocytes/macrophages of an inflamed lung [38]. Supporting this, it was shown that retroviral injection of H-PGDS expressing fibroblasts in the lung attenuated bleomycin-induced lung injury and fibrosis [39]. Though the direct role of PGD2 was not shown in PF, the impact of PGDS and PGD receptor has been identified. Genetic depletion of chemoattractant receptor homologous with T-helper cell type 2 (CRTH2), a receptor for PGD2, in mice aggravated bleomycin-induced PF, as seen by prolonged inflammation and delayed resolution of fibrosis [40]. γδT cells expressing CRTH2 were found to be important in imparting protection against bleomycin-induced PF, when compared to splenocytes and other hematopoietic cells [40]. PGD2 also regulates fibroblast activity by attenuating TGFβ-induced collagen secretion via the DP receptor and the c-AMP pathway [41]. PGD2 was also found to mediate IPF by triggering of MUC5B gene expression in airway epithelial cells through the activation of ERK MAPK/RSK1/CREB pathway by binding to D-prostanoid receptor (DP1) [42]. MUC5B plays a key role in the development of honeycomb cysts seen in the lungs of IPF patients. Thus, activation of H-PGDS or depletion/inhibition of CRTH2 in specific cell types may be a therapeutic approach in ameliorating experimental PF in animal models. It is unclear if blocking PGDS or CRTH2 in IPF patients has any beneficial impact on human lung fibrosis.

2.5. PGE2 and PGE2 Signaling in the Pathogenesis of Pulmonary Fibrosis

Prostaglandin E2 (PGE2), a COX-2-derived eicosanoid, plays an important role in regulating homeostatic signaling between AECs and lung fibroblasts; however, evidence for its profibrotic role in animal models is controversial. PGE2 levels in BAL fluids from IPF patients and bleomycin-treated mouse lungs are lower compared to normal subjects and mice not challenged with bleomycin. While most of the in vitro studies show that COX-2 and PGE2 are anti-fibrotic, the in vivo data have been inconsistent in animal models of lung fibrosis. In a vanadium pentoxide-induced model of PF, mice lacking COX-2 exhibited higher inflammatory responses and lung fibrosis compared to controls or mice lacking COX-1 [43]. However, in another study, COX-2 deficient mice had exacerbated lung dysfunction but not fibrosis to bleomycin challenge [44], while a different study showed that $COX-2^{-/-}$ mice developed both losses of pulmonary function and severe lung fibrosis to bleomycin challenge [45]. Reasons for these discrepancies are unclear but could be due to differences in the sex and genetic background of the mice, as well as the dose and route of bleomycin administration. Further, there have been controversies over the efficacy of PGE2 vs. PGI2 in conferring protection against bleomycin-induced PF in animal models. Treatment with PGE2 in the lung was found to improve the bleomycin-induced decline in the lung function and the inflammatory responses in the lung caused by bleomycin. However, there was no change in the extent of bleomycin-induced fibrosis in the mice treated with PGE2 [45]. A robust protection against bleomycin-induced lung injury and fibrosis was achieved by a nanostructured lipid carrier-based delivery of PGE2 specifically to the lungs along with siRNA targeted against CCL2, HIF1α, and MMP2. This was found to reduce the tissue damage, inflammation, fibrotic markers in the lung, in addition to improving the mortality [46]. Liposomal instillation of PGE2 through inhalation, but not intravenous injection protected mice against bleomycin-induced inflammation, reduction in body weight, extent of fibrosis, and mortality rates [47]. Losartan, a selective AT1 receptor antagonist, which increases PGE2 levels in the lung was found to improve bleomycin-induced inflammation and also reduce the hydroxyproline content in the lungs [48]. However, using several in vivo genetic models it was demonstrated that prostacyclin (PGI2), but not PGE2, protected against the development of fibrosis and decline in lung function in response to bleomycin treatment [49]. Derivatives of PGE2 have been tested for protection against lung fibrosis. Administration of 16, 16-dimethy PGE2 (DM PGE2) protected mice against bleomycin-induced lung inflammation and PF [50]; however, the effect of DM PGE2 on pulmonary function was not examined. In addition to the direct role of PGE2 in modulating fibrosis, the PGE2 transporter, PGT/SLCO2A1, may also play a pathophysiological role in the development and progression of fibrosis. Extracellular PGE2 is taken up by cells via the high-affinity transporter *SLCO2A1*, which is expressed in the vascular endothelium, human and mouse airway bronchial and alveolar Type I and Type II epithelial cells [51]. Mice deficient in *Slco2a1* exhibited more severe fibrosis to bleomycin administration as characterized by exacerbated collagen deposition as compared to WT mice [51]. The mechanism for the protection conferred by SLCO2A1 is unclear and it is not known why blocking PGE2 transport from outside to inside the cell exacerbated the fibrosis in bleomycin-challenged mice. Several unanswered questions such as expression of PGE2 receptors (EP2 and EP4), levels of PGE2 and TGF-β, and PGE2 signaling in

Sclco2a1 deficient cells in mouse lung need to be addressed to define the potential mechanism(s) of regulation by the transporter.

The current understanding of potential mechanisms involved in PGE2-mediated attenuation of lung fibrosis and fibroblast to myofibroblast differentiation could involve but not limited to: (1) PGE2 deficiency in fibrotic lungs; (2) modulation of PGE2 signaling via EP1-4 receptors regulating apoptosis and proliferation; (3) epigenetic regulation of EP receptors in fibrotic lungs; (4) Interaction between plasminogen activation and PGE2 generation; and (5) Modulation of cross-talk between PGE2/EP2/EP4 and TGF-β/TGF-βR1/TGF-βR2 signal transduction (Figure 1).

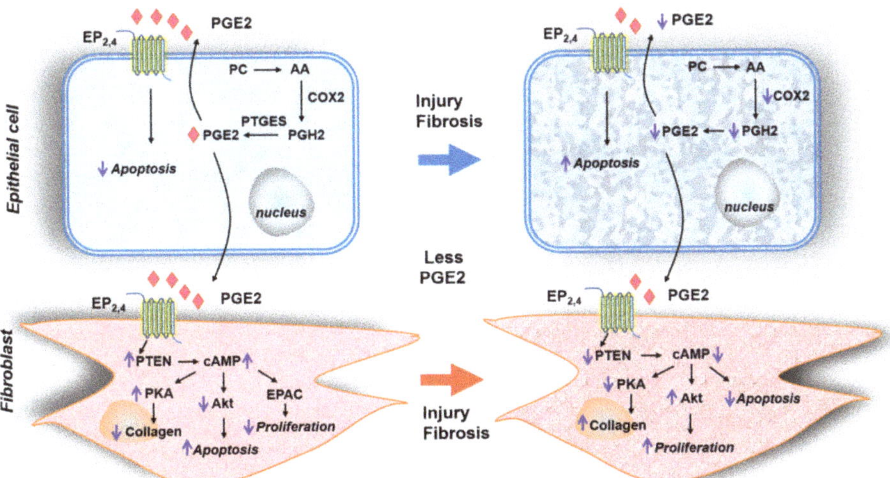

Figure 1. Prostaglandin E2 signaling via EP2/EP4 in epithelial cells and fibroblasts for the development of pulmonary fibrosis. Schema depicts PGE2 biosynthesis from arachidonic acid by COX2 and prostaglandin E synthases in alveolar epithelial cells. The autacoid function of PGE2 via its receptors EP2 and EP4 regulates homeostatic signaling between the alveolar epithelial cells (AECs) and pulmonary fibroblasts. Fibrotic signaling is specified by enhanced apoptosis and diminished secretion of PGE2 by AECs due to injury, which in turn promotes fibroblast proliferation, collagen deposition, and myofibroblast differentiation, distinct in pulmonary fibrosis. AA—Arachidonic acid, cAMP—cyclic adenosine monophosphate, COX2—Cyclooxygenase-2, EP$_{2,4}$—Prostaglandin EP$_2$,EP$_4$ receptor, EPAC—Exchange protein directly activated by cAMP, PC—Phosphatidyl choline, PGE2—Prostaglandin E2, PGH2—Prostaglandin H2, PKA—Protein kinase A, PTGES—Prostaglandin E Synthase, PTEN—Phosphatase and tensin homolog.

2.5.1. PGE2 Deficiency in Fibrotic Lungs

Reduced PGE2 levels have been reported in bronchoalveolar lavage fluid and conditioned culture media of alveolar macrophages (AMs) from IPF patients [52,53], which are consistent with reduced COX-2 expression in IPF lungs [54–56]. Reduced PGE2 synthesis has been observed in fibroblasts isolated from bleomycin-challenged rat lungs [57–59] due to diminished expression of COX-2. This reduction in COX-2 expression in fibroblasts from IPF lungs was attributed to the inability of transcriptional factors such as NF-κB/p65, CEBPb, and CREB-1 to bind to COX-2 promoter due to a defective H3 and H4 histone acetylation resulting from increased recruitment of HDACs and decreased HATs [54]. Injury to the AECs during lung fibrogenesis resulted in an elevated release of chemokine CCL2 [59], which reduced PGE2 production and stimulated fibrogenesis [60].

2.5.2. Modulation of PGE2 Signaling via EP1-4 Receptors Regulating Apoptosis and Proliferation

The cellular membrane receptors for PGE2 are termed EP receptors that consist of four different isoforms namely EP1, EP2, EP3, and EP4 generated from a primitive PGE receptor by gene duplication [61]. These four EP receptor isoforms show varying degrees of binding to PGE2 with EP3 and EP4 exhibiting high-affinity binding to the ligand [62]. In the normal lung, PGE2 secreted by AECs inhibits apoptosis of the cells in an autocrine fashion via EP2. However, PGE2 secreted by AECs limit fibroblast proliferation via EP2/EP4, activate PTEN, increase cAMP levels via adenylate cyclase that result in diminished fibroblast proliferation. Further, the ability of PGE2 to promote normal fibroblast apoptosis requires signaling through EP2/EP4 and reduction in protein kinase B (Akt) [63]. Akt is also negatively regulated by PGE2 by increased cAMP levels generated by PGE2 via EP2/EP4 and PTEN. PGE2 also limits myofibroblast differentiation in normal lungs by amplifying the inhibitory cAMP and PTEN [64]. cAMP generated by adenylate cyclase is rapidly degraded in cells by phosphodiesterase (PDE) 4, and blocking PDE4 that results in accumulation of cAMP induced by PGE2 also limits TGF-β-induced myofibroblast differentiation [65,66]. In contrast to the anti-proliferative effect of PGE2 in lung fibroblasts via EP2/EP4, PGE2 can promote proliferation of NIH 3T3 fibroblasts by promoting calcium mobilization [67]. Similarly, PGE2 stimulated neonatal rat ventricular fibroblast proliferation that was mimicked by sulprostone, an antagonist of EP1/EP3 [68]. Thus, PGE2 is a double-edged sword that can be antifibrotic or proliferative depending upon the type of EP receptor signal transduction and nature of the fibroblast investigated.

2.5.3. Epigenetic Regulation of EP Receptors in Fibrotic Lungs

In addition to diminished PGE2 production and signaling in fibrotic lungs, fibroblasts from IPF lungs [69], and mice with experimental fibrosis [70] showed resistance to antifibrotic activities of PGE2 due to decreased expression of EP2 (PTGER2). The decreased expression of EP2 could be due to increased degradation, decreased synthesis, or aberrations in DNA methylation. Studies carried out with IPF patients and two animal models of PF identified hypermethylation of the human *PTGER2* and mouse *Ptger2* promoters, respectively, containing abundant CpG dinucleotides susceptible to methylation [71]. Further, inhibition of DNA methylation with 5-aza-2′-deoxycytidine restored EP2 mRNA and protein expression and responsiveness to PGE2 in IPF lung fibroblasts. The increased *PTGER2* promoter methylation was mediated by increased Akt signaling and *PTEN* suppression in PGE2 resistant IPF lung fibroblasts [71]. Thus, epigenetic hypermethylation of EP2 provides a novel mechanism of conferring resistance to PGE2 signaling in IPF lung fibroblasts, and in experimental animal models of fibrosis.

2.5.4. Inter-relationship between Plasminogen Activation and PGE2 Production in Pulmonary Fibrosis

Plasminogen activation to plasmin protects lungs from fibrosis and patients with PF exhibit fibrin accumulation in the lungs due to increased expression of plasminogen activation inhibitor-1 (PAI-1) [72,73]. One potential mechanism underlying the antifibrotic effect of plasmin is through PGE2. Plasminogen activation upregulates PGE2 in AECs, fibroblasts, and fibrocytes from control and bleomycin-treated mice, and PGE2 production was exaggerated in lung fibroblasts, fibrocytes and AECs from *Pai1$^{-/-}$* mice compared to the cells from the control group. Further, it has been shown that plasmin stimulated PGE2 production in AECs, and fibroblasts involved the enzymatic release of HGF by plasmin and subsequent HGF-mediated upregulation of COX-2 [74]. Interestingly, PGE2 also modulates expression of the plasminogen activation system such as PAI-1 in non-lung cells [75,76]. These findings suggest an important inter-relationship between plasminogen activation to PGE2 and PGE2 stimulation of PAI-1 regulates lung fibrosis process.

2.5.5. Crosstalk between PGE2 and TGF-β Signaling in Fibroblast Differentiation

PGE2 is an antifibrotic lipid mediator and TGF-β is a multifunctional cytokine that drives fibrosis in IPF and experimental models of lung fibrosis. Evidence strongly suggests potential interaction and crosstalk between PGE2/EP2/EP4 and TGF-β/TGF-βR1/TGF-βR2 signaling pathways in regulating anti- and profibrotic cascades in lung fibroblasts. It has been reported that TGF-β1 induces PGE2, but not procollagen synthesis in human fetal lung fibroblasts that was pertussis-toxin sensitive [77]; however, other studies have demonstrated that TGF-β1 downregulates COX-2 expression leading to decreased PGE2 in human lung cancer A549 epithelial-like cells, which is involved in fibrotic response to TGF-β1 [78]. Several studies have shown that PGE2 antagonizes TGF-β signaling and responses in lung fibroblasts. PGE2 inhibited TGF-β1-induced fibroblast-to-myofibroblast differentiation that was SMAD-independent but involved cell shape and adhesion-dependent binding. PGE2 had no effect on TGF-β1-mediated SMAD phosphorylation or its translocation to the nucleus but diminished phosphorylation of paxillin, STAT-3, and FAK and limited activation of PKB/Akt pathway [79]. Interestingly, PGE2 not only prevented but also reversed the TGF-β1-induced myofibroblast differentiation as characterized by the ability of PGE2 to reverse expression of 368 genes upregulated and 345 genes down-regulated by TGF-β1 [80]. Moreover, PGE2 inhibited TGF-β-induced mesenchymal-epithelial transition [81] suggesting modulation of the injury-repair process. Many of the opposing effects of PGE2 to TGF-β response seem to involve changes in intracellular calcium. In fibroblasts isolated from normal lungs, PGE2 inhibited TGF-β-promoted [Ca^{2+}] oscillations and prevented the activation of Akt and Ca^{2+}/calmodulin-dependent protein kinase-II (CaMK-II) but did not prevent activation of Smad-2 or ERK. PGE2 also eliminated TGF-β-stimulated expression of collagen A1, and α-smooth muscle actin (α-SMA) [82]. However, as fibroblasts isolated from IPF lungs show resistance to TGF-β and PGE2 signaling, it is unclear if the expression of EP2/EP4 and TGF-β R1/TGF-βR2 are both modulated in these differentiated myofibroblasts. Further studies are necessary to define the precise interaction between these two signaling pathways in the regulation of fibrogenesis.

2.6. PGF2α and PGF2α Receptor in Pulmonary Fibrosis

Prostaglandin F2α (PGF2α) has been shown to be a pro-fibrotic eicosanoid that stimulated fibroblasts proliferation and collagen production in a TGF-β independent manner. In vivo studies revealed that signaling via its cognate receptor *Ptgfr* is involved in the development of lung fibrosis, with attenuation of bleomycin-induced fibrosis but not inflammation in *Ptgfr*$^{-/-}$ mice and was independent of the TGF-β pathway [83]. Pharmacological inhibition of TGF-βR1 kinase in *Ptgfr*$^{-/-}$ mice further inhibited lung fibrosis suggesting that PGF2α/PF pathway was signaling in a TGF-β independent manner. Further, PGF2α and TGF-β signaling were synergistic in mediating increases in proliferation and collagen synthesis by murine and human fibroblast cell lines. PGF2α-induced collagen gene transcription was Rho kinase-dependent; however, Rho-kinase was not involved in TGF-β-mediated collagen transcription. Additionally, bronchoalveolar lavage (BAL) fluid from IPF patients had high levels of PGF2α when compared to BAL fluid from patients with sarcoidosis [83].

The clinical relevance of PGF2α in PF was investigated by measuring levels of 15-keto-dihydro PGF2α, a major and stable metabolite of PGF2α in plasma of IPF patients. Plasma concentrations of 15-keto-dihydro PGF2α were significantly higher in IPF patients than controls, which correlated with forced expiratory volume in 1 s, forced vital capacity, diffusing capacity for carbon, the composite physiologic index, 6 min walk distance, and end-exercise oxygen saturation [84]. Thus, an association of plasma PGF2α metabolite, 15-keto-dihydro-PGF2α with disease severity of IPF and prognosis, supports a potential pathogenic role for PGF2α in human IPF.

2.7. Leukotrienes and Its Role in Pulmonary Fibrosis

Leukotrienes are immuno-regulatory lipid mediators primarily derived from arachidonic acid by 5-lipoxygenase (5-LO) [85]. 5-LO converts arachidonic acid to 5-hydroperoxy eicosatetraenoic acid

(5-HPETE), and then to leukotriene LTA_4. LTA_4 is converted to the dihydroxy fatty acid leukotriene LTB_4 or conjugated to glutathione to generate LTC_4. LTC_4 is converted to LTD_4 and LTE_4 by sequential peptidolytic cleavage [85]. The leukotrienes LTC_4, LTD_4, and LTE_4 are collectively known as cysteinyl leukotrienes (cysLT) and play an important role in airway diseases. Leukotrienes signal through G-protein coupled receptors (GPCRs), namely BLT1 and BLT2 for LTB_4, and CysLT1, CysLT2, and CysLTE, also known as gpr99, for cysLTs [86]. Leukotrienes, initially identified to be generated by leukocytes, are also produced by mast cells, eosinophils, macrophages, and inflammatory cells [85]. LTB_4 and LTC_4 levels are elevated in BAL and lung tissue lysates from IPF patients, suggesting constitutive activation of 5-LO in IPF [85,87]. AMs were the major source of increased LTB_4 and LTC_4 levels in IPF lungs [88]. In animal models such as bleomycin- and silica-induced PF, the tissue and BAL fluid content of cysLTs and LTB_4 was significantly elevated [89]. Furthermore, targeted disruption of 5-LO also attenuated bleomycin-induced injury as determined by reduction in the lung content of hydroxyproline [90]. Additionally, in the silica-induced model of PF, the expression of the LTB_4 receptor was increased while the expression of CysLT type 2 receptor was downregulated in lung tissue. Furthermore, strong immunohistochemical staining for the CysLT type 1 receptor, but not CysLT type 2receptor, was observed in pathological lesions [89]. These findings suggest that an increase in LT production in the lung and modulation of the cysLT receptors may contribute to the progression of PF. The importance of cysLT receptors in the pathogenesis of PF was further revealed by downregulation of LTC_4 synthase, a key enzyme in cysLT biosynthesis. Genetic deletion of LTC_4 synthase protected mice from bleomycin-induced alveolar septal thickening by macrophages and fibroblasts and collagen deposition [91]. In contrast, knockdown of the cysLT1 receptor significantly increased both, the concentration of cysLTs in BAL and the magnitude of septal thickening. These findings provide strong evidence for cysLT1 in regulating bleomycin-mediated lung fibrosis and a shift in the homeostatic balance from cysLT1 to cysLT2 to drive the fibrogenesis [92]. However, the clinical relevance of these pathways in patients with IPF and other forms of PF are unknown.

Cellular Senescence and Leukotriene Metabolism in Pulmonary Fibrosis

IPF is a lung disorder of the elderly population, and there is compelling evidence for aging-associated factors such as cellular senescence, telomerase attrition, and dysregulated metabolism in the pathogenesis of lung fibrosis. Accumulation of senescent cells in the fibrotic tissue has been linked to the severity of IPF, and earlier studies have shown that modulation of 5-LO and COX-2 in senescent fibroblasts [93,94]. In vitro, induction of senescence in human lung fibroblasts (IMR-90) using irradiation increased expression of phospho-5-LO/total 5-LO ratio and secretion of cysLTs [95]. Further, the cysLT-rich conditional medium of senescent IMR-90 fibroblasts induced pro-fibrotic signaling in naïve fibroblasts, which was abrogated by inhibition of 5-LO. In a preclinical model of lung fibrosis, pre-treatment with ABT-263 that selectively eliminates senescent cells attenuated bleomycin-induced PF [96]. Interestingly, senescent fibroblast from IPF lungs, but not normal lungs, secreted cysLTs and not the antifibrotic PGs_1 suggesting a role for senescent lung fibroblasts in contributing to a pool of cysLTs regulating the development of fibrosis [95].

3. Sphingolipids, Sphingolipid Metabolizing Enzymes, and S1p Receptors in Pulmonary Fibrosis

Sphingolipids are present across all eukaryotic cells and serve as structural and signaling lipids that regulate a variety of cellular functions under normal and pathological conditions. Advances in comprehensive lipid profiling techniques using LC-MS/MS and LC-ESI-MS/MS have identified close to 400 sphingolipidome in human plasma and tissues [97] that provide a better understanding of the functions, regulation and complex network of sphingolipids and sphingolipid-derived mediators in human health and diseases. Further, the sphingolipids and sphingolipid metabolites signal, both, intracellularly and extracellularly via G-protein coupled receptors. All sphingolipids share a common structural feature comprising of a long-chain sphingoid base such as sphingosine [(2S, 3R, 4E)-2-amino octadecc-4-ene-1,3 diol]. The -NH2 (amino) group on C2 is either free or linked to long-chain (C16–C18) or very long-chain (C20–C24) fatty acids with or without double bonds to generate ceramides. Ceramides can be further derivatized by the addition of a head group such as phosphorus, phosphocholine, glucose, galactose, or several sugar residues. Small amounts of sphingoid base containing a head group such as phosphorus, phosphocholine, or sugar units without the amide-linked fatty acid termed as "lyso sphingolipids" have been identified in biological fluids and tissues. As several excellent reviews describe the chemistry and biochemistry of sphingoid bases and sphingolipids [98–100], only a brief outline of the metabolism of sphingolipids is presented here.

Sphingomyelin (SM), the most abundant sphingolipid in mammalian cells, is hydrolyzed by three major sphingomyelinases (SMases), namely acidic, alkaline, and neutral to generate ceramide and phosphocholine [101]. Ceramide, a pro-apoptotic lipid molecule, promotes cell cycle arrest and regulates epithelial and endothelial apoptosis in the lung tissue [102]. Ceramide can also be phosphorylated to ceramide-1-phosphate (C1P) by ceramide kinase, which in contrast to ceramide has pro-survival function and also plays a role in the inflammatory process [103]. Ceramidases (acidic, alkaline and neutral) catabolize ceramide to sphingosine, a precursor of sphingosine-1-phosphate (S1P) [104,105].

Sphingosine, derived from ceramide or dihydrosphingosine, obtained by biosynthesis from palmitoyl CoA and serine in the de novo pathway, is phosphorylated by sphingosine kinase (SPHK) 1 or 2 to generate S1P [106]. S1P is dephosphorylated to sphingosine by S1P phosphatases (SPP1 or -2, encoded by SGPP1-2) [107] or lipid phosphate phosphatases (LPPs) [107]. However, S1P can be irreversibly degraded by S1P Lyase (encoded by *SGPL1*) to ethanolamine phosphate and Δ2-hexadecenal (Δ2-HDE) [106]. In addition to S1P, S1P lyase also hydrolyzes dihydro-S1P (DH-S1P) to hexadecanal and ethanolamine phosphate (Figure 2).

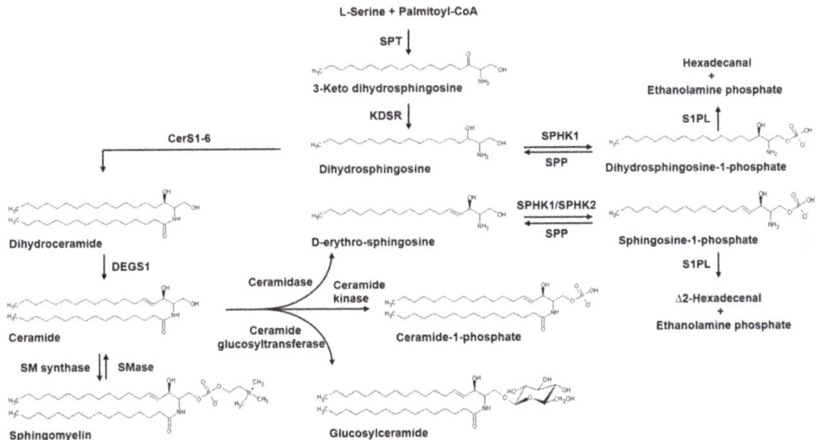

Figure 2. Sphingolipids implicated in pulmonary fibrosis, their structure, and metabolizing enzymes. An overview of de novo pathways of sphingolipid biosynthesis. Formation of the precursor 3-keto dihydrosphingosine from palmitoyl-CoA and L-serine catalyzed by SPT is the initial rate-limiting step, followed by generation of complex sphingolipids catalyzed by specific metabolic enzymes resulting in the generation of ceramides, sphingosine, sphingosine-1-phosphate (S1P), ceramide-1-phosphate and glucosyl ceramide. S1P or Dihydro S1P is hydrolyzed by S1PL to Δ2-hexadecenal or hexadecanal, respectively, and ethanolamine phosphate. S1P or dihydro S1P is also converted back to sphingosine by SPP. Dysregulation of the metabolic pathway intermediates is involved in the pathogenesis of pulmonary fibrosis. CerS1-6—Ceramide synthase 1-6, DEGS1—Delta 4-Desaturase Sphingolipid 1, KDSR—Keto dihydrosphingosine reductase, SPT—Serine palmitoyltransferase, SMase—Sphingomyelinase, SM synthase—Sphingomyelin synthase, SPHK1—Sphingosine kinase 1, SPHK2—Sphingosine kinase 2, S1P—Sphingosine-1-phosphate, SPP—Sphingosine-1-phosphate phosphatase, S1PL—Sphingosine-1-phosphate lyase.

In cells, Δ2-HDE is subsequently oxidized to *trans*-2-hexadecenoic acid followed by CoA addition to generate *trans*-Δ2-hexadecenoyl CoA, which is reduced to palmitoyl CoA by *Trans*-2-enoyl-CoA reductase (TER) in mammalian cells [108]. The palmitoyl CoA generated from Δ2-HDE is channeled to glycerophospholipids [108]; thus S1P lyase is a key enzyme in connecting the sphingolipid catabolism to glycerophospholipid metabolism. S1P is a pleotropic bioactive sphingolipid that is angiogenic and involved in several cellular functions and signals both intracellularly and extracellularly via G-protein coupled $S1P_{1-5}$ receptors on the plasma membrane of cells [109]. Ceramide can also be converted to glucosylceramide by ceramide glucosyltransferase [110] and other glycosphingolipids [111], and to ceramide-1-phosphate catalyzed by ceramide kinase [112,113]. Ceramide-1-phosphate is a naturally occurring bioactive sphingophospholipid involved in non-receptor mediated intracellular actions of cell proliferation, migration, and inflammation [114]. It is evident that sphingolipid metabolites generated from ceramide regulate a variety of cellular functions extracellularly via G-protein coupled receptors and intracellularly by non-receptor mechanisms.

3.1. Sphingomyelin and Sphingomyelinase in Pulmonary Fibrosis

Sphingomyelin, once considered to be an inert but essential membrane component, is now recognized as an important bioactive lipid and precursor for ceramide and sphingosylphosphocholine in mammalian cells [115]. Sphingomyelin is hydrolyzed by sphingomyelinase (SMase), a phosphodiesterase, to ceramide, which regulates a variety of cellular processes such as apoptosis, autophagy, senescence, infection, and inflammation [104]. At least five types of SMases have been identified and classified according to their cation dependence and pH optima. The five types of SMases are: lysosomal acid SMase; secreted zinc-dependent SMase; Mg^{2+}-dependent neutral SMase; Mg^{2+}-independent SMase; and alkaline SMase [116]. Among the SMases, the acid SMase (ASMase) has been widely studied as it is activated by stress and specific developmental cues that result in rapid generation of ceramide in the plasma membrane. Importantly, the ASMase exists in complex with acid ceramidase that cleaves ceramide to sphingosine. ASMase and its role in several human pathologies such as Farber disease [117], and cystic fibrosis [118] are known. Recent advances in the development of small molecule inhibitors to block ASMase to reduce inflammation in preclinical models of cystic fibrosis are encouraging but unclear for IPF and further studies are necessary to explore the therapeutic potential in lung fibrosis [119,120]. ASMase plays a role in experimental PF. Mice challenged with bleomycin to induce inflammation and PF showed increased activity of ASMase and acid ceramidase in lung tissue lysates, and deletion of ASMase in mice reduced bleomycin-induced lung inflammation, collagen deposition, and development of lung fibrosis [121]. Although these data suggest a role for ASMase in bleomycin-induced PF, it is unclear if ASMase expression and activity are altered in IPF lungs and in cells from IPF lungs, which require further investigation. Additionally, *SMPD1* the gene that encodes ASMase is subject to epigenetic regulation through methylation [122], which requires further investigation in IPF.

3.2. Ceramide Metabolism and Signaling in Pulmonary Fibrosis

In cells, ceramide could be generated by de novo biosynthesis, sphingomyelin degradation, synthesis from sphingosine and fatty acid, degradation of glucosyl- and galactosyl-ceramide and ceramide-1-phosphate [103]; however, the SMase mediated sphingomyelin degradation probably represents the major mechanism for intracellular ceramide production by cellular stimulation and stress [123]. Ceramide is central to sphingolipid pathways involved in several human diseases and accumulation of ceramide has been shown in pathologies such as COPD, cystic fibrosis, PF, ischemia/reperfusion injury and acute inflammation. Unlike cystic fibrosis [124] and COPD [125] definitive studies relating to ceramide levels in BAL fluid, plasma, or lung tissue to collagen deposition or fibrosis in IPF lungs have not been performed. In bleomycin-induced PF, changes in ceramide levels in BAL fluid and lung tissue were not significantly different from control mice, although S1P levels were elevated [126]. However, in the radiation model of pneumonitis and fibrosis, ceramide levels in the lung tissues were decreased at one-week post-irradiation but significantly increased at six weeks, which was also seen in the BAL fluid in late stages of radiation [127]. More importantly, the ratio of ceramide to S1P levels might be a better indicator of sphingolipid involvement in the pathology. In another study, it was found that mice exposed to ^{56}Fe radiation, the total lung ceramide levels were found to be elevated predominantly by the increase in palmitoyl fatty acid (C16:0)-containing ceramide molecular species [128]. While limited data are available on the sphingolipidome of lung tissues, plasma, or BAL fluid, metabolomics of IPF lungs showed dysregulated *SMPD1*, *SMPD4*, and *DEGS1* mRNA expression that point to aberrant ceramide production, whereas reduced mRNA expression of *ACER3* suggested dysregulated ceramide metabolism [129]. However, changes in mRNA levels of the dysregulated sphingolipid metabolizing enzymes were not validated by determining the protein levels in this study, as mRNA levels may not reflect true metabolomics change.

3.3. S1P Signaling Axis in the Pathophysiology of IPF and Animal Models of Pulmonary Fibrosis

S1P is the simplest bioactive sphingophospholipid that is present in circulating cells in the blood, biological fluids including plasma, BAL fluid, and in all the cell types in various organs. Cellular S1P levels are much lower (<0.5 μM) compared to plasma (0.5–1 μM) and tightly regulated by synthesis and degradation. About two-thirds of plasma S1P is bound to apolipoprotein M (apoM), which is a minor component of the High-density lipoprotein (HDL) particle [130–132]. In mammalian cells, S1P is generated by phosphorylation of sphingosine catalyzed by the lipid kinase sphingosine kinase (SPHK) 1 and 2 [133]. In addition to sphingosine, SPHK2 can also phosphorylate FTY720 a structural analog of sphingosine [134]. S1P is degraded to sphingosine by S1P phosphatases 1 and 2 [135], lipid phosphate phosphatases, and by S1P lyase to Δ2-hexadecenal and ethanolamine phosphate [109,136,137]. Platelets and erythrocytes have much higher levels of S1P as they lack S1P lyase [138]. S1P that is generated inside the cell is transported to outside by ABC transporters [139] and spinster homolog 2 (SPNS2) transporter [139]. S1P is a potent angiogenic factor and exhibits a plethora of effects on cellular functions [134] by binding to a family of G-protein coupled receptors S1P$_{1-5}$ present on the plasma membrane of cells. S1P signals intracellularly, independent of S1P$_{1-5}$, by binding to target proteins such as telomerase and HDACs [109], and modulates calcium homeostasis, regulates mitochondrial assembly and function by binding to prohibitin 2 [140], and is a modulator of BACE1 activity in Alzheimer's disease [141]. Additionally, S1P has been identified as a missing cofactor required for the E3 ligase activity of TNF receptor-associated factor 2 (TRAF2) [142]. However, in keratinocytes, deletion of *Traf2*, but not *Sphk1*, disrupted TNF-α-mediated NF-kB and MAPK signaling causing skin inflammation [143]. There is overwhelming evidence for protection as well as the detrimental effects of S1P in human diseases. In lipopolysaccharide (LPS)-induced and ventilator-induced lung inflammatory injury in mice, S1P levels in plasma [144], BAL fluid, and lung tissues were significantly lower compared to controls [106], and infusion of S1P was found to be beneficial in mouse and canine models of sepsis [145,146]. However, in lung disorders such as bronchopulmonary dysplasia (BPD) [147], pulmonary arterial hypertension (PAH) [148], asthma [149], experimental models of PF [150], and IPF [151] circulating and lung tissue levels of S1P were significantly elevated compared to controls, and reducing S1P levels by genetic deletion of *Sphk1* in mice or inhibition of SPHK1 by small molecule inhibitors conferred protection. Several excellent reviews have dealt with the role of S1P signaling in sepsis [152], asthma [153], and BPD [154]; therefore in this section, the role of S1P signaling in IPF and animal models of PF involving dysregulation of S1P metabolizing enzymes will be considered.

3.3.1. S1P Levels Are Altered in IPF and Animal Models of Pulmonary Fibrosis

There is only one study that describes S1P levels in IPF. S1P levels were increased in BAL fluid and serum from IPF patients [151], which correlated with lung function parameters such as diffusion capacity, forced expiratory volume, and forced vital capacity. In bleomycin- and radiation-induced mouse models of PF, sphingolipid levels were altered compared to control groups. S1P and DH-S1P levels were increased up to 4-fold in mouse lungs after 3, 7, and 14 days of post-bleomycin challenge [126,150]. The increase in S1P and DH-S1P levels in lung tissues on day 21 after bleomycin challenge were partly restored after administration of a SPHK1 inhibitor, SKI-II [126]. This decrease in S1P and DH-S1P levels correlated with protection against bleomycin-induced PF and mortality. Similarly, an increase in S1P and DH-S1P levels was detected in plasma, lung tissue, and BAL fluids 18 weeks after a 20 Gy thoracic irradiation of mice [155]. In the radiation model of PF, inhibition of sphingolipid de novo biosynthesis by targeting serine palmitoyltransferase with myriocin decreased levels of S1P and DH-S1P in mouse lung and plasma and delayed the onset of radiation-induced PF [155]. These studies show that S1P and DH-S1P levels are upregulated in human IPF and animal models of PF and blocking their production or enhancing their catabolism can be a therapeutic approach for fibrotic lung diseases.

3.3.2. SPHK1/S1P Signaling Promotes Lung Inflammation and Pulmonary Fibrosis

Increased S1P levels observed in plasma, BAL fluid, and lung tissues of IPF patients and in animal models of lung fibrosis could be attributed to modulation of its metabolism mediated by SPHKs, S1P lyase, and S1P phosphatases/lipid phosphate phosphatases. SPHK1, but not SPHK2, protein expression was increased in lung tissue lysates from IPF patients compared to control subjects, as well as in the murine model of bleomycin-induced lung inflammation and PF [126,151]. Microarray analysis of peripheral blood mononuclear cells (PBMCs) for mRNA expression of S1P synthesizing enzymes (SPHK1/2) negatively correlated with DLCO (diffusing capacity of the lung for carbon monoxide) in IPF, and Kaplan-Meier survival analysis comparing IPF with control groups demonstrated significantly reduced survival of patients with high expression of *SPHK1* or *SPHK2* [126]. The causative role of SPHK1 in PF was confirmed in genetically engineered mice lacking *Sphk1*. Bleomycin upregulated the expression of SPHK1 in lungs compared to WT mice, and genetic deletion of *Sphk1*, but not *Sphk2*, attenuated bleomycin-induced mortality, lung injury and collagen deposition in the lungs. Further, administration of SPHK1 inhibitor, SKI-II to mice attenuated bleomycin-induced lung inflammation and collagen deposition in lungs confirming a role for SPHK1-mediated S1P in the development of PF [126]. The development and progression of IPF and experimental PF show the involvement of both immune- and non-immune cells. Among the several lung cell types, AECs, fibroblasts, AMs, and endothelial cells have been implicated in the pathogenesis of lung fibrosis [15,156,157]. Genetic deletion of *Sphk1* in fibroblasts and AECs, but not endothelial cells, protected mice from bleomycin-induced lung fibrosis [158]. Additionally, a role for SPHK1 was also shown using two alternative models, namely the radiation-induced lung injury/pulmonary fibrosis (RILI/PF) and asbestos-induced lung fibrosis (AIPF) models. In the RILI/PF model, the expression of both SPHK1 and SPHK2 was elevated at 6 weeks in lung tissues [127]. Further, in this model, simvastatin augmented the expression of both the isoforms of SPHK and conferred protection from radiation-induced lung injury. Although the mechanisms underlying the differential effects of simvastatin on lung SPHK1 and SPHK2 expression are unclear, evidence of these changes in RILI/PF supports the idea that SPHK1 and SPHK2 could potentially serve as useful clinical biomarkers of lung inflammatory injury. In the asbestos-induced PF model, the SPHK1 inhibitor, PF-543 mitigated collagen deposition, and development of PF in mice [159]. Thus, SPHK1 appears to be a viable target to ameliorate the development of lung fibrosis in preclinical models; however, clinical trials to determine the efficacy of SPHK1 inhibitor(s) in treating IPF are required.

3.3.3. Dihydro S1P Signaling in Pulmonary Fibrosis

DH-S1P levels, similar to S1P are increased in lung tissues of bleomycin-challenged mice [150] and plasma, BAL fluid, and lung tissues of thoracic radiated mice [126,127]; however, DH-S1P levels in IPF patients have not been reported. While S1P is profibrotic and activates fibroblasts, DH-S1P seems to exhibit an antifibrotic property in fibroblasts. DH-S1P inhibited TGF-β-induced SMAD3 signaling and collagen upregulation in human foreskin fibroblasts through a PTEN/PPM1A-dependent pathway [160]. Additionally, S1P and DH-S1P showed opposing roles in the regulation of the MMP1/TIMP1 pathway in dermal fibroblasts [161,162]. DH-S1P is antifibrotic in scleroderma fibroblasts wherein PTEN protein levels were low that correlated with elevated levels of collagen and phospho-Smad3 and reduced levels of MMP1. DH-S1P treatment restored PTEN levels and normalized collagen and MMP1 expression, as well as SMAD3 phosphorylation. The distribution and function of S1P receptors differ in scleroderma and healthy fibroblasts, suggesting that alteration in sphingolipid signaling pathway may contribute to scleroderma fibrosis [163].

3.3.4. S1P lyase/S1P Signaling in IPF and Pulmonary Fibrosis

S1P lyase irreversibly hydrolyzes S1P to Δ2-HDE and ethanolamine phosphate and thus regulates intracellular S1P levels [109,136]. S1P lyase plays an important role in sepsis-induced inflammatory lung injury, where circulating and lung tissue S1P levels are lower compared to controls. Inhibition of S1P lyase in vivo increased circulating S1P levels and mitigated LPS-induced lung inflammation [144] and in vitro restored LPS-induced endothelial dysfunction. In contrast to the sepsis model, inhibition of *Sgpl1* or S1P lyase activity had an opposite effect on fibrogenesis. S1P lyase expression was upregulated in IPF lung tissues, primary lung fibroblasts isolated from patients with IPF and bleomycin-challenged mice. Knockdown of S1P lyase ($Sgpl1^{+/-}$) in mice augmented bleomycin-induced PF, and patients with IPF had reduced *Sgpl1* mRNA expression in PBMCs, exhibited higher severity of fibrosis and lower survival rate [164]. Thus, the sphingolipid metabolizing enzyme S1P lyase may be a potential target that would require in vivo activator(s) to reduce S1P levels in IPF patients.

3.3.5. Autophagy and S1P lyase/S1P Pathway in IPF and Pulmonary Fibrosis

Autophagy, an intracellular catabolic process triggered to remove aggregated/misfolded protein(s) and damaged organelles, plays a role in IPF. Lung tissues from IPF patients demonstrated decreased autophagic activity as assessed by LC3, p62 protein expression and immunofluorescence, and numbers of autophagosomes [165]. This inhibition of autophagy was attributed to TGF-β action on fibroblasts via activation of mTORC1 and increased expression of TIGAR. However, the role of S1P and S1P signaling in autophagy is controversial. Earlier studies indicate that S1P is an inducer [166–170] or inhibitor of autophagy [171,172]. Studies carried out in fibroblasts isolated from $Sgpl1^{+/-}$ mouse lung and overexpression of h*SGPL1* in HLFs clearly established a role for S1P in bleomycin-induced autophagy. Expression of beclin1, LC3, and the total number of autophagosomes in lung fibroblasts isolated from $Sgpl1^{+/-}$ mice were significantly lower than in WT controls, and chloroquine treatment increased autophagosome numbers in lung fibroblasts isolated from WT mice compared with those from $Sgpl1^{+/-}$ [164]. Similarly, transfection of HLFs with adenoviral construct of *hSGPL1* enhanced the expression of LC3 and beclin 1, and reversed TGF-β-induced decrease of LC3 expression and autophagosome formation [164]. Similar to TGF-β signaling, the S1P-induced mRNA and protein expression of FN, α-SMA, were also suppressed by overexpression of S1P lyase in HLFs. S1P challenge also attenuated LC3 expression and autophagosome formation, and overexpression of S1P lyase blocked S1P-induced attenuation of LC3 mRNA and protein expression [164]. Thus, it has been suggested that increased expression of S1P lyase in IPF lungs could represent a compensatory mechanism to partly counterbalance the TGF-β- and S1P-induced inhibition of autophagy, and the enhanced expression of S1P lyase serves as an endogenous suppressor of PF [164]. Autophagy inhibition might also mediate epithelial-mesenchymal transition and lung myofibroblast differentiation in IPF [173]. Thus, increased S1P levels could account for EMT of AEC in IPF [151].

3.3.6. Serine palmitoyltransferase Modulation of S1P Signaling and Pulmonary Fibrosis

The first and rate-limiting enzyme in the de novo biosynthesis of sphingolipids is serine palmitoyltransferase (SPT), a pyridoxal phosphate-dependent enzyme, which condenses serine and palmitoyl CoA to generate 3-ketosphinganine (3-keto dihydrosphingosine) that is subsequently converted to ceramide and sphingomyelin [174] (Figure 2). SPT is composed of two major subunits SPTLC1 and SPTLC2 that encode 53- and 63-kDa proteins, respectively [175]. Fungal metabolites such as sulfamisterin [176] and myriocin [177,178] have been employed to block SPT and interrogate the role of this enzyme on the overall sphingolipid metabolism in animals and pathologies. In a major study by Gorshkova et al., a single dose of myriocin decreased radiation-induced pulmonary inflammation and fibrosis and alleviated the dysregulated lung gene expression at 18 weeks post-radiation [155]. Additionally, myriocin inhibited the upregulation of S1P/DH-S1P levels and modified ceramide-sphingoid base molecular species levels in the irradiated animals. Myriocin also modulated the expression and/or activity of S1P metabolizing enzymes in the lung tissue. Myriocin treatment attenuated the radiation-induced increase in expression of SPHK1, SPT, and SPGL1, but not SPHK2; however, the activity of S1P lyase was not modulated compared to control animals. Myriocin elicited its effect by attenuating TGF-β-induced α-SMA and SPT2 expression and myofibroblast differentiation in HLFs [155]. Thus, the ability of myriocin to ameliorate radiation-induced lung inflammation and fibrosis suggests that SPT might be a novel therapeutic target in radiation-induced lung fibrosis.

3.3.7. S1P Receptors in Pulmonary Fibrosis

S1P signals intracellularly and extracellularly via five G-protein coupled receptors $S1P_{1-5}$. Additionally, S1P generated in the nucleus signals within the nucleus independent of $S1P_{1-5}$ and is involved in epigenetic regulation of pro-inflammatory genes activated by *Pseudomonas aeruginosa* in the lung epithelium [155]. The $S1P_{1-5}$ are coupled to G_i, G_α, G_o, G_q, and $G_{12/13}$ and this differential coupling dictates S1P signaling, in part, to various downstream effectors such as MAPKs, PI3K, Src, nMLCK, adenylate cyclase, phospholipase C (PLC), phospholipase D that result in a plethora of cellular responses [134,137]. A growing body of evidence suggests a role for $S1P_{2,3}$ in PF; however, the role of $S1P_1$, the predominant S1P receptor expressed in many mammalian cells, is unclear as complete deletion of *S1p1* is embryonically lethal. However, some light has been shed on the $S1P_1$ role in a vascular leak/lung fibrosis. In the bleomycin murine model, prolonged exposure of FTY720 (a non-selective $S1P_{1,3}$ modulator) and AUY954 (an $S1P_1$ selective modulator) caused a pulmonary leak in mouse lungs while low doses of bleomycin did not induce lung fibrosis. However, administration of either FTY720 or AUY954 along with low doses of bleomycin exacerbated vascular leak that was accompanied by intra-alveolar coagulation and development of extensive lung fibrosis in mice [179]. To understand the mechanism of FTY720 or AUY954 + bleomycin effect on the development of lung fibrosis in the context of a vascular leak, an in vivo thrombin coagulation-vascular leak model was investigated in the presence of bleomycin. Inhibition of thrombin-induced coagulopathy with an anticoagulant dabigatran, but not warfarin, attenuated lung fibrosis mediated by FTY720+low doses of bleomycin [180]. The FTY720 + bleomycin-induced vascular leak correlated with increased $\alpha_v\beta_6$ expression in the lung and thrombin inhibition with dabigatran diminished $\alpha_v\beta_6$ expression and activation of the TGF-β canonical signaling [180]. However, this in vivo FTY720 + bleomycin model does not explain the ability of FTY720 to enhance endothelial barrier function independent of $S1P_1$ in lung endothelial cells [181]. In contrast to FTY720, another analog, FTY720 (*S*)-phosphonate, that is non-hydrolyzable by lipid phosphate phosphatases, significantly inhibited bleomycin-induced alveolar-capillary leakage and inflammatory cell recruitment while FTY720 failed to confer protection against bleomycin-mediated lung inflammatory injury [182]. The mechanism of protection by FTY720 (*S*)-phosphonate is unclear, but it preserved expression of $S1P_1$ on the cell surface while FTY720 allowed $S1P_1$ internalization and recycling in endothelial cells.

$S1P_2$ and $S1P_3$ have been shown to be pro-inflammatory and profibrotic in the bleomycin model of lung fibrosis. Genetic deletion of *S1pr2* or pharmacological inhibition of S1PR2 alleviated bleomycin-induced PF [183]. Bone marrow chimera experiments showed that bone marrow-derived cells contributed to the development of lung fibrosis, and depletion of macrophages also reduced bleomycin-induced lung fibrosis. Further, bleomycin challenge increased expression of pro-inflammatory cytokines such as IL-4 and IL-13, which were diminished in *S1pr2*-deleted mice [183]. A similar role for *S1pr2* in the development of bleomycin-induced lung fibrosis was demonstrated using

S1pr2 KO mice and JTE-013 [184], a pharmacological inhibitor of S1PR2 in mice [184]. Knockdown of *S1pr3* attenuated bleomycin-induced lung inflammation and PF in mice without changing TGF-β levels, but by reducing connect tissue growth factor (CTGF) [185]. S1PR1 and S1PR3 agonists such as FTY720-phosphate and Ponesimod, but not SEW2871, caused a robust stimulation of ECM synthesis and expression of pro-fibrotic genes including CTGF [186]. Depletion of *S1PR2*, or *S1PR3*, but not *S1PR1*, in HLFs attenuated Rho activity that is closely associated with fibrosis and differentiation of myofibroblasts [187]. Thus, both *S1PR1* and *S1PR3* could be therapeutic targets in PF. Interestingly, Fingolimod (FTY720) at 1.25 mg and 5.0 mg daily dose given to patients with relapsing multiple sclerosis showed a reduction in pulmonary function [188] as observed in IPF patients.

3.3.8. Interaction between TGF-β and S1P Signaling in the Pathogenesis of Pulmonary Fibrosis

TGF-β is a critical cytokine that drives the development of IPF and PF in animal models. TGF-β promotes EMT in AECs [189,190], fibroblast to myofibroblast transdifferentiation [191], and EndMT, which are key pathways involved in the development of lung fibrosis. TGF-β expression is increased in the lungs of IPF patients and preclinical models of lung fibrosis [17,192,193]. TGF-β binds to TGF-βRII and initiates binding to and phosphorylation of TGF-βRI. This triggers recruitment of SMAD2/3 to the cytoplasmic domain of TGF-βRI, phosphorylates SMAD2/3, SMAD2/3 forms a trimer with SMAD4, which translocates to the nucleus where it binds to SMAD-binding elements in promoter regions to modulate gene transcription [194]. Conditional deletion of TGF-βRII in lung epithelial cells protected mice from bleomycin-induced fibrosis, but these mice developed emphysema-like phenotypes [195,196]. Similarly, *Smad3*-deficient mice showed alveolar destruction resembling emphysema but developed lung fibrosis in response to bleomycin [197]. In addition to the canonical pathway of signaling via SMAD, TGF-β can signal via the non-canonical pathways such as MAPKs, PI3K/Akt, and Rho-GTPase to modulate AECs and fibroblasts to stimulate a fibrotic phenotype [18,198]. Interactions between TGF-β/TGF-βR signaling with serotonin, integrins, sGC-cGMP-PKG, S1P/S1PRs, and lysophosphatidic acid (LPA)/ LPA receptors (LPARs) have been demonstrated. There is overwhelming evidence indicating the importance of SPHK1/S1P signaling in TGF-β-mediated fibroblast to myofibroblast differentiation. In HLFs, downregulation of *Sphk1* with siRNA or inhibition of SPHK1 activity with inhibitors decreased α-SMA and fibronectin expression upregulated by TGF-β [126,150,187]. Further, knocking down $S1P_2$ and $S1P_3$, but not $S1P_1$, with siRNA reduced TGF-β-induced α-SMA expression, and blocking SPHK1 had no effect on SMAD2/3 phosphorylation [187,199]. In the murine model of lung fibrosis, bleomycin-induced TGF-β secretion and phosphorylation of SMAD2/3, AKT, JNK1, and p38 MAPK were mitigated in *Sphk1* deleted or SPHK1 inhibitor administered mice, demonstrating the interaction between the TGF-β and SPHK1/S1P signaling axis in vivo [126,150]. Both in vivo and in vitro, TGF-β upregulated SPHK1 and enhanced S1P levels and exogenous addition of S1P antibody to fibroblasts prevented α-SMA and fibronectin increase [126,150], further confirming the release of S1P from the cell for its extracellular responses. Similarly, TGF-β-induced EMT in A549 epithelial cells was partially dependent on SPHK1/$S1P_{2,3}$ signaling axis [151]. TGF-β increased the expression of both SPHK1 and S1P lyase in HLF, which was blocked by anti-TGF-β neutralizing antibody and *SMAD3* siRNA [164]. The transcriptional activity of *SMAD3* in regulating *SGPL1* expression induced by TGF-β was verified by ChIP assay as well as *hSGPL1* luciferase promoter activity in HLF [164]. The transcriptional regulation of *SPHK1* by SMAD3 or other transcriptional factors in response to TGF-β in normal and IPF lung fibroblasts needs to be further investigated. The TGF-β-mediated fibroblast differentiation to myofibroblast was attenuated by overexpression of *SGPL1* in HLF and is attributed to the decrease in intracellular S1P level, increase in expression of LC3 and Beclin1, and autophagy [164]. Thus, understanding the balance in expression of SPHK1 and S1P lyase in lung tissues of IPF and animal models will provide a handle to S1P signaling in the progression or resolution of PF (Figure 3).

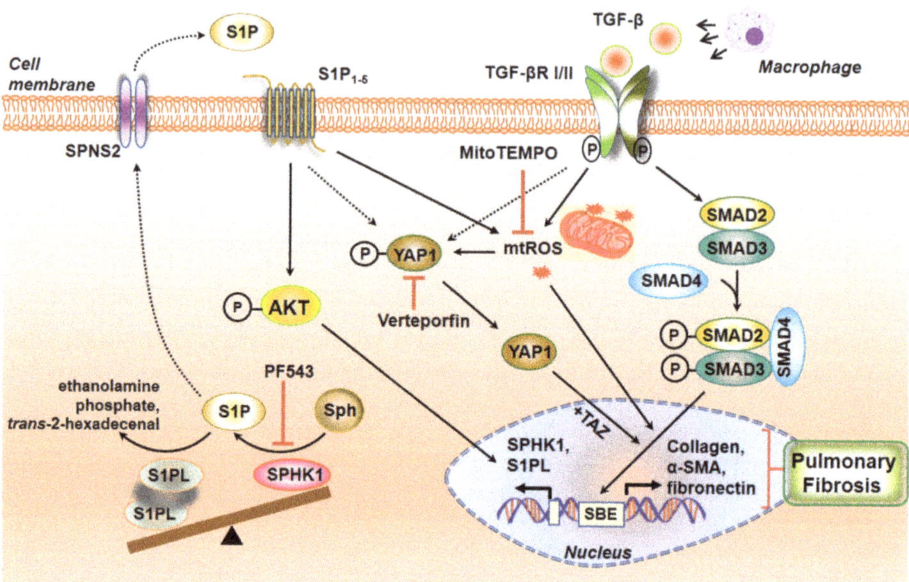

Figure 3. Crosstalk between S1P and TGF-β signaling cascade in pulmonary fibrosis. Profibrotic signaling by the inflammatory cytokine TGF-β is affected through the dimerization of TGF-β1 and TGF-βII receptors and its phosphorylation and activation of SMADs to induce a fibrogenic transcriptional program. Transport of S1P generated by SPHK1 activation from the cell to the extracellular milieu is enabled by SPNS2 transporter. S1P binding to S1P receptors stimulates mitochondrial reactive oxygen species (mtROS) generation and YAP1 translocation to the nucleus to influence myofibroblast transdifferentiation and matrix remodeling. AKT—Protein kinase B, αSMA—smooth muscle αActin, MitoTEMPO—(2-(2,2,6,6-Tetramethylpiperidin-1-oxyl-4-ylamino) -2-oxoethyl)triphenylphosphonium chloride, PF543—SPHK1 inhibitor, SBE—SMAD-Binding Element, SPNS2—spinster homolog 2, Sph—Sphingosine, SPHK1—Sphingosine kinase 1, S1P—Sphingosine -1-phosphate, S1PL—Sphingosine-1-phosphate lyase, S1P$_{1-5}$—Sphingosine-1-phosphate receptors 1–5, SMAD—Mothers Against Decapentaplegic Homolog, TGF-β—Transforming growth factor beta, TGF-β RI/II—Transforming growth factor beta receptor I/II, TAZ—Tafazzin, YAP1—Yes-associated Protein 1.

3.3.9. SPHK1/S1P Signaling and Mitochondrial ROS in Pulmonary Fibrosis

The pathophysiology of IPF is characterized by increased ROS production and an imbalance in redox status of the lung tissue [200,201]. The two major sources of ROS in cells are from mitochondrial oxidative phosphorylation and activation of NADPH oxidases (NOXs) [202,203]. Altered mitochondrial homeostasis that includes increased mitochondrial ROS (mtROS), impaired respiration, mtDNA damage, compromised mitochondrial dynamics and mitophagy have been reported in epithelial cells and fibroblasts from healthy aged lungs, IPF lungs and lungs from murine models of bleomycin- and asbestos-induced PF, resulting from upregulation of NADPH Oxidase (NOX) 4, and increased mtROS and mtDNA damage [204–207]. TGF-β has been shown to induce mtROS through decreased Complex IV activity in senescent cells [208,209], and recent studies suggest that SPHK1/S1P signaling stimulates NOX2-dependent ROS generation in lung endothelial cells [137,210–212], and mtROS in lung fibroblasts [158]. TGF-β stimulated mtROS in HLF that was dependent on SPHK1 expression and activity as well as YAP activation. Inhibition or downregulation of SPHK1, and YAP1 activity or expression reduced TGF-β mediated mtROS and scavenging mtROS with MitoTEMPO attenuated TGF-β-dependent expression of fibronectin and α-SMA, demonstrating a definitive role for mtROS in fibroblast differentiation. Further, genetic deletion of Sphk1 in mouse lung fibroblasts or inhibition of

SPHK1 with PF543 (a specific inhibitor of SPHK1) reduced bleomycin-induced YAP1 co-localization with FSP-1 in fibrotic foci rich in fibroblasts [158]. This study demonstrated a role for SPHK1/S1P signaling in TGF-β-induced YAP1 activation that is essential for mtROS generation and expression of fibronectin and α-SMA in fibroblasts. How SPHK1 activates YAP1 in HLFs is unclear. It is important to determine if both SPHK1 and YAP1 are translocated to the mitochondrial outer membrane to initiate the process of mtROS production in response to a stimulus such as TGF-β. In confluent cells, YAP1 is primarily phosphorylated and localized in the cytosol, and upon activation gets dephosphorylated by phosphatases including PTPN14 [213] and translocates to the nucleus and functions as a co-transcriptional regulator with TAZ [214,215] (Figure 3).

4. Phospholipids and Phospholipid Metabolizing Enzymes in Pulmonary Fibrosis

Phospholipids are essential components of all biological membranes and metabolic dysregulation of phospholipids have been shown to contribute to the pathogenesis of several pulmonary disorders including IPF and experimental PF. Lipidomics of IPF lungs or lungs of animal models have not been performed; however, phospholipid content in the BAL fluids was reduced in IPF lungs and mouse lungs of animal models. Aberrant phospholipid metabolism by phospholipid metabolizing enzymes such as PLA_2, PLC, PLD, and lyso PLD or autotaxin (ATX) generate bioactive lipid molecules that play key roles in the development and progression of lung fibrosis [216]. Further, metabolism of cardiolipin, a major mitochondrial phospholipid, also seems to contribute to the pathogenesis of IPF. The role of arachidonic acid-derived prostanoids and leukotrienes on lung fibrosis have been dealt with in the earlier section and here we will review the recent developments on fatty acid, PLD/PA, ATX/LPA, cardiolipin metabolism by lysocardiolipin acyltransferase (LYCAT) and oxidized phospholipids in the pathophysiology of lung fibrosis.

4.1. Surfactant Lipids and Surfactant Proteins in IPF and Pulmonary Fibrosis

The lung surfactants, in addition to dipalmitoyl phosphatidylcholine, dipalmitoyl phosphatidylglycerol and phosphatidylinositol also include four surfactant proteins, SP-A, SP-B, SP-C and SP-D, which act as a defense mechanism against toxic pathogens or microbes trying to invade the lung. SP-A and SP-D serve as biomarkers of IPF, as seen by their ability to predict the survival rates in IPF patients. In the BAL fluid, the levels of phosphatidylglycerol were lower while the contents of phosphatidylinositol and sphingomyelin were elevated, and the dysregulated phospholipid levels correlated with the severity of the disease [217–220]. Administration of a natural bovine lung extract neutral lipids, phospholipids enriched with phosphatidylethanolamine, fatty acids, and surfactant proteins attenuated bleomycin-induced lung fibrosis and soluble collagen levels in mouse lungs [221]. This protective role of the administered surfactant was attributed to the inclusion of phosphatidylethanolamine in the surfactant preparation. SP-D levels were found to be elevated in the serum of patients with radiographic abnormalities during IPF [222], whereas elevated levels of SP-A and SP-D were seen in IPF patients and also systemic sclerosis patients when compared to normal individuals [223]. Interestingly, another study showed that SP-D levels were increased only in patients who were diagnosed in late stages of IPF and could be used as a biomarker for progression of disease during anti-fibrotic treatments [222]. SP-B levels were increased in patients with IPF [224,225], and BAL fluid analysis showed that SP-A/phospholipid ratio in the BALF was lower in IPF patients and it could be used a predictive marker for survival in these patients [225]. Mutations in SP-C were seen in familial and sporadic cases of PF. The levels of SP-B were decreased, and SP-D was increased during bleomycin-induced fibrosis in the mice, but there was no change in the levels of SP-A and SP-C [226]. SP-C deficient mice were found to have increased infiltration of inflammatory cells, lung architectural distortion, and increased collagen deposition. There was also delayed resolution of fibrosis in these mice as seen by sustained apoptosis of the lung parenchymal cells and prominent fibrosis in the centriacinar and sub-pleural regions containing fibroblasts, collagen fibrils, and damaged interalveolar septa. Mitochondrial fusion related proteins such as Mitofusin 1 (Mfn1) and Mitofusin 2 (Mfn2) regulate surfactant production in alveolar type II epithelial cells by playing a key role in lipid metabolism which in turn regulates fibrosis in the lungs. Loss of Mfn1 and Mfn2 in AEC in mice was found to promote bleomycin-induced lung fibrosis in a recent study [227]. More interestingly, fibrosis was spontaneously induced in mice with both, Mfn1 and Mfn2 deletion in AECs. AEC cells from *Mfn1*$^{-/-}$ and *Mfn2*$^{-/-}$ mice challenged with bleomycin were found to have altered lipid metabolism of cholesterol, ceramides, phosphatidic acids, phosphatidylcholine, phosphatidylethanolamine, phosphatidylserine, plasmalogen phosphatidylethanolamine, and decreased SP-B, SP-C gene expression [227]. Mfn1/2-deficient AECs showed no changes in the SP-B, SP-C gene expression, but with altered lipid profile by changes in mono-acylglycerol, diacylglycerol, acylcarnitine, cholesterol, phosphatidylserine, and phosphatidylglycerol levels. Thus, mitochondrial dynamics seems to have an impact on the altered lipid metabolism in AECs due to bleomycin-induced lung injury.

4.2. Phospholipase D/Phosphatidic acid Signaling Axis in Development of Pulmonary Fibrosis

PLD hydrolyzes phosphatidylcholine to phosphatidic acid (PA), and choline [228–231]. It can degrade other phospholipids such as phosphatidylethanolamine and phosphatidylserine to PA and ethanolamine or serine, respectively. In addition to the phosphohydrolase activity, it has a transphosphatidylation activity where the PA is transferred to primary short-chain alcohols such as methanol, ethanol, butanol, and propanol, but not secondary or tertiary alcohols, generating the corresponding phosphatidylalcohol [230]. There are six isoforms of PLD, PLD1-6, of which PLD1 and PLD2 isoforms exhibit the ability to hydrolyze phospholipids and these two isoforms have been widely recognized in several human pathophysiologies including cancer, hypertension, neurodisorders, diabetes, and acute lung injury [228–231]. PA is a bioactive lipid second messenger, which is further converted to diacylglycerol (DAG) or lyso-PA (LPA) by PA phosphatase [232,233] or PA-specific PLA$_1$/PLA$_2$ [234,235], respectively. PLD mediated PA generation is involved in the regulation of various cellular processes including cell survival, cell migration, cell proliferation, differentiation, cytoskeletal changes, membrane trafficking, and autophagy [230,236,237]. PLD was activated by bleomycin in lung endothelial cells and led to reactive oxygen species generation [238,239]. PLD activation and PA generation induced by bleomycin in bovine lung ECs were significantly attenuated by the thiol protectant (*N*-acetyl-L-cysteine), antioxidants, and iron chelators suggesting the role of ROS, lipid peroxidation, and iron in the process. This study revealed a novel mechanism of the bleomycin-induced redox-sensitive activation of PLD that led to the generation of PA, which was cytotoxic to lung ECs, thus suggesting a possible bioactive lipid-signaling mechanism of microvascular disorders encountered in PF [238]. A more recent study showed that PLD2, but not PLD1, expression was elevated in lung tissues of IPF patients compared to control subjects and in lung tissues of mice with bleomycin-induced PF [239]. Both *Pld1*$^{-/-}$ and *Pld2*$^{-/-}$ deficient mice were protected against bleomycin-induced lung inflammation and fibrosis, thereby establishing the role of PLD in fibrogenesis. Further, bleomycin stimulated mitochondrial superoxide production, mtDNA damage, and apoptosis, which was attenuated by the catalytically inactive mutants of PLD1 or PLD2, downregulation of PLD2 expression with siRNA or inhibition of PLD1 and PLD2 with specific small molecule inhibitors [239]. The downstream targets of PLD/PA signaling in the activation of mtROS and fibroblasts are unclear; however, PA is known to stimulate SPHK1 [239], IQGAP1 via RAC1 [240], mTOR [241], and modulates mitochondrial function and dynamics [242]. In addition to PLD1 and PLD2, a novel role for PLD4 in the development of kidney fibrosis has been reported. Genetic deletion of PLD4, a transmembrane glycoprotein and an isoform of PLD that lacks any enzyme activity, in kidney tubular epithelial cells attenuated development of kidney fibrosis [243]; however, the mechanism of protection in the absence of PA generation is unknown.

4.3. Diacylglycerol Kinase in Radiation-Induced Fibrosis

Diacylglycerols are lipid intermediates generated de novo in cells for the biosynthesis of triglycerides (TGs) and phospholipids and a product of PLC and PLD pathways. DAGs are physiological and endogenous activators of protein kinase C (PKC) conventional (α, β, γ) and novel (δ, ε, θ, η) isoforms that are known regulators of pleiotropic downstream signaling cascades in cells. DAGs do not accumulate and are converted to PA by DAG kinase (DAGK) or TGs by acyltransferases. Of the ten isoforms of DAGK, DAGK-α has been shown to promote radiation-induced fibrosis [244]. Radiation-induced transcription of DAGK-α in cells was facilitated by profibrotic transcription factor early growth response 1 and DNA methylation profiling showed a hypomethylation pattern [245]. In dermal fibroblasts isolated from breast cancer patients, DAGK-α regulated TGF-β-induced profibrotic activation and lipid signaling as evidenced by siRNA downregulation of the enzyme or inhibition by R59949 [246]. In addition, activation of Collagen 1A mRNA expression in fibroblasts by ionizing radiation or bleomycin was attenuated by DAGK-α siRNA or the inhibitor, R59949. Inhibition of DAGK-α increased the accumulation of 7 molecular species of DAG and reduced the accumulation of specific PA and LPA molecular species [245]. In contrast to the radiation model of fibrosis, a 4-fold upregulation of C18:1/C24:2 DAG species was observed in lung tissues from bleomycin-challenged mice; however DAGK-α expression and levels of PA or LPA were not determined and the VEGF inhibitor CBO-P11 had no effect on the DAG levels due to bleomycin treatment [247]. Thus, DAGK-α is an epigenetically regulated lipid kinase involved in radiation-induced fibrosis and may serve as a marker and therapeutic target in radiation therapy.

5. Lysophospholipids and Lysophospholipids Metabolizing Enzymes in Pulmonary Fibrosis

Lysophospholipids such as lysophosphatidylcholine (LPC) and lysophosphatidylethanolamine (LPE) are generated from PC and PE, respectively by PLA_1/PLA_2 while 1-acyl or 2-acyl lysophosphatidic acid (LPA) is formed from 1-acyl- or 2-acyl LPC by the action of lyso PLD or ATX [248] (Figure 4).

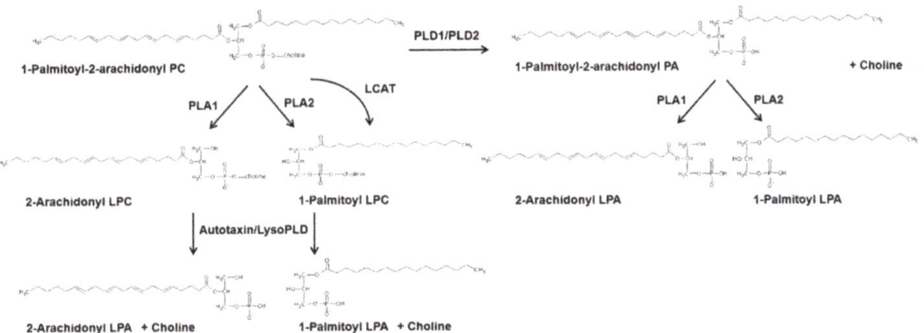

Figure 4. Generation of lysophosphatidic acid by two different pathways in mammalian cells. Schema showing the conversion of 1-Palmitoyl-2-arachidonyl phosphatidylcholine to LPA by the metabolizing enzymes, PLD, PLA_1, PLA_2, and autotaxin or lyso PLD. LCAT—lecithin-cholesterol acyltransferase, LPA—lysophosphatidic acid, LPC—lysophosphatidylcholine, lyso PLD—lysophospholipase D, PA—Phosphatidic acid, PC—Phosphatidylcholine, PLA1/2—Phospholipase A1/2, PLD1/2—Phospholipase D 1/2.

Lysophospholipids are not major components of normal cellular lipids but their levels are increased in several human pathologies. LPC levels were elevated in serum samples from IPF patients as determined by ultra-high-performance liquid chromatography coupled to high-resolution mass spectrometry, and it was suggested that LPC could serve as a biomarker for IPF [249]. However, metabolomics analysis of peripheral blood samples from abnormal interstitial lung patients revealed downregulation of 1-acyl LPC but upregulation of PC, PA, and PE suggesting potential activation of lipid metabolizing enzymes during the development of interstitial lung abnormalities [250]. However, LPA, the major lysolipid linked to development of IPF, was not identified in the analysis. Interestingly, LPA levels were significantly elevated in exhaled breath condensate (EBC) of IPF patients compared to normal subjects with LPA 22:4 as the predominant molecular species [251]. In contrast to EBC, plasma LPA levels were not significantly different between the IPF and normal subjects. LPC levels were not measured in EBCs in this study. LPA levels were also elevated in BAL fluids from segmental allergen-challenged asthmatics compared to control subjects with specific upregulation of LPA 22:5 and LPA 22:6 molecular species [252]. The presence of these unusual polyunsaturated LPA molecular species in EBC, and BAL fluids, but not in plasma, suggests that these could serve as potential biomarkers in IPF and other lung diseases. Cardiolipin is metabolized by mitochondrial PLA$_2$ to mono- and dilyso-cardiolipin during normal and oxidative stress and the lysocardiolipin is converted back to cardiolipin by two remodeling enzymes, tafazzin [253], and LYCAT [254]. Mutations in the X-linked tafazzin gene result in the accumulation of mono-lysocardiolipin and development of Barth syndrome [255]. LYCAT plays an important role in IPF and bleomycin-induced PF in mice [254].

5.1. Lysocardiolipin Acyltransferase in Cardiolipin Remodeling

Cardiolipin (CL), a major phospholipid of mitochondria, plays an important role in the structural organization, energy metabolism, and functioning of the mitochondria [256,257]. It is located mainly in the inner mitochondrial membrane (IMM), where it interacts with a number of mitochondrial proteins and enzymes. CL has been identified as an integral component of mitochondrial electron transport complex III, IV, and the ADP/ATP carrier and is essential for the stability of the quaternary protein structure. There is evidence that changes in composition and distribution of CL molecular species may be involved in the impairment of oxidative phosphorylation [258,259]. Conversion of CL to monolyso CL (MLCL) by mitochondrial phospholipase A2 has been implicated in the process of apoptosis through its interaction with a number of death-inducing proteins including cytochrome c, t-Bid, and caspase-8 [260–262]. Recent studies suggest the importance of CL in the severity of lung injury in experimental pneumonia [263] and role of CL fatty acid composition in mitochondrial cell function [258]. LYCAT is a key enzyme involved in the remodeling of mitochondrial CL from monounsaturated to polyunsaturated fatty acyl chains (>60% linoleic acid (C18:2) levels), and LYCAT transfers poly-unsaturated fatty acyl CoAs to mono- and dilyso-CL resulting in mitochondrial CL remodeling [254,258](Figure 5).

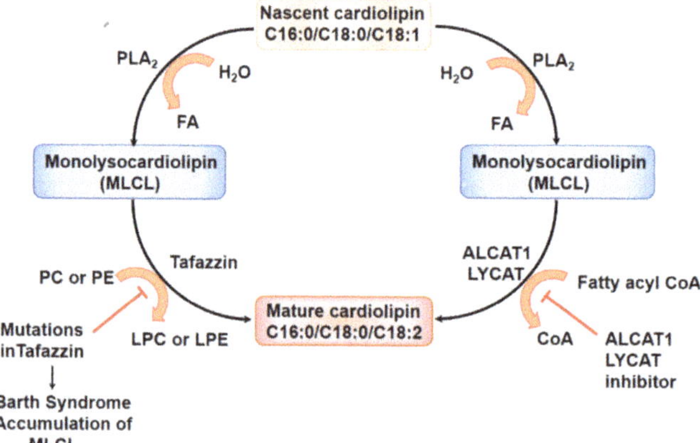

Figure 5. Cardiolipin remodeling by lysocardiolipin acyltransferase in pulmonary fibrosis. The mitochondrial phospholipid cardiolipin is converted by iPLA$_2$ or cPLA$_2$ to monolysocardiolipin (MLCL), a regulator of apoptosis, in the mitochondria. Dysregulation of LYCAT function, which remodels the fatty acid composition of cardiolipin, has been implicated in pulmonary fibrosis and Barth syndrome. ALCAT1—Acyl-CoA: lysocardiolipin acyltransferase, CoA—CoenzymeA, FA—Fatty acid,

LYCAT—lysocardiolipin acyltransferase, LPC—lysophosphatidylcholine, LPE—lysophosphatidylethanolamine, PC—Phosphatidylcholine, PE—Phosphatidylethanolamine, iPLA$_2$—calcium-independent Phospholipase A2, cPLA$_2$—cytosolic phospholipase A$_2$.

Bleomycin challenge significantly decreased the cardiolipin level and mol% of C18:1 and C18:2 fatty acids in mouse lung compared to the control group. In contrast to C18:1 and C18:2, the mol% of C18:0 was higher in the lung of bleomycin-challenged mice. Bleomycin challenge shifted the unsaturated to saturated fatty acid ratio and unsaturation index were lower in bleomycin-challenged animals, indicating loss of unsaturated fatty acids, especially C18:1 and C18:2 fatty acids in the cardiolipin. Overexpression of *hLYCAT* in bleomycin-treated mouse lung restored the cardiolipin level and unsaturated to saturated fatty acid ratio and unsaturation index [254].

Lysocardiolipin Acyltransferase and Pulmonary Fibrosis

LYCAT or Acyl-CoA cardiolipin acyltransferase (ALCAT1) is one of the key enzymes involved in remodeling the fatty acid composition of cardiolipin from saturated and monounsaturated to polyunsaturated (~60% linoleic acid) fatty acids in the mitochondria [264]. Defective cardiolipin remodeling and loss of linoleic acid cause dilated cardiomyopathy and Barth syndrome, a genetic disorder characterized by mitochondrial dysfunction, growth retardation, and neutropenia [265]. LYCAT mRNA expression was significantly reduced in PBMCs of IPF patients; however, in lung tissues from patients with IPF, and in two preclinical murine models of IPF, bleomycin- and radiation-induced PF, the LYCAT protein expression was significantly higher compared to controls. LYCAT mRNA expression in PBMCs directly and significantly correlated with carbon monoxide diffusion capacity. In murine models, *hLYCAT* overexpression reduced several indices of lung fibrosis induced by bleomycin and radiation, whereas downregulation of native LYCAT expression by siRNA accentuated fibrogenesis in the preclinical bleomycin- and radiation mice models. In vitro, LYCAT overexpression attenuated bleomycin-induced cardiolipin remodeling, mitochondrial membrane potential, ROS generation, and apoptosis of AECs. Thus, modulating LYCAT expression could offer a novel approach to ameliorate the progression of lung fibrosis.

LYCAT expressed in AECs of IPF lungs [254] and in isolated fibroblasts from fibrotic IPF lung specimens [266]. TGF-β stimulated LYCAT expression in HLF, which was dependent on SMAD3 and mutation of the SMAD2/3 binding sites (−179/−183 and −540/−544) reduced TGF-β-stimulated LYCAT promoter activity. Overexpression of LYCAT attenuated TGF-β-induced mitochondrial and intracellular oxidative stress, NOX4 expression, and differentiation of HLFs. MitoTEMPO, a mitochondrial ROS scavenger, blocked TGF-β-induced mitochondrial superoxide, NOX4 expression, and differentiation of HLFs. Treatment of HLF with NOX1/NOX4 inhibitor, GKT137831, also attenuated TGF-β induced fibroblast differentiation and mitochondrial oxidative stress. These results suggest that TGF-β stimulates LYCAT expression that negatively regulates TGF-β-induced lung fibroblast differentiation by modulation of mitochondrial ROS. Thus, TGF-β could function as a pro-fibrotic and anti-fibrotic cytokine in lung fibroblasts and fibrogenesis.

5.2. Autotaxin (Lysophospholipase D), LPA and LPARs

LPA is the simplest naturally occurring glycerophospholipid and consists of a glycerol backbone attached to a long-chain fatty acid of varying chain length and saturation/unsaturation and a free polar phosphate group. LPA is found in all cells and biological fluids including plasma, serum, and BAL, and its levels in biological fluids and tissues are elevated in several human pathologies including IPF and other forms of PF [248,252,267–269]. Plasma levels of LPA (~0.1–1.0 μM) are much lower compared to serum (>1.0 μM) due to release of LPC, a precursor for LPA production, from activated platelets and other circulating cells [248,270].

5.2.1. LPA Production in Cells by Autotaxin, Phospholipase D, and Acylglycerol Kinase

At least two major pathways have been identified for LPA production in biological systems. The first pathway is mediated by autotaxin (ATX) or lysophospholipase D that utilizes extracellular LPC as a substrate to generate LPA and the majority of plasma LPA is produced via this mechanism [248,269, 271]. Intra- and extra-cellular LPA levels are also regulated by membrane-associated lipid phosphate phosphatases [107,272]. ATX is an ectonucleotide pyrophaphatase-phosphodiesterase 2 (ENPP2) and encoded by *ENPP2* gene in mammalian cells [273,274]. It is secreted as an active protein from cells and is not a transmembrane protein such as other ENPPs [275]. It was later discovered that plasma ATX has lyso PLD activity [276] and uses LPC or lysophosphatidylserine as the substrate [248,271]. LPC is abundant in plasma and is associated with albumin and lipoproteins [248,277,278]. LPC is generated through the hydrolysis of PC by PLA_1 or PLA_2 and lecithin cholesterol acyltransferase (LCAT) enzyme. LCAT is a transacylase that transfers fatty acid from *sn*-2 position of PC to cholesterol to generate LPC and cholesteryl ester that is predominantly C81:1 (oleic acid) [248]. ATX can also utilize sphingosylphosphorylcholine as a substrate to generate S1P but the physiological relevance of this pathway is unclear. ATX is secreted by macrophages, AECs, and adipocytes, and increased ATX expression and activity has been reported in various human pathologies and experimental models mimicking human diseases [252,267,271].

The second pathway of intracellular LPA generation in cells involves phospholipase A1/A2 mediated hydrolysis of PA derived from either the de novo biosynthesis or PLD1/PLD2 signaling axis that uses PC, PS, and PE as substrates. While it will be difficult to distinguish between the two pools of PA generated inside the cell, specific PLA_1 or PLA_2 can hydrolyze PA to 1-acyl or 2-acyl-*sn*-glycero-3-phosphate within the cell [234,235]. The fatty acid composition of LPA derived from ATX/LPC vs. PLA_1-PLA_2/PA pathways may vary based on the fatty acid composition of the substrate. The third minor pathway of intracellular LPA production involves phosphorylation of 1-acyl- or 2-acyl-monoglycerol by acylglycerol kinase (AGK) [279,280]. AGK is a mitochondrial lipid kinase and PA generated in the mitochondria by AGK can serve as a substrate for the biosynthesis of CL in the mitochondria. Further, AGK is a subunit of the mitochondrial TIM22 protein import complex. Mutations in *AGK*, independent of its kinase activity, dysregulates mitochondrial protein import leading to Sengers syndrome [281]. The role of AGK in lung pathologies has not been investigated.

5.2.2. LPA Signals via LPARs

The numerous physiological and pathophysiological responses of LPA are mediated through six 7-transmembrane G-protein coupled receptors (GPCRs) collectively termed as LPA_{1-6}. The six receptors are widely distributed with overlapping specificities and different affinity for LPA [282]. The LPARs are coupled differentially to G_i, $G_{12/13}$, $G_{q/11}$, and G_s that initiate numerous signaling cascades within the cell. Several studies have identified stimulation of MAPKs, PI3K, Rho, Rac, and phospholipase activation by LPA through differential coupling to various G_α proteins. Of the six LPARs, LPA_6 has sequence homology with P2Y receptor and higher affinity to 2-acyl LPA than 1-acyl LPA [283]. These receptors are highly expressed in lung cells and some of the LPARs are actively involved in the development and progression of several lung pathologies.

5.2.3. PPARγ is an Intracellular Receptor of LPA

The transcriptional factor, PPARγ regulates a wide range of physiological and pathophysiological activities including energy metabolism, inflammation, atherogenesis, and fibrosis [284,285]. LPA and 1-alkyl LPA (alkyl group instead of acyl group) have been shown to be agonists for PPARγ in mammalian cells [286]. PPARγ is also activated by oxidized phospholipids, fatty acids, eicosanoids, and oxidized-LDL [287] and activation of PPARγ inhibits the differentiation of fibroblast to myofibroblast [32]. In contrast to LPA, cyclic PA (cPA), the structural analog of LPA, is generated by PLD2 mediated hydrolysis of 1-acyl LPC [288] and has been shown to be a physiological inhibitor of PPARγ [289,290]. The pathophysiological role of LPA and cPA in modulation of PPARγ and development of lung fibrosis is unclear.

5.2.4. ATX/LPA Signaling Axis in Pulmonary Fibrosis

Elevated LPA levels were found in BAL fluids of IPF patients and the bleomycin murine model [268,291]. The increased LPA levels in BAL fluid from IPF patients mediated fibroblast recruitment and vascular leak [268]. LPA stimulated migration of fibroblasts [292] and induced expression of TGF-β, fibronectin, α-SMA, and collagen via AKT, SMAD3, and MAPK pathways in HLF [293]. LPA signaling via LPARs is known to transactivate PDGFR via PLD2 [294] and EGFR [295] in primary human bronchial epithelial cells. However, in fibroblast activation, LPA increased TGF-β expression that signaled via TGF-βRI/TGF-βRII and this was blocked by anti-TGF-β antibody. This suggests the crosstalk between LPA signaling and TGF-β signaling in fibroblast differentiation.

ENPP2 was identified as a candidate gene regulating lung formation, development, and remodeling by genome-wide linkage analysis coupled with expression profiling [296]. Based on this, the deletion of both alleles of *ENPP2* in mouse ($ENPP2^{-/-}$) was embryonically lethal [297–299]; however, knockdown of a single allele ($ENPP2^{+/-}$) in mouse had no major phenotypic effect in lungs. Overexpression of *ENPP2* in bronchial epithelium or liver resulted in a 2-fold increase of ATX expression in plasma without any phenotype suggesting ATX/LPA signaling axis per se has no consequence on lung morphology and function [300]. ATX is constitutively expressed in bronchial epithelium, AMs and other cells in both humans and rodents, and increased ATX staining mainly localized in bronchial epithelial cells around fibroblastic foci was seen in lung tissues from IPF patients [291,301]. The upregulation of ATX directly correlated with disease progression and irreversible PF development. A similar increase in ATX staining and protein expression was observed in the lungs of bleomycin-challenged mice. Conditional deletion of *ENPP2* in bronchial epithelial cells and macrophages reduced ATX levels in BAL fluid and disease severity confirming a pathophysiological role for ATX in lung fibrosis [291]. A number of small molecular weight inhibitors have been developed to ameliorate IPF and reverse fibrosis in patients. ATX inhibitors GWJ-A-23 [291], PF-8380 [216], BBT-877 [302], and GLPG1690 [303] have been shown to confer protection against bleomycin-induced PF in mice. Among the various ATX inhibitors, the Galapagos compound GLPG1690 has advanced to Phase III clinical trials and awaiting outcome on safety and side effects. The ATX inhibitor PAT-048 (Bristol Myers Squibb) seems to have a different mechanism in attenuating PF mediated by bleomycin [301]. PAT-048 only blocked ATX-dependent LPA production in plasma and had no effect on bleomycin-induced lung fibrosis and BAL fluid LPA levels. Additionally, there was discordance in LPC and LPA molecular species from plasma and BAL fluid. In the BAL fluid, the predominant LPA molecular species were long-chain polyunsaturated C22:5 and C22:6 fatty acids whereas the BAL fluid LPC molecular species contained shorter and saturated C16:0 and C18:0 fatty acids. It was concluded that an alternate ATX-independent pathway(s) was most likely contributed for the local generation of highly polyunsaturated LPA species in the bleomycin-injured [301]. Lipidomic analysis revealed the presence of highly unsaturated fatty acids in phospholipids of lung bronchial epithelial cells, alveolar type II epithelial cells, and AMs [304], indicating the phospholipids in lung cells could be a substrate for PLD to generate PA with polyunsaturated fatty acids that could generate LPA in the lung.

5.2.5. LPA Signaling via LPA Receptors in the Development of IPF and Pulmonary Fibrosis

LPA signaling via LPA$_{1-3}$ has been identified to be involved in the development of PF. Fibroblasts isolated from IPF lungs had elevated expression of LPA$_1$, and inhibition of LPA$_1$ reduced fibroblast responses to chemotactic activity [268]. The number of apoptotic cells present in alveolar and bronchial epithelia was significantly reduced in *Lpar1* deficient mice that were exposed to bleomycin, and LPA signaling through LPA$_1$ induced apoptosis of normal bronchial epithelial cells [305]. In contrast to epithelial cells, LPA signaling through LPA$_1$ promoted resistance of lung fibroblasts to apoptosis. An antagonist of LPA$_1$, BMS-986020, was found to lessen the extent of the decline in the forced vital capacity of lungs in IPF patients [306] while another LPA$_1$ antagonist AM966 attenuated bleomycin-induced vascular leakage, lung injury, inflammation and fibrosis [307]. Thus, LPA signaling via LPA$_1$ promotes apoptotic and profibrotic responses in lung epithelial cells and fibroblasts, respectively. Other LPA$_1$ antagonists such as AM152, SAR100842 and ONO-7300243 have been shown to be beneficial in vivo against bleomycin mediated lung inflammation and fibrosis and in vitro on epithelial cell apoptosis and fibroblast migration and differentiation; however, none have advanced to or beyond clinical trials. Similar to LPA$_1$, deficiency of *Lpar2* in mice conferred protection from bleomycin-induced lung inflammatory injury and PF in mice [293]. Reduced number of TUNEL$^+$ apoptotic alveolar and bronchial epithelial cells was observed in the lung tissues of bleomycin-challenged *Lpar2* mice compared to the WT mice. Further, LPA-induced expression of TGF-β and differentiation of HLFs was reduced in lung fibroblasts deficient of *Lpar2*. LPA was found to signal through *Lpar2* to regulate the ERK1/2, SMAD3, AKT, and p38 MAPK pathways, but not SMAD2 and JNK pathways during PF [293] (Figure 6). The role of LPA$_{3-6}$ in IPF and experimental lung fibrosis is yet to be established.

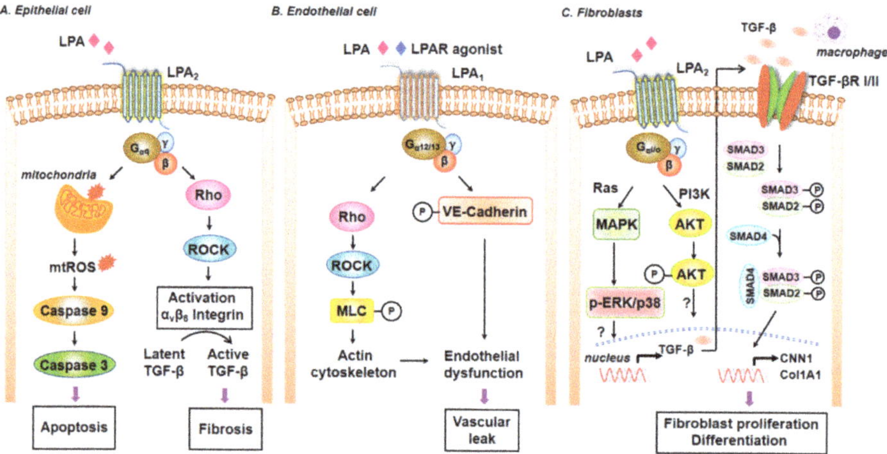

Figure 6. Role of lysophosphatidic signaling via lysophosphatidic acid receptors in pulmonary fibrosis. (**A**), Alveolar epithelial cell stimulation by lysophosphatidic acid (LPA) activates the G-protein coupled receptor LPA$_2$ and generation of mitochondrial reactive species (mtROS) promotes caspase-3 mediated apoptotic pathway. Rho/ROCK activation by LPA directs the fibrogenic signaling via α$_v$β$_6$ and TGF-β. (**B**), In the lung vascular endothelial cells, the LPA/LPA$_1$ activation leads to cytoskeletal changes and VE-cadherin-mediated junctional disruption. (**C**), LPA/LPA2 signaling in the fibroblasts stimulates the MAPK/ERK and PI3K/AKT pathways, resulting in enhanced TGF-β production and secretion. This results in activation of the TGF-β dependent profibrotic signaling mechanisms via TGF-βI/TGF-βII receptors and activation of SMAD proteins leading to myofibroblast differentiation and collagen production in PF. AKT—Protein kinase B, CNN1—Calponin1, Col1A1—Collagen Type I Alpha 1 chain, ERK—Extracellular signal-regulated kinase, G$_α$q,12/13,i/o—GTP-binding protein complex alpha subunit, LPA—lysophosphatidic acid, LPA1/2—lysophosphatidic acid receptor type1/2, LPAR—lysophosphatidic acid receptor, mtROS—mitochondrial reactive oxygen species, MAPK—mitogen-activated protein kinase, MLC—myosin light chain, PI3K—Phosphatidylinositol-3-kinase, Ras—RAS family of GTPases, Rho—Rho family of small GTPase, ROCK—Rho-associated coiled-coil containing protein kinase, SMAD2/3/4—Mothers Against Decapentaplegic Homolog 2/3/4, TGF-β—transforming growth factor beta, TGF-β RI/II—transforming growth factor beta receptor I/II, VE—Cadherin vascular endothelial cadherin.

5.3. Lipid Peroxidation and Oxidized Phospholipids in Pulmonary Fibrosis

Membrane lipids are rich in polyunsaturated fatty acids, which are susceptible to oxidation and peroxidation by hydroxyl ($^{\bullet}$OH), hydroperoxyl ($^{\bullet}$OOH) radicals, and ROS to generate lipid peroxidation end products, mildly oxidized and truncated oxidized phospholipids [308]. Lipids can also be oxidized by LPs, COXs, and cytochrome P450 to yield lipid peroxidation products, and oxidized phospholipids. Cellular antioxidant defense mechanisms maintain the normal redox status of the cell by scavenging, neutralizing, and repairing the free radicals and oxidized lipids; however, in pathological conditions, these lipid oxidation products accumulate and affect cell function and viability. Elevated levels of lipid peroxidation products such as 4-hydroxynonenal, oxidized phospholipids, or their protein adducts have been identified in BAL fluids and lung tissues of patients with IPF, COPD, and ARDS [309–311]. Generation and role for lipid peroxidation products have been shown in bleomycin-induced PF [312] and antioxidants such as α-tocopherol protected the mice against bleomycin-induced lung injury [313] suggesting a role for lipid peroxidation in pulmonary toxicity. Resveratrol, a phytochemical and antioxidant, also alleviated bleomycin-induced lung injury in rats [314] and the thiol protectant, *N*-acetylcysteine, and the iron chelator, deferoxamine attenuated the bleomycin-mediated oxidative stress and lung injury [315] supporting the findings that bleomycin-induced oxidative stress, altered thiol redox status, induced lipid peroxidation, activated PLD, and caused cytotoxicity in a redox-dependent pathway in lung endothelial cells [238]. Glutathione peroxidase (GPx) is an important antioxidant enzyme that reduces phospholipid hydroperoxides and modulation of GPx expression or activity would tilt the redox balance towards oxidation. Reduced expression of GPx4 and increased 4-hydroxynonenal immunostaining was seen in IPF lungs, and genetic deletion of GPx4 ($Gpx4^{+/-}$) mice showed enhanced bleomycin-induced lung fibrosis and TGF-β-mediated myofibroblast differentiation in vitro [316]. Thus, changes in the redox status due to tress could play a critical role in the pathophysiology of IPF.

There is compelling evidence for monocyte-derived macrophages and not residential lung macrophages to be involved in the development of PF in experimental murine models [317]. Accumulation of lipid-laden AMs, also known as foam cells, has been characterized in experimental models of fibrosis and histological lung specimens from IPF patients [318,319]. A significant increase in oxidized-PC (Ox-PC) was observed in AMs and BAL fluid after bleomycin injury and Ox-PC, but not native PC, enhanced the expression of *Tgfβ1*, *Ym1*, *cd36, cd63,* and *cd56* transcripts indicating polarization of macrophages to M2 phenotype. Instillation of Ox-PC directly into the mouse lung induced foam cell formation and accumulation and development of PF. The increased accumulation of lipids in macrophages occurs through efferocytosis whereby macrophages engulf apoptotic alveolar type II epithelial cells after injury via CD36 on the macrophages [319–321]. Targeted deletion of the lipid efflux transporter, ATP-binding cassette sub-family G member 1 (*Abcg1*) increased foam cell formation and exacerbated lung fibrosis in mice from bleomycin challenge [320]. Thus, a pneumocytes-macrophage-CD36-oxidized lipid signaling axis has been proposed to play a key role in the development of lung fibrosis in the bleomycin model [319,322]. Another potential mechanism of Ox-PC in development of PF is through PAI-1. Administration of 1-palmitoyl-2-(5-oxo-valeroyl)-*sn*-glycero-3-phosphocholine (POVPC), an Ox-PC, to *Pai-1*$^{+/+}$ increased hydroxyproline in the lung and reduced serum SP-D levels in contrast to *Pai-1*$^{-/-}$ mice treated with POVPC [323,324]. These results suggest that PAI-1 promotes fibrosis in response to oxidized phospholipid administration in mouse lungs.

6. Conclusions

IPF is a heterogeneous interstitial lung disease caused by abnormal host-defense, activation of immune and non-immune cells, and dysfunctional wound, male gender, cigarette-smoking, and mutations in *MUC5B* and *Sftpc* gene confer a predisposition to the development of IPF. In the last three decades, several target genes and proteins have been identified as key players in the pathophysiology healing and lung repair that result in lung fibrosis. With the advancement in mass spectrometry and lipidomics, mediators derived from fatty acids, glycerolipids, phospholipids, and sphingolipids have been identified in lung tissues and biological fluids of IPF patients and experimental models of fibrosis. These bioactive lipid mediators exhibit pro- or anti-fibrotic effects in vivo and *in vitro*. PLA$_2$ catalyzed hydrolysis of a phospholipid such as PC enriched in arachidonic acid generates free AA that is subsequently converted by COX2/PG synthase to prostanoids such as PGD2, PGE$_2$, PGF2α, and by 5-LO to LTs. While PGD2 and PGF2α and LTs have pro-fibrotic properties, PGE2 exhibits anti-fibrotic effects in the experimental bleomycin model of PF and TGF-β-induced fibroblast activation. The SPHK1/S1P/S1PR, PLD2/PA, and ATX/LPA/LPAR signaling axes have been identified as significant contributors to fibrosis in IPF patients and bleomycin/radiation models of PF. Blocking SPHK1 with SKI-II attenuated lung inflammation and development of lung fibrosis 3 weeks post-bleomycin challenge indicating a long-term beneficial effect of this SPHK1 inhibitor. However, PF543 is yet to be evaluated for its efficacy in clinical trials. As S1P signals via S1PRs, and deletion of *S1PR2* or *S1PR3* ameliorated bleomycin-induced PF, use of specific receptor antagonists for S1PR2 or S1PR3 might be beneficial, which needs further investigation.

There has been tremendous advancement in developing ATX inhibitor(s) as a potential therapy in IPF (Table 1). The ATX inhibitor from Galapagos Pharmaceuticals, GLPG1690 exhibited a good PK/PD profile in experimental animal model of PF and has entered Phase III clinical trials after completing the Phase II studies in IPF patients. Blocking LPA$_1$ with antagonists such as BMS-976278, AM152, SAR100842, and ONO-7300243 conferred protection against bleomycin-mediated lung inflammation and fibrosis in vivo and in vitro on epithelial cell apoptosis and fibroblast migration and differentiation; BMS-976278 has shown promise in clinical trials. Another promising target is LYCAT, a cardiolipin remodeling enzyme, which is upregulated in the lungs of IPF patients and bleomycin treated animals. LYCAT overexpression in mouse lung reduced bleomycin-induced PF and activation of LYCAT activity or enhances its expression in lung epithelial cells and fibroblasts might be a novel approach to reduce PF in pre-clinical models before advancing to human studies. Other potential targets for drug development for IPF include cysLTs and its receptors. Montelukast, a cysLT type I receptor antagonist, approved by the FDA for the management of asthma. Although Montelukast showed some beneficial effects against IPF, and its potential as a therapeutic agent for lung fibrosis is inconclusive and require more trials. Another class of mediators known as lipoxins, resolvins, protectins, and maresins, which are derived from polyunsaturated fatty acids may modulate the resolution of inflammation and fibrosis. Resolvin D1 is one lipid mediator that is derived from eicosapentaenoic acid and docosahexaenoic acid attenuates lung inflammation and interstitial fibrosis as well as inhibits TGF-β-induced EMT suggesting it as a potential therapeutic agent against IPF. Thus, understanding the mechanisms of generation, and signaling pathways of lipid mediators generated in PF provides an opportunity to develop therapies to attenuate the development of lung fibrosis and facilitate the process resolution of fibrotic injury.

Table 1. Clinical studies to evaluate new drugs targeting lipid-mediated pathways in the treatment of pulmonary fibrosis.

Drug	Target Specificity	Identifier [a]	Stage
BMS-986278	LPA$_1$ receptor antagonist	NCT04308681	Phase II
BMS-986020	LPA$_1$ receptor antagonist	NCT01766817	Phase II
18F-BMS-986327	LPA$_1$ ligand for PET	NCT04069143	Phase I
Tipelukast/MN-001	Orphan drug (LT receptor, PDE inhibitor, 5-LO pathway)	NCT02503657	Phase II
GLPG1690	ATX inhibitor	NCT03733444	Phase III
BBT-877	ATX inhibitor	NCT03830125	Phase I
X-165	ATX inhibitor	Preclinical, IND	-
GSK2126458	PI3K/mTOR inhibitor	NCT01725139	Phase I
Sirolimus	mTOR inhibitor	NCT01462006	Phase I
HEC68498	PI3K/mTOR inhibitor	NCT03502902	Phase I

[a] Unique identifier from https://clinicaltrials.gov/; ATX—Autotaxin; IND—Investigational new drug; LPA$_1$—Lysophosphatidic acid receptor type1; 5-LO—5-lipoxygenase; LT—Leukotriene; mTOR—mammalian target of rapamycin; PDE—Phosphodiesterase; PET—Positron emission tomography; PI3K—Phosphoinositide 3-kinase.

Author Contributions: Conceptualization and supervision—V.S., R.R., and V.N.; Writing—Original Draft Preparation—V.S., R.R., and V.N.; Writing, Review and Editing—V.N., V.S., D.W.K., and R.R.; Graphics—V.N. and R.R. All authors have read and agreed to the published version of the manuscript.

Funding: This work was supported by grants from national Institutes of Health HLBI P01 HL060678 Project4), P01 HL126609 (Project 3), R01 HL127342 (Multi PI) and R01 HL147942 (Co-I) to VN; Veterans Affairs Merit Award 2IO1BX000786-05A2 to DWK.

Acknowledgments: The authors wish to thank Longshuang Huang for his contribution to the pulmonary fibrosis project during his tenure in VN's laboratory at UIC. All publications of Huang related to pulmonary fibrosis have been cited. The technical assistance of Alison Ha is appreciated. The authors also wish to dedicate this review article to Andrew Tager, a pulmonologist and physician-scientist from Massachusetts General Hospital, Boston who passed away in August 2017. Andy made several seminal contributions to the field of PF, and his original work related to LPA/LPARs signaling in the development of IPF and PF in experimental models of lung inflammation are internationally recognized by researchers. Andy's contributions to the field of IPF will be remembered by all the researchers working in PF.

Conflicts of Interest: The authors declare no conflict of interest. The funders had no role in the design of the study; in the collection, analyses or interpretation of the data; in the writing of the manuscript or in the decision to publish this review article.

Abbreviations

5-HPETE	5-hydroperoxy eicosatetraenoic acid
5-LO	5-lipoxygenase
Abcg1	ATP-binding cassette sub-family G member 1
AEC	Alveolar epithelial cells
ALCAT1	Acyl-CoA lysocardiolipin acyltransferase
AM	Alveolar macrophages
ApoM	Apolipoprotein M
ASMase	Acid sphingomyelinase
ATX	Autotaxin
BAL	Bronchoalveolar lavage
BPD	Bronchopulmonary dysplasia
C1P	Ceramide-1-phosphate
CL	Cardiolipin
CRTH2	Chemoattractant receptor homologous with T-helper cell type 2
COPD	Chronic obstructive pulmonary disease

COX	Cyclooxygenases
CTGF	Connective tissue growth factor
cysLT	Cysteinyl leukotrienes
DAG	Diacylglycerol
DAGK	Diacylglycerol kinase
DM PGE2	16,16-dimethy PGE2
DP1	D-prostanoid receptor
EBC	Exhaled breath condensate
EGR1	Early growth response 1
EndMT	Endothelial-mesenchymal transdifferentiation
ENPP2	Ectonucleotide pyrophaphatase-phosphodiesterase 2
EP2, EP4	PGE2 receptors
ER	Endoplasmic reticulum
GPCRs	G-protein coupled receptors
GSH	Glutathione
GPx	Glutathione peroxidase
H-PGDS	Hematopoietic PGD synthase
HLF	Human lung fibroblast
IMM	Inner mitochondrial membrane
IPF	Idiopathic pulmonary fibrosis
LNO2	12-nitrolinoleic acid
5-LO	5-lipoxygenase
LPA	Lysophosphatidic acid
LPAR	Lysophosphatidic acid receptor
LPC	Lysophosphatidylcholine
LPE	Lysophosphatidylethanolamine
LPP	Lipid phosphate phosphatases
LPS	Lipopolysachharide
LT	Leukotriene
LYCAT	Lysocardiolipin acyltransferase
Mfn1	Mitofusin 1
Mfn2	Mitofusin 2
MLCL	Monolyso CL
mtROS	Mitochondrial ROS
NFA	Nitrated fatty acids
NO	Nitric oxide
NOX	NADPH oxidase
OA-NO2	10-nitro-oleic acid
Ox-PC	Oxidized-PC
PA	Phosphatidic acid
PBMC	Peripheral blood mononuclear cells (PBMCs)
PC	Phosphatidylcholine
PPARγ	Peroxisome-activated receptor γ
PG	Prostaglandin
PGD2	Prostaglandin D2
PGE2	Prostaglandin E2
PGI2	Prostacyclin
PGF2α	Prostaglandin F2α
PGET	PGE2 transporter
PF	Pulmonary fibrosis
PKC	Protein kinase C
PLD	Phospholipase D

PTEN	Phosphatase and tensin homolog
ROS	Reactive oxygen species
S1P	Sphingosine-1-phosphate
Sgpl1	S1P lyase
SM	Sphingomyelin
SMAD2/3/4	Mothers Against Decapentaplegic Homolog 2/3/4
SMase	Sphingomyelinase
SPHK	Sphingosine kinase
SPNS2	Spinster homolog 2
SPP	S1P phosphatase
SPT	Serine palmitoyltransferase
TRAF2	TNF receptor-associated factor 2

References

1. Abrass, C.K. Cellular lipid metabolism and the role of lipids in progressive renal disease. *Am. J. Nephrol.* **2004**, *24*, 46–53. [CrossRef] [PubMed]
2. Adibhatla, R.M.; Hatcher, J.F. Role of Lipids in Brain Injury and Diseases. *Future Lipidol.* **2007**, *2*, 403–422. [CrossRef] [PubMed]
3. Burnett, J.R.; Hooper, A.J.; Hegele, R.A. Lipids and cardiovascular disease. *Pathology* **2019**, *51*, 129–130. [CrossRef] [PubMed]
4. Sulciner, M.L.; Gartung, A.; Gilligan, M.M.; Serhan, C.N.; Panigrahy, D. Targeting lipid mediators in cancer biology. *Cancer Metastasis Rev.* **2018**, *37*, 557–572. [CrossRef]
5. Higenbottam, T. Lung lipids and disease. *Respiration* **1989**, *55*, 14–27. [CrossRef] [PubMed]
6. Luo, X.; Zhao, X.; Cheng, C.; Li, N.; Liu, Y.; Cao, Y. The implications of signaling lipids in cancer metastasis. *Exp. Mol. Med.* **2018**, *50*, 127. [CrossRef]
7. Raghu, G.; Collard, H.R.; Egan, J.J.; Martinez, F.J.; Behr, J.; Brown, K.K.; Colby, T.V.; Cordier, J.F.; Flaherty, K.R.; Lasky, J.A.; et al. An official ATS/ERS/JRS/ALAT statement: Idiopathic pulmonary fibrosis: Evidence-based guidelines for diagnosis and management. *Am. J. Respir. Crit. Care Med.* **2011**, *183*, 788–824. [CrossRef]
8. Loomis-King, H.; Flaherty, K.R.; Moore, B.B. Pathogenesis, current treatments and future directions for idiopathic pulmonary fibrosis. *Curr. Opin. Pharmacol.* **2013**, *13*, 377–385. [CrossRef]
9. Selman, M.; King, T.E.; Pardo, A.; American Thoracic, S.; European Respiratory, S.; American College of Chest, P. Idiopathic pulmonary fibrosis: Prevailing and evolving hypotheses about its pathogenesis and implications for therapy. *Ann. Intern. Med.* **2001**, *134*, 136–151. [CrossRef]
10. Sontake, V.; Gajjala, P.R.; Kasam, R.K.; Madala, S.K. New therapeutics based on emerging concepts in pulmonary fibrosis. *Expert Opin. Ther. Targets* **2019**, *23*, 69–81. [CrossRef]
11. Richeldi, L.; Collard, H.R.; Jones, M.G. Idiopathic pulmonary fibrosis. *Lancet* **2017**, *389*, 1941–1952. [CrossRef]
12. Du Bois, R.M. Strategies for treating idiopathic pulmonary fibrosis. *Nat. Rev. Drug Discov.* **2010**, *9*, 129–140. [CrossRef] [PubMed]
13. Desai, O.; Winkler, J.; Minasyan, M.; Herzog, E.L. The Role of Immune and Inflammatory Cells in Idiopathic Pulmonary Fibrosis. *Front. Med.* **2018**, *5*, 43. [CrossRef] [PubMed]
14. Barkauskas, C.E.; Noble, P.W. Cellular mechanisms of tissue fibrosis. 7. New insights into the cellular mechanisms of pulmonary fibrosis. *Am. J. Physiol. Cell Physiol.* **2014**, *306*, C987–C996. [CrossRef]
15. Sakai, N.; Tager, A.M. Fibrosis of two: Epithelial cell-fibroblast interactions in pulmonary fibrosis. *Biochim. Biophys. Acta (BBA)-Mol. Basis Dis.* **2013**, *1832*, 911–921. [CrossRef]
16. Massague, J. TGFbeta in Cancer. *Cell* **2008**, *134*, 215–230. [CrossRef]
17. Fernandez, I.E.; Eickelberg, O. The impact of TGF-beta on lung fibrosis: From targeting to biomarkers. *Proc. Am. Thorac. Soc.* **2012**, *9*, 111–116. [CrossRef]
18. Derynck, R.; Zhang, Y.E. Smad-dependent and Smad-independent pathways in TGF-beta family signalling. *Nature* **2003**, *425*, 577–584. [CrossRef]
19. Walton, K.L.; Johnson, K.E.; Harrison, C.A. Targeting TGF-beta Mediated SMAD Signaling for the Prevention of Fibrosis. *Front. Pharmacol.* **2017**, *8*, 461. [CrossRef]

20. Daniels, C.E.; Wilkes, M.C.; Edens, M.; Kottom, T.J.; Murphy, S.J.; Limper, A.H.; Leof, E.B. Imatinib mesylate inhibits the profibrogenic activity of TGF-beta and prevents bleomycin-mediated lung fibrosis. *J. Clin. Investig.* **2004**, *114*, 1308–1316. [CrossRef]

21. Mackinnon, A.C.; Gibbons, M.A.; Farnworth, S.L.; Leffler, H.; Nilsson, U.J.; Delaine, T.; Simpson, A.J.; Forbes, S.J.; Hirani, N.; Gauldie, J.; et al. Regulation of transforming growth factor-beta1-driven lung fibrosis by galectin-3. *Am. J. Respir. Crit. Care Med.* **2012**, *185*, 537–546. [CrossRef] [PubMed]

22. Blackwell, T.S.; Tager, A.M.; Borok, Z.; Moore, B.B.; Schwartz, D.A.; Anstrom, K.J.; Bar-Joseph, Z.; Bitterman, P.; Blackburn, M.R.; Bradford, W.; et al. Future directions in idiopathic pulmonary fibrosis research. An NHLBI workshop report. *Am. J. Respir. Crit. Care Med.* **2014**, *189*, 214–222. [CrossRef] [PubMed]

23. Yan, F.; Wen, Z.; Wang, R.; Luo, W.; Du, Y.; Wang, W.; Chen, X. Identification of the lipid biomarkers from plasma in idiopathic pulmonary fibrosis by Lipidomics. *BMC Pulm. Med.* **2017**, *17*, 174. [CrossRef] [PubMed]

24. Chu, S.G.; Villalba, J.A.; Liang, X.; Xiong, K.; Tsoyi, K.; Ith, B.; Ayaub, E.A.; Tatituri, R.V.; Byers, D.E.; Hsu, F.F.; et al. Palmitic Acid-Rich High-Fat Diet Exacerbates Experimental Pulmonary Fibrosis by Modulating Endoplasmic Reticulum Stress. *Am. J. Respir. Cell Mol. Biol.* **2019**, *61*, 737–746. [CrossRef]

25. Sunaga, H.; Matsui, H.; Ueno, M.; Maeno, T.; Iso, T.; Syamsunarno, M.R.; Anjo, S.; Matsuzaka, T.; Shimano, H.; Yokoyama, T.; et al. Deranged fatty acid composition causes pulmonary fibrosis in Elovl6-deficient mice. *Nat. Commun.* **2013**, *4*, 2563. [CrossRef]

26. Romero, F.; Hong, X.; Shah, D.; Kallen, C.B.; Rosas, I.; Guo, Z.; Schriner, D.; Barta, J.; Shaghaghi, H.; Hoek, J.B.; et al. Lipid Synthesis Is Required to Resolve Endoplasmic Reticulum Stress and Limit Fibrotic Responses in the Lung. *Am. J. Respir. Cell Mol. Biol.* **2018**, *59*, 225–236. [CrossRef]

27. Baker, P.R.; Schopfer, F.J.; Sweeney, S.; Freeman, B.A. Red cell membrane and plasma linoleic acid nitration products: Synthesis, clinical identification, and quantitation. *Proc. Natl. Acad. Sci. USA* **2004**, *101*, 11577–11582. [CrossRef]

28. Baker, P.R.; Lin, Y.; Schopfer, F.J.; Woodcock, S.R.; Groeger, A.L.; Batthyany, C.; Sweeney, S.; Long, M.H.; Iles, K.E.; Baker, L.M.; et al. Fatty acid transduction of nitric oxide signaling: Multiple nitrated unsaturated fatty acid derivatives exist in human blood and urine and serve as endogenous peroxisome proliferator-activated receptor ligands. *J. Biol. Chem.* **2005**, *280*, 42464–42475. [CrossRef]

29. Delmastro-Greenwood, M.; Freeman, B.A.; Wendell, S.G. Redox-dependent anti-inflammatory signaling actions of unsaturated fatty acids. *Annu. Rev. Physiol.* **2014**, *76*, 79–105. [CrossRef]

30. O'Donnell, V.B.; Eiserich, J.P.; Bloodsworth, A.; Chumley, P.H.; Kirk, M.; Barnes, S.; Darley-Usmar, V.M.; Freeman, B.A. Nitration of unsaturated fatty acids by nitric oxide-derived reactive species. *Methods Enzymol.* **1999**, *301*, 454–470. [CrossRef]

31. Michalik, L.; Wahli, W. Involvement of PPAR nuclear receptors in tissue injury and wound repair. *J. Clin. Investig.* **2006**, *116*, 598–606. [CrossRef] [PubMed]

32. Burgess, H.A.; Daugherty, L.E.; Thatcher, T.H.; Lakatos, H.F.; Ray, D.M.; Redonnet, M.; Phipps, R.P.; Sime, P.J. PPARgamma agonists inhibit TGF-beta induced pulmonary myofibroblast differentiation and collagen production: Implications for therapy of lung fibrosis. *Am. J. Physiol. Lung Cell. Mol. Physiol.* **2005**, *288*, L1146–L1153. [CrossRef]

33. Milam, J.E.; Keshamouni, V.G.; Phan, S.H.; Hu, B.; Gangireddy, S.R.; Hogaboam, C.M.; Standiford, T.J.; Thannickal, V.J.; Reddy, R.C. PPAR-gamma agonists inhibit profibrotic phenotypes in human lung fibroblasts and bleomycin-induced pulmonary fibrosis. *Am. J. Physiol. Lung Cell. Mol. Physiol.* **2008**, *294*, L891–L901. [CrossRef]

34. Reddy, A.T.; Lakshmi, S.P.; Zhang, Y.; Reddy, R.C. Nitrated fatty acids reverse pulmonary fibrosis by dedifferentiating myofibroblasts and promoting collagen uptake by alveolar macrophages. *FASEB J.* **2014**, *28*, 5299–5310. [CrossRef]

35. Olman, M.A. Beyond TGF-beta: A prostaglandin promotes fibrosis. *Nat. Med.* **2009**, *15*, 1360–1361. [CrossRef] [PubMed]

36. Nagase, T.; Uozumi, N.; Ishii, S.; Kita, Y.; Yamamoto, H.; Ohga, E.; Ouchi, Y.; Shimizu, T. A pivotal role of cytosolic phospholipase A(2) in bleomycin-induced pulmonary fibrosis. *Nat. Med.* **2002**, *8*, 480–484. [CrossRef] [PubMed]

37. Shimizu, H.; Ito, A.; Sakurada, K.; Nakamura, J.; Tanaka, K.; Komatsu, M.; Takeda, M.; Saito, K.; Endo, Y.; Kozaki, T.; et al. AK106-001616, a Potent and Selective Inhibitor of Cytosolic Phospholipase A2: In Vivo Efficacy for Inflammation, Neuropathic Pain, and Pulmonary Fibrosis. *J. Pharmacol. Exp. Ther.* **2019**, *369*, 511–522. [CrossRef] [PubMed]

38. Kida, T.; Ayabe, S.; Omori, K.; Nakamura, T.; Maehara, T.; Aritake, K.; Urade, Y.; Murata, T. Prostaglandin D2 Attenuates Bleomycin-Induced Lung Inflammation and Pulmonary Fibrosis. *PLoS ONE* **2016**, *11*, e0167729. [CrossRef]

39. Ando, M.; Murakami, Y.; Kojima, F.; Endo, H.; Kitasato, H.; Hashimoto, A.; Kobayashi, H.; Majima, M.; Inoue, M.; Kondo, H.; et al. Retrovirally introduced prostaglandin D2 synthase suppresses lung injury induced by bleomycin. *Am. J. Respir. Cell Mol. Biol.* **2003**, *28*, 582–591. [CrossRef]

40. Ueda, S.; Fukunaga, K.; Takihara, T.; Shiraishi, Y.; Oguma, T.; Shiomi, T.; Suzuki, Y.; Ishii, M.; Sayama, K.; Kagawa, S.; et al. Deficiency of CRTH2, a Prostaglandin D2 Receptor, Aggravates Bleomycin-induced Pulmonary Inflammation and Fibrosis. *Am. J. Respir. Cell Mol. Biol.* **2019**, *60*, 289–298. [CrossRef]

41. Ayabe, S.; Kida, T.; Hori, M.; Ozaki, H.; Murata, T. Prostaglandin D2 inhibits collagen secretion from lung fibroblasts by activating the DP receptor. *J. Pharmacol. Sci.* **2013**, *121*, 312–317. [CrossRef]

42. Zhang, Q.; Wang, Y.; Qu, D.; Yu, J.; Yang, J. The Possible Pathogenesis of Idiopathic Pulmonary Fibrosis considering MUC5B. *Biomed Res. Int.* **2019**, *2019*, 9712464. [CrossRef] [PubMed]

43. Bonner, J.C.; Rice, A.B.; Ingram, J.L.; Moomaw, C.R.; Nyska, A.; Bradbury, A.; Sessoms, A.R.; Chulada, P.C.; Morgan, D.L.; Zeldin, D.C.; et al. Susceptibility of cyclooxygenase-2-deficient mice to pulmonary fibrogenesis. *Am. J. Pathol.* **2002**, *161*, 459–470. [CrossRef]

44. Card, J.W.; Voltz, J.W.; Carey, M.A.; Bradbury, J.A.; Degraff, L.M.; Lih, F.B.; Bonner, J.C.; Morgan, D.L.; Flake, G.P.; Zeldin, D.C. Cyclooxygenase-2 deficiency exacerbates bleomycin-induced lung dysfunction but not fibrosis. *Am. J. Respir. Cell Mol. Biol.* **2007**, *37*, 300–308. [CrossRef]

45. Dackor, R.T.; Cheng, J.; Voltz, J.W.; Card, J.W.; Ferguson, C.D.; Garrett, R.C.; Bradbury, J.A.; DeGraff, L.M.; Lih, F.B.; Tomer, K.B.; et al. Prostaglandin E(2) protects murine lungs from bleomycin-induced pulmonary fibrosis and lung dysfunction. *Am. J. Physiol. Lung Cell. Mol. Physiol.* **2011**, *301*, L645–L655. [CrossRef] [PubMed]

46. Garbuzenko, O.B.; Ivanova, V.; Kholodovych, V.; Reimer, D.C.; Reuhl, K.R.; Yurkow, E.; Adler, D.; Minko, T. Combinatorial treatment of idiopathic pulmonary fibrosis using nanoparticles with prostaglandin E and siRNA(s). *Nanomedicine* **2017**, *13*, 1983–1992. [CrossRef]

47. Ivanova, V.; Garbuzenko, O.B.; Reuhl, K.R.; Reimer, D.C.; Pozharov, V.P.; Minko, T. Inhalation treatment of pulmonary fibrosis by liposomal prostaglandin E2. *Eur. J. Pharm. Biopharm.* **2013**, *84*, 335–344. [CrossRef] [PubMed]

48. Molina-Molina, M.; Serrano-Mollar, A.; Bulbena, O.; Fernandez-Zabalegui, L.; Closa, D.; Marin-Arguedas, A.; Torrego, A.; Mullol, J.; Picado, C.; Xaubet, A. Losartan attenuates bleomycin induced lung fibrosis by increasing prostaglandin E2 synthesis. *Thorax* **2006**, *61*, 604–610. [CrossRef]

49. Lovgren, A.K.; Jania, L.A.; Hartney, J.M.; Parsons, K.K.; Audoly, L.P.; Fitzgerald, G.A.; Tilley, S.L.; Koller, B.H. COX-2-derived prostacyclin protects against bleomycin-induced pulmonary fibrosis. *Am. J. Physiol. Lung Cell. Mol. Physiol.* **2006**, *291*, L144–L156. [CrossRef]

50. Failla, M.; Genovese, T.; Mazzon, E.; Fruciano, M.; Fagone, E.; Gili, E.; Barera, A.; La Rosa, C.; Conte, E.; Crimi, N.; et al. 16,16-Dimethyl prostaglandin E2 efficacy on prevention and protection from bleomycin-induced lung injury and fibrosis. *Am. J. Respir. Cell Mol. Biol.* **2009**, *41*, 50–58. [CrossRef]

51. Nakanishi, T.; Hasegawa, Y.; Mimura, R.; Wakayama, T.; Uetoko, Y.; Komori, H.; Akanuma, S.; Hosoya, K.; Tamai, I. Prostaglandin Transporter (PGT/SLCO2A1) Protects the Lung from Bleomycin-Induced Fibrosis. *PLoS ONE* **2015**, *10*, e0123895. [CrossRef] [PubMed]

52. Borok, Z.; Gillissen, A.; Buhl, R.; Hoyt, R.F.; Hubbard, R.C.; Ozaki, T.; Rennard, S.I.; Crystal, R.G. Augmentation of functional prostaglandin E levels on the respiratory epithelial surface by aerosol administration of prostaglandin E. *Am. Rev. Respir. Dis.* **1991**, *144*, 1080–1084. [CrossRef] [PubMed]

53. Ozaki, T.; Moriguchi, H.; Nakamura, Y.; Kamei, T.; Yasuoka, S.; Ogura, T. Regulatory effect of prostaglandin E2 on fibronectin release from human alveolar macrophages. *Am. Rev. Respir. Dis.* **1990**, *141*, 965–969. [CrossRef]

54. Coward, W.R.; Watts, K.; Feghali-Bostwick, C.A.; Knox, A.; Pang, L. Defective histone acetylation is responsible for the diminished expression of cyclooxygenase 2 in idiopathic pulmonary fibrosis. *Mol. Cell. Biol.* **2009**, *29*, 4325–4339. [CrossRef]
55. Petkova, D.K.; Clelland, C.A.; Ronan, J.E.; Lewis, S.; Knox, A.J. Reduced expression of cyclooxygenase (COX) in idiopathic pulmonary fibrosis and sarcoidosis. *Histopathology* **2003**, *43*, 381–386. [CrossRef]
56. Maher, T.M.; Evans, I.C.; Bottoms, S.E.; Mercer, P.F.; Thorley, A.J.; Nicholson, A.G.; Laurent, G.J.; Tetley, T.D.; Chambers, R.C.; McAnulty, R.J. Diminished prostaglandin E2 contributes to the apoptosis paradox in idiopathic pulmonary fibrosis. *Am. J. Respir. Crit. Care Med.* **2010**, *182*, 73–82. [CrossRef]
57. Ogushi, F.; Endo, T.; Tani, K.; Asada, K.; Kawano, T.; Tada, H.; Maniwa, K.; Sone, S. Decreased prostaglandin E2 synthesis by lung fibroblasts isolated from rats with bleomycin-induced lung fibrosis. *Int. J. Exp. Pathol.* **1999**, *80*, 41–49. [CrossRef]
58. Vancheri, C.; Sortino, M.A.; Tomaselli, V.; Mastruzzo, C.; Condorelli, F.; Bellistri, G.; Pistorio, M.P.; Canonico, P.L.; Crimi, N. Different expression of TNF-alpha receptors and prostaglandin E(2)Production in normal and fibrotic lung fibroblasts: Potential implications for the evolution of the inflammatory process. *Am. J. Respir. Cell Mol. Biol.* **2000**, *22*, 628–634. [CrossRef]
59. Wilborn, J.; Crofford, L.J.; Burdick, M.D.; Kunkel, S.L.; Strieter, R.M.; Peters-Golden, M. Cultured lung fibroblasts isolated from patients with idiopathic pulmonary fibrosis have a diminished capacity to synthesize prostaglandin E2 and to express cyclooxygenase-2. *J. Clin. Investig.* **1995**, *95*, 1861–1868. [CrossRef]
60. Moore, B.B.; Peters-Golden, M.; Christensen, P.J.; Lama, V.; Kuziel, W.A.; Paine, R., 3rd; Toews, G.B. Alveolar epithelial cell inhibition of fibroblast proliferation is regulated by MCP-1/CCR2 and mediated by PGE2. *Am. J. Physiol. Lung Cell. Mol. Physiol.* **2003**, *284*, L342–L349. [CrossRef]
61. Toh, H.; Ichikawa, A.; Narumiya, S. Molecular evolution of receptors for eicosanoids. *FEBS Lett.* **1995**, *361*, 17–21. [CrossRef]
62. Dey, I.; Lejeune, M.; Chadee, K. Prostaglandin E2 receptor distribution and function in the gastrointestinal tract. *Br. J. Pharmacol.* **2006**, *149*, 611–623. [CrossRef] [PubMed]
63. Huang, S.K.; White, E.S.; Wettlaufer, S.H.; Grifka, H.; Hogaboam, C.M.; Thannickal, V.J.; Horowitz, J.C.; Peters-Golden, M. Prostaglandin E(2) induces fibroblast apoptosis by modulating multiple survival pathways. *FASEB J.* **2009**, *23*, 4317–4326. [CrossRef]
64. White, E.S.; Atrasz, R.G.; Dickie, E.G.; Aronoff, D.M.; Stambolic, V.; Mak, T.W.; Moore, B.B.; Peters-Golden, M. Prostaglandin E(2) inhibits fibroblast migration by E-prostanoid 2 receptor-mediated increase in PTEN activity. *Am. J. Respir. Cell Mol. Biol.* **2005**, *32*, 135–141. [CrossRef]
65. Dunkern, T.R.; Feurstein, D.; Rossi, G.A.; Sabatini, F.; Hatzelmann, A. Inhibition of TGF-beta induced lung fibroblast to myofibroblast conversion by phosphodiesterase inhibiting drugs and activators of soluble guanylyl cyclase. *Eur. J. Pharmacol.* **2007**, *572*, 12–22. [CrossRef]
66. Togo, S.; Liu, X.; Wang, X.; Sugiura, H.; Kamio, K.; Kawasaki, S.; Kobayashi, T.; Ertl, R.F.; Ahn, Y.; Holz, O.; et al. PDE4 inhibitors roflumilast and rolipram augment PGE2 inhibition of TGF-{beta}1-stimulated fibroblasts. *Am. J. Physiol. Lung Cell. Mol. Physiol.* **2009**, *296*, L959–L969. [CrossRef]
67. Watanabe, T.; Satoh, H.; Togoh, M.; Taniguchi, S.; Hashimoto, Y.; Kurokawa, K. Positive and negative regulation of cell proliferation through prostaglandin receptors in NIH-3T3 cells. *J. Cell. Physiol.* **1996**, *169*, 401–409. [CrossRef]
68. Harding, P.; LaPointe, M.C. Prostaglandin E2 increases cardiac fibroblast proliferation and increases cyclin D expression via EP1 receptor. *Leukot. Essent. Fat. Acids* **2011**, *84*, 147–152. [CrossRef]
69. Huang, S.K.; Wettlaufer, S.H.; Hogaboam, C.M.; Flaherty, K.R.; Martinez, F.J.; Myers, J.L.; Colby, T.V.; Travis, W.D.; Toews, G.B.; Peters-Golden, M. Variable prostaglandin E2 resistance in fibroblasts from patients with usual interstitial pneumonia. *Am. J. Respir. Crit. Care Med.* **2008**, *177*, 66–74. [CrossRef]
70. Moore, B.B.; Ballinger, M.N.; White, E.S.; Green, M.E.; Herrygers, A.B.; Wilke, C.A.; Toews, G.B.; Peters-Golden, M. Bleomycin-induced E prostanoid receptor changes alter fibroblast responses to prostaglandin E2. *J. Immunol.* **2005**, *174*, 5644–5649. [CrossRef]
71. Huang, S.K.; Fisher, A.S.; Scruggs, A.M.; White, E.S.; Hogaboam, C.M.; Richardson, B.C.; Peters-Golden, M. Hypermethylation of PTGER2 confers prostaglandin E2 resistance in fibrotic fibroblasts from humans and mice. *Am. J. Pathol.* **2010**, *177*, 2245–2255. [CrossRef]
72. Chapman, H.A.; Allen, C.L.; Stone, O.L. Abnormalities in pathways of alveolar fibrin turnover among patients with interstitial lung disease. *Am. Rev. Respir. Dis.* **1986**, *133*, 437–443. [CrossRef]

73. Eitzman, D.T.; McCoy, R.D.; Zheng, X.; Fay, W.P.; Shen, T.; Ginsburg, D.; Simon, R.H. Bleomycin-induced pulmonary fibrosis in transgenic mice that either lack or overexpress the murine plasminogen activator inhibitor-1 gene. *J. Clin. Investig.* **1996**, *97*, 232–237. [CrossRef]

74. Bauman, K.A.; Wettlaufer, S.H.; Okunishi, K.; Vannella, K.M.; Stoolman, J.S.; Huang, S.K.; Courey, A.J.; White, E.S.; Hogaboam, C.M.; Simon, R.H.; et al. The antifibrotic effects of plasminogen activation occur via prostaglandin E2 synthesis in humans and mice. *J. Clin. Investig.* **2010**, *120*, 1950–1960. [CrossRef]

75. Allan, E.H.; Martin, T.J. Prostaglandin E2 regulates production of plasminogen activator isoenzymes, urokinase receptor, and plasminogen activator inhibitor-1 in primary cultures of rat calvarial osteoblasts. *J. Cell. Physiol.* **1995**, *165*, 521–529. [CrossRef]

76. Pai, R.; Nakamura, T.; Moon, W.S.; Tarnawski, A.S. Prostaglandins promote colon cancer cell invasion; signaling by cross-talk between two distinct growth factor receptors. *FASEB J.* **2003**, *17*, 1640–1647. [CrossRef]

77. Lim, X.; Bless, D.M.; Munoz-Del-Rio, A.; Welham, N.V. Changes in cytokine signaling and extracellular matrix production induced by inflammatory factors in cultured vocal fold fibroblasts. *Ann. Otol. Rhinol. Laryngol.* **2008**, *117*, 227–238. [CrossRef]

78. Takai, E.; Tsukimoto, M.; Kojima, S. TGF-beta1 downregulates COX-2 expression leading to decrease of PGE2 production in human lung cancer A549 cells, which is involved in fibrotic response to TGF-beta1. *PLoS ONE* **2013**, *8*, e76346. [CrossRef]

79. Thomas, P.E.; Peters-Golden, M.; White, E.S.; Thannickal, V.J.; Moore, B.B. PGE(2) inhibition of TGF-beta1-induced myofibroblast differentiation is Smad-independent but involves cell shape and adhesion-dependent signaling. *Am. J. Physiol. Lung Cell. Mol. Physiol.* **2007**, *293*, L417–L428. [CrossRef]

80. Wettlaufer, S.H.; Scott, J.P.; McEachin, R.C.; Peters-Golden, M.; Huang, S.K. Reversal of the Transcriptome by Prostaglandin E2 during Myofibroblast Dedifferentiation. *Am. J. Respir. Cell Mol. Biol.* **2016**, *54*, 114–127. [CrossRef] [PubMed]

81. Remes Lenicov, F.; Paletta, A.L.; Gonzalez Prinz, M.; Varese, A.; Pavillet, C.E.; Lopez Malizia, A.; Sabatte, J.; Geffner, J.R.; Ceballos, A. Prostaglandin E2 Antagonizes TGF-beta Actions During the Differentiation of Monocytes Into Dendritic Cells. *Front. Immunol.* **2018**, *9*, 1441. [CrossRef]

82. Mukherjee, S.; Sheng, W.; Michkov, A.; Sriarm, K.; Sun, R.; Dvorkin-Gheva, A.; Insel, P.A.; Janssen, L.J. Prostaglandin E2 inhibits profibrotic function of human pulmonary fibroblasts by disrupting Ca(2+) signaling. *Am. J. Physiol. Lung Cell. Mol. Physiol.* **2019**, *316*, L810–L821. [CrossRef]

83. Oga, T.; Matsuoka, T.; Yao, C.; Nonomura, K.; Kitaoka, S.; Sakata, D.; Kita, Y.; Tanizawa, K.; Taguchi, Y.; Chin, K.; et al. Prostaglandin F(2alpha) receptor signaling facilitates bleomycin-induced pulmonary fibrosis independently of transforming growth factor-beta. *Nat. Med.* **2009**, *15*, 1426–1430. [CrossRef] [PubMed]

84. Aihara, K.; Handa, T.; Oga, T.; Watanabe, K.; Tanizawa, K.; Ikezoe, K.; Taguchi, Y.; Sato, H.; Chin, K.; Nagai, S.; et al. Clinical relevance of plasma prostaglandin F2alpha metabolite concentrations in patients with idiopathic pulmonary fibrosis. *PLoS ONE* **2013**, *8*, e66017. [CrossRef] [PubMed]

85. Radmark, O.; Werz, O.; Steinhilber, D.; Samuelsson, B. 5-Lipoxygenase, a key enzyme for leukotriene biosynthesis in health and disease. *Biochim. Biophys. Acta (BBA)-Mol. Cell Biol. Lipids* **2015**, *1851*, 331–339. [CrossRef] [PubMed]

86. Haeggstrom, J.Z. Leukotriene biosynthetic enzymes as therapeutic targets. *J. Clin. Investig.* **2018**, *128*, 2680–2690. [CrossRef] [PubMed]

87. Wardlaw, A.J.; Hay, H.; Cromwell, O.; Collins, J.V.; Kay, A.B. Leukotrienes, LTC4 and LTB4, in bronchoalveolar lavage in bronchial asthma and other respiratory diseases. *J. Allergy Clin. Immunol.* **1989**, *84*, 19–26. [CrossRef]

88. Wilborn, J.; Bailie, M.; Coffey, M.; Burdick, M.; Strieter, R.; Peters-Golden, M. Constitutive activation of 5-lipoxygenase in the lungs of patients with idiopathic pulmonary fibrosis. *J. Clin. Investig.* **1996**, *97*, 1827–1836. [CrossRef]

89. Shimbori, C.; Shiota, N.; Okunishi, H. Involvement of leukotrienes in the pathogenesis of silica-induced pulmonary fibrosis in mice. *Exp. Lung Res.* **2010**, *36*, 292–301. [CrossRef]

90. Peters-Golden, M.; Bailie, M.; Marshall, T.; Wilke, C.; Phan, S.H.; Toews, G.B.; Moore, B.B. Protection from pulmonary fibrosis in leukotriene-deficient mice. *Am. J. Respir. Crit. Care Med.* **2002**, *165*, 229–235. [CrossRef]

91. Beller, T.C.; Friend, D.S.; Maekawa, A.; Lam, B.K.; Austen, K.F.; Kanaoka, Y. Cysteinyl leukotriene 1 receptor controls the severity of chronic pulmonary inflammation and fibrosis. *Proc. Natl. Acad. Sci. USA* **2004**, *101*, 3047–3052. [CrossRef] [PubMed]

92. Yokomizo, T.; Nakamura, M.; Shimizu, T. Leukotriene receptors as potential therapeutic targets. *J. Clin. Investig.* **2018**, *128*, 2691–2701. [CrossRef] [PubMed]

93. Zdanov, S.; Bernard, D.; Debacq-Chainiaux, F.; Martien, S.; Gosselin, K.; Vercamer, C.; Chelli, F.; Toussaint, O.; Abbadie, C. Normal or stress-induced fibroblast senescence involves COX-2 activity. *Exp. Cell Res.* **2007**, *313*, 3046–3056. [CrossRef] [PubMed]

94. Catalano, A.; Rodilossi, S.; Caprari, P.; Coppola, V.; Procopio, A. 5-Lipoxygenase regulates senescence-like growth arrest by promoting ROS-dependent p53 activation. *EMBO J.* **2005**, *24*, 170–179. [CrossRef]

95. Wiley, C.D.; Brumwell, A.N.; Davis, S.S.; Jackson, J.R.; Valdovinos, A.; Calhoun, C.; Alimirah, F.; Castellanos, C.A.; Ruan, R.; Wei, Y.; et al. Secretion of leukotrienes by senescent lung fibroblasts promotes pulmonary fibrosis. *JCI Insight* **2019**, *4*. [CrossRef]

96. Lagares, D.; Santos, A.; Grasberger, P.E.; Liu, F.; Probst, C.K.; Rahimi, R.A.; Sakai, N.; Kuehl, T.; Ryan, J.; Bhola, P.; et al. Targeted apoptosis of myofibroblasts with the BH3 mimetic ABT-263 reverses established fibrosis. *Sci. Transl. Med.* **2017**, *9*. [CrossRef]

97. Haynes, C.A.; Allegood, J.C.; Park, H.; Sullards, M.C. Sphingolipidomics: Methods for the comprehensive analysis of sphingolipids. *J. Chromatogr. B* **2009**, *877*, 2696–2708. [CrossRef]

98. Goins, L.; Spassieva, S. Sphingoid bases and their involvement in neurodegenerative diseases. *Adv. Biol. Regul.* **2018**, *70*, 65–73. [CrossRef]

99. Bieberich, E. Sphingolipids and lipid rafts: Novel concepts and methods of analysis. *Chem. Phys. Lipids* **2018**, *216*, 114–131. [CrossRef] [PubMed]

100. Hannun, Y.A.; Obeid, L.M. Sphingolipids and their metabolism in physiology and disease. *Nat. Rev. Mol. Cell Biol.* **2018**, *19*, 175–191. [CrossRef] [PubMed]

101. Pavoine, C.; Pecker, F. Sphingomyelinases: Their regulation and roles in cardiovascular pathophysiology. *Cardiovasc. Res.* **2009**, *82*, 175–183. [CrossRef] [PubMed]

102. Noe, J.; Petrusca, D.; Rush, N.; Deng, P.; VanDemark, M.; Berdyshev, E.; Gu, Y.; Smith, P.; Schweitzer, K.; Pilewsky, J.; et al. CFTR regulation of intracellular pH and ceramides is required for lung endothelial cell apoptosis. *Am. J. Respir. Cell Mol. Biol.* **2009**, *41*, 314–323. [CrossRef] [PubMed]

103. Arana, L.; Gangoiti, P.; Ouro, A.; Trueba, M.; Gomez-Munoz, A. Ceramide and ceramide 1-phosphate in health and disease. *Lipids Health Dis.* **2010**, *9*, 15. [CrossRef] [PubMed]

104. Testai, F.D.; Xu, H.L.; Kilkus, J.; Suryadevara, V.; Gorshkova, I.; Berdyshev, E.; Pelligrino, D.A.; Dawson, G. Changes in the metabolism of sphingolipids after subarachnoid hemorrhage. *J. Neurosci. Res.* **2015**, *93*, 796–805. [CrossRef] [PubMed]

105. Shea, B.S.; Tager, A.M. Sphingolipid regulation of tissue fibrosis. *Open Rheumatol. J.* **2012**, *6*, 123–129. [CrossRef]

106. Suryadevara, V.; Fu, P.; Ebenezer, D.L.; Berdyshev, E.; Bronova, I.A.; Huang, L.S.; Harijith, A.; Natarajan, V. Sphingolipids in Ventilator Induced Lung Injury: Role of Sphingosine-1-Phosphate Lyase. *Int. J. Mol. Sci.* **2018**, *19*. [CrossRef]

107. Tang, X.; Benesch, M.G.; Brindley, D.N. Lipid phosphate phosphatases and their roles in mammalian physiology and pathology. *J. Lipid Res.* **2015**, *56*, 2048–2060. [CrossRef]

108. Wakashima, T.; Abe, K.; Kihara, A. Dual functions of the trans-2-enoyl-CoA reductase TER in the sphingosine 1-phosphate metabolic pathway and in fatty acid elongation. *J. Biol. Chem.* **2014**, *289*, 24736–24748. [CrossRef] [PubMed]

109. Ebenezer, D.L.; Fu, P.; Suryadevara, V.; Zhao, Y.; Natarajan, V. Epigenetic regulation of pro-inflammatory cytokine secretion by sphingosine 1-phosphate (S1P) in acute lung injury: Role of S1P lyase. *Adv. Biol. Regul.* **2017**, *63*, 156–166. [CrossRef]

110. Ichikawa, S.; Sakiyama, H.; Suzuki, G.; Hidari, K.I.; Hirabayashi, Y. Expression cloning of a cDNA for human ceramide glucosyltransferase that catalyzes the first glycosylation step of glycosphingolipid synthesis. *Proc. Natl. Acad. Sci. USA* **1996**, *93*, 12654. [CrossRef]

111. D'Angelo, G.; Capasso, S.; Sticco, L.; Russo, D. Glycosphingolipids: Synthesis and functions. *FEBS J.* **2013**, *280*, 6338–6353. [CrossRef] [PubMed]

112. Sugiura, M.; Kono, K.; Liu, H.; Shimizugawa, T.; Minekura, H.; Spiegel, S.; Kohama, T. Ceramide kinase, a novel lipid kinase. Molecular cloning and functional characterization. *J. Biol. Chem.* **2002**, *277*, 23294–23300. [CrossRef] [PubMed]

113. Bornancin, F. Ceramide kinase: The first decade. *Cell. Signal.* **2011**, *23*, 999–1008. [CrossRef] [PubMed]

114. Hoeferlin, L.A.; Wijesinghe, D.S.; Chalfant, C.E. The role of ceramide-1-phosphate in biological functions. *Handb. Exp. Pharmacol.* **2013**, 153–166. [CrossRef]

115. Berger, A.; Rosenthal, D.; Spiegel, S. Sphingosylphosphocholine, a signaling molecule which accumulates in Niemann-Pick disease type A, stimulates DNA-binding activity of the transcription activator protein AP-1. *Proc. Natl. Acad. Sci. USA* **1995**, *92*, 5885–5889. [CrossRef] [PubMed]

116. Jenkins, R.W.; Canals, D.; Hannun, Y.A. Roles and regulation of secretory and lysosomal acid sphingomyelinase. *Cell. Signal.* **2009**, *21*, 836–846. [CrossRef]

117. Beckmann, N.; Becker, K.A.; Kadow, S.; Schumacher, F.; Kramer, M.; Kuhn, C.; Schulz-Schaeffer, W.J.; Edwards, M.J.; Kleuser, B.; Gulbins, E.; et al. Acid Sphingomyelinase Deficiency Ameliorates Farber Disease. *Int. J. Mol. Sci.* **2019**, *20*. [CrossRef]

118. Teichgraber, V.; Ulrich, M.; Endlich, N.; Riethmuller, J.; Wilker, B.; De Oliveira-Munding, C.C.; van Heeckeren, A.M.; Barr, M.L.; von Kurthy, G.; Schmid, K.W.; et al. Ceramide accumulation mediates inflammation, cell death and infection susceptibility in cystic fibrosis. *Nat. Med.* **2008**, *14*, 382–391. [CrossRef]

119. Becker, K.A.; Riethmuller, J.; Luth, A.; Doring, G.; Kleuser, B.; Gulbins, E. Acid sphingomyelinase inhibitors normalize pulmonary ceramide and inflammation in cystic fibrosis. *Am. J. Respir. Cell Mol. Biol.* **2010**, *42*, 716–724. [CrossRef]

120. Riethmuller, J.; Anthonysamy, J.; Serra, E.; Schwab, M.; Doring, G.; Gulbins, E. Therapeutic efficacy and safety of amitriptyline in patients with cystic fibrosis. *Cell. Physiol. Biochem.* **2009**, *24*, 65–72. [CrossRef]

121. Dhami, R.; He, X.; Schuchman, E.H. Acid sphingomyelinase deficiency attenuates bleomycin-induced lung inflammation and fibrosis in mice. *Cell. Physiol. Biochem.* **2010**, *26*, 749–760. [CrossRef] [PubMed]

122. Simonaro, C.M.; Park, J.H.; Eliyahu, E.; Shtraizent, N.; McGovern, M.M.; Schuchman, E.H. Imprinting at the SMPD1 locus: Implications for acid sphingomyelinase-deficient Niemann-Pick disease. *Am. J. Hum. Genet.* **2006**, *78*, 865–870. [CrossRef] [PubMed]

123. Berdyshev, E.V.; Gorshkova, I.; Skobeleva, A.; Bittman, R.; Lu, X.; Dudek, S.M.; Mirzapoiazova, T.; Garcia, J.G.; Natarajan, V. FTY720 inhibits ceramide synthases and up-regulates dihydrosphingosine 1-phosphate formation in human lung endothelial cells. *J. Biol. Chem.* **2009**, *284*, 5467–5477. [CrossRef] [PubMed]

124. Scholte, B.J.; Horati, H.; Veltman, M.; Vreeken, R.J.; Garratt, L.W.; Tiddens, H.; Janssens, H.M.; Stick, S.M. Australian Respiratory Early Surveillance Team for Cystic, F. Oxidative stress and abnormal bioactive lipids in early cystic fibrosis lung disease. *J. Cyst. Fibros.* **2019**, *18*, 781–789. [CrossRef] [PubMed]

125. Tibboel, J.; Reiss, I.; de Jongste, J.C.; Post, M. Ceramides: A potential therapeutic target in pulmonary emphysema. *Respir. Res* **2013**, *14*, 96. [CrossRef] [PubMed]

126. Huang, L.S.; Berdyshev, E.; Mathew, B.; Fu, P.; Gorshkova, I.A.; He, D.; Ma, W.; Noth, I.; Ma, S.F.; Pendyala, S.; et al. Targeting sphingosine kinase 1 attenuates bleomycin-induced pulmonary fibrosis. *FASEB J.* **2013**, *27*, 1749–1760. [CrossRef]

127. Mathew, B.; Jacobson, J.R.; Berdyshev, E.; Huang, Y.; Sun, X.; Zhao, Y.; Gerhold, L.M.; Siegler, J.; Evenoski, C.; Wang, T.; et al. Role of sphingolipids in murine radiation-induced lung injury: Protection by sphingosine 1-phosphate analogs. *FASEB J.* **2011**, *25*, 3388–3400. [CrossRef]

128. Christofidou-Solomidou, M.; Pietrofesa, R.A.; Arguiri, E.; Schweitzer, K.S.; Berdyshev, E.V.; McCarthy, M.; Corbitt, A.; Alwood, J.S.; Yu, Y.; Globus, R.K.; et al. Space radiation-associated lung injury in a murine model. *Am. J. Physiol. Lung Cell. Mol. Physiol.* **2015**, *308*, L416–L428. [CrossRef]

129. Zhao, Y.D.; Yin, L.; Archer, S.; Lu, C.; Zhao, G.; Yao, Y.; Wu, L.; Hsin, M.; Waddell, T.K.; Keshavjee, S.; et al. Metabolic heterogeneity of idiopathic pulmonary fibrosis: A metabolomic study. *BMJ Open Respir. Res.* **2017**, *4*, e000183. [CrossRef]

130. Okajima, F. Plasma lipoproteins behave as carriers of extracellular sphingosine 1-phosphate: Is this an atherogenic mediator or an anti-atherogenic mediator? *Biochim. Biophys. Acta (BBA)-Mol. Cell Biol. Lipids* **2002**, *1582*, 132–137. [CrossRef]

131. Xu, N.; Dahlback, B. A novel human apolipoprotein (apoM). *J. Biol. Chem.* **1999**, *274*, 31286–31290. [CrossRef] [PubMed]

132. Christoffersen, C.; Obinata, H.; Kumaraswamy, S.B.; Galvani, S.; Ahnstrom, J.; Sevvana, M.; Egerer-Sieber, C.; Muller, Y.A.; Hla, T.; Nielsen, L.B.; et al. Endothelium-protective sphingosine-1-phosphate provided by HDL-associated apolipoprotein M. *Proc. Natl. Acad. Sci. USA* **2011**, *108*, 9613–9618. [CrossRef] [PubMed]

133. Ebenezer, D.L.; Berdyshev, E.V.; Bronova, I.A.; Liu, Y.; Tiruppathi, C.; Komarova, Y.; Benevolenskaya, E.V.; Suryadevara, V.; Ha, A.W.; Harijith, A.; et al. Pseudomonas aeruginosa stimulates nuclear sphingosine-1-phosphate generation and epigenetic regulation of lung inflammatory injury. *Thorax* **2019**, *74*, 579–591. [CrossRef] [PubMed]

134. Natarajan, V.; Dudek, S.M.; Jacobson, J.R.; Moreno-Vinasco, L.; Huang, L.S.; Abassi, T.; Mathew, B.; Zhao, Y.; Wang, L.; Bittman, R.; et al. Sphingosine-1-phosphate, FTY720, and sphingosine-1-phosphate receptors in the pathobiology of acute lung injury. *Am. J. Respir. Cell Mol. Biol.* **2013**, *49*, 6–17. [CrossRef] [PubMed]

135. Fu, P.; Ebenezer, D.L.; Ha, A.W.; Suryadevara, V.; Harijith, A.; Natarajan, V. Nuclear lipid mediators: Role of nuclear sphingolipids and sphingosine-1-phosphate signaling in epigenetic regulation of inflammation and gene expression. *J. Cell. Biochem.* **2018**, *119*, 6337–6353. [CrossRef] [PubMed]

136. Saba, J.D. Fifty years of lyase and a moment of truth: Sphingosine phosphate lyase from discovery to disease. *J. Lipid Res.* **2019**, *60*, 456–463. [CrossRef]

137. Ebenezer, D.L.; Fu, P.; Natarajan, V. Targeting sphingosine-1-phosphate signaling in lung diseases. *Pharmacol. Ther.* **2016**, *168*, 143–157. [CrossRef]

138. Ksiazek, M.; Chacinska, M.; Chabowski, A.; Baranowski, M. Sources, metabolism, and regulation of circulating sphingosine-1-phosphate. *J. Lipid Res.* **2015**, *56*, 1271–1281. [CrossRef]

139. Nagahashi, M.; Takabe, K.; Terracina, K.P.; Soma, D.; Hirose, Y.; Kobayashi, T.; Matsuda, Y.; Wakai, T. Sphingosine-1-phosphate transporters as targets for cancer therapy. *Biomed Res. Int.* **2014**, *2014*, 651727. [CrossRef]

140. Strub, G.M.; Paillard, M.; Liang, J.; Gomez, L.; Allegood, J.C.; Hait, N.C.; Maceyka, M.; Price, M.M.; Chen, Q.; Simpson, D.C.; et al. Sphingosine-1-phosphate produced by sphingosine kinase 2 in mitochondria interacts with prohibitin 2 to regulate complex IV assembly and respiration. *FASEB J.* **2011**, *25*, 600–612. [CrossRef]

141. Takasugi, N.; Sasaki, T.; Suzuki, K.; Osawa, S.; Isshiki, H.; Hori, Y.; Shimada, N.; Higo, T.; Yokoshima, S.; Fukuyama, T.; et al. BACE1 activity is modulated by cell-associated sphingosine-1-phosphate. *J. Neurosci.* **2011**, *31*, 6850–6857. [CrossRef] [PubMed]

142. Alvarez, S.E.; Harikumar, K.B.; Hait, N.C.; Allegood, J.; Strub, G.M.; Kim, E.Y.; Maceyka, M.; Jiang, H.; Luo, C.; Kordula, T.; et al. Sphingosine-1-phosphate is a missing cofactor for the E3 ubiquitin ligase TRAF2. *Nature* **2010**, *465*, 1084–1088. [CrossRef] [PubMed]

143. Etemadi, N.; Chopin, M.; Anderton, H.; Tanzer, M.C.; Rickard, J.A.; Abeysekera, W.; Hall, C.; Spall, S.K.; Wang, B.; Xiong, Y.; et al. TRAF2 regulates TNF and NF-kappaB signalling to suppress apoptosis and skin inflammation independently of Sphingosine kinase 1. *Elife* **2015**, *4*. [CrossRef] [PubMed]

144. Zhao, Y.; Gorshkova, I.A.; Berdyshev, E.; He, D.; Fu, P.; Ma, W.; Su, Y.; Usatyuk, P.V.; Pendyala, S.; Oskouian, B.; et al. Protection of LPS-induced murine acute lung injury by sphingosine-1-phosphate lyase suppression. *Am. J. Respir. Cell Mol. Biol.* **2011**, *45*, 426–435. [CrossRef]

145. McVerry, B.J.; Peng, X.; Hassoun, P.M.; Sammani, S.; Simon, B.A.; Garcia, J.G. Sphingosine 1-phosphate reduces vascular leak in murine and canine models of acute lung injury. *Am. J. Respir. Crit. Care Med.* **2004**, *170*, 987–993. [CrossRef]

146. Peng, X.; Hassoun, P.M.; Sammani, S.; McVerry, B.J.; Burne, M.J.; Rabb, H.; Pearse, D.; Tuder, R.M.; Garcia, J.G. Protective effects of sphingosine 1-phosphate in murine endotoxin-induced inflammatory lung injury. *Am. J. Respir. Crit. Care Med.* **2004**, *169*, 1245–1251. [CrossRef]

147. Harijith, A.; Pendyala, S.; Reddy, N.M.; Bai, T.; Usatyuk, P.V.; Berdyshev, E.; Gorshkova, I.; Huang, L.S.; Mohan, V.; Garzon, S.; et al. Sphingosine kinase 1 deficiency confers protection against hyperoxia-induced bronchopulmonary dysplasia in a murine model: Role of S1P signaling and Nox proteins. *Am. J. Pathol.* **2013**, *183*, 1169–1182. [CrossRef]

148. Chen, J.; Tang, H.; Sysol, J.R.; Moreno-Vinasco, L.; Shioura, K.M.; Chen, T.; Gorshkova, I.; Wang, L.; Huang, L.S.; Usatyuk, P.V.; et al. The sphingosine kinase 1/sphingosine-1-phosphate pathway in pulmonary arterial hypertension. *Am. J. Respir. Crit. Care Med.* **2014**, *190*, 1032–1043. [CrossRef]

149. Ammit, A.J.; Hastie, A.T.; Edsall, L.C.; Hoffman, R.K.; Amrani, Y.; Krymskaya, V.P.; Kane, S.A.; Peters, S.P.; Penn, R.B.; Spiegel, S.; et al. Sphingosine 1-phosphate modulates human airway smooth muscle cell functions that promote inflammation and airway remodeling in asthma. *FASEB J.* **2001**, *15*, 1212–1214. [CrossRef]

150. Huang, L.S.; Natarajan, V. Sphingolipids in pulmonary fibrosis. *Adv. Biol. Regul.* **2015**, *57*, 55–63. [CrossRef]

151. Milara, J.; Navarro, R.; Juan, G.; Peiro, T.; Serrano, A.; Ramon, M.; Morcillo, E.; Cortijo, J. Sphingosine-1-phosphate is increased in patients with idiopathic pulmonary fibrosis and mediates epithelial to mesenchymal transition. *Thorax* **2012**, *67*, 147–156. [CrossRef] [PubMed]

152. Wang, L.; Dudek, S.M. Regulation of vascular permeability by sphingosine 1-phosphate. *Microvasc. Res.* **2009**, *77*, 39–45. [CrossRef] [PubMed]

153. Lai, W.Q.; Wong, W.S.; Leung, B.P. Sphingosine kinase and sphingosine 1-phosphate in asthma. *Biosci. Rep.* **2011**, *31*, 145–150. [CrossRef] [PubMed]

154. Sudhadevi, T.; Ha, A.W.; Ebenezer, D.L.; Fu, P.; Putherickal, V.; Natarajan, V.; Harijith, A. Advancements in understanding the role of lysophospholipids and their receptors in lung disorders including bronchopulmonary dysplasia. *Biochim. Biophys. Acta Mol. Cell Biol. Lipids* **2020**, *1865*, 158685. [CrossRef]

155. Gorshkova, I.; Zhou, T.; Mathew, B.; Jacobson, J.R.; Takekoshi, D.; Bhattacharya, P.; Smith, B.; Aydogan, B.; Weichselbaum, R.R.; Natarajan, V.; et al. Inhibition of serine palmitoyltransferase delays the onset of radiation-induced pulmonary fibrosis through the negative regulation of sphingosine kinase-1 expression. *J. Lipid Res.* **2012**, *53*, 1553–1568. [CrossRef]

156. Morishima, Y.; Nomura, A.; Uchida, Y.; Noguchi, Y.; Sakamoto, T.; Ishii, Y.; Goto, Y.; Masuyama, K.; Zhang, M.J.; Hirano, K.; et al. Triggering the induction of myofibroblast and fibrogenesis by airway epithelial shedding. *Am. J. Respir. Cell Mol. Biol.* **2001**, *24*, 1–11. [CrossRef]

157. Leach, H.G.; Chrobak, I.; Han, R.; Trojanowska, M. Endothelial cells recruit macrophages and contribute to a fibrotic milieu in bleomycin lung injury. *Am. J. Respir. Cell Mol. Biol.* **2013**, *49*, 1093–1101. [CrossRef]

158. Huang, L.S.; Sudhadevi, T.; Fu, P.; Punathil-Kannan, P.K.; Ebenezer, D.L.; Ramchandran, R.; Putherickal, V.; Cheresh, P.; Zhou, G.; Ha, A.W.; et al. Sphingosine Kinase 1/S1P Signaling Contributes to Pulmonary Fibrosis by Activating Hippo/YAP Pathway and Mitochondrial Reactive Oxygen Species in Lung Fibroblasts. *Int. J. Mol. Sci.* **2020**, *21*. [CrossRef]

159. Kim, S.-J.; Cheresh, P.; Huang, L.; Watanabe, S.; Joshi, N.; Williams, K.; Piseaux-Allion, R.; Chi, M.; Yeldandi, A.; Lam, A. The Sphingosine Kinase 1 Inhibitor, PF543, Mitigates Asbestos-Induced Pulmonary Fibrosis and Lung mtDNA Damage in Mice in *D107. Mitochondria and er Stress in Homeostasis and Repair.* Presented at American Thoracis Society 2019 International Conference, Dallas, TX, USA, 17–22 May 2019; p. A7220.

160. Bu, S.; Kapanadze, B.; Hsu, T.; Trojanowska, M. Opposite effects of dihydrosphingosine 1-phosphate and sphingosine 1-phosphate on transforming growth factor-beta/Smad signaling are mediated through the PTEN/PPM1A-dependent pathway. *J. Biol. Chem.* **2008**, *283*, 19593–19602. [CrossRef]

161. Bu, S.; Yamanaka, M.; Pei, H.; Bielawska, A.; Bielawski, J.; Hannun, Y.A.; Obeid, L.; Trojanowska, M. Dihydrosphingosine 1-phosphate stimulates MMP1 gene expression via activation of ERK1/2-Ets1 pathway in human fibroblasts. *FASEB J.* **2006**, *20*, 184–186. [CrossRef]

162. Yamanaka, M.; Shegogue, D.; Pei, H.; Bu, S.; Bielawska, A.; Bielawski, J.; Pettus, B.; Hannun, Y.A.; Obeid, L.; Trojanowska, M. Sphingosine kinase 1 (SPHK1) is induced by transforming growth factor-beta and mediates TIMP-1 up-regulation. *J. Biol. Chem.* **2004**, *279*, 53994–54001. [CrossRef]

163. Bu, S.; Asano, Y.; Bujor, A.; Highland, K.; Hant, F.; Trojanowska, M. Dihydrosphingosine 1-phosphate has a potent antifibrotic effect in scleroderma fibroblasts via normalization of phosphatase and tensin homolog levels. *Arthritis Rheum.* **2010**, *62*, 2117–2126. [CrossRef] [PubMed]

164. Huang, L.S.; Berdyshev, E.V.; Tran, J.T.; Xie, L.; Chen, J.; Ebenezer, D.L.; Mathew, B.; Gorshkova, I.; Zhang, W.; Reddy, S.P.; et al. Sphingosine-1-phosphate lyase is an endogenous suppressor of pulmonary fibrosis: Role of S1P signalling and autophagy. *Thorax* **2015**, *70*, 1138–1148. [CrossRef] [PubMed]

165. Patel, A.S.; Lin, L.; Geyer, A.; Haspel, J.A.; An, C.H.; Cao, J.; Rosas, I.O.; Morse, D. Autophagy in idiopathic pulmonary fibrosis. *PLoS ONE* **2012**, *7*, e41394. [CrossRef] [PubMed]

166. Huang, Y.L.; Chang, C.L.; Tang, C.H.; Lin, Y.C.; Ju, T.K.; Huang, W.P.; Lee, H. Extrinsic sphingosine 1-phosphate activates S1P5 and induces autophagy through generating endoplasmic reticulum stress in human prostate cancer PC-3 cells. *Cell. Signal.* **2014**, *26*, 611–618. [CrossRef] [PubMed]

167. Huang, Y.L.; Huang, W.P.; Lee, H. Roles of sphingosine 1-phosphate on tumorigenesis. *World J. Biol. Chem.* **2011**, *2*, 25–34. [CrossRef]

168. Lepine, S.; Allegood, J.C.; Park, M.; Dent, P.; Milstien, S.; Spiegel, S. Sphingosine-1-phosphate phosphohydrolase-1 regulates ER stress-induced autophagy. *Cell Death Differ.* **2011**, *18*, 350–361. [CrossRef]

169. Sheng, R.; Zhang, T.T.; Felice, V.D.; Qin, T.; Qin, Z.H.; Smith, C.D.; Sapp, E.; Difiglia, M.; Waeber, C. Preconditioning stimuli induce autophagy via sphingosine kinase 2 in mouse cortical neurons. *J. Biol. Chem.* **2014**, *289*, 20845–20857. [CrossRef] [PubMed]

170. Chang, C.L.; Ho, M.C.; Lee, P.H.; Hsu, C.Y.; Huang, W.P.; Lee, H. S1P(5) is required for sphingosine 1-phosphate-induced autophagy in human prostate cancer PC-3 cells. *Am. J. Physiol. Cell Physiol.* **2009**, *297*, C451–G458. [CrossRef]

171. Slattum, G.; Gu, Y.; Sabbadini, R.; Rosenblatt, J. Autophagy in oncogenic K-Ras promotes basal extrusion of epithelial cells by degrading S1P. *Curr. Biol.* **2014**, *24*, 19–28. [CrossRef]

172. Taniguchi, M.; Kitatani, K.; Kondo, T.; Hashimoto-Nishimura, M.; Asano, S.; Hayashi, A.; Mitsutake, S.; Igarashi, Y.; Umehara, H.; Takeya, H.; et al. Regulation of autophagy and its associated cell death by "sphingolipid rheostat": Reciprocal role of ceramide and sphingosine 1-phosphate in the mammalian target of rapamycin pathway. *J. Biol. Chem.* **2012**, *287*, 39898–39910. [CrossRef] [PubMed]

173. Hill, C.; Li, J.; Liu, D.; Conforti, F.; Brereton, C.J.; Yao, L.; Zhou, Y.; Alzetani, A.; Chee, S.J.; Marshall, B.G.; et al. Autophagy inhibition-mediated epithelial-mesenchymal transition augments local myofibroblast differentiation in pulmonary fibrosis. *Cell Death Dis.* **2019**, *10*, 591. [CrossRef] [PubMed]

174. Park, K.H.; Ye, Z.W.; Zhang, J.; Hammad, S.M.; Townsend, D.M.; Rockey, D.C.; Kim, S.H. 3-ketodihydrosphingosine reductase mutation induces steatosis and hepatic injury in zebrafish. *Sci. Rep.* **2019**, *9*, 1138. [CrossRef]

175. Weiss, B.; Stoffel, W. Human and murine serine-palmitoyl-CoA transferase–cloning, expression and characterization of the key enzyme in sphingolipid synthesis. *Eur. J. Biochem.* **1997**, *249*, 239–247. [CrossRef] [PubMed]

176. Yamaji-Hasegawa, A.; Takahashi, A.; Tetsuka, Y.; Senoh, Y.; Kobayashi, T. Fungal metabolite sulfamisterin suppresses sphingolipid synthesis through inhibition of serine palmitoyltransferase. *Biochemistry* **2005**, *44*, 268–277. [CrossRef]

177. Wadsworth, J.M.; Clarke, D.J.; McMahon, S.A.; Lowther, J.P.; Beattie, A.E.; Langridge-Smith, P.R.; Broughton, H.B.; Dunn, T.M.; Naismith, J.H.; Campopiano, D.J. The chemical basis of serine palmitoyltransferase inhibition by myriocin. *J. Am. Chem. Soc.* **2013**, *135*, 14276–14285. [CrossRef]

178. Watson, M.L.; Coghlan, M.; Hundal, H.S. Modulating serine palmitoyl transferase (SPT) expression and activity unveils a crucial role in lipid-induced insulin resistance in rat skeletal muscle cells. *Biochem. J.* **2009**, *417*, 791–801. [CrossRef]

179. Shea, B.S.; Brooks, S.F.; Fontaine, B.A.; Chun, J.; Luster, A.D.; Tager, A.M. Prolonged exposure to sphingosine 1-phosphate receptor-1 agonists exacerbates vascular leak, fibrosis, and mortality after lung injury. *Am. J. Respir. Cell Mol. Biol.* **2010**, *43*, 662–673. [CrossRef]

180. Shea, B.S.; Probst, C.K.; Brazee, P.L.; Rotile, N.J.; Blasi, F.; Weinreb, P.H.; Black, K.E.; Sosnovik, D.E.; Van Cott, E.M.; Violette, S.M.; et al. Uncoupling of the profibrotic and hemostatic effects of thrombin in lung fibrosis. *JCI Insight* **2017**, *2*. [CrossRef]

181. Dudek, S.M.; Camp, S.M.; Chiang, E.T.; Singleton, P.A.; Usatyuk, P.V.; Zhao, Y.; Natarajan, V.; Garcia, J.G. Pulmonary endothelial cell barrier enhancement by FTY720 does not require the S1P1 receptor. *Cell. Signal.* **2007**, *19*, 1754–1764. [CrossRef]

182. Wang, L.; Sammani, S.; Moreno-Vinasco, L.; Letsiou, E.; Wang, T.; Camp, S.M.; Bittman, R.; Garcia, J.G.; Dudek, S.M. FTY720 (s)-phosphonate preserves sphingosine 1-phosphate receptor 1 expression and exhibits superior barrier protection to FTY720 in acute lung injury. *Crit. Care Med.* **2014**, *42*, e189–e199. [CrossRef] [PubMed]

183. Zhao, J.; Okamoto, Y.; Asano, Y.; Ishimaru, K.; Aki, S.; Yoshioka, K.; Takuwa, N.; Wada, T.; Inagaki, Y.; Takahashi, C.; et al. Sphingosine-1-phosphate receptor-2 facilitates pulmonary fibrosis through potentiating IL-13 pathway in macrophages. *PLoS ONE* **2018**, *13*, e0197604. [CrossRef] [PubMed]

184. Park, S.J.; Im, D.S. Deficiency of Sphingosine-1-Phosphate Receptor 2 (S1P2) Attenuates Bleomycin-Induced Pulmonary Fibrosis. *Biomol. Ther.* **2019**, *27*, 318–326. [CrossRef] [PubMed]

185. Murakami, K.; Kohno, M.; Kadoya, M.; Nagahara, H.; Fujii, W.; Seno, T.; Yamamoto, A.; Oda, R.; Fujiwara, H.; Kubo, T.; et al. Knock out of S1P3 receptor signaling attenuates inflammation and fibrosis in bleomycin-induced lung injury mice model. *PLoS ONE* **2014**, *9*, e106792. [CrossRef]

186. Sobel, K.; Menyhart, K.; Killer, N.; Renault, B.; Bauer, Y.; Studer, R.; Steiner, B.; Bolli, M.H.; Nayler, O.; Gatfield, J. Sphingosine 1-phosphate (S1P) receptor agonists mediate pro-fibrotic responses in normal human lung fibroblasts via S1P2 and S1P3 receptors and Smad-independent signaling. *J. Biol. Chem.* **2013**, *288*, 14839–14851. [CrossRef]

187. Kono, Y.; Nishiuma, T.; Nishimura, Y.; Kotani, Y.; Okada, T.; Nakamura, S.; Yokoyama, M. Sphingosine kinase 1 regulates differentiation of human and mouse lung fibroblasts mediated by TGF-beta1. *Am. J. Respir. Cell Mol. Biol.* **2007**, *37*, 395–404. [CrossRef] [PubMed]

188. Kappos, L.; Antel, J.; Comi, G.; Montalban, X.; O'Connor, P.; Polman, C.H.; Haas, T.; Korn, A.A.; Karlsson, G.; Radue, E.W.; et al. Oral fingolimod (FTY720) for relapsing multiple sclerosis. *N. Engl. J. Med.* **2006**, *355*, 1124–1140. [CrossRef]

189. Willis, B.C.; Borok, Z. TGF-beta-induced EMT: Mechanisms and implications for fibrotic lung disease. *Am. J. Physiol. Lung Cell. Mol. Physiol.* **2007**, *293*, L525–L534. [CrossRef]

190. Wynn, T.A. Cellular and molecular mechanisms of fibrosis. *J. Pathol.* **2008**, *214*, 199–210. [CrossRef]

191. Wynn, T.A. Integrating mechanisms of pulmonary fibrosis. *J. Exp. Med.* **2011**, *208*, 1339–1350. [CrossRef]

192. Hoyt, D.G.; Lazo, J.S. Alterations in pulmonary mRNA encoding procollagens, fibronectin and transforming growth factor-beta precede bleomycin-induced pulmonary fibrosis in mice. *J. Pharmacol. Exp. Ther.* **1988**, *246*, 765–771. [PubMed]

193. Yi, E.S.; Bedoya, A.; Lee, H.; Chin, E.; Saunders, W.; Kim, S.J.; Danielpour, D.; Remick, D.G.; Yin, S.; Ulich, T.R. Radiation-induced lung injury in vivo: Expression of transforming growth factor-beta precedes fibrosis. *Inflammation* **1996**, *20*, 339–352. [CrossRef] [PubMed]

194. Massague, J. TGFbeta signalling in context. *Nat. Rev. Mol. Cell Biol.* **2012**, *13*, 616–630. [CrossRef] [PubMed]

195. Chen, H.; Zhuang, F.; Liu, Y.H.; Xu, B.; Del Moral, P.; Deng, W.; Chai, Y.; Kolb, M.; Gauldie, J.; Warburton, D.; et al. TGF-beta receptor II in epithelia versus mesenchyme plays distinct roles in the developing lung. *Eur. Respir. J.* **2008**, *32*, 285–295. [CrossRef]

196. Li, M.; Krishnaveni, M.S.; Li, C.; Zhou, B.; Xing, Y.; Banfalvi, A.; Li, A.; Lombardi, V.; Akbari, O.; Borok, Z.; et al. Epithelium-specific deletion of TGF-beta receptor type II protects mice from bleomycin-induced pulmonary fibrosis. *J. Clin. Investig.* **2011**, *121*, 277–287. [CrossRef]

197. Zhao, J.; Shi, W.; Wang, Y.L.; Chen, H.; Bringas, P., Jr.; Datto, M.B.; Frederick, J.P.; Wang, X.F.; Warburton, D. Smad3 deficiency attenuates bleomycin-induced pulmonary fibrosis in mice. *Am. J. Physiol. Lung Cell. Mol. Physiol.* **2002**, *282*, L585–L593. [CrossRef] [PubMed]

198. Gyorfi, A.H.; Matei, A.E.; Distler, J.H.W. Targeting TGF-beta signaling for the treatment of fibrosis. *Matrix Biol.* **2018**, *68–69*, 8–27. [CrossRef]

199. Cencetti, F.; Bernacchioni, C.; Nincheri, P.; Donati, C.; Bruni, P. Transforming growth factor-beta1 induces transdifferentiation of myoblasts into myofibroblasts via up-regulation of sphingosine kinase-1/S1P3 axis. *Mol. Biol. Cell* **2010**, *21*, 1111–1124. [CrossRef]

200. Cheresh, P.; Kim, S.J.; Tulasiram, S.; Kamp, D.W. Oxidative stress and pulmonary fibrosis. *Biochim. Biophys. Acta (BBA)-Mol. Basis Dis.* **2013**, *1832*, 1028–1040. [CrossRef]

201. Kurundkar, A.; Thannickal, V.J. Redox mechanisms in age-related lung fibrosis. *Redox Biol.* **2016**, *9*, 67–76. [CrossRef]

202. Dan Dunn, J.; Alvarez, L.A.; Zhang, X.; Soldati, T. Reactive oxygen species and mitochondria: A nexus of cellular homeostasis. *Redox Biol.* **2015**, *6*, 472–485. [CrossRef] [PubMed]

203. Dikalov, S. Cross talk between mitochondria and NADPH oxidases. *Free Radic. Biol. Med.* **2011**, *51*, 1289–1301. [CrossRef] [PubMed]

204. Liu, X.; Chen, Z. The pathophysiological role of mitochondrial oxidative stress in lung diseases. *J. Transl. Med.* **2017**, *15*, 207. [CrossRef] [PubMed]

205. Carnesecchi, S.; Deffert, C.; Donati, Y.; Basset, O.; Hinz, B.; Preynat-Seauve, O.; Guichard, C.; Arbiser, J.L.; Banfi, B.; Pache, J.C.; et al. A key role for NOX4 in epithelial cell death during development of lung fibrosis. *Antioxid. Redox Signal.* **2011**, *15*, 607–619. [CrossRef]

206. Hecker, L.; Vittal, R.; Jones, T.; Jagirdar, R.; Luckhardt, T.R.; Horowitz, J.C.; Pennathur, S.; Martinez, F.J.; Thannickal, V.J. NADPH oxidase-4 mediates myofibroblast activation and fibrogenic responses to lung injury. *Nat. Med.* **2009**, *15*, 1077–1081. [CrossRef] [PubMed]

207. Turn, C.S.; Lockey, R.F.; Kolliputi, N. Putting the brakes on age-related idiopathic pulmonary fibrosis: Can Nox4 inhibitors suppress IPF? *Exp. Gerontol.* **2015**, *63*, 81–82. [CrossRef]

208. Yoon, Y.S.; Lee, J.H.; Hwang, S.C.; Choi, K.S.; Yoon, G. TGF beta1 induces prolonged mitochondrial ROS generation through decreased complex IV activity with senescent arrest in Mv1Lu cells. *Oncogene* **2005**, *24*, 1895–1903. [CrossRef]

209. Zhao, R.Z.; Jiang, S.; Zhang, L.; Yu, Z.B. Mitochondrial electron transport chain, ROS generation and uncoupling (Review). *Int. J. Mol. Med.* **2019**, *44*, 3–15. [CrossRef]

210. Fu, P.; Usatyuk, P.V.; Jacobson, J.; Cress, A.E.; Garcia, J.G.; Salgia, R.; Natarajan, V. Role played by paxillin and paxillin tyrosine phosphorylation in hepatocyte growth factor/sphingosine-1-phosphate-mediated reactive oxygen species generation, lamellipodia formation, and endothelial barrier function. *Pulm. Circ.* **2015**, *5*, 619–630. [CrossRef] [PubMed]

211. Harijith, A.; Pendyala, S.; Ebenezer, D.L.; Ha, A.W.; Fu, P.; Wang, Y.T.; Ma, K.; Toth, P.T.; Berdyshev, E.V.; Kanteti, P.; et al. Hyperoxia-induced p47phox activation and ROS generation is mediated through S1P transporter Spns2, and S1P/S1P1&2 signaling axis in lung endothelium. *Am. J. Physiol. Lung Cell. Mol. Physiol.* **2016**, *311*, L337–L351. [CrossRef]

212. Fu, P.; Shaaya, M.; Harijith, A.; Jacobson, J.R.; Karginov, A.; Natarajan, V. Sphingolipids Signaling in Lamellipodia Formation and Enhancement of Endothelial Barrier Function. *Curr. Top. Membr.* **2018**, *82*, 1–31. [CrossRef] [PubMed]

213. Michaloglou, C.; Lehmann, W.; Martin, T.; Delaunay, C.; Hueber, A.; Barys, L.; Niu, H.; Billy, E.; Wartmann, M.; Ito, M.; et al. The tyrosine phosphatase PTPN14 is a negative regulator of YAP activity. *PLoS ONE* **2013**, *8*, e61916. [CrossRef] [PubMed]

214. Boggiano, J.C.; Fehon, R.G. Growth control by committee: Intercellular junctions, cell polarity, and the cytoskeleton regulate Hippo signaling. *Dev. Cell* **2012**, *22*, 695–702. [CrossRef] [PubMed]

215. Zhao, B.; Tumaneng, K.; Guan, K.L. The Hippo pathway in organ size control, tissue regeneration and stem cell self-renewal. *Nat. Cell Biol.* **2011**, *13*, 877–883. [CrossRef]

216. Ninou, I.; Kaffe, E.; Muller, S.; Budd, D.C.; Stevenson, C.S.; Ullmer, C.; Aidinis, V. Pharmacologic targeting of the ATX/LPA axis attenuates bleomycin-induced pulmonary fibrosis. *Pulm. Pharmacol. Ther.* **2018**, *52*, 32–40. [CrossRef]

217. Robinson, P.C.; Watters, L.C.; King, T.E.; Mason, R.J. Idiopathic pulmonary fibrosis. Abnormalities in bronchoalveolar lavage fluid phospholipids. *Am. Rev. Respir. Dis.* **1988**, *137*, 585–591. [CrossRef]

218. Han, S.; Mallampalli, R.K. The Role of Surfactant in Lung Disease and Host Defense against Pulmonary Infections. *Ann. Am. Thorac. Soc.* **2015**, *12*, 765–774. [CrossRef]

219. Low, R.B. Bronchoalveolar lavage lipids in idiopathic pulmonary fibrosis. *Chest* **1989**, *95*, 3–5. [CrossRef]

220. Gunther, A.; Schmidt, R.; Nix, F.; Yabut-Perez, M.; Guth, C.; Rosseau, S.; Siebert, C.; Grimminger, F.; Morr, H.; Velcovsky, H.G.; et al. Surfactant abnormalities in idiopathic pulmonary fibrosis, hypersensitivity pneumonitis and sarcoidosis. *Eur. Respir. J.* **1999**, *14*, 565–573. [CrossRef]

221. Vazquez-de-Lara, L.G.; Tlatelpa-Romero, B.; Romero, Y.; Fernandez-Tamayo, N.; Vazquez-de-Lara, F.; M Justo-Janeiro, J.; Garcia-Carrasco, M.; de-la-Rosa Paredes, R.; Cisneros-Lira, J.G.; Mendoza-Milla, C.; et al. Phosphatidylethanolamine Induces an Antifibrotic Phenotype in Normal Human Lung Fibroblasts and Ameliorates Bleomycin-Induced Lung Fibrosis in Mice. *Int. J. Mol. Sci.* **2018**, *19*. [CrossRef]

222. El Nady, M.; Kaddah, S.; El Hinnawy, Y.; Halim, R.M.; Kandeel, R. Plasma surfactant protein-D as a potential biomarker in idiopathic pulmonary fibrosis. *Egypt. J. Bronchol.* **2019**, *13*, 214–218. [CrossRef]

223. Greene, K.E.; King, T.E., Jr.; Kuroki, Y.; Bucher-Bartelson, B.; Hunninghake, G.W.; Newman, L.S.; Nagae, H.; Mason, R.J. Serum surfactant proteins-A and -D as biomarkers in idiopathic pulmonary fibrosis. *Eur. Respir. J.* **2002**, *19*, 439–446. [CrossRef] [PubMed]

224. Ruppert, C.; Hirschbach, L.; Nef, H.; Seeger, W.; Guenther, A.; Markart, P. Surfactant protein B proforms as potential new biomarkers for idiopathic pulmonary fibrosis. *Eur. Respir. J.* **2014**, *44*, P772.

225. McCormack, F.X.; King, T.E., Jr.; Bucher, B.L.; Nielsen, L.; Mason, R.J. Surfactant protein A predicts survival in idiopathic pulmonary fibrosis. *Am. J. Respir. Crit. Care Med.* **1995**, *152*, 751–759. [CrossRef] [PubMed]

226. Lawson, W.E.; Polosukhin, V.V.; Stathopoulos, G.T.; Zoia, O.; Han, W.; Lane, K.B.; Li, B.; Donnelly, E.F.; Holburn, G.E.; Lewis, K.G.; et al. Increased and prolonged pulmonary fibrosis in surfactant protein C-deficient mice following intratracheal bleomycin. *Am. J. Pathol.* **2005**, *167*, 1267–1277. [CrossRef]

227. Chung, K.P.; Hsu, C.L.; Fan, L.C.; Huang, Z.; Bhatia, D.; Chen, Y.J.; Hisata, S.; Cho, S.J.; Nakahira, K.; Imamura, M.; et al. Mitofusins regulate lipid metabolism to mediate the development of lung fibrosis. *Nat. Commun.* **2019**, *10*, 3390. [CrossRef]

228. Frohman, M.A. The phospholipase D superfamily as therapeutic targets. *Trends Pharmacol. Sci.* **2015**, *36*, 137–144. [CrossRef]

229. Gomez-Cambronero, J. New concepts in phospholipase D signaling in inflammation and cancer. *Sci. World J.* **2010**, *10*, 1356–1369. [CrossRef]

230. Cummings, R.; Parinandi, N.; Wang, L.; Usatyuk, P.; Natarajan, V. Phospholipase D/phosphatidic acid signal transduction: Role and physiological significance in lung. *Mol. Cell. Biochem.* **2002**, *234–235*, 99–109. [CrossRef]

231. Brown, H.A.; Thomas, P.G.; Lindsley, C.W. Targeting phospholipase D in cancer, infection and neurodegenerative disorders. *Nat. Rev. Drug Discov.* **2017**, *16*, 351–367. [CrossRef]

232. Carman, G.M.; Han, G.S. Phosphatidic acid phosphatase, a key enzyme in the regulation of lipid synthesis. *J. Biol. Chem.* **2009**, *284*, 2593–2597. [CrossRef] [PubMed]

233. Siniossoglou, S. Phospholipid metabolism and nuclear function: Roles of the lipin family of phosphatidic acid phosphatases. *Biochim. Biophys. Acta (BBA)-Mol. Cell Biol. Lipids* **2013**, *1831*, 575–581. [CrossRef] [PubMed]

234. Sonoda, H.; Aoki, J.; Hiramatsu, T.; Ishida, M.; Bandoh, K.; Nagai, Y.; Taguchi, R.; Inoue, K.; Arai, H. A novel phosphatidic acid-selective phospholipase A1 that produces lysophosphatidic acid. *J. Biol. Chem.* **2002**, *277*, 34254–34263. [CrossRef] [PubMed]

235. Ito, M.; Tchoua, U.; Okamoto, M.; Tojo, H. Purification and properties of a phospholipase A2/lipase preferring phosphatidic acid, bis(monoacylglycerol) phosphate, and monoacylglycerol from rat testis. *J. Biol. Chem.* **2002**, *277*, 43674–43681. [CrossRef]

236. Gomez-Cambronero, J.; Kantonen, S. A river runs through it: How autophagy, senescence, and phagocytosis could be linked to phospholipase D by Wnt signaling. *J. Leukoc. Biol.* **2014**, *96*, 779–784. [CrossRef]

237. Henkels, K.M.; Miller, T.E.; Ganesan, R.; Wilkins, B.A.; Fite, K.; Gomez-Cambronero, J. A Phosphatidic Acid (PA) conveyor system of continuous intracellular transport from cell membrane to nucleus maintains EGF receptor homeostasis. *Oncotarget* **2016**, *7*, 47002–47017. [CrossRef]

238. Patel, R.B.; Kotha, S.R.; Sherwani, S.I.; Sliman, S.M.; Gurney, T.O.; Loar, B.; Butler, S.O.; Morris, A.J.; Marsh, C.B.; Parinandi, N.L. Pulmonary fibrosis inducer, bleomycin, causes redox-sensitive activation of phospholipase D and cytotoxicity through formation of bioactive lipid signal mediator, phosphatidic acid, in lung microvascular endothelial cells. *Int. J. Toxicol.* **2011**, *30*, 69–90. [CrossRef]

239. Suryadevara, V.; Huang, L.; Kim, S.J.; Cheresh, P.; Shaaya, M.; Bandela, M.; Fu, P.; Feghali-Bostwick, C.; Di Paolo, G.; Kamp, D.W.; et al. Role of phospholipase D in bleomycin-induced mitochondrial reactive oxygen species generation, mitochondrial DNA damage, and pulmonary fibrosis. *Am. J. Physiol. Lung Cell. Mol. Physiol.* **2019**, *317*, L175–L187. [CrossRef]

240. Usatyuk, P.V.; Gorshkova, I.A.; He, D.; Zhao, Y.; Kalari, S.K.; Garcia, J.G.; Natarajan, V. Phospholipase D-mediated activation of IQGAP1 through Rac1 regulates hyperoxia-induced p47phox translocation and reactive oxygen species generation in lung endothelial cells. *J. Biol. Chem.* **2009**, *284*, 15339–15352. [CrossRef]

241. Foster, D.A. Regulation of mTOR by phosphatidic acid? *Cancer Res.* **2007**, *67*, 1–4. [CrossRef]

242. Kameoka, S.; Adachi, Y.; Okamoto, K.; Iijima, M.; Sesaki, H. Phosphatidic Acid and Cardiolipin Coordinate Mitochondrial Dynamics. *Trends Cell Biol.* **2018**, *28*, 67–76. [CrossRef] [PubMed]

243. Trivedi, P.; Kumar, R.K.; Iyer, A.; Boswell, S.; Gerarduzzi, C.; Dadhania, V.P.; Herbert, Z.; Joshi, N.; Luyendyk, J.P.; Humphreys, B.D.; et al. Targeting Phospholipase D4 Attenuates Kidney Fibrosis. *J. Am. Soc. Nephrol.* **2017**, *28*, 3579–3589. [CrossRef] [PubMed]

244. Liu, C.S.; Schmezer, P.; Popanda, O. Diacylglycerol Kinase Alpha in Radiation-Induced Fibrosis: Potential as a Predictive Marker or Therapeutic Target. *Front. Oncol.* **2020**, *10*, 737. [CrossRef] [PubMed]

245. Weigel, C.; Veldwijk, M.R.; Oakes, C.C.; Seibold, P.; Slynko, A.; Liesenfeld, D.B.; Rabionet, M.; Hanke, S.A.; Wenz, F.; Sperk, E.; et al. Epigenetic regulation of diacylglycerol kinase alpha promotes radiation-induced fibrosis. *Nat. Commun.* **2016**, *7*, 10893. [CrossRef] [PubMed]

246. Sato, M.; Liu, K.; Sasaki, S.; Kunii, N.; Sakai, H.; Mizuno, H.; Saga, H.; Sakane, F. Evaluations of the selectivities of the diacylglycerol kinase inhibitors R59022 and R59949 among diacylglycerol kinase isozymes using a new non-radioactive assay method. *Pharmacology* **2013**, *92*, 99–107. [CrossRef]

247. Kulkarni, Y.M.; Dutta, S.; Iyer, A.K.; Venkatadri, R.; Kaushik, V.; Ramesh, V.; Wright, C.A.; Semmes, O.J.; Yakisich, J.S.; Azad, N. A proteomics approach to identifying key protein targets involved in VEGF inhibitor mediated attenuation of bleomycin-induced pulmonary fibrosis. *Proteomics* **2016**, *16*, 33–46. [CrossRef]

248. Aoki, J.; Inoue, A.; Okudaira, S. Two pathways for lysophosphatidic acid production. *Biochim. Biophys. Acta (BBA)-Mol. Cell Biol. Lipids* **2008**, *1781*, 513–518. [CrossRef]

249. Rindlisbacher, B.; Schmid, C.; Geiser, T.; Bovet, C.; Funke-Chambour, M. Serum metabolic profiling identified a distinct metabolic signature in patients with idiopathic pulmonary fibrosis—A potential biomarker role for LysoPC. *Respir Res* **2018**, *19*, 7. [CrossRef]

250. Tan, Y.; Jia, D.; Lin, Z.; Guo, B.; He, B.; Lu, C.; Xiao, C.; Liu, Z.; Zhao, N.; Bian, Z.; et al. Potential Metabolic Biomarkers to Identify Interstitial Lung Abnormalities. *Int. J. Mol. Sci.* **2016**, *17*. [CrossRef]

251. Montesi, S.B.; Mathai, S.K.; Brenner, L.N.; Gorshkova, I.A.; Berdyshev, E.V.; Tager, A.M.; Shea, B.S. Docosatetraenoyl LPA is elevated in exhaled breath condensate in idiopathic pulmonary fibrosis. *BMC Pulm. Med.* **2014**, *14*, 5. [CrossRef]

252. Park, G.Y.; Lee, Y.G.; Berdyshev, E.; Nyenhuis, S.; Du, J.; Fu, P.; Gorshkova, I.A.; Li, Y.; Chung, S.; Karpurapu, M.; et al. Autotaxin production of lysophosphatidic acid mediates allergic asthmatic inflammation. *Am. J. Respir. Crit. Care Med.* **2013**, *188*, 928–940. [CrossRef] [PubMed]

253. Houtkooper, R.H.; Turkenburg, M.; Poll-The, B.T.; Karall, D.; Perez-Cerda, C.; Morrone, A.; Malvagia, S.; Wanders, R.J.; Kulik, W.; Vaz, F.M. The enigmatic role of tafazzin in cardiolipin metabolism. *Biochim. Biophys. Acta (BBA)-Biomembr.* **2009**, *1788*, 2003–2014. [CrossRef] [PubMed]

254. Huang, L.S.; Mathew, B.; Li, H.; Zhao, Y.; Ma, S.F.; Noth, I.; Reddy, S.P.; Harijith, A.; Usatyuk, P.V.; Berdyshev, E.V.; et al. The mitochondrial cardiolipin remodeling enzyme lysocardiolipin acyltransferase is a novel target in pulmonary fibrosis. *Am. J. Respir. Crit. Care Med.* **2014**, *189*, 1402–1415. [CrossRef] [PubMed]

255. Saric, A.; Andreau, K.; Armand, A.S.; Moller, I.M.; Petit, P.X. Barth Syndrome: From Mitochondrial Dysfunctions Associated with Aberrant Production of Reactive Oxygen Species to Pluripotent Stem Cell Studies. *Front. Genet.* **2015**, *6*, 359. [CrossRef]

256. Dudek, J. Role of Cardiolipin in Mitochondrial Signaling Pathways. *Front. Cell Dev. Biol.* **2017**, *5*, 90. [CrossRef] [PubMed]

257. Paradies, G.; Paradies, V.; Ruggiero, F.M.; Petrosillo, G. Role of Cardiolipin in Mitochondrial Function and Dynamics in Health and Disease: Molecular and Pharmacological Aspects. *Cells* **2019**, *8*. [CrossRef]

258. Baile, M.G.; Sathappa, M.; Lu, Y.W.; Pryce, E.; Whited, K.; McCaffery, J.M.; Han, X.; Alder, N.N.; Claypool, S.M. Unremodeled and remodeled cardiolipin are functionally indistinguishable in yeast. *J. Biol. Chem.* **2014**, *289*, 1768–1778. [CrossRef]

259. Schenkel, L.C.; Bakovic, M. Formation and regulation of mitochondrial membranes. *Int. J. Cell Biol.* **2014**, *2014*, 709828. [CrossRef]

260. Schug, Z.T.; Gottlieb, E. Cardiolipin acts as a mitochondrial signalling platform to launch apoptosis. *Biochim. Biophys. Acta (BBA)-Biomembr.* **2009**, *1788*, 2022–2031. [CrossRef]

261. Paradies, G.; Paradies, V.; De Benedictis, V.; Ruggiero, F.M.; Petrosillo, G. Functional role of cardiolipin in mitochondrial bioenergetics. *Biochim. Biophys. Acta (BBA)-Biomembr.* **2014**, *1837*, 408–417. [CrossRef]

262. Borisenko, G.G. Mitochondrial phospholipid cardiolipin and its triggering functions in the cells. *Lipid Technol.* **2016**, *28*, 40–43. [CrossRef]

263. Ray, N.B.; Durairaj, L.; Chen, B.B.; McVerry, B.J.; Ryan, A.J.; Donahoe, M.; Waltenbaugh, A.K.; O'Donnell, C.P.; Henderson, F.C.; Etscheidt, C.A.; et al. Dynamic regulation of cardiolipin by the lipid pump Atp8b1 determines the severity of lung injury in experimental pneumonia. *Nat. Med.* **2010**, *16*, 1120–1127. [CrossRef] [PubMed]

264. Raja, V.; Greenberg, M.L. The functions of cardiolipin in cellular metabolism-potential modifiers of the Barth syndrome phenotype. *Chem. Phys. Lipids* **2014**, *179*, 49–56. [CrossRef] [PubMed]

265. Kiebish, M.A.; Yang, K.; Liu, X.; Mancuso, D.J.; Guan, S.; Zhao, Z.; Sims, H.F.; Cerqua, R.; Cade, W.T.; Han, X.; et al. Dysfunctional cardiac mitochondrial bioenergetic, lipidomic, and signaling in a murine model of Barth syndrome. *J. Lipid Res.* **2013**, *54*, 1312–1325. [CrossRef]

266. Huang, L.S.; Jiang, P.; Feghali-Bostwick, C.; Reddy, S.P.; Garcia, J.G.N.; Natarajan, V. Lysocardiolipin acyltransferase regulates TGF-beta mediated lung fibroblast differentiation. *Free Radic. Biol. Med.* **2017**, *112*, 162–173. [CrossRef]

267. Ackerman, S.J.; Park, G.Y.; Christman, J.W.; Nyenhuis, S.; Berdyshev, E.; Natarajan, V. Polyunsaturated lysophosphatidic acid as a potential asthma biomarker. *Biomark. Med.* **2016**, *10*, 123–135. [CrossRef]

268. Tager, A.M.; LaCamera, P.; Shea, B.S.; Campanella, G.S.; Selman, M.; Zhao, Z.; Polosukhin, V.; Wain, J.; Karimi-Shah, B.A.; Kim, N.D.; et al. The lysophosphatidic acid receptor LPA1 links pulmonary fibrosis to lung injury by mediating fibroblast recruitment and vascular leak. *Nat. Med.* **2008**, *14*, 45–54. [CrossRef]

269. Federico, L.; Jeong, K.J.; Vellano, C.P.; Mills, G.B. Autotaxin, a lysophospholipase D with pleomorphic effects in oncogenesis and cancer progression. *J. Lipid Res.* **2016**, *57*, 25–35. [CrossRef]

270. Van Meeteren, L.A.; Moolenaar, W.H. Regulation and biological activities of the autotaxin-LPA axis. *Prog. Lipid Res.* **2007**, *46*, 145–160. [CrossRef] [PubMed]

271. Ninou, I.; Magkrioti, C.; Aidinis, V. Autotaxin in Pathophysiology and Pulmonary Fibrosis. *Front. Med.* **2018**, *5*, 180. [CrossRef]

272. Benesch, M.G.; Tang, X.; Venkatraman, G.; Bekele, R.T.; Brindley, D.N. Recent advances in targeting the autotaxin-lysophosphatidate-lipid phosphate phosphatase axis in vivo. *J. Biomed. Res.* **2016**, *30*, 272–284. [CrossRef] [PubMed]

273. Stracke, M.L.; Krutzsch, H.C.; Unsworth, E.J.; Arestad, A.; Cioce, V.; Schiffmann, E.; Liotta, L.A. Identification, purification, and partial sequence analysis of autotaxin, a novel motility-stimulating protein. *J. Biol. Chem.* **1992**, *267*, 2524–2529. [PubMed]

274. Murata, J.; Lee, H.Y.; Clair, T.; Krutzsch, H.C.; Arestad, A.A.; Sobel, M.E.; Liotta, L.A.; Stracke, M.L. cDNA cloning of the human tumor motility-stimulating protein, autotaxin, reveals a homology with phosphodiesterases. *J. Biol. Chem.* **1994**, *269*, 30479–30484. [PubMed]

275. Stefan, C.; Jansen, S.; Bollen, M. NPP-type ectophosphodiesterases: Unity in diversity. *Trends Biochem. Sci.* **2005**, *30*, 542–550. [CrossRef]

276. Tokumura, A.; Majima, E.; Kariya, Y.; Tominaga, K.; Kogure, K.; Yasuda, K.; Fukuzawa, K. Identification of human plasma lysophospholipase D, a lysophosphatidic acid-producing enzyme, as autotaxin, a multifunctional phosphodiesterase. *J. Biol. Chem.* **2002**, *277*, 39436–39442. [CrossRef]

277. Aikawa, S.; Hashimoto, T.; Kano, K.; Aoki, J. Lysophosphatidic acid as a lipid mediator with multiple biological actions. *J. Biochem.* **2015**, *157*, 81–89. [CrossRef]

278. Zhao, Y.; Natarajan, V. Lysophosphatidic acid (LPA) and its receptors: Role in airway inflammation and remodeling. *Biochim. Biophys. Acta (BBA)-Mol. Cell Biol. Lipids* **2013**, *1831*, 86–92. [CrossRef] [PubMed]

279. Bektas, M.; Payne, S.G.; Liu, H.; Goparaju, S.; Milstien, S.; Spiegel, S. A novel acylglycerol kinase that produces lysophosphatidic acid modulates cross talk with EGFR in prostate cancer cells. *J. Cell Biol.* **2005**, *169*, 801–811. [CrossRef] [PubMed]

280. Kalari, S.; Zhao, Y.; Spannhake, E.W.; Berdyshev, E.V.; Natarajan, V. Role of acylglycerol kinase in LPA-induced IL-8 secretion and transactivation of epidermal growth factor-receptor in human bronchial epithelial cells. *Am. J. Physiol. Lung Cell. Mol. Physiol.* **2009**, *296*, L328–L336. [CrossRef] [PubMed]

281. Kang, Y.; Stroud, D.A.; Baker, M.J.; De Souza, D.P.; Frazier, A.E.; Liem, M.; Tull, D.; Mathivanan, S.; McConville, M.J.; Thorburn, D.R.; et al. Sengers Syndrome-Associated Mitochondrial Acylglycerol Kinase Is a Subunit of the Human TIM22 Protein Import Complex. *Mol. Cell* **2017**, *67*, 457–470.e455. [CrossRef] [PubMed]

282. Yung, Y.C.; Stoddard, N.C.; Chun, J. LPA receptor signaling: Pharmacology, physiology, and pathophysiology. *J. Lipid Res.* **2014**, *55*, 1192–1214. [CrossRef] [PubMed]

283. Webb, T.E.; Kaplan, M.G.; Barnard, E.A. Identification of 6H1 as a P2Y purinoceptor: P2Y5. *Biochem. Biophys. Res. Commun.* **1996**, *219*, 105–110. [CrossRef]

284. Issemann, I.; Green, S. Activation of a member of the steroid hormone receptor superfamily by peroxisome proliferators. *Nature* **1990**, *347*, 645–650. [CrossRef] [PubMed]

285. Ricote, M.; Huang, J.; Fajas, L.; Li, A.; Welch, J.; Najib, J.; Witztum, J.L.; Auwerx, J.; Palinski, W.; Glass, C.K. Expression of the peroxisome proliferator-activated receptor gamma (PPARgamma) in human atherosclerosis and regulation in macrophages by colony stimulating factors and oxidized low density lipoprotein. *Proc. Natl. Acad. Sci. USA* **1998**, *95*, 7614–7619. [CrossRef]

286. McIntyre, T.M.; Pontsler, A.V.; Silva, A.R.; St Hilaire, A.; Xu, Y.; Hinshaw, J.C.; Zimmerman, G.A.; Hama, K.; Aoki, J.; Arai, H.; et al. Identification of an intracellular receptor for lysophosphatidic acid (LPA): LPA is a transcellular PPARgamma agonist. *Proc. Natl. Acad. Sci. USA* **2003**, *100*, 131–136. [CrossRef] [PubMed]

287. Lakatos, H.F.; Thatcher, T.H.; Kottmann, R.M.; Garcia, T.M.; Phipps, R.P.; Sime, P.J. The Role of PPARs in Lung Fibrosis. *PPAR Res.* **2007**, *2007*, 71323. [CrossRef] [PubMed]

288. Fujiwara, Y. Cyclic phosphatidic acid—A unique bioactive phospholipid. *Biochim. Biophys. Acta (BBA)-Mol. Cell Biol. Lipids* **2008**, *1781*, 519–524. [CrossRef] [PubMed]

289. Murakami-Murofushi, K.; Uchiyama, A.; Fujiwara, Y.; Kobayashi, T.; Kobayashi, S.; Mukai, M.; Murofushi, H.; Tigyi, G. Biological functions of a novel lipid mediator, cyclic phosphatidic acid. *Biochim. Biophys. Acta (BBA)-Mol. Cell Biol. Lipids* **2002**, *1582*, 1–7. [CrossRef]

290. Tsukahara, T. PPAR gamma Networks in Cell Signaling: Update and Impact of Cyclic Phosphatidic Acid. *J. Lipids* **2013**, *2013*, 246597. [CrossRef]

291. Oikonomou, N.; Mouratis, M.A.; Tzouvelekis, A.; Kaffe, E.; Valavanis, C.; Vilaras, G.; Karameris, A.; Prestwich, G.D.; Bouros, D.; Aidinis, V. Pulmonary autotaxin expression contributes to the pathogenesis of pulmonary fibrosis. *Am. J. Respir. Cell Mol. Biol.* **2012**, *47*, 566–574. [CrossRef]

292. Sakai, T.; Peyruchaud, O.; Fassler, R.; Mosher, D.F. Restoration of beta1A integrins is required for lysophosphatidic acid-induced migration of beta1-null mouse fibroblastic cells. *J. Biol. Chem.* **1998**, *273*, 19378–19382. [CrossRef] [PubMed]

293. Huang, L.S.; Fu, P.; Patel, P.; Harijith, A.; Sun, T.; Zhao, Y.; Garcia, J.G.; Chun, J.; Natarajan, V. Lysophosphatidic acid receptor-2 deficiency confers protection against bleomycin-induced lung injury and fibrosis in mice. *Am. J. Respir. Cell Mol. Biol.* **2013**, *49*, 912–922. [CrossRef] [PubMed]

294. Wang, L.; Cummings, R.; Zhao, Y.; Kazlauskas, A.; Sham, J.K.; Morris, A.; Georas, S.; Brindley, D.N.; Natarajan, V. Involvement of phospholipase D2 in lysophosphatidate-induced transactivation of platelet-derived growth factor receptor-beta in human bronchial epithelial cells. *J. Biol. Chem.* **2003**, *278*, 39931–39940. [CrossRef] [PubMed]

295. He, D.; Natarajan, V.; Stern, R.; Gorshkova, I.A.; Solway, J.; Spannhake, E.W.; Zhao, Y. Lysophosphatidic acid-induced transactivation of epidermal growth factor receptor regulates cyclo-oxygenase-2 expression and prostaglandin E(2) release via C/EBPbeta in human bronchial epithelial cells. *Biochem. J.* **2008**, *412*, 153–162. [CrossRef]

296. Ganguly, K.; Stoeger, T.; Wesselkamper, S.C.; Reinhard, C.; Sartor, M.A.; Medvedovic, M.; Tomlinson, C.R.; Bolle, I.; Mason, J.M.; Leikauf, G.D.; et al. Candidate genes controlling pulmonary function in mice: Transcript profiling and predicted protein structure. *Physiol. Genom.* **2007**, *31*, 410–421. [CrossRef]

297. Van Meeteren, L.A.; Ruurs, P.; Stortelers, C.; Bouwman, P.; van Rooijen, M.A.; Pradere, J.P.; Pettit, T.R.; Wakelam, M.J.; Saulnier-Blache, J.S.; Mummery, C.L.; et al. Autotaxin, a secreted lysophospholipase D, is essential for blood vessel formation during development. *Mol. Cell. Biol.* **2006**, *26*, 5015–5022. [CrossRef]

298. Tanaka, M.; Okudaira, S.; Kishi, Y.; Ohkawa, R.; Iseki, S.; Ota, M.; Noji, S.; Yatomi, Y.; Aoki, J.; Arai, H. Autotaxin stabilizes blood vessels and is required for embryonic vasculature by producing lysophosphatidic acid. *J. Biol. Chem.* **2006**, *281*, 25822–25830. [CrossRef]

299. Ferry, G.; Giganti, A.; Coge, F.; Bertaux, F.; Thiam, K.; Boutin, J.A. Functional invalidation of the autotaxin gene by a single amino acid mutation in mouse is lethal. *FEBS Lett.* **2007**, *581*, 3572–3578. [CrossRef]

300. Mouratis, M.A.; Magkrioti, C.; Oikonomou, N.; Katsifa, A.; Prestwich, G.D.; Kaffe, E.; Aidinis, V. Autotaxin and Endotoxin-Induced Acute Lung Injury. *PLoS ONE* **2015**, *10*, e0133619. [CrossRef]

301. Black, K.E.; Berdyshev, E.; Bain, G.; Castelino, F.V.; Shea, B.S.; Probst, C.K.; Fontaine, B.A.; Bronova, I.; Goulet, L.; Lagares, D.; et al. Autotaxin activity increases locally following lung injury, but is not required for pulmonary lysophosphatidic acid production or fibrosis. *FASEB J.* **2016**, *30*, 2435–2450. [CrossRef]

302. Lee, G.; Kang, S.U.; Ryou, J.-H.; Lim, J.-J.; Lee, Y.-H. Late Breaking Abstract—BBT-877, a Potent Autotaxin Inhibitor in Clinical Development to Treat Idiopathic Pulmonary Fibrosis. *Eur. Respir. J.* **2019**, *54*, PA1293. [CrossRef]

303. Maher, T.M.; van der Aar, E.M.; Van de Steen, O.; Allamassey, L.; Desrivot, J.; Dupont, S.; Fagard, L.; Ford, P.; Fieuw, A.; Wuyts, W. Safety, tolerability, pharmacokinetics, and pharmacodynamics of GLPG1690, a novel autotaxin inhibitor, to treat idiopathic pulmonary fibrosis (FLORA): A phase 2a randomised placebo-controlled trial. *Lancet Respir. Med.* **2018**, *6*, 627–635. [CrossRef]

304. Zemski Berry, K.A.; Murphy, R.C.; Kosmider, B.; Mason, R.J. Lipidomic characterization and localization of phospholipids in the human lung. *J. Lipid Res.* **2017**, *58*, 926–933. [CrossRef] [PubMed]

305. Funke, M.; Zhao, Z.; Xu, Y.; Chun, J.; Tager, A.M. The lysophosphatidic acid receptor LPA1 promotes epithelial cell apoptosis after lung injury. *Am. J. Respir. Cell Mol. Biol.* **2012**, *46*, 355–364. [CrossRef] [PubMed]

306. Palmer, S.M.; Snyder, L.; Todd, J.L.; Soule, B.; Christian, R.; Anstrom, K.; Luo, Y.; Gagnon, R.; Rosen, G. Randomized, Double-Blind, Placebo-Controlled, Phase 2 Trial of BMS-986020, a Lysophosphatidic Acid Receptor Antagonist for the Treatment of Idiopathic Pulmonary Fibrosis. *Chest* **2018**, *154*, 1061–1069. [CrossRef]

307. Swaney, J.S.; Chapman, C.; Correa, L.D.; Stebbins, K.J.; Bundey, R.A.; Prodanovich, P.C.; Fagan, P.; Baccei, C.S.; Santini, A.M.; Hutchinson, J.H.; et al. A novel, orally active LPA(1) receptor antagonist inhibits lung fibrosis in the mouse bleomycin model. *Br. J. Pharmacol.* **2010**, *160*, 1699–1713. [CrossRef]

308. Gianazza, E.; Brioschi, M.; Fernandez, A.M.; Banfi, C. Lipoxidation in cardiovascular diseases. *Redox Biol.* **2019**, *23*, 101119. [CrossRef]

309. Thimmulappa, R.K.; Gang, X.; Kim, J.H.; Sussan, T.E.; Witztum, J.L.; Biswal, S. Oxidized phospholipids impair pulmonary antibacterial defenses: Evidence in mice exposed to cigarette smoke. *Biochem. Biophys. Res. Commun.* **2012**, *426*, 253–259. [CrossRef] [PubMed]

310. Paliogiannis, P.; Fois, A.G.; Collu, C.; Bandinu, A.; Zinellu, E.; Carru, C.; Pirina, P.; Mangoni, A.A.; Zinellu, A. Oxidative stress-linked biomarkers in idiopathic pulmonary fibrosis: A systematic review and meta-analysis. *Biomark. Med.* **2018**, *12*, 1175–1184. [CrossRef] [PubMed]

311. Martinez, F.J.; Collard, H.R.; Pardo, A.; Raghu, G.; Richeldi, L.; Selman, M.; Swigris, J.J.; Taniguchi, H.; Wells, A.U. Idiopathic pulmonary fibrosis. *Nat. Rev. Dis. Primers* **2017**, *3*, 17074. [CrossRef] [PubMed]

312. Sato, K.; Tashiro, Y.; Chibana, S.; Yamashita, A.; Karakawa, T.; Kohrogi, H. Role of lipid-derived free radical in bleomycin-induced lung injury in mice: Availability for ESR spin trap method with organic phase extraction. *Biol. Pharm. Bull.* **2008**, *31*, 1855–1859. [CrossRef] [PubMed]

313. Suntres, Z.E.; Shek, P.N. Protective effect of liposomal alpha-tocopherol against bleomycin-induced lung injury. *Biomed. Environ. Sci.* **1997**, *10*, 47–59. [PubMed]

314. Sener, G.; Topaloglu, N.; Sehirli, A.O.; Ercan, F.; Gedik, N. Resveratrol alleviates bleomycin-induced lung injury in rats. *Pulm. Pharmacol. Ther.* **2007**, *20*, 642–649. [CrossRef] [PubMed]

315. Teixeira, K.C.; Soares, F.S.; Rocha, L.G.; Silveira, P.C.; Silva, L.A.; Valenca, S.S.; Dal Pizzol, F.; Streck, E.L.; Pinho, R.A. Attenuation of bleomycin-induced lung injury and oxidative stress by N-acetylcysteine plus deferoxamine. *Pulm. Pharmacol. Ther.* **2008**, *21*, 309–316. [CrossRef] [PubMed]

316. Tsubouchi, K.; Araya, J.; Yoshida, M.; Sakamoto, T.; Koumura, T.; Minagawa, S.; Hara, H.; Hosaka, Y.; Ichikawa, A.; Saito, N.; et al. Involvement of GPx4-Regulated Lipid Peroxidation in Idiopathic Pulmonary Fibrosis Pathogenesis. *J. Immunol.* **2019**, *203*, 2076–2087. [CrossRef]

317. Misharin, A.V.; Morales-Nebreda, L.; Reyfman, P.A.; Cuda, C.M.; Walter, J.M.; McQuattie-Pimentel, A.C.; Chen, C.I.; Anekalla, K.R.; Joshi, N.; Williams, K.J.N.; et al. Monocyte-derived alveolar macrophages drive lung fibrosis and persist in the lung over the life span. *J. Exp. Med.* **2017**, *214*, 2387–2404. [CrossRef]

318. Yasuda, K.; Sato, A.; Nishimura, K.; Chida, K.; Hayakawa, H. Phospholipid analysis of alveolar macrophages and bronchoalveolar lavage fluid following bleomycin administration to rabbits. *Lung* **1994**, *172*, 91–102. [CrossRef]

319. Romero, F.; Shah, D.; Duong, M.; Penn, R.B.; Fessler, M.B.; Madenspacher, J.; Stafstrom, W.; Kavuru, M.; Lu, B.; Kallen, C.B.; et al. A pneumocyte-macrophage paracrine lipid axis drives the lung toward fibrosis. *Am. J. Respir. Cell Mol. Biol.* **2015**, *53*, 74–86. [CrossRef]

320. Kim, K.K.; Dotson, M.R.; Agarwal, M.; Yang, J.; Bradley, P.B.; Subbotina, N.; Osterholzer, J.J.; Sisson, T.H. Efferocytosis of apoptotic alveolar epithelial cells is sufficient to initiate lung fibrosis. *Cell Death Dis.* **2018**, *9*, 1056. [CrossRef]

321. Parks, B.W.; Black, L.L.; Zimmerman, K.A.; Metz, A.E.; Steele, C.; Murphy-Ullrich, J.E.; Kabarowski, J.H. CD36, but not G2A, modulates efferocytosis, inflammation, and fibrosis following bleomycin-induced lung injury. *J. Lipid Res.* **2013**, *54*, 1114–1123. [CrossRef]

322. Bradley, P.; Subbotina, N.; Dotson, M.; Teitz-Tennenbaum, S.; Roussey, J.; Osterholzer, J.; Sisson, T. CD36 Scavenger Receptor Promotes Pulmonary Fibrosis in Response to Oropharyngeal Administration of Oxidized Phospholipid. In *C72. Pulmonary Fibrosis: Mechanisms and Models*. Presented at American Thoracic Society 2018 Conference, San Diego, CA, USA, 18–23 May 2018; p. A5759.

323. Boorjian, S. Commentary on "Conditional survival of patients with metastatic renal cell carcinoma treated with VEGF-targeted therapy: A population-based study." Harshman LC, Xie W, Bjarnason GA, Knox JJ, MacKenzie M, Wood L, Srinivas S, Vaishampayan UN, Tan MH, Rha SY, Donskov F, Agarwal N, Kollmannsberger C, North S, Rini BI, Heng DY, Choueiri TK, Stanford Cancer Institute, Stanford University School of Medicine, Stanford, CA: Lancet Oncol 2012;13(9):927-35 (Epub 2012 Aug 8). *Urol. Oncol.* **2013**, *31*, 127–128. [CrossRef] [PubMed]

324. Osterholzer, J.J.; Subbotina, N.; Dotson, M.; Teitz-Tennenbaum, S.; Sisson, T. The Plasminogen Activator Inhibitor-1 Promotes Pulmonary Fibrosis In Response To Intratracheal Administration Of Oxidized Phospholipid. In *C78. Fibrosis: Mediators and Modulators*. Presented at American Thoracic Society 2017 International Conference, Washington, DC, USA, 19–24 May 2017; p. A6438.

International Journal of
Molecular Sciences

Article

Circulating CRP Levels Are Associated with Epicardial and Visceral Fat Depots in Women with Metabolic Syndrome Criteria

Federico Carbone [1,2,*], Maria Stefania Lattanzio [3,*], Silvia Minetti [1,2], Anna Maria Ansaldo [1], Daniele Ferrara [1], Emilio Molina-Molina [3], Anna Belfiore [3], Edoardo Elia [1], Stefania Pugliese [3], Vincenzo Ostilio Palmieri [3], Fabrizio Montecucco [2,4,†] and Piero Portincasa [3,†]

[1] First Clinic of Internal Medicine, Department of Internal Medicine, University of Genoa, 6 viale Benedetto XV, 16132 Genoa, Italy; silvia.minetti@unige.it (S.M.); annamaria.ansaldo@gmail.com (A.M.A.); daniele.ferrara0292@gmail.com (D.F.); edoardo.elia93@gmail.com (E.E.)

[2] IRCCS Ospedale Policlinico San Martino Genoa–Italian Cardiovascular Network, 10 Largo Benzi, 16132 Genoa, Italy; fabrizio.montecucco@unige.it

[3] Clinica Medica "A. Murri", Department of Biomedical Sciences & Human Oncology, University of Bari Medical School, Piazza Giulio Cesare 11, 70124 Bari, Italy; emmolin.ugr@gmail.com (E.M.-M.); belfiore.murri@gmail.com (A.B.); drstefaniapugliese@libero.it (S.P.); vincenzoostilio.palmieri@uniba.it (V.O.P.); piero.portincasa@uniba.it (P.P.)

[4] First Clinic of Internal Medicine, Department of Internal Medicine and Centre of Excellence for Biomedical Research (CEBR), University of Genoa, 6 viale Benedetto XV, 16132 Genoa, Italy

[*] Correspondence: federico.carbone@unige.it (F.C.); stefanialattanzio@hotmail.it (M.S.L.); Tel.: +39-010-33-51054 (F.C.); Fax: +39-010-353-8686 (F.C.)

[†] These authors contributed equally to this work.

Received: 14 November 2019; Accepted: 23 November 2019; Published: 27 November 2019

Abstract: Sexual dimorphism accounts for significant differences in adipose tissue mass and distribution. However, how the crosstalk between visceral and ectopic fat depots occurs and which are the determinants of ectopic fat expansion and dysfunction remains unknown. Here, we focused on the impact of gender in the crosstalk between visceral and epicardial fat depots and the role of adipocytokines and high-sensitivity C-reactive protein (hs-CRP). A total of 141 outward patients (both men and women) with one or more defining criteria for metabolic syndrome (MetS) were consecutively enrolled. For all patients, demographic and clinical data were collected and ultrasound assessment of visceral adipose tissue (VFth) and epicardial fat (EFth) thickness was performed. Hs-CRP and adipocytokine levels were assessed by enzyme-linked immunosorbent assay (ELISA). Men were characterized by increased VFth and EFth (p-value < 0.001 and 0.014, respectively), whereas women showed higher levels of adiponectin and leptin (p-value < 0.001 for both). However, only in women VFth and EFth significantly correlated between them ($p = 0.013$) and also with leptin ($p < 0.001$ for both) and hs-CRP ($p = 0.005$ and $p = 0.028$, respectively). Linear regression confirmed an independent association of both leptin and hs-CRP with VFth in women, also after adjustment for age and MetS ($p = 0.012$ and 0.007, respectively). In conclusion, men and women present differences in epicardial fat deposition and systemic inflammation. An intriguing association between visceral/epicardial fat depots and chronic low-grade inflammation also emerged. In women Although a further validation in larger studies is needed, these findings suggest a critical role of sex in stratification of obese/dysmetabolic patients.

Keywords: epicardial fat thickness; visceral fat thickness; high-sensitivity c-reactive protein; leptin; gender; female

1. Introduction

The classical paradigm of obesity, defined by body mass index (BMI) >30 kg/m^2, is no longer considered representative of this heterogeneous condition, characterized by many phenotypes [1]. Rather, the term "adiposopathy" has been recently coined to describe the pathological response of adipose tissue to positive caloric balance, behavioral changes and environmental factors in susceptible individuals [2]. Alongside the shift to a visceral adipose tissue (VAT) distribution and pro-inflammatory adipocytokine unbalance, the deposition of ectopic fat depots (e.g., within liver, pancreas, heart, kidney and skeletal muscle) is now considered a defining feature of adiposopathy. The growing evidence of individual variation in body fat distribution also raised the interest toward the susceptibility of visceral fat storing to genetic factors [3], including racial and sex differences [4,5]. Especially, sexual dimorphism accounts for significant differences in visceral fat mass and distribution, women being characterized by greater BMI with prevalent subcutaneous distribution. This has important clinical implications as visceral and subcutaneous fat greatly differ in terms of function and response to weight gain. Sex differences were also reported in the chronic low-grade inflammatory response underlying obesity. Full adipose tissue profiling in both experimental and clinical studies demonstrated sex difference in insulin resistance and inflammatory response. More specifically, women seem to be protected from macrophage infiltration into the adipose tissue [6], and show a stronger association between adiposity and C-reactive protein (CRP) [7].

Conversely less is known about the role of epicardial fat tissue thickness (EFth) and function. EFth correlates with metabolic syndrome and its relative clinical features, but the strength of this association is less than half of that with visceral adipose tissue [8]. Sex-related differences in the crosstalk between visceral, ectopic fat and circulating inflammatory molecules are also poorly investigated [9]. Nevertheless, this is a very critical point because epicardial fat is increasingly described as a metabolic active organ, with detrimental effects on surrounding tissues (i.e., coronary arteries and myocardium) [10]. There is indeed an urgent need to understand, which are the determinants of epicardial fat expansion and dysfunction, whether sex-related differences in epicardial fat deposition exist and how the crosstalk between visceral and ectopic fat depots occurs. With this aim, a cohort of outward patients with one or more defining criteria for MetS have been analyzed here. We particularly focused on the potential impact of sex in visceral and epicardial fat deposition, also considering if any correlation with adipocytokines and high-sensitivity CRP (hs-CRP) exists.

2. Results

2.1. Men and Women Have a Similar Metabolic Profile

The characteristics of the overall cohort are shown in Table 1.

Table 1. Baseline characteristics in the overall cohort.

Parameters	Overall Cohort (*n* = 125)
Clinical	
Age, yr. [IQR]	56 (49–62)
Men, no. (%)	58 (46.4)
Active smokers, no. (%)	21 (16.8)
Hypertension, no (%)	100 (80.0)
IFG, no (%)	1 (0.8)
T2DM, no (%)	17 (13.6)
sBP, mmHg [IQR]	130 (120–135)
dBP, mmHg [IQR]	80 (72–85)

Table 1. *Cont.*

Parameters	Overall Cohort (*n* = 125)
Waist circumference, cm [IQR]	99 (93–108)
Weight, Kg [IQR]	77 (68–90)
BMI, Kg/m^2 [IQR]	27.2 (25.1–30.3)
MetS criteria	
1	41 (32.7)
2	49 (39.2)
3	27 (21.6)
4	7 (5.6)
5	1 (0.8)
MetS	35 (28.0)
Ultrasound assessment	
EFth, mm [IQR]	5.6 (4.8–6.5)
VFth, mm [IQR]	61 (46–76)
Hepatic steatosis, no. (%)	85 (68.0)
Biochemistry	
Serum total-c, mg/dL [IQR]	194 (168–219)
Serum LDL-c, mg/dL [IQR]	111 (83–129)
Serum HDL-c, mg/dL [IQR]	60 (50–71)
Serum TAG, mg/dL [IQR]	101 (71–139)
Fasting glycaemia, mg/dL [IQR]	89 (83–99)
VAI, *n* [IQR]	1.3 (0.8–1.9)

sBP: systolic blood pressure; dBP: diastolic blood pressure; BMI: body mass index; MetS: metabolic syndrome; EFth: epicardial fat thickness; VFth: visceral fat thickness; LDL-c: low-density lipoprotein cholesterol; HDL-c: high-density lipoprotein cholesterol; TAG: triglycerides; VAI: visceral adiposity index.

Median age of patients was 56 years with almost equal distribution across sex (men 46.4%). Concerning metabolic syndrome criteria, hypertension was highly frequent (80.0%), whereas the prevalence of glucose intolerance was lower (impaired fasting glucose 0.8% and diabetes 13.6%). Overall, more than half of patients had 1 or 2 MetS defining criteria (32.7% and 39.2%, respectively), whereas MetS was recorded in 28.0% of patients. Once categorized for sex, men showed a weak increase of blood pressure values, waist circumference/weight measurement and glycolipid profile, but not BMI (Table 2).

Table 2. Baseline clinical/biochemical characteristics across sex.

Clinical Data	Men (*n* = 66)	Women (*n* = 75)	*p*-Value
Age, yr. (IQR)	56 (49–60)	57 (47–63)	0.335
Active smokers, no. (%)	11 (19.0)	10 (14.9)	0.634
sBP, mmHg (IQR)	130 (125–135)	125 (115–136)	0.017
dBP, mmHg (IQR)	80 (80–89)	80 (70–85)	0.010
Waist circumference, cm (IQR)	103 (95–110)	97 (90–102)	0.005
Weight, Kg (IQR)	86 (76–93)	70 (63–80)	<0.001

Table 2. *Cont.*

Clinical Data	Men ($n = 66$)	Women ($n = 75$)	p-Value
BMI, kg/m^2 (IQR)	28.4 (25.7–30.4)	26.7 (24.8–30.0)	0.216
Biochemistry			
Serum total-c, mg/dL (IQR)	186 (161–210)	197 (170–225)	0.042
Serum LDL-c, mg/dL (IQR)	110 (81–127)	116 (89–133)	0.259
Serum HDL-c, mg/dL (IQR)	53 (43–59)	69 (60–77)	<0.001
Serum TAG, mg/dL (IQR)	118 (85–169)	90 (65–126)	0.001
Fasting glycaemia, mg/dL (IQR)	92 (86–104)	87 (82–98)	0.013
VAI, n (IQR)	1.4 (0.9–2.2)	1.3 (0.8–1.7)	0.130

sBP: systolic blood pressure; dBP: diastolic blood pressure; BMI: body mass index; total-c: total cholesterol; LDL-c: low-density lipoprotein cholesterol; HDL-c: high-density lipoprotein cholesterol; TAG: triglycerides; VAI: visceral adiposity index.

Of interest, the prevalence of MetS did not differ across sex (Figure 1A).

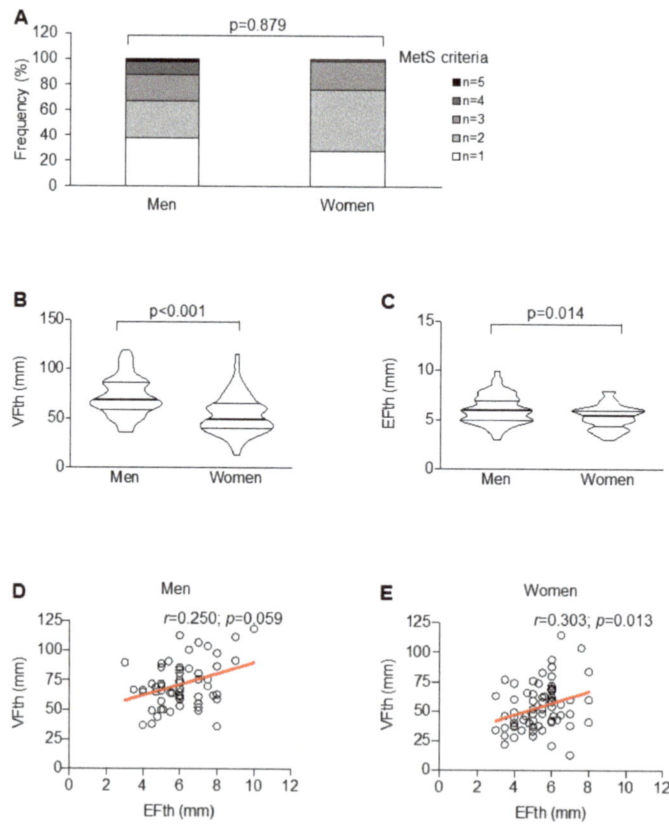

Figure 1. Metabolic differences and adipose tissue distribution across sex. Sex distribution in term of numbers of metabolic syndrome (MetS) criteria (**A**). Sex-related differences in the extent of visceral fat thickness (VFth) (**B**) and epicardial fat thickness (EFth) (**C**) and their correlation in men (**D**) and women (**E**).

2.2. Men Have Increased Ectopic Fat Depots but Reduced Circulating Adipocytokine Levels as Compared to Women

In our cohort of patients with at least one MetS criterion, men were characterized by greater VFth and EFth (p-value < 0.001 and 0.014, respectively) as compared to women (Figure 1B,C). However, only in women, VFth and EFth significantly correlated with each other (r = 0.303; p = 0.013) (Figure 1D,E).

Conversely, women showed increased levels of adiponectin and leptin (p-value < 0.001 for both; Figure 2A,B), whereas the concentrations of hs-CRP were similar across sex (Figure 2C).

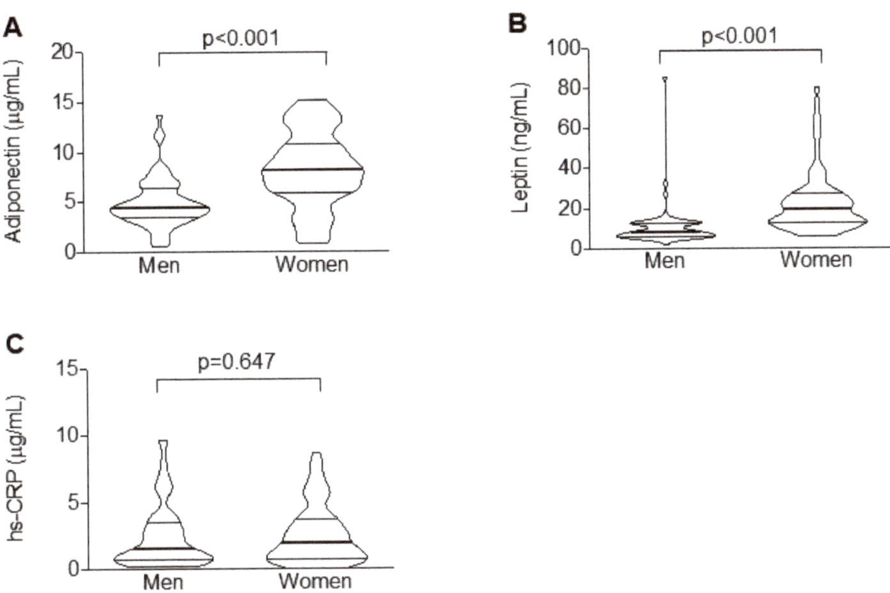

Figure 2. Women are characterized by greater serum levels of adipocytokines but not hs-CRP. Violin plots illustrating the median values across sex of adiponectin (**A**), leptin (**B**), and high-sensitivity C-reactive protein (hs-CRP) (**C**).

2.3. Only in Women, Circulating Levels of Leptin and CRP are Associated with Visceral and Ectopic Fat Depots

Considering potential association between circulating levels of inflammatory molecules and fat distribution in men and women, no correlation was found between adiponectin and VFth (Figure 3A,B).

Conversely, the adipocytokine leptin correlated with VFth both in men and women (p = 0.043 and < 0.001, respectively) (Figure 3C,D). Hs-CRP correlated with VFth only in women (r = 0.341; p = 0.005) (Figure 3E,F). Considering EFth, no correlation between serum adiponectin and this variable was observed in both men and women (Figure 4A,B).

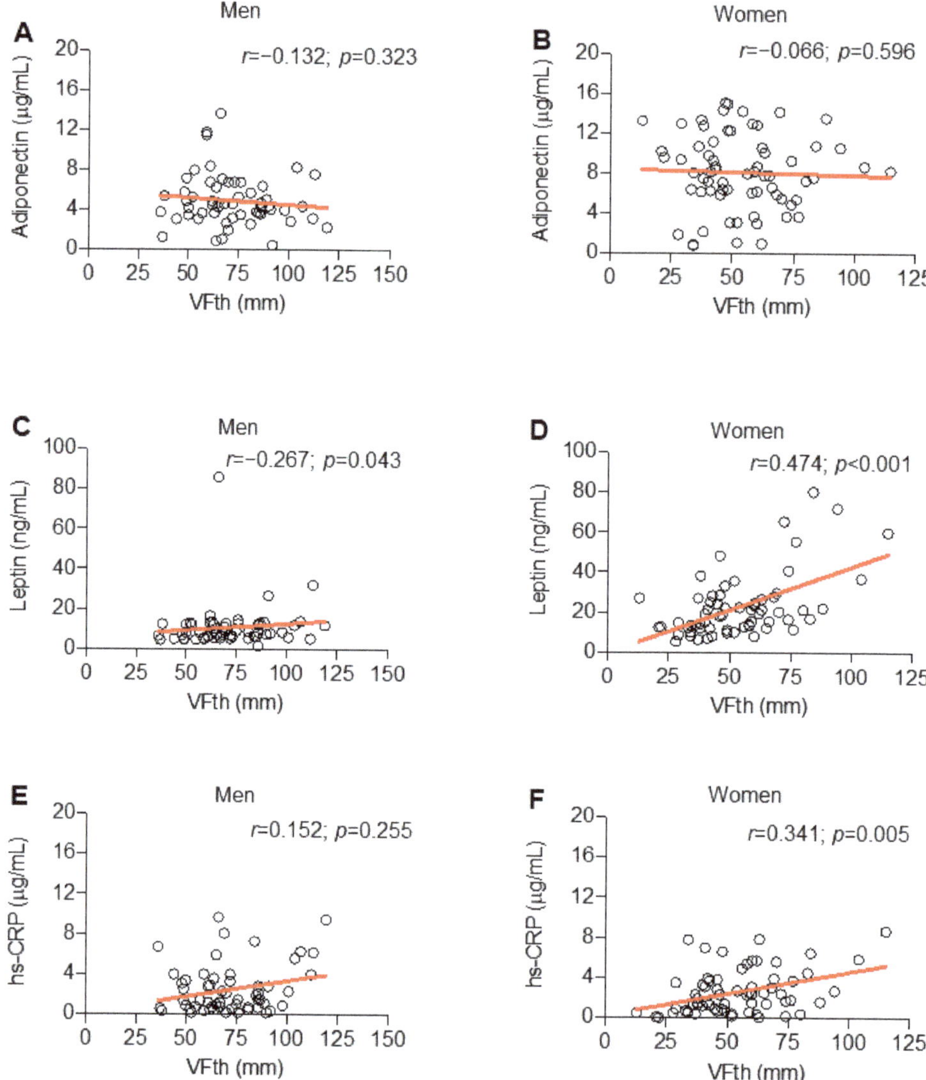

Figure 3. Only in women, hs-CRP correlates with the extent of visceral fat thickness (VFth). Scatter plot illustrating the correlation of VFth with serum biomarkers across sex: adiponectin (**A**,**B**), leptin (**C**,**D**) and hs-CRP (**E**,**F**).

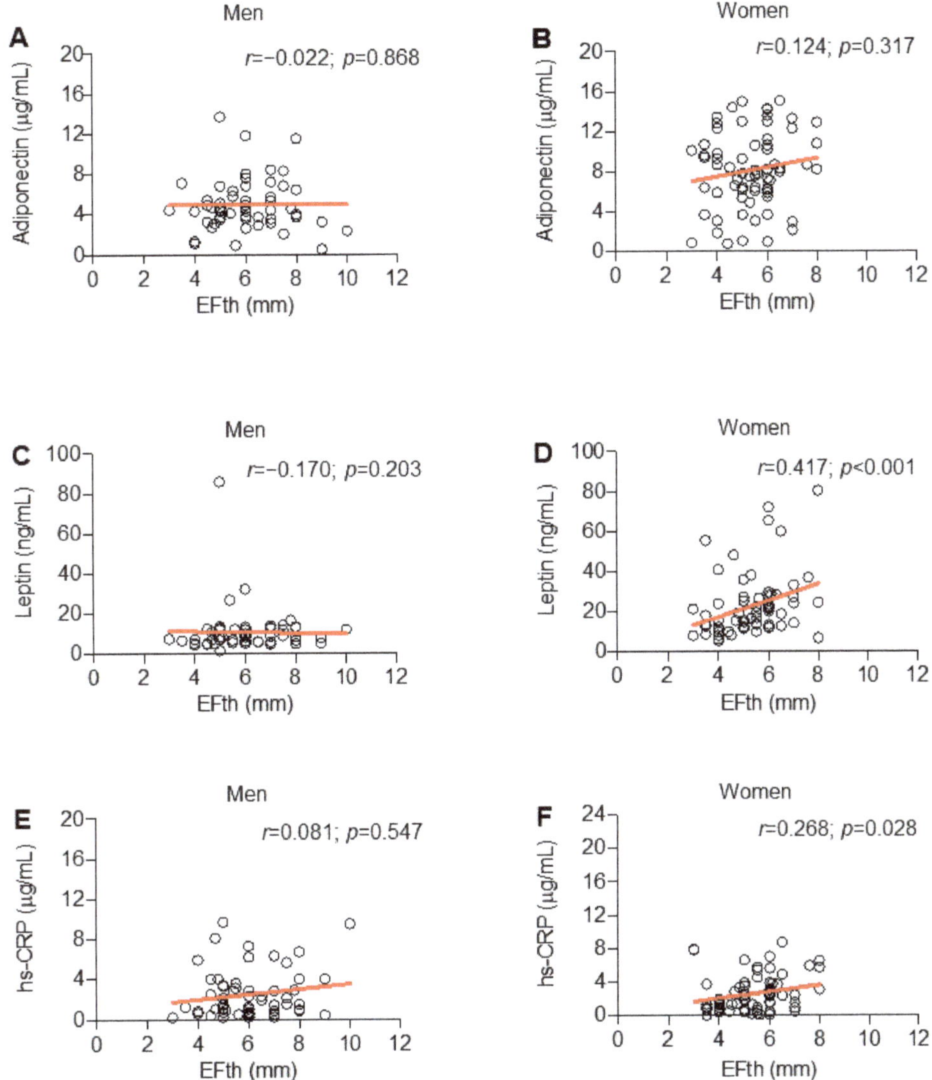

Figure 4. Only in women, leptin and high-sensitivity C-reactive protein (hs-CRP) correlate with the extent of epicardial fat thickness (EFth). Scatter plot illustrating the correlation of EFth with serum biomarkers across sex: adiponectin (**A**,**B**), leptin (**C**,**D**) and hs-CRP (**E**,**F**).

Only in women, significant correlations between EFth and leptin ($r = 0.417$; $p < 0.001$) and hs-CRP ($r = 0.268$; $p = 0.028$) were demonstrated (Figure 4C–F).

As difference in systemic biomarkers occurred only in women, linear regression analyses to test the independent associations were performed only in this sub-group. Both serum leptin and hs-CRP were independently associated with VFth also after adjustment for age and MetS criteria (Table 3).

Table 3. Linear regression showing the association of adipose tissue depots with inflammatory biomarkers in women.

Variables	Univariate		Adjusted	
VFth	**B (95% CI)**	***p*-Value**	**B (95% CI)**	***p*-Value**
Age	0.69 (0.22–1.15)	0.004	0.33 (–0.09–0.76)	0.122
MetS criteria	13.37 (7.84–18.91)	< 0.001	9.46 (3.78–15)	0.001
Leptin	0.15 (0.02–0.28)	0.025	0.14 (0.03–0.24)	0.012
hs-CRP	3.52 (1.47–5.57)	0.001	2.61 (0.75–4.47)	0.007
EFth	B (95% CI)	*p*–value	B (95% CI)	*p*–value
Age	0.05 (0.02–0.08)	< 0.001	0.03 (0.01–0.06)	0.006
MetS criteria	0.85 (0.51–1.18)	< 0.001	0.61 (0.30–0.12)	0.001
Leptin	0.01 (–0.00–0.01)	0.116	-	-
hs-CRP	0.12 (0.02–0.25)	0.026	0.05 (–0.06–0.17)	0.390

Vth: visceral fat thickness; MetS: metabolic syndrome; EFth: epicardial fat thickness; hs-CRP: high-sensitivity C-reactive protein.

Conversely, hs-CRP was independently associated with EFth in univariate analysis (B 0.12 [95%CI 0.02–0.25]; p = 0.026) but not in the adjusted analysis (Table 3).

3. Discussion

The present study further supports current evidence about the role of sex in determining adipose tissue distribution and inflammatory status. Although the extent of both visceral and epicardial fat depots was greater in men, only in women a correlation between serum inflammatory molecules and fat depots was observed. This may be not surprising as sexual dimorphism is already known to influence adipose distribution, systemic inflammation, cardio-metabolic disorders and clinical response to bariatric surgery as well [7,11,12]. Such differences in adipose tissue distribution may be ascribed to a different lipoprotein lipase activity and estrogen receptor expression [5]. Estrogens also modulate food intake and energy expenditure, potentially inducing anti-obese effects. Due to the high aromatase content, especially subcutaneous adipose tissue (SAT) contributes to this cardio-metabolic protection, which is lost in post-menopausal women [13]. Concerning sex-related differences in ectopic fat depots, the higher EFth observed in men may be ascribed to the well-known association existing between VAT extent and ectopic fat deposition. Sexual dimorphism in epicardial fat occurs in aging rats and this may be ascribed to differences in adipocytokines and obesity-related genes [14,15]. Also in human beings, the SWAN Cardiovascular Fat Ancillary Study reported increased EFth and coronary artery calcification in post-menopausal women, in association with decreased estradiol levels [16,17].

Viewed from an allostatic perspective, ectopic fat deposition occurs when storage capacity is exceeded over a defined limit. A genetic control of this limit would exist at both population and individual levels, with a significant role of age and sex [18–20]. However, VAT extent by itself is unlikely the only determinant of ectopic fat distribution and then is not sufficient to explain the new concepts of "obesity paradox" and "adiposopathy" [1,21]. In line with other studies, our results rather suggest a sex-specific correlation of chronic low-grade systemic inflammation with visceral and ectopic fat depots. Although preliminary, this finding may be considered as a strength of this study and could be a flywheel for future researches. More specifically, hs-CRP showed the strongest association with adipose depots. Overall, those data further claim for a role of hs-CRP in cardio-metabolic risk stratification, whereas it remains largely unexplained whether and how adipocytokines contribute to adipose tissue extent and function. Many researches on different models have attempted to address this point. A stronger association between obesity and inflammation has been already observed in women than men, especially for hs-CRP. A recent meta-analysis enrolling more than eighty thousand obese patients from 51 cross-sectional studies confirmed a correlation between BMI and serum hs-CRP, which was stronger in Caucasian women [22]. High inflammation in women was later reported in the HERMEX study [23]. Our findings are in line with this sex-specific link between visceral and epicardial fat through chronic low-grade inflammatory processes. We might speculate on pathophysiological mechanisms underlying such associations. Estrogens are widely associated with anti-inflammatory effects [24], so that insulin resistance and ectopic fat deposition are more frequently observed in post- than pre-menopausal women [25]. Nevertheless, even in post-menopausal women, hormone replacement therapy is associated with higher levels of hs-CRP [26,27]. Similarly, sexual dimorphism also modulates adipocytokine metabolism [28,29]. In line with our results, circulating levels of both leptin and adiponectin increased in women as compared to men [30,31]. Biological reasons for those differences are still unknown, but they seem human-specific as in rodents they are less apparent or even opposite [32].

This study has several limitations. Firstly, due the relatively small size of our cohort, the study may not be representative of the general population and thereby does not take into account all potential variables influencing EFth. In addition, the cross-sectional design of the study does not only allow to provide any relationship with long-term outcomes. Secondly, we have to acknowledge that the reproductive status of patients was not presented in this study, as we did not collect information about physiological/pharmacological history and hormonal background. Thirdly, computerized tomography and magnetic resonance would be the gold standard method to quantify adipose tissue depots. However, ultrasound is increasingly being validated and not burdened by high cost and limitation due to accessibility and radiation exposure. In our context, ultrasonography offers a wide range of opportunities in clinical setting, by providing accurate, non-invasive, and reproducible measurement of adipose tissue depots in different districts.

Finally, here we focused only of EFth, without considering other ectopic fat depots, especially hepatic steatosis, for which a sex-related differences have been widely reported. However, an extensive evaluation of whole-body fat depots—which should also include intramuscular, pancreatic and renal ones among others—is out of scope for the present study. Too many variables should have been considered, due to the differences between ectopic fat tissues in terms of vascularization, metabolic function and tissue environment. Future larger studies can discriminate between the unicity of each tissue and the common underlying pathways leading to ectopic fat deposition, including the sex-related ones.

4. Material and Methods

4.1. Patients

In this cross-sectional observational monocentric study, we enrolled 141 consecutive patients aged 30 to 65 years referred for one or more diagnostic criteria for MetS to the Clinica Medica "A. Murri", Department of Biomedical Sciences and Human Oncology, University of Bari Medical School (Bari, Italy). Exclusion criteria were systolic blood pressure (sBP) > 200 mm Hg or diastolic blood pressure (dBP) ≥ 120 mm Hg; reduction of left ventricular ejection fraction; cardiovascular disease (myocardial infarction or stroke within 6 months, angina, coronary revascularization, systolic heart failure); atrial fibrillation or flutter as well as other dysrhythmias; inflammatory bowel disease; severe psychiatric disorders; chronic kidney disease (glomerular filtration rate < 60 mL/min); secondary hypertension; peripheral arterial disease; alcoholism or use of illicit drugs; severe hepatic diseases; history of cancer; use of immunosuppressive drugs, chemotherapy, or radiotherapy; or inability to understand or adhere to study procedures.

In a second step we have further excluded patients with serum levels of hs-CRP above 10 μg/dL, in order to avoid potential bias due to other concomitant inflammatory conditions. The Ethics Committee of University Hospital Policlinico in Bari approved this protocol, performed in accordance to the guidelines of the Declaration of Helsinki. Patients gave informed consent before entering in the study (study number 5408, protocol number 0013869; approved by AOUCPG23/COMET/P on 7 July 2017).

4.2. Study Endpoints and Statistical Power Calculation

The primary outcome was to determine sex-related differences in EFth. According with the power study calculation based also on previous studies [33] our sample size (n = 125) was able to detect differences in EFth across sex with a power of 80% and a two-sided alpha error of 5%. Secondary endpoints include the potential impact of sex in the inflammatory crosstalk between visceral and epicardial adipose tissues (i.e., serum levels of adipocytokines and hs-CRP).

4.3. Data Collection and Assessment

For all patients, clinical data were collected, including, comorbidities and anthropometric measures. Waist circumference was measured at the midpoint between the lower border of the rib cage and the iliac crest. Visceral obesity was then diagnosed in the presence of waist circumference ≥ 96 cm in men and ≥ 80 cm in women, as recommended for the Mediterranean/European population [34]. The diagnosis of arterial hypertension was based on sBP ≥ 140 mmHg and/or dBP ≥ 90 mmHg (measured three times within 30 min, in the sitting position and using a brachial sphygmomanometer), or use of blood-pressure-lowering agents. The diagnosis of impaired fasting glucose and of type 2 diabetes (T2DM) was based on the revised criteria of the American Diabetes Association, using a value of fasting blood glucose ≥ 100 to < 126 mg/dL, and ≥126 mg/dL or the use of insulin or oral hypoglycemic agents, respectively [35]. A 12-h overnight fasting blood sample was drawn to determine hematological and biochemical profile. As surrogate marker of visceral adipose tissue dysfunction, we calculated the visceral adiposity index (VAI), through a validated mathematical model that uses both anthropometric (body mass index [BMI] and waist circumference) and functional (triglycerides [TAG] and high-density lipoprotein cholesterol [HDL-c]) simple parameters [36,37]. Finally, diagnosis of MetS was made based on the harmonized MetS criteria [34].

4.4. Ultrasound Assessment of Fat Depots

Visceral fat was measured with the Hitachi Noblus-E ecocolordoppler (Hitachi Medical, Tokyo, Japan) and Logiq E9 (GE, Healthcare, Chicago, IL, USA) ultrasound equipment equipped with a 3.5 MHz convex probe.

Epicardial fat thickness was measured with the M5S 7.5 MHz convex probe while the ML6-15 linear probe was used for the measurement of carotid IMT and plaques, Logiq E9 (GE, Healthcare,). All ultrasound examinations and diagnoses were performed by a trained internal medicine specialist. The visceral fat thickness was measured from the center of the left hepatic lobe [38,39].

4.5. Statistical Analysis

Analyses were performed with IBM SPSS Statistics for Windows, Version 23.0 (IBM CO. Armonk, NY, USA). Categorical data are presented as absolute (and relative) frequencies and analyzed by Chi square or Fisher's exact test, as appropriate. Continuous variables have been expressed as median and interquartile range (IQR) as the normality assumption was not demonstrated. Intergroup comparisons were then drawn by Mann–Whitney test, whereas correlations were investigated through Spearman's rank correlation coefficient. Adjusted linear and logistic regressions have been used to evaluate the independency of the association between inflammatory biomarkers and adipose tissue depots. For all statistical analyses, a 2-sided p-value < 0.05 has been considered as statistically significant.

5. Conclusions

In conclusion, our study supports a close association between visceral and epicardial fat depots in the subset of women with at least one MetS criterion. This crosstalk would be mediated by a chronic low-grade inflammatory status. Since ectopic lipid accumulation precedes the metabolic and atherosclerotic complications, identifying fat-related inflammatory biomarkers might have a clinical utility in stratifying obese/dysmetabolic patients and eventually in CV risk estimation. This approach might then optimize intervention outcomes and produce a more personalized approach [40].

Author Contributions: P.P. and F.M. contribute to conceptualization of the study, methodology, project administration, funding acquisition and supervision; E.M.-M. also contribute to funding acquisition. M.S.L., E.M.-M., A.B., S.P., and V.O.P., performed investigation and data curation. F.C., S.M., A.M.A., D.F., and E.E. performed formal analysis. F.C. took also care of analysis with statistical software, original draft preparation, review, editing and visualization.

Funding: This study was supported by a grant from the Rete Cardiologica of Italian Ministry of Health (#2754291) to F. Montecucco. This study was supported by a grant from Fondazione Carige to F. Montecucco. The present paper is written in the context of the project FOIE GRAS, which has received funding from the European Union's Horizon 2020 Research and Innovation program under the Marie Skłodowska-Curie Grant Agreement (#722619; recipient Emilio Molina-Molina).

Conflicts of Interest: The authors declare no conflict of interest.

References

1. Vecchie, A.; Dallegri, F.; Carbone, F.; Bonaventura, A.; Liberale, L.; Portincasa, P.; Fruhbeck, G.; Montecucco, F. Obesity phenotypes and their paradoxical association with cardiovascular diseases. *Eur. J. Intern. Med.* **2018**, *48*, 6–17. [CrossRef] [PubMed]

2. Bays, H.E. Adiposopathy is "sick fat" a cardiovascular disease? *J. Am. Coll Cardiol* **2011**, *57*, 2461–2473. [CrossRef] [PubMed]

3. Bouchard, C.; Tremblay, A.; Despres, J.P.; Nadeau, A.; Lupien, P.J.; Theriault, G.; Dussault, J.; Moorjani, S.; Pinault, S.; Fournier, G. The response to long-term overfeeding in identical twins. *N. Engl. J. Med.* **1990**, *322*, 1477–1482. [CrossRef] [PubMed]

4. Nazare, J.A.; Smith, J.D.; Borel, A.L.; Haffner, S.M.; Balkau, B.; Ross, R.; Massien, C.; Almeras, N.; Despres, J.P. Ethnic influences on the relations between abdominal subcutaneous and visceral adiposity, liver fat, and cardiometabolic risk profile: The International Study of Prediction of Intra-Abdominal Adiposity and Its Relationship With Cardiometabolic Risk/Intra-Abdominal Adiposity. *Am. J. Clin. Nutr* **2012**, *96*, 714–726. [PubMed]

5. Chella Krishnan, K.; Mehrabian, M.; Lusis, A.J. Sex differences in metabolism and cardiometabolic disorders. *Curr Opin Lipidol* **2018**, *29*, 404–410. [CrossRef] [PubMed]

6. Singer, K.; Maley, N.; Mergian, T.; DelProposto, J.; Cho, K.W.; Zamarron, B.F.; Martinez-Santibanez, G.; Geletka, L.; Muir, L.; Wachowiak, P.; et al. Differences in Hematopoietic Stem Cells Contribute to Sexually Dimorphic Inflammatory Responses to High Fat Diet-induced Obesity. *J. Biol. Chem.* **2015**, *290*, 13250–13262. [CrossRef]

7. Fall, T.; Hagg, S.; Ploner, A.; Magi, R.; Fischer, K.; Draisma, H.H.; Sarin, A.P.; Benyamin, B.; Ladenvall, C.; Akerlund, M.; et al. Age- and sex-specific causal effects of adiposity on cardiovascular risk factors. *Diabetes* **2015**, *64*, 1841–1852. [CrossRef]

8. Rabkin, S.W. The relationship between epicardial fat and indices of obesity and the metabolic syndrome: A systematic review and meta-analysis. *Metab. Syndr. Relat. Disord.* **2014**, *12*, 31–42. [CrossRef]

9. Bredella, M.A. Sex Differences in Body Composition. *Adv. Exp. Med. Biol* **2017**, *1043*, 9–27.

10. Ferrara, D.; Montecucco, F.; Dallegri, F.; Carbone, F. Impact of different ectopic fat depots on cardiovascular and metabolic diseases. *J. Cell. Physiol.* **2019**, *234*, 21630–21641. [CrossRef]

11. Thorand, B.; Baumert, J.; Doring, A.; Herder, C.; Kolb, H.; Rathmann, W.; Giani, G.; Koenig, W.; Group, K. Sex differences in the relation of body composition to markers of inflammation. *Atherosclerosis* **2006**, *184*, 216–224. [CrossRef]

12. Carbone, F.; Nulli Migliola, E.; Bonaventura, A.; Vecchie, A.; De Vuono, S.; Ricci, M.A.; Vaudo, G.; Boni, M.; Dallegri, F.; Montecucco, F.; et al. High serum levels of C-reactive protein (CRP) predict beneficial decrease of visceral fat in obese females after sleeve gastrectomy. *Nutr Metab Cardiovasc Dis* **2018**, *28*, 494–500. [CrossRef] [PubMed]

13. Belanger, C.; Luu-The, V.; Dupont, P.; Tchernof, A. Adipose tissue intracrinology: Potential importance of local androgen/estrogen metabolism in the regulation of adiposity. *Horm Metab Res.* **2002**, *34*, 737–745. [CrossRef] [PubMed]

14. Fei, J.; Cook, C.; Blough, E.; Santanam, N. Age and sex mediated changes in epicardial fat adipokines. *Atherosclerosis* **2010**, *212*, 488–494. [CrossRef]

15. Kocher, C.; Christiansen, M.; Martin, S.; Adams, C.; Wehner, P.; Gress, T.; Santanam, N. Sexual dimorphism in obesity-related genes in the epicardial fat during aging. *J. Physiol Biochem* **2017**, *73*, 215–224. [CrossRef]

16. El Khoudary, S.R.; Shields, K.J.; Janssen, I.; Hanley, C.; Budoff, M.J.; Barinas-Mitchell, E.; Everson-Rose, S.A.; Powell, L.H.; Matthews, K.A. Cardiovascular Fat, Menopause, and Sex Hormones in Women: The SWAN Cardiovascular Fat Ancillary Study. *J. Clin. Endocrinol Metab* **2015**, *100*, 3304–3312. [CrossRef]

17. El Khoudary, S.R.; Shields, K.J.; Janssen, I.; Budoff, M.J.; Everson-Rose, S.A.; Powell, L.H.; Matthews, K.A. Postmenopausal Women With Greater Paracardial Fat Have More Coronary Artery Calcification Than Premenopausal Women: The Study of Women's Health Across the Nation (SWAN) Cardiovascular Fat Ancillary Study. *J. Am. Heart Assoc.* **2017**, *6*, 2. [CrossRef]

18. Chu, A.Y.; Deng, X.; Fisher, V.A.; Drong, A.; Zhang, Y.; Feitosa, M.F.; Liu, C.T.; Weeks, O.; Choh, A.C.; Duan, Q.; et al. Multiethnic genome-wide meta-analysis of ectopic fat depots identifies loci associated with adipocyte development and differentiation. *Nat. Genet.* **2017**, *49*, 125–130. [CrossRef]

19. Tamura, Y. Ectopic fat, insulin resistance and metabolic disease in non-obese Asians: Investigating metabolic gradation. *Endocr. J.* **2019**, *66*, 1–9. [CrossRef]

20. Mancuso, P.; Bouchard, B. The Impact of Aging on Adipose Function and Adipokine Synthesis. *Front. Endocrinol.* **2019**, *10*, 137. [CrossRef]

21. Neeland, I.J.; Poirier, P.; Despres, J.P. Cardiovascular and Metabolic Heterogeneity of Obesity: Clinical Challenges and Implications for Management. *Circulation* **2018**, *137*, 1391–1406. [CrossRef] [PubMed]

22. Choi, J.; Joseph, L.; Pilote, L. Obesity and C-reactive protein in various populations: A systematic review and meta-analysis. *Obes Rev.* **2013**, *14*, 232–244. [CrossRef] [PubMed]

23. Soriano-Maldonado, A.; Aparicio, V.A.; Felix-Redondo, F.J.; Fernandez-Berges, D. Severity of obesity and cardiometabolic risk factors in adults: Sex differences and role of physical activity. The HERMEX study. *Int J. Cardiol* **2016**, *223*, 352–359. [CrossRef] [PubMed]

24. Camporez, J.P.; Lyu, K.; Goldberg, E.L.; Zhang, D.; Cline, G.W.; Jurczak, M.J.; Dixit, V.D.; Petersen, K.F.; Shulman, G.I. Anti-inflammatory effects of oestrogen mediate the sexual dimorphic response to lipid-induced insulin resistance. *J. Physiol.* **2019**, *597*, 3885–3903. [CrossRef] [PubMed]

25. Abildgaard, J.; Danielsen, E.R.; Dorph, E.; Thomsen, C.; Juul, A.; Ewertsen, C.; Pedersen, B.K.; Pedersen, A.T.; Ploug, T.; Lindegaard, B. Ectopic Lipid Deposition Is Associated With Insulin Resistance in Postmenopausal Women. *J. Clin. Endocrinol Metab* **2018**, *103*, 3394–3404. [CrossRef]

26. Ridker, P.M.; Hennekens, C.H.; Rifai, N.; Buring, J.E.; Manson, J.E. Hormone replacement therapy and increased plasma concentration of C-reactive protein. *Circulation* **1999**, *100*, 713–716. [CrossRef]

27. Rexrode, K.M.; Pradhan, A.; Manson, J.E.; Buring, J.E.; Ridker, P.M. Relationship of total and abdominal adiposity with CRP and IL-6 in women. *Ann. Epidemiol.* **2003**, *13*, 674–682. [CrossRef]

28. White, U.A.; Tchoukalova, Y.D. Sex dimorphism and depot differences in adipose tissue function. *Biochim. Biophys. Acta* **2014**, *1842*, 377–392. [CrossRef]

29. Valencak, T.G.; Osterrieder, A.; Schulz, T.J. Sex matters: The effects of biological sex on adipose tissue biology and energy metabolism. *Redox Biol.* **2017**, *12*, 806–813. [CrossRef]

30. Hickey, M.S.; Israel, R.G.; Gardiner, S.N.; Considine, R.V.; McCammon, M.R.; Tyndall, G.L.; Houmard, J.A.; Marks, R.H.; Caro, J.F. Gender differences in serum leptin levels in humans. *Biochem. Mol. Med.* **1996**, *59*, 1–6. [CrossRef]

31. Kennedy, A.; Gettys, T.W.; Watson, P.; Wallace, P.; Ganaway, E.; Pan, Q.; Garvey, W.T. The metabolic significance of leptin in humans: Gender-based differences in relationship to adiposity, insulin sensitivity, and energy expenditure. *J. Clin. Endocrinol Metab* **1997**, *82*, 1293–1300. [CrossRef] [PubMed]

32. Shen, W.; Punyanitya, M.; Silva, A.M.; Chen, J.; Gallagher, D.; Sardinha, L.B.; Allison, D.B.; Heymsfield, S.B. Sexual dimorphism of adipose tissue distribution across the lifespan: A cross-sectional whole-body magnetic resonance imaging study. *Nutr. Metab.* **2009**, *6*, 17. [CrossRef] [PubMed]

33. Salami, S.S.; Tucciarone, M.; Bess, R.; Kolluru, A.; Szpunar, S.; Rosman, H.; Cohen, G. Race and epicardial fat: The impact of anthropometric measurements, percent body fat and sex. *Ethn. Dis.* **2013**, *23*, 281–285. [PubMed]
34. Alberti, K.G.; Eckel, R.H.; Grundy, S.M.; Zimmet, P.Z.; Cleeman, J.I.; Donato, K.A.; Fruchart, J.C.; James, W.P.; Loria, C.M.; Smith, S.C.; et al. Harmonizing the metabolic syndrome: A joint interim statement of the International Diabetes Federation Task Force on Epidemiology and Prevention; National Heart, Lung, and Blood Institute; American Heart Association; World Heart Federation; International Atherosclerosis Society; and International Association for the Study of Obesity. *Circulation* **2009**, *120*, 1640–1645. [PubMed]
35. American Diabetes, A. 2. Classification and Diagnosis of Diabetes. *Diabetes Care* **2016**, *39*, S13–S22.
36. Amato, M.C.; Giordano, C.; Galia, M.; Criscimanna, A.; Vitabile, S.; Midiri, M.; Galluzzo, A.; AlkaMeSy Study, G. Visceral Adiposity Index: A reliable indicator of visceral fat function associated with cardiometabolic risk. *Diabetes Care* **2010**, *33*, 920–922. [CrossRef]
37. Amato, M.C.; Giordano, C.; Pitrone, M.; Galluzzo, A. Cut-off points of the visceral adiposity index (VAI) identifying a visceral adipose dysfunction associated with cardiometabolic risk in a Caucasian Sicilian population. *Lipids Health Dis.* **2011**, *10*, 183. [CrossRef]
38. Stolk, R.P.; Meijer, R.; Mali, W.P.; Grobbee, D.E.; van der Graaf, Y.; Secondary Manifestations of Arterial Disease (SMART) Study Group. Ultrasound measurements of intraabdominal fat estimate the metabolic syndrome better than do measurements of waist circumference. *Am. J. Clin. Nutr* **2003**, *77*, 857–860. [CrossRef]
39. Kim, S.K.; Kim, H.J.; Hur, K.Y.; Choi, S.H.; Ahn, C.W.; Lim, S.K.; Kim, K.R.; Lee, H.C.; Huh, K.B.; Cha, B.S. Visceral fat thickness measured by ultrasonography can estimate not only visceral obesity but also risks of cardiovascular and metabolic diseases. *Am. J. Clin. Nutr* **2004**, *79*, 593–599. [CrossRef]
40. Trouwborst, I.; Bowser, S.M.; Goossens, G.H.; Blaak, E.E. Ectopic Fat Accumulation in Distinct Insulin Resistant Phenotypes; Targets for Personalized Nutritional Interventions. *Front. Nutr.* **2018**, *5*, 77. [CrossRef]

International Journal of
Molecular Sciences

Concept Paper

Lipids and Lipid-Processing Pathways in Drug Delivery and Therapeutics

Milica Markovic [1], Shimon Ben-Shabat [1], Aaron Aponick [2], Ellen M. Zimmermann [3] and Arik Dahan [1,*]

[1]   Department of Clinical Pharmacology, School of Pharmacy, Faculty of Health Sciences,
    Ben-Gurion University of the Negev, Beer-Sheva 8410501, Israel; milica@post.bgu.ac.il (M.M.);
    sbs@bgu.ac.il (S.B.-S.)
[2]   Department of Chemistry, University of Florida, Gainesville, FL 32603, USA; aponick@chem.ufl.edu
[3]   Department of Medicine, Division of Gastroenterology, University of Florida, Gainesville, FL 32610, USA;
    Ellen.Zimmermann@medicine.ufl.edu
*   Correspondence: arikd@bgu.ac.il

Received: 13 April 2020; Accepted: 2 May 2020; Published: 4 May 2020

Abstract: The aim of this work is to analyze relevant endogenous lipid processing pathways, in the context of the impact that lipids have on drug absorption, their therapeutic use, and utilization in drug delivery. Lipids may serve as biomarkers of some diseases, but they can also provide endogenous therapeutic effects for certain pathological conditions. Current uses and possible clinical benefits of various lipids (fatty acids, steroids, triglycerides, and phospholipids) in cancer, infectious, inflammatory, and neurodegenerative diseases are presented. Lipids can also be conjugated to a drug molecule, accomplishing numerous potential benefits, one being the improved treatment effect, due to joined influence of the lipid carrier and the drug moiety. In addition, such conjugates have increased lipophilicity relative to the parent drug. This leads to improved drug pharmacokinetics and bioavailability, the ability to join endogenous lipid pathways and achieve drug targeting to the lymphatics, inflamed tissues in certain autoimmune diseases, or enable overcoming different barriers in the body. Altogether, novel mechanisms of the lipid role in diseases are constantly discovered, and new ways to exploit these mechanisms for the optimal drug design that would advance different drug delivery/therapy aspects are continuously emerging.

Keywords: lipid; fatty acid; glyceride; steroid; phospholipid; oral drug absorption; prodrug; phospholipase A_2 (PLA_2)

---

1. Introduction

Lipids are hydrophobic biomolecules, which include fatty acids (FA), glycerides, phospholipids (PL), sterols, sphingolipids, and prenol lipids (Figure 1) [1]. Lipids play an important role in energy metabolism and storage, as structural components, in signaling, and as hormones. The disruption of lipid metabolic enzymes and pathways occurs in many disease such as cancer, diabetes, infectious, neurodegenerative, and inflammatory diseases [2]. The aim of this work is to elucidate the effect of lipids, and lipid excipients on drug absorption, to describe the metabolic lipid pathways and to demonstrate the role that lipids have in many pathological conditions, as well as their endogenous pharmacological activity. An additional section is dedicated to lipidic prodrugs that can exploit lipid processing pathways in order to achieve their effect. The presence of dietary lipids or lipids from drug formulations/lipidic prodrugs can influence drug absorption by incorporating to the natural lipid metabolic pathways. Hereinafter, we provide a small overview of the lipid influence on drug absorption.

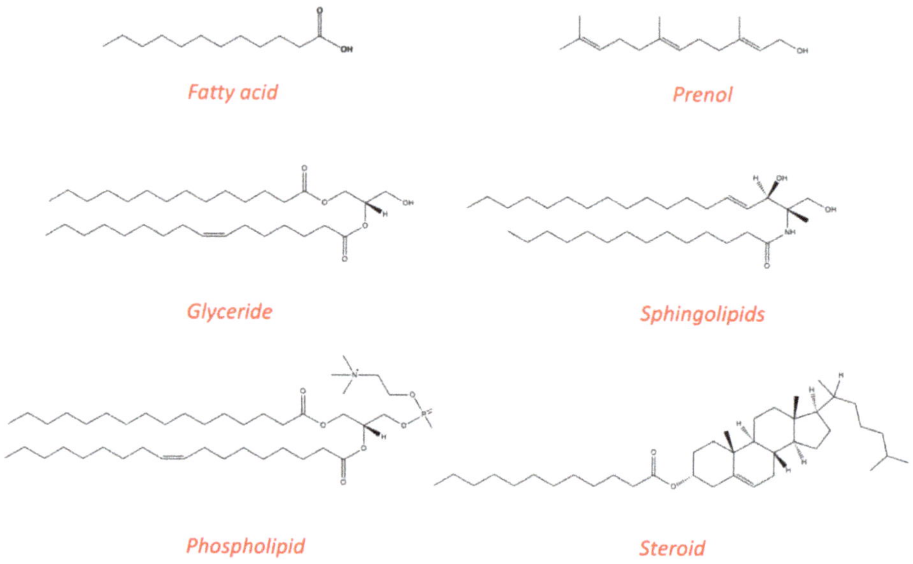

Figure 1. Main lipid categories: examples of chemical structures.

Molecular revolution, such as development of in-vitro high-throughput screening methods and combinatorial chemistry, resulted in a high number of poor aqueous solubility molecules to be selected as drug candidates. Nearly 40% of all novel drug candidates are lipophilic and demonstrate low water solubility [3,4]. Following oral administration, drugs encounter various obstacles on their way to the blood circulation. Absorption is a process, in which orally administered compounds travel from the gastrointestinal (GI) lumen into the intestinal membrane and enter systemic circulation/become bioavailable. Following ingestion and prior to permeation, the drug has to be dissolved in the GI milieu and turn into a molecular form close to the intestinal membrane. This can be difficult for lipophilic compounds with poor solubility in water, thereby presenting a limiting step in the absorption process. The presence of lipids derived from food or lipid-based formulations in the intestinal lumen can influence the oral absorption of highly lipophilic drugs in many different ways. The solubility of the drug can be increased, due to the creation of various colloidal formations (vesicles, micelles). Drug solubilization can be influenced by lipid presence itself and by the simulation of physiological lipid processing pathways, leading to increased secretion of bile-salts and phospholipids [5,6]. The lipids can influence intestinal metabolism and influx/efflux transport. Studies showed that, in some cases, lipid excipients could improve drug absorption through the influence on the P-glycoprotein (P-gp) functioning [7]. Additionally, nuclear hormone receptors (NHR) were shown to play an important role in lipid trafficking and metabolism and, thus, in intestinal lipid and drug absorption, since they control a number of proteins (e.g., fatty-acid-binding proteins) that are involved in lipid/drug transport and metabolism [8–10]. Following oral administration, in many cases drugs pass through the hepatic vein on their way to the systemic blood, whereas highly lipophilic compounds may be transported through the intestinal lymphatic system. Lipids can also stimulate intestinal lymphatic drug transport, in which the drugs can bypass the first-pass hepatic metabolism and go directly to the systemic circulation [11]. After solubilization, drugs permeate through the intestinal membrane via passive diffusion/active transport through the enterocytes. For hydrophilic drugs with poor solubility in lipids, this step can be the rate-limiting in the absorption cascade, whereas, for lipophilic drugs, the unstirred water layer (UWL) in the proximity of the intestinal membrane is an obstacle for effective permeability. The diffusion of FA, monoglycerides (MG), and many other lipophilic molecules (including drugs and prodrugs) through UWL is significantly increased through micellar solubilization, prior to arriving to the enterocytes. It is likely these lipophilic molecules dissociate from the micelles prior to going into enterocytes or through binding to the transporter or vesicular-mediated transport of micelles. The following stage of drug absorption is leaving the enterocyte into the lamina propria, from where the drug is usually absorbed into the hepatic blood flow, unless it undertakes lymphatic transport, which is particularly important for lipophilic molecules, and it will be discussed in detail in part 2. The following section describes lipid processing pathways subsequent to oral ingestion. In addition to lipid processing pathways, this work provides an overview of lipid balance disruption in disease, pharmacological roles of certain lipids, and their role as lipid carriers in the drug delivery.

2. Lipid Processing Pathways

The digestion of exogenous lipids begins in the mouth via enzyme lingual lipase. It is followed by the gastric lipase in the stomach, combined with mechanical mixing. This is where PL and TG, alongside initial amphiphilic digestion products, such as FA, diglycerides (DG), and lysophospholipids (LPL), form crude emulsion [12,13]. Digestion continues in the small intestine, where pancreatic lipase, alongside its cofactor co-lipase, finalizes the disintegration of TG to DG, MG, and FA [14]. This enzyme mainly hydrolyses *sn*-1 and *sn*-3 positions of the TG, producing FA and MG [15]. PL digestion in the small intestine occurs via pancreatic phospholipase A_2 (PLA_2) enzyme, which hydrolyses the FA in the *sn*-2 position of the PL, producing a FA and LPL [16]. Exogenous lipids stimulate the secretion of biliary lipids, such as bile salts, PL, and unesterified cholesterol, which in turn stimulates the formation of vesicles and micelles that incorporate these exogenous lipids. This step allows for lipophilic drugs, as well as lipidic prodrugs, to incorporate into micelles, thereby increasing the solubilization of lipids/lipid-drug conjugates/lipophilic drugs. After the micelles break down, lipid digestion products become available for absorption to the enterocytes. Many molecules pass through the basolateral enterocyte membrane into the lamina propria on their way to hepatic blood flow. Others, such as lipophilic digestion products (FA, MG), are reassembled into TG/PL in the enterocyte, followed by incorporation to lipoproteins (rich with TG). Lipoproteins (LP), in this case, chylomicrons (CM) also then undergo exocytosis to the lamina propria, however they do not enter blood vessels, due to its large size that obstructs the passage through tight junctions between capillary endothelial cells; rather, LP enter leaky lymphatic capillary, lacteal [17]. The contents of lacteal are released to the systemic circulation via bigger lymph vessels, hence avoiding the hepatic blood flow. This intestinal lymphatic system allows the transport of absorbed lipids and other lipophilic molecules (drugs, liposoluble vitamins, prodrugs) from the small intestine to the blood circulation. The main advantages of designing a drug/prodrug that undergoes lymphatic transport is bypassing the portal blood flow and, thus, hepatic metabolism, and achieving lymphatic targeting in cases when its desired (e.g., treating cancer metastases).

Another important class of lipids is sterols, with cholesterol being the most familiar human sterol. Cholesterol is transferred from the intestine to the liver as a part of LP complex [18]. Figure 2 demonstrates the structure of LP. The role of LP is to transfer various lipids to the tissues [19]. According to its density, LP can be categorized by their density as ultralow density LP, chylomicrons (CM), very low-density lipoproteins (VLDL), low-density lipoproteins (LDL), and high-density lipoproteins (HDL) [20]. Another species of LP formed through the degradation of the VLDL and HDL are intermediate-density lipoproteins (IDL). CM is synthesized in the intestinal tissue and it is responsible for carrying the cholesterol and TG throughout the body. VLDL distribute the TG to various tissues. It is produced in the liver are transformed to LDL via lipoprotein lipase. LDL is responsible for getting cholesterol from the liver to other parts of the body; it contains apolipoprotein (apo B-100), which permits LDL binding to the LDL receptors on the cell surfaces. Lastly, the role of HDL is to carry cholesterol from the tissues back to the liver. Steroid hormones, mineralocorticoids, and glucocorticoids are synthesized from cholesterol, as well as bile salts.

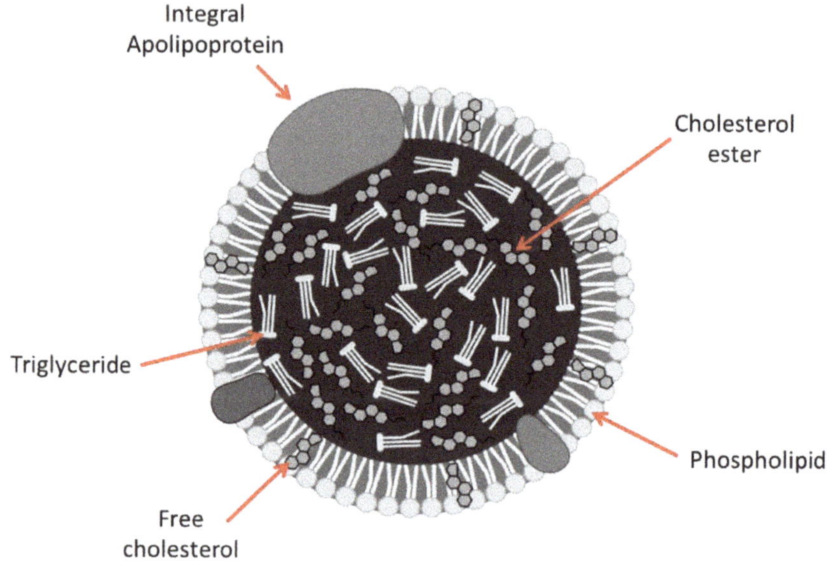

Figure 2. Lipoprotein structure.

The misbalance in the presence of different lipid moieties and their oxidation products can be a sign of pathological condition. The role of lipids in pathological diseases and their therapeutic potential, as well as basic principles of lipidic prodrugs, are discussed in the following section.

3. Lipid Role in Pathological Conditions and Their Therapeutic/Drug Delivery Use

The application of lipidomics can help us to study lipid metabolism and it can help us to understand the underlying mechanisms of various metabolic diseases and the role of lipids in modulating homeostasis [21]. Figure 3 presents several diseases caused by disruption in lipid metabolism. The identification of specific lipid biomarkers and mechanism-based knowledge of particular metabolic lipid processing pathways in connection to different pathologies is basis for developing innovative therapeutic approach for treating different pathologies. In addition, we mention the use of lipid carriers in drug delivery, particularly the use of lipidic prodrugs. Lipidic prodrugs contain the drug covalently attached to the lipid carrier (e.g., FA, glyceride, PL, or steroid). This conjugation leads to higher lipophilicity when compared to the drug alone, and can thus lead to better pharmacokinetic profile and provide other significant benefits, such as improved absorption through biological barriers, extended blood half-life, selective distribution to tissues (e.g., brain, intestine), decreased liver first-pass metabolism, and overall improved bioavailability of the drug. Drug-lipid conjugates can also join the physiological lipid trafficking pathways and accomplish drug targeting to specific sites [22]. Exploiting the lymphatic transport is one of the mechanisms of lipidic prodrug incorporation to the lipid processing pathways, and it is highlighted in this paragraph.

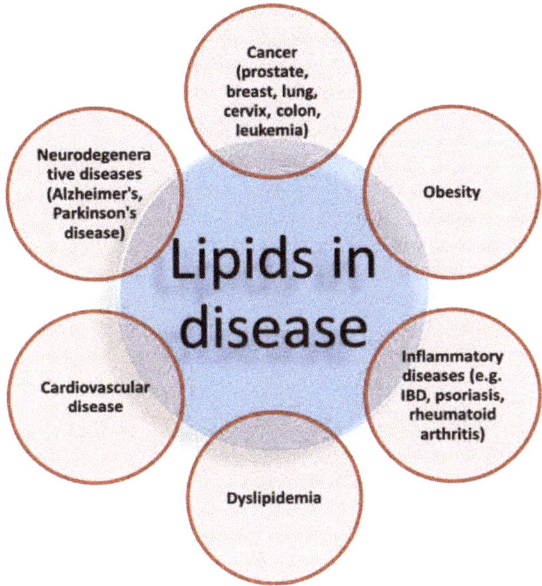

Figure 3. The representation of various diseases caused by disruption in the physiological lipid metabolism.

3.1. Fatty Acids (FA)

FA contains hydrocarbon chains of various lengths and degrees of desaturation. Numerous lipids are synthesized from FA; for instance, PL and glycerides contain hydrophobic FA tail, as well as triacylglycerides, which are synthesized and stored in the state of elevated nutrient availability [23]. In the physiological conditions, de novo FA synthesis in humans originates in the liver, breasts during lactation, and adipose tissues [24]. However, in pathological conditions, such as cancer, it was demonstrated that FA biosynthesis plays an important role [25,26]. Fatty acid synthase was identified as the tumor antigen OA-519 in invasive breast cancer, in retinoblastoma [27], and it was also found in proliferating fetal tissues [24], signifying that the reactivation of cancer FA synthesis could mean a retreat to a less-differentiated embryonic state [23]. The elevated presence of FA in cancerous tissues could also be due to the increased metabolic demand in the cancer tissues or adaptation to decreased presence of serum lipids in the tumor environment [23].

The ratio of free FA turnover is reduced in the growing white adipose tissue. Adipose tissues and dietary saturated FA are in correlation with increased fat cell size and number. Adipose TG lipase inhibition is responsible for TG buildup, while the inhibition of hormone-sensitive lipase. the buildup of DG. Surplus triacylglycerols, sterols, and sterol esters are enclosed by the PL monolayer and create lipid droplets. The size and number of lipid droplet distribution is linked to obesity [28].

Unsaturated FA (UFA) are present in the human fat and they contain at least one unsaturated double bond. They could be monounsaturated (MUFA), if only one unsaturated double bond is present (oleic, elaidic acid), or polyunsaturated (PUFA), where a number of double bonds are present in the alkyl chain (i.e., ω-3 position: docosahexaenoic, eicosapentaenoic acid; or, ω-6 position: arachidonic, linoleic acid). PUFA are considered as essential FA, due to their significant physiological function, and due to the fact that they only come from exogenouse sourses (diet). UFA in general, have numerous physiological functions, such as maintaining functional biomembrane, promote cardiovascular wellbeing through lowering cholesterol/TG [29], anti-inflammatory effect, synthesis of prostaglandins/thromboxanes, and they promote brain functioning [30].

Omega-3 PUFA (4g a day) were shown to decrease the TG levels by 30%, respective of the baseline levels, and can thus be used in the treatment of pancreatitis [31]. However meta-analysis revealed that omega-3 PUFA supplementation did not decrease the risk of stroke, cardiac sudden death, myocardial infarction, and all-cause mortality [32].

Additionally, UFA were also found to have some endogenous anticancer properties, and some capability of promoting chemotherapeutic effect of anticancer drugs [33]. UFA have good biocompatibility, innate tumor-targeting features, and likely pharmacological activity in cancer treatment; therefore, they present good lipid carriers for prodrug development in cancer therapy. The conjugation of UFA with chemotherapeutic agent was shown to increase the lipophilicity of the drugs, which might enable the cellular uptake of anticancer drugs through passive transport (which is particularly important for water-soluble nucleoside drugs). Anti-cancer agent-UFA conjugates that entered clinical trials gemcitabine-elaidic acid (CP-4126), cytarabine-elaidic acid (CP-4055), and paclitaxel-DHA (PTXDHA) prodrug [33].

Lymphatic transport with FA is variable; whereas, short/medium-chain FA might be absorbed via hepatic blood, long-chain FA have a tendency to be reacylated into TG, integrated in CM, and transported through the lymphatics. FA-drug conjugates usually do not undergo lymphatic transport, due to extensive hydrolysis prior to absorption process, although some FA-drug conjugates have indeed shown successful lymphatic transport [34]. Testosterone is an androgen hormone that is subjected to high first-pass hepatic metabolism, which leads to very low systemic bioavailability following oral absorption; thus, testosterone by itself cannot treat male androgen deficiency syndrome. However when conjugated with an undecanoic acid, it showed an increase in systemic bioavailability up to 7% [35]. Testosterone-undecanoate undergoes lymphatic transport, thus avoiding the portal blood and first-pass metabolism (Table 1). In the blood, free testosterone is released from the prodrug and it is free to demonstrate its therapeutic effect [34–36].

3.2. Glycerides

Elevated plasma TG and TG-rich LP in non-fasting conditions have a significant role in various cardiovascular diseases, such as myocardial infarction, ischemic heart disease, and even death [37,38]. CM residues and LDL are entrapped by the arterial wall cells; in the general population, in people who do not have familial hyperlipoproteinemia, it was shown that atherogenesis can occur after the postprandial period [39]. High TG can be a marker for several types of atherogenic LP. TG-rich LP, such as IDL and VLDL, can be trapped within the arterial wall, while nascent CM and VLDL are too big to penetrate the wall [40]. It was suggested that lifelong high plasma TG-rich LP or their remnants are related to greater risk of ischemic heart disease, regardless of suboptimal HDL levels [41].

On the other hand, TG use in drug administration is important; TG can be used to enhance formulations of highly lipophilic drugs. They can be TG oils, various combinations of TG, DG, and MG. The sort of oil that is used in the formulation has significant impact on the formulation capacity to improve absorption [42]. Non-digestable lipids, such as sucrose polyesters, are not absorbed from the intestinal membrane, whereas digestable lipids, such as DG, TG, PL, FA, cholesterol, and some synthetic derivatives, are suitable components of drug formulation. They can be classified according to the degree of saturation, interaction with water, and the length of their carbon chain, which can be long-chain TG (LCT), medium-chain TG (MCT), as well as DG, MG, FA, PL, and others [42]. The effect of peanut oil (LCT), miglyol (MCT), and paraffin oil was studied on the absorption of probucol, hypocholesterolemic drug with high lipophilicity, with log P around 10 [43]. The highest bioavailability was obtained with LCT, peanut oil, then MCT, and lastly no absorption was evident for paraffin oil solution. However each drug formulation should be evaluated on a case-by-case basis.

Glycerides have an important role in the development of novel prodrugs with improved targeting features when compared to the parent drug. Oftentimes, the TG-drug conjugates undergo lymphatic transport and lymphatic drug delivery. In one instance, the immunosuppressant drug, mycophenolic acid (MPA) was conjugated to the sn-2 position of a glyceride [44]. This prodrug (2-MPA-TG) was designed to incorporate MPA to the TG deacylation–reacylation pathway, and increase the lymphatic transport of MPA following oral administration (Table 1). MPA lymphatic transport and systemic bioavailability was evaluated in dogs, following oral administration of MPA alone and MPA-prodrug [45].MPA levels in the lymph nodes and lymph residing lymphocytes were studied in order to determine the extent of MPA targeting to the sites of action within the lymph. The TG-MPA prodrug significantly increased lymphatic transport 288-fold, in comparison to the MPA; 36.4% of the dose was recovered in the lymph. The MPA level in the lymph nodes was increased 5- to 6-fold and in the lymph lymphocytes 21-fold *vs.* MPA administration alone. This study demonstrated that the TG-MPA prodrug could incorporate deacylation–reacylation pathway of the TG and successfully target the lymphatics through improved MPA uptake within lymph residing lymphocytes, resulting in the more effective immunomodulation of MPA [45].

3.3. Phospholipids (PL)

Lipids are strongly linked with amyloid precursor protein metabolism, resulting in amyloid-beta peptide (Aβ) formation, one of the key component of senile plaques, which characterize the pathological hallmark of Alzheimer disease [46]. Certain choline PLs were proposed as likely biomarkers of Alzheimer disease. It was shown that lysophosphatidylcholine, lyso-platelet, and choline plasmalogen activating factor levels during normal aging increase meaningfully; comparable but even more pronounced alterations were found in people with probable Alzheimer disease. Thus, higher choline-containing phospholipids in the plasma may be representative of a quicker aging process [47].

On the other hand, PL therapeutic effects were demonstrated in several diseases. It was shown in that certain PL moieties have therapeutic activity in ulcerative colitis [48,49]; some clinical trials exposed that the adding phosphatidylcholine (PC) to the colonic mucosa alleviates inflammatory activity [50]. One study, phase IIA, double blind, randomized, placebo controlled study including 60 patients with chronic active, non-steroid dependent, ulcerative colitis, showed that 6 g of retarded release phosphatidylcholine rich PL during three months period alleviates inflammatory activity that is caused by ulcerative colitis [50]. Another study showed that polyunsaturated PC is beneficial supplementary treatment for patient management in HBsAg negative chronic active hepatitis, where the condition is inefficiently controlled with standard doses of immunosuppressive therapy [51]. Dietary PC was also shown to alleviate the orotic acid-induced fatty liver in rats OA-, mostly through the reduction of TG synthesis in the liver and improvement of FA β-oxidation [52]. Dietary lecithin (mixture of phosphatidylcholine, phosphatidylethanolamine, phosphatidylinositol, phosphatidylserine, and phosphatidic acid) was shown to be protective in cholestatic liver disease in cholic acid-Fed Abcb4-Deficient Mice, through the drastic mitigation of the hepatic damage [53].

Oxidized PL are present in the inflamed tissues and they are taught to have an important role in the immune response modulation [54]. In most studies, oxidized PL have proinflammatory properties, however it was shown that particular PL oxidation products can display anti-inflammatory features. The derivatives of oxidized PL may, in fact, be a new treatment option for immune diseases (i.e., atherosclerosis, psoriasis, multiple sclerosis, and rheumatoid arthritis) [55].

Since PL have the innate ability to reduce inflammatory activity in certain disease, they make an interesting carrier for lipidic prodrug design. Our group investigated the PLA_2-mediated activation of the PL-drug conjugate, and its potential use in inflammatory diseases, such as inflammatory bowel disease (IBD); in IBD intestinal tissues, the levels of PLA_2 expression are elevated [56,57]. The aim of our work is to target the inflamed tissues with elevated PLA_2 by oral PL-based prodrugs. Linking the drug directly to the sn-2 PL position showed the absence of PLA_2-mediated activation [58]; however, once the linker was introduced the activation of the prodrug was possible [59–61]. Novel computational analysis was used to optimize the prodrug design (linker length) [62–64]. For instance, PL-indomethacin prodrug was orally administered to rats and the prodrug with 5-carbon linker (DP-155, Table 1) showed a 20-fold increase in free drug vs. the PL-indomethacin prodrug with the 2-carbon linker. Free drug was liberated in the intestinal lumen via PLA_2-mediated hydrolysis [59]. Linker design is an essential parameter in the prodrug activation through PLA_2. This was also demonstrated on an antiangiogenesis agent, fumagillin, by adding a 7-carbon acyl linker in the sn-2 position of the PL, the resulting PL-fumagilin prodrug was activated by local PLA_2, and free fumagillin, free drug was released, demonstrating reduced angiogenesis in-vivo [65]. Through the smart design of PL-drug conjugates, we can enable the exploitation of endogenous PL processing pathway, through activation with the enzyme PLA_2 [66].

3.4. Steroids

Cholesterol is the main sterol synthesized in humans. It can originate from the diet or can be synthesized *de novo* in the body [67]. As mentioned earlier, it has a variety of roles, from being a structural component of the cell membrane, part of different classes of LP, to being a precursor of steroid hormones, bile acids, and vitamin D [68,69]. Besides these physiological roles, cholesterol is found to play an essential role in the pathogenesis of some cancer types [70]; low cholesterol serum levels were found to be in connection with lung, cervix, breast, colon cancer, and leukemia, whereas high levels of cholesterol were associated with brain tumors [71]. For instance, prostate cancer is recognized as a lipid-rich tumor [72], with androgen receptor playing a main role in the development of this type of cancer. Genes for lipogenic enzymes can be regulated through androgen: the FA synthesis pathway and cholesterol pathway are highly influenced in this way, resulting in elevated synthesis of FA and cholesterol. Increased lipogenesis, as a result leads to elevated production of PL, cholesterol, and other cell membrane components, all characteristics of cancer cells. Additionally, bioactive lipids are also involved in prostate cancer progression [73]. Cholesterol is a precursor of estrogen, and elevated estrogen levels are linked with a greater risk of breast cancer [74]. In the growing cancerous cell, cholesterol is needed to make up cell membranes, which are synthesized faster in these tissues; the cholesterol is obtained from *de novo* sources or through LDL particles via high affinity receptor-mediated uptake. In many cases, cancer cells show high affinity to LDL particles vs. normal cells [74,75]. This increased demand for LDL by malignant cells and the overexpression of LDL receptors can be used for developing a novel targeted drug delivery system, among which are cholesterol-based prodrugs, as described in Section 4.

Cholesterol can be enzymatically transformed or oxygenated by free radicals to form oxysterols, 27-carbon derivatives of cholesterol [76]. They are connected to several diseases, e.g., atherosclerosis, Alzheimer's disease, Parkinson's disease, as well as cancer progression. Oxysterols also impact vascular ageing, by gathering in the of ageing blood vessel walls, they can stimulate monocytes and endothelial cells, and enable smooth muscle cells of blood vessels to proliferate, migrate, and act as fibroblast-like cells [77].

Atherosclerosis is a widely spread human pathology, associated with cholesterol and lipid metabolism [19]. The main risk factor for developing atherosclerosis is high cholesterol levels in the body, leading to deposits of fat in the arterial walls. Atherosclerosis therapy is dedicated to the reduction of cholesterol ester buildup. Prospective treatment to decrease cholesterol esters is targeted cholesterol ester hydrolase (CEH) delivery to hepatocytes using galactose-functionalized polyamidoamine dendrimer generation 5. The upward regulation of CEH prompts an increase in the cholesterol ester hydrolysis towards free cholesterol, which is consequently secreted as bile acids. This approach has been utilized for easing the accumulation of cholesterol esters in patients suffering from atherosclerosis [78].

Differences in the cholesterol uptake between normal and cancerous tissues could be used for developing cholesterol-based conjugates as a drug delivery system for targeting malignant diseases [79]. Cancer cells demonstrate an increase in LDL receptor activity, due to the high demand of cancerous tissue for cholesterol. This is due to fast growth of these cells or due to cell transformation mechanism. Hence, LDL is exploited as a carrier of anticancer agents, as a novel drug delivery approach for cancer therapy [79]. This method is based on the fact that, LDL is the endogenous transporter of cholesterol. Cholesterol in the body is mainly obtained through the LDL receptor-mediated endocytosis prevalently as cholesterol ester. Phosphotyrosine-cholesterol prodrug demonstrated efficacy against platinum-resistant ovarian cancer cells [80]. Ursodeoxycholic acid (UDCA) and zidovudine conjugation yielded a prodrug (UDCA−AZT), which was able to enter central nervous system, in comparison to free zidovudine (Table 1). Other advantages of this prodrug included enhanced antiviral activity, since it decreased zidovudine hydrolysis in the plasma, and improved the multi-drug resistance of zidovudine [81].

Table 1. Molecular structures of lipid-based prodrugs.

Drug	Lipid-Drug Conjugate
Testosterone [82]	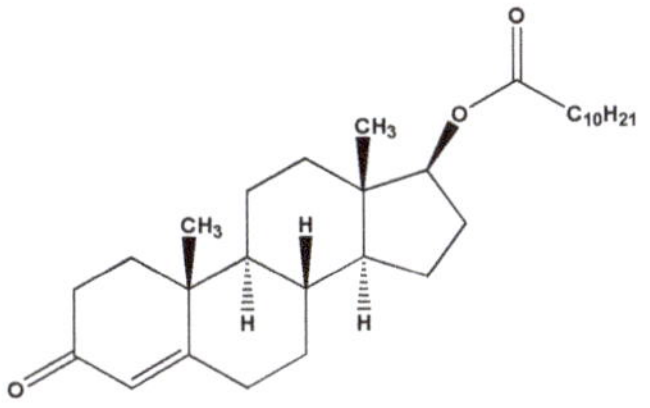 Testosterone undecanoat
Mycophenolic acid (MPA) [45]	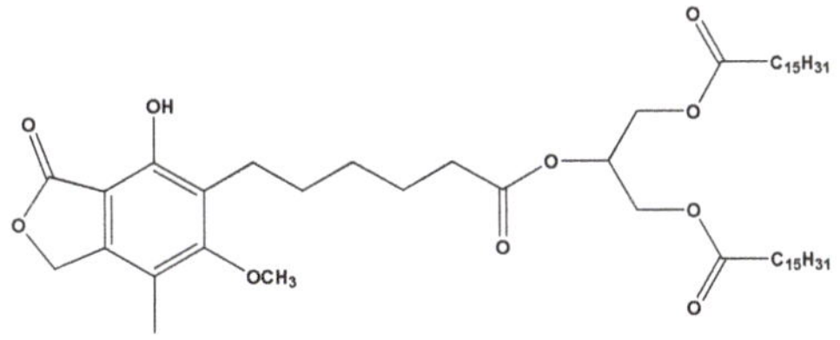 2-MPA-TG
Indomethacin [59]	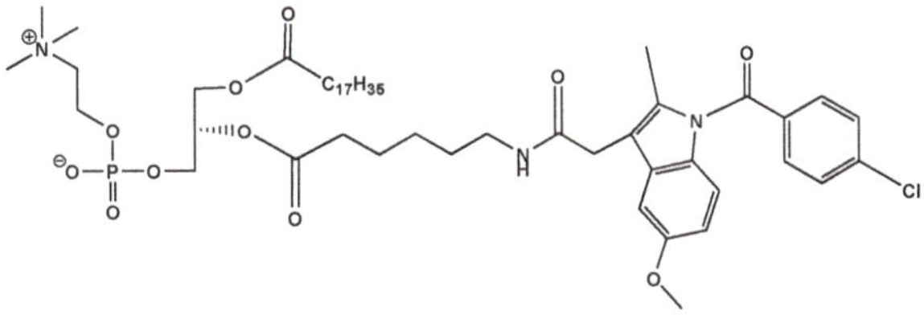 DP-155
Zidovudine [81]	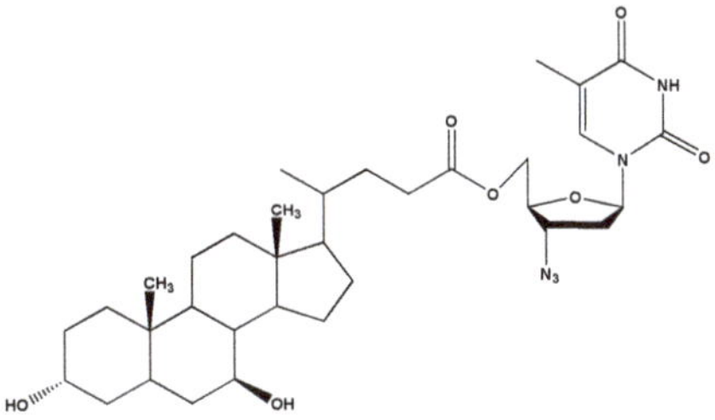 UDCA−AZT

4. Discussion

Lipids are not only storage of body energy; they are fundamentally involved in the preservation and regulation of the cell function. Modifications in the lipid metabolism are linked with human diseases, such as cancer, diabetes, atherosclerosis, dyslipidemia, and neurodegenerative and infectious diseases. A number of drugs were developed to target specific lipid metabolic and signaling pathways, such as cyclooxygenase (COX) inhibitors or statins for the lowering of cholesterol. Researchers are pursuing particular regulators of lipid targets, such as nuclear hormone receptors (peroxisome proliferator-activated receptors, liver X receptor), phosphatidylinositol 3-kinases, ceramide kinases, and sphingosine [2]. An immunosuppressant drug targeting sphingosine-1-phosphate was approved for the treatment of multiple sclerosis, due to its targeting ability towards sphingosine-1-phosphate receptors, but not towards serine palmitoyl transferase, in contrast to its parent compound myriocin [83]. However, aside from glycolipids, very few antibodies are developed to recognize specific lipids; this is an important field that should be further investigated.

Lipids have vast potential in disease therapy, from being disease markers that enable drug targeted therapy, to being natively used as a supplement, as well as being part of the lipid drug formulation and used as a carrier for a particular drug. Some lipidic prodrugs (mainly FA-drug conjugates) were shown to be very successful in preclinical studies and, consequently, underwent clinical investigations. It was mentioned earlier that some UFA derivatives with chemotherapeutic agents entered clinical trials, most importantly gemcitabine-eladic acid [84], cytarabine-elaidic [85], paclitaxel-docosahexaenoic acid [86], and lipophilic docetaxel prodrug (MNK-010) [87]. In addition, cholesterol-siRNA prodrug (ARC-520) entered clinical trial for the treatment of chronic hepatitis B infection [88]. Commercially available lipidic prodrug product is TU for oral or intramuscular application in androgen deficiency [82]. A recent lipidic prodrug that reached the market is aripiprazole lauroxil and paliperidone palmitate an extended-release, long-acting intramuscular injection for the treatment of adults with schizophrenia [89,90]. Hence, lipidic prodrugs are a promising strategy for developing novel commercial products in the future [91]. Prospective direction for development of novel lipidic prodrugs containing glyceride/PL carriers could include even greater improved prodrug design, with linkers/spacers that would allow for controlled release of the parent drug from the prodrug complex [59–61,92].

Formulation lipids, such as LCT, MCT, as well as DG, MG, FA, and PL, are used to improve drug absorption. Their success in doing so is determined by several parameters, such as the amount of digestion products, digestion rate, and extent, as well as level of distribution of such products. The importance of combining lipidic prodrugs and lipid formulation design was recently demonstrated with lipidic prodrug, methotrexate-DG ester that was incorporated in the lipid bilayer of liposomes composed of egg (PC)/yeast phosphatidylinositol and demonstrated lower toxicity and slower lymphoma growth in mice when compared with methotrexate alone [93]. Lipids, both as carriers and formulation components (permeability enhancers), have a promising role as particularly diverse materials, with a vast space for discovery, development, and optimization in drug delivery.

5. Conclusions

Lipids may serve as biomarkers of diseases, as targets for drug molecules and as lipid carriers in the prodrug approach. Some lipids can be used as therapeutic supplementation to the drug therapy due to their innate pharmacological effect. All of this makes them valuable and diverse materials, with an encouraging role in future discovery, development, and optimization in drug delivery and therapeutics.

Author Contributions: M.M., S.B.-S., A.A., E.M.Z. and A.D. worked on conceptualization, analyzed the data, and outlined the manuscript. M.M. and A.D. wrote the skeleton of the paper, and S.B.-S., A.A. and E.M.Z. contributed to the writing-review and editing of the full version. All authors have read and agreed to the published version of the manuscript.

Funding: This work was funded through the United States-Israel Binational Science Foundation (BSF), grant number 2015365.

Acknowledgments: This work is a part of Milica Markovic Ph.D. dissertation.

Conflicts of Interest: The authors declare no conflict of interest.

References

1. Fahy, E.; Cotter, D.; Sud, M.; Subramaniam, S. Lipid classification, structures and tools. *Biochim. Biophys. Acta* **2011**, *1811*, 637–647. [CrossRef] [PubMed]

2. Wenk, M.R. The emerging field of lipidomics. *Nat. Rev. Drug Discov.* **2005**, *4*, 594–610. [CrossRef] [PubMed]

3. Lipinski, C.A.; Lombardo, F.; Dominy, B.W.; Feeney, P.J. Experimental and computational approaches to estimate solubility and permeability in drug discovery and development settings. *Adv. Drug Deliv. Rev.* **2001**, *46*, 3–26. [CrossRef]

4. van De Waterbeemd, H.; Smith, D.A.; Beaumont, K.; Walker, D.K. Property-based design: Optimization of drug absorption and pharmacokinetics. *J. Med. Chem.* **2001**, *44*, 1313–1333. [CrossRef]

5. Kalantzi, L.; Persson, E.; Polentarutti, B.; Abrahamsson, B.; Goumas, K.; Dressman, J.B.; Reppas, C. Canine intestinal contents vs. simulated media for the assessment of solubility of two weak bases in the human small intestinal contents. *Pharm. Res.* **2006**, *23*, 1373–1381. [CrossRef]

6. Mudie, D.M.; Samiei, N.; Marshall, D.J.; Amidon, G.E.; Bergstrom, C.A.S. Selection of In Vivo Predictive Dissolution Media Using Drug Substance and Physiological Properties. *Aaps. J.* **2020**, *22*, 34. [CrossRef]

7. Cornaire, G.; Woodley, J.; Hermann, P.; Cloarec, A.; Arellano, C.; Houin, G. Impact of excipients on the absorption of P-glycoprotein substrates in vitro and in vivo. *Int. J. Pharm.* **2004**, *278*, 119–131. [CrossRef]

8. Kliewer, S.A.; Xu, H.E.; Lambert, M.H.; Willson, T.M. Peroxisome proliferator-activated receptors: From genes to physiology. *Recent Prog. Horm. Res.* **2001**, *56*, 239–263. [CrossRef]

9. Poirier, H.; Niot, I.; Monnot, M.C.; Braissant, O.; Meunier-Durmort, C.; Costet, P.; Pineau, T.; Wahli, W.; Willson, T.M.; Besnard, P. Differential involvement of peroxisome-proliferator-activated receptors alpha and delta in fibrate and fatty-acid-mediated inductions of the gene encoding liver fatty-acid-binding protein in the liver and the small intestine. *Biochem. J.* **2001**, *355*, 481–488. [CrossRef]

10. Trevaskis, N.L.; Nguyen, G.; Scanlon, M.J.; Porter, C.J.H. Fatty acid binding proteins: Potential chaperones of cytosolic drug transport in the enterocyte? *Pharm. Res.* **2011**, *28*, 2176–2190. [CrossRef]

11. Porter, C.J.; Trevaskis, N.L.; Charman, W.N. Lipids and lipid-based formulations: Optimizing the oral delivery of lipophilic drugs. *Nat. Rev. Drug Discov.* **2007**, *6*, 231–248. [CrossRef] [PubMed]

12. Carriere, F.; Barrowman, J.A.; Verger, R.; Laugier, R. Secretion and contribution to lipolysis of gastric and pancreatic lipases during a test meal in humans. *Gastroenterology* **1993**, *105*, 876–888. [CrossRef]

13. Mead, J.F. Lipid Metabolism. *Annu. Rev. Biochem.* **1963**, *32*, 241–268. [CrossRef] [PubMed]

14. Lowe, M.E. Structure and function of pancreatic lipase and colipase. *Annu. Rev. Nutr.* **1997**, *17*, 141–158. [CrossRef] [PubMed]

15. Borgstrom, B. On the mechanism of pancreatic lipolysis of glycerides. *Biochim. Biophys. Acta* **1954**, *13*, 491–504. [CrossRef]

16. van den Bosch, H.; Postema, N.M.; de Haas, G.H.; van Deenen, L.L. On the positional specificity of phospholipase A from pancreas. *Biochim. Biophys. Acta* **1965**, *98*, 657–659. [CrossRef]

17. Porter, C.J.; Charman, W.N. Intestinal lymphatic drug transport: An update. *Adv. Drug Deliv. Rev.* **2001**, *50*, 61–80. [CrossRef]

18. Vance, D.E.; Van den Bosch, H. Cholesterol in the year 2000. *Biochim. Biophys. Acta* **2000**, *1529*, 1–8. [CrossRef]

19. Maxfield, F.R.; Tabas, I. Role of cholesterol and lipid organization in disease. *Nature* **2005**, *438*, 612–621. [CrossRef]

20. Schumaker, V.N.; Adams, G.H. Circulating lipoproteins. *Annu Rev Biochem* **1969**, *38*, 113–136. [CrossRef]

21. Stephenson, D.J.; Hoeferlin, L.A.; Chalfant, C.E. Lipidomics in translational research and the clinical significance of lipid-based biomarkers. *Transl. Res.* **2017**, *189*, 13–29. [CrossRef] [PubMed]

22. Markovic, M.; Ben-Shabat, S.; Keinan, S.; Aponick, A.; Zimmermann, E.M.; Dahan, A. Lipidic prodrug approach for improved oral drug delivery and therapy. *Med. Res. Rev.* **2019**, *39*, 579–607. [CrossRef] [PubMed]

23. Röhrig, F.; Schulze, A. The multifaceted roles of fatty acid synthesis in cancer. *Nat. Rev. Cancer* **2016**, *16*, 732–749. [CrossRef] [PubMed]

24. Kusakabe, T.; Maeda, M.; Hoshi, N.; Sugino, T.; Watanabe, K.; Fukuda, T.; Suzuki, T. Fatty acid synthase is expressed mainly in adult hormone-sensitive cells or cells with high lipid metabolism and in proliferating fetal cells. *J. Histochem. Cytochem. Off. J. Histochem. Soc.* **2000**, *48*, 613–622. [CrossRef]

25. Currie, E.; Schulze, A.; Zechner, R.; Walther, T.C.; Farese, R.V., Jr. Cellular fatty acid metabolism and cancer. *Cell Metab.* **2013**, *18*, 153–161. [CrossRef]

26. Santos, C.R.; Schulze, A. Lipid metabolism in cancer. *FEBS J.* **2012**, *279*, 2610–2623. [CrossRef]

27. Camassei, F.D.; Cozza, R.; Acquaviva, A.; Jenkner, A.; Ravà, L.; Gareri, R.; Donfrancesco, A.; Bosman, C.; Vadalà, P.; Hadjistilianou, T.; et al. Expression of the Lipogenic Enzyme Fatty Acid Synthase (FAS) in Retinoblastoma and Its Correlation with Tumor Aggressiveness. *Investig. Ophthalmol. Vis. Sci.* **2003**, *44*, 2399–2403. [CrossRef]

28. Hafidi, M.E.; Buelna-Chontal, M.; Sanchez-Munoz, F.; Carbo, R. Adipogenesis: A Necessary but Harmful Strategy. *Int. J. Mol. Sci.* **2019**, *20*, 3657. [CrossRef]

29. Yang, W.-S.; Chen, Y.-Y.; Chen, P.-C.; Hsu, H.-C.; Su, T.-C.; Lin, H.-J.; Chen, M.-F.; Lee, Y.-T.; Chien, K.-L. Association between Plasma N-6 Polyunsaturated Fatty Acids Levels and the Risk of Cardiovascular Disease in a Community-based Cohort Study. *Sci. Rep.* **2019**, *9*, 19298. [CrossRef]

30. Bazinet, R.P.; Laye, S. Polyunsaturated fatty acids and their metabolites in brain function and disease. *Nat. Rev. Neurosci.* **2014**, *15*, 771–785. [CrossRef]

31. Yuan, G.; Al-Shali, K.Z.; Hegele, R.A. Hypertriglyceridemia: Its etiology, effects and treatment. *CMAJ* **2007**, *176*, 1113–1120. [CrossRef] [PubMed]

32. Rizos, E.C.; Ntzani, E.E.; Bika, E.; Kostapanos, M.S.; Elisaf, M.S. Association between omega-3 fatty acid supplementation and risk of major cardiovascular disease events: A systematic review and meta-analysis. *JAMA* **2012**, *308*, 1024–1033. [CrossRef] [PubMed]

33. Sun, B.; Luo, C.; Cui, W.; Sun, J.; He, Z. Chemotherapy agent-unsaturated fatty acid prodrugs and prodrug-nanoplatforms for cancer chemotherapy. *J. Control. Release: Off. J. Control. Release Soc.* **2017**, *264*, 145–159. [CrossRef] [PubMed]

34. Shackleford, D.M.; Faassen, W.A.; Houwing, N.; Lass, H.; Edwards, G.A.; Porter, C.J.; Charman, W.N. Contribution of lymphatically transported testosterone undecanoate to the systemic exposure of testosterone after oral administration of two andriol formulations in conscious lymph duct-cannulated dogs. *J. Pharmacol. Exp. Ther.* **2003**, *306*, 925–933. [CrossRef] [PubMed]

35. Tauber, U.; Schroder, K.; Dusterberg, B.; Matthes, H. Absolute bioavailability of testosterone after oral administration of testosterone-undecanoate and testosterone. *Eur. J. Drug Metab. Pharmacokinet.* **1986**, *11*, 145–149. [CrossRef] [PubMed]

36. Nieschlag, E.; Mauss, J.; Coert, A.; Kicovic, P. Plasma androgen levels in men after oral administration of testosterone or testosterone undecanoate. *Acta Endocrinol.* **1975**, *79*, 366–374. [CrossRef] [PubMed]

37. Bansal, S.; Buring, J.E.; Rifai, N.; Mora, S.; Sacks, F.M.; Ridker, P.M. Fasting compared with nonfasting triglycerides and risk of cardiovascular events in women. *JAMA* **2007**, *298*, 309–316. [CrossRef]

38. Nordestgaard, B.G.; Benn, M.; Schnohr, P.; Tybjaerg-Hansen, A. Nonfasting triglycerides and risk of myocardial infarction, ischemic heart disease, and death in men and women. *JAMA* **2007**, *298*, 299–308. [CrossRef]

39. Zilversmit, D.B. Atherogenesis: A postprandial phenomenon. *Circulation* **1979**, *60*, 473–485. [CrossRef]

40. Mamo, J.C.; Proctor, S.D.; Smith, D. Retention of chylomicron remnants by arterial tissue; importance of an efficient clearance mechanism from plasma. *Atherosclerosis* **1998**. [CrossRef]

41. Hegele, R.A.; Ginsberg, H.N.; Chapman, M.J.; Nordestgaard, B.G.; Kuivenhoven, J.A.; Averna, M.; Borén, J.; Bruckert, E.; Catapano, A.L.; Descamps, O.S.; et al. The polygenic nature of hypertriglyceridaemia: Implications for definition, diagnosis, and management. *Lancet Diabetes Endocrinol.* **2014**, *2*, 655–666. [CrossRef]

42. Dahan, A.; Hoffman, A. The effect of different lipid based formulations on the oral absorption of lipophilic drugs: The ability of in vitro lipolysis and consecutive ex vivo intestinal permeability data to predict in vivo bioavailability in rats. *Eur. J. Pharm. Biopharm. Off. J. Arb. Fur Pharm. Verfahr. E.V.* **2007**, *67*, 96–105. [CrossRef] [PubMed]

43. Palin, K.J.; Wilson, C.G. The effect of different oils on the absorption of probucol in the rat. *J. Pharm. Pharmacol.* **1984**, *36*, 641–643. [CrossRef] [PubMed]

44. Han, S.; Hu, L.; Quach, T.; Simpson, J.S.; Trevaskis, N.L.; Porter, C.J.H. Constitutive Triglyceride Turnover into the Mesenteric Lymph Is Unable to Support Efficient Lymphatic Transport of a Biomimetic Triglyceride Prodrug. *J. Pharm. Sci.* **2015**. [CrossRef] [PubMed]

45. Han, S.; Hu, L.; Gracia; Quach, T.; Simpson, J.S.; Edwards, G.A.; Trevaskis, N.L.; Porter, C.J. Lymphatic Transport and Lymphocyte Targeting of a Triglyceride Mimetic Prodrug Is Enhanced in a Large Animal Model: Studies in Greyhound Dogs. *Mol. Pharm.* **2016**, *13*, 3351–3361. [CrossRef] [PubMed]

46. Kosicek, M.; Hecimovic, S. Phospholipids and Alzheimer's disease: Alterations, mechanisms and potential biomarkers. *Int. J. Mol. Sci.* **2013**, *14*, 1310–1322. [CrossRef]

47. Dorninger, F.; Moser, A.B.; Kou, J.; Wiesinger, C.; Forss-Petter, S.; Gleiss, A.; Hinterberger, M.; Jungwirth, S.; Fischer, P.; Berger, J. Alterations in the Plasma Levels of Specific Choline Phospholipids in Alzheimer's Disease Mimic Accelerated Aging. *J. Alzheimers Dis.* **2018**, *62*, 841–854. [CrossRef]

48. Schneider, H.; Braun, A.; Füllekrug, J.; Stremmel, W.; Ehehalt, R. Lipid based therapy for ulcerative colitis-modulation of intestinal mucus membrane phospholipids as a tool to influence inflammation. *Int. J. Mol. Sci.* **2010**, *11*, 4149–4164. [CrossRef]

49. Stremmel, W.; Ehehalt, R.; Staffer, S.; Stoffels, S.; Mohr, A.; Karner, M.; Braun, A. Mucosal protection by phosphatidylcholine. *Dig. Dis. (Basel Switz.)* **2012**, *30*, 85–91. [CrossRef]

50. Stremmel, W.; Merle, U.; Zahn, A.; Autschbach, F.; Hinz, U.; Ehehalt, R. Retarded release phosphatidylcholine benefits patients with chronic active ulcerative colitis. *Gut* **2005**, *54*, 966–971. [CrossRef]

51. Jenkins, P.J.; Portmann, B.P.; Eddleston, A.L.W.F.; Williams, R. Use of polyunsaturated phosphatidyl choline in HBsAg negative chronic active hepatitis: Results of prospective double-blind controlled trial. *Liver* **1982**, *2*, 77–81. [CrossRef] [PubMed]

52. Buang, Y.; Wang, Y.M.; Cha, J.Y.; Nagao, K.; Yanagita, T. Dietary phosphatidylcholine alleviates fatty liver induced by orotic acid. *Nutr. (BurbankLos Angeles Cty. Calif.)* **2005**, *21*, 867–873. [CrossRef] [PubMed]

53. Lamireau, T.; Bouchard, G.; Yousef, I.M.; Clouzeau-Girard, H.; Rosenbaum, J.; Desmoulière, A.; Tuchweber, B. Dietary Lecithin Protects Against Cholestatic Liver Disease in Cholic Acid–Fed Abcb4– Deficient Mice. *Pediatric Res.* **2007**, *61*, 185–190. [CrossRef] [PubMed]

54. Deigner, H.P.; Hermetter, A. Oxidized phospholipids: Emerging lipid mediators in pathophysiology. *Curr. Opin. Lipidol.* **2008**, *19*, 289–294. [CrossRef]

55. Feige, E.; Mendel, I.; George, J.; Yacov, N.; Harats, D. Modified phospholipids as anti-inflammatory compounds. *Curr. Opin. Lipidol.* **2010**, *21*, 525–529. [CrossRef]

56. Minami, T.; Shinomura, Y.; Miyagawa, J.; Tojo, H.; Okamoto, M.; Matsuzawa, Y. Immunohistochemical localization of group II phospholipase A2 in colonic mucosa of patients with inflammatory bowel disease. *Am. J. Gastroenterol.* **1997**, *92*, 289–292.

57. Minami, T.; Tojo, H.; Shinomura, Y.; Matsuzawa, Y.; Okamoto, M. Increased group II phospholipase A2 in colonic mucosa of patients with Crohn's disease and ulcerative colitis. *Gut* **1994**, *35*, 1593–1598. [CrossRef]

58. Dahan, A.; Duvdevani, R.; Shapiro, I.; Elmann, A.; Finkelstein, E.; Hoffman, A. The oral absorption of phospholipid prodrugs: In vivo and in vitro mechanistic investigation of trafficking of a lecithin-valproic acid conjugate following oral administration. *J. Control. Release Off. J. Control. Release Soc.* **2008**, *126*, 1–9. [CrossRef]

59. Dahan, A.; Duvdevani, R.; Dvir, E.; Elmann, A.; Hoffman, A. A novel mechanism for oral controlled release of drugs by continuous degradation of a phospholipid prodrug along the intestine: In-vivo and in-vitro evaluation of an indomethacin-lecithin conjugate. *J. Control. Release Off. J. Control. Release Soc.* **2007**, *119*, 86–93. [CrossRef]

60. Dahan, A.; Markovic, M.; Epstein, S.; Cohen, N.; Zimmermann, E.M.; Aponick, A.; Ben-Shabat, S. Phospholipid-drug conjugates as a novel oral drug targeting approach for the treatment of inflammatory bowel disease. *Eur. J. Pharm. Sci. Off. J. Eur. Fed. Pharm. Sci.* **2017**, *108*, 78–85. [CrossRef]

61. Markovic, M.; Dahan, A.; Keinan, S.; Kurnikov, I.; Aponick, A.; Zimmermann, E.M.; Ben-Shabat, S. Phospholipid-Based Prodrugs for Colon-Targeted Drug Delivery: Experimental Study and In-Silico Simulations. *Pharmaceutics* **2019**, *11*, 186. [CrossRef] [PubMed]

62. Dahan, A.; Ben-Shabat, S.; Cohen, N.; Keinan, S.; Kurnikov, I.; Aponick, A.; Zimmermann, E.M. Phospholipid-Based Prodrugs for Drug Targeting in Inflammatory Bowel Disease: Computational Optimization and In-Vitro Correlation. *Curr. Top. Med. Chem.* **2016**, *16*, 2543–2548. [CrossRef] [PubMed]

63. Dahan, A.; Markovic, M.; Keinan, S.; Kurnikov, I.; Aponick, A.; Zimmermann, E.M.; Ben-Shabat, S. Computational modeling and in-vitro/in-silico correlation of phospholipid-based prodrugs for targeted drug delivery in inflammatory bowel disease. *J. Comput.-Aided Mol. Des.* **2017**, *31*, 1021–1028. [CrossRef] [PubMed]

64. Markovic, M.; Ben-Shabat, S.; Keinan, S.; Aponick, A.; Zimmermann, E.M.; Dahan, A. Molecular Modeling-Guided Design of Phospholipid-Based Prodrugs. *Int. J. Mol. Sci.* **2019**, *20*, 2210. [CrossRef]

65. Pan, D.; Pham, C.T.N.; Weilbaecher, K.N.; Tomasson, M.H.; Wickline, S.A.; Lanza, G.M. Contact-facilitated drug delivery with Sn2 lipase labile prodrugs optimize targeted lipid nanoparticle drug delivery. *Wiley Interdiscip. Rev. Nanomed. Nanobiotechnol.* **2016**, *8*, 85–106. [CrossRef]

66. Markovic, M.; Ben-Shabat, S.; Keinan, S.; Aponick, A.; Zimmermann, E.M.; Dahan, A. Prospects and Challenges of Phospholipid-Based Prodrugs. *Pharmaceutics* **2018**, *10*, 210. [CrossRef]

67. Hume, R.; Boyd, G.S. Cholesterol metabolism and steroid-hormone production. *Biochem. Soc. Trans.* **1978**, *6*, 893–898. [CrossRef]

68. Janowski, B.A.; Willy, P.J.; Devi, T.R.; Falck, J.R.; Mangelsdorf, D.J. An oxysterol signalling pathway mediated by the nuclear receptor LXR alpha. *Nature* **1996**, *383*, 728–731. [CrossRef]

69. Umetani, M.; Domoto, H.; Gormley, A.K.; Yuhanna, I.S.; Cummins, C.L.; Javitt, N.B.; Korach, K.S.; Shaul, P.W.; Mangelsdorf, D.J. 27-Hydroxycholesterol is an endogenous SERM that inhibits the cardiovascular effects of estrogen. *Nat. Med.* **2007**, *13*, 1185–1192. [CrossRef]

70. Hu, J.; La Vecchia, C.; de Groh, M.; Negri, E.; Morrison, H.; Mery, L. Dietary cholesterol intake and cancer. *Ann. Oncol. Off. J. Eur. Soc. Med Oncol.* **2012**, *23*, 491–500. [CrossRef]

71. Kambach, D.M.; Halim, A.S.; Cauer, A.G.; Sun, Q.; Tristan, C.A.; Celiku, O.; Kesarwala, A.H.; Shankavaram, U.; Batchelor, E.; Stommel, J.M. Disabled cell density sensing leads to dysregulated cholesterol synthesis in glioblastoma. *Oncotarget* **2017**, *8*, 14860–14875. [CrossRef] [PubMed]

72. Wu, X.; Daniels, G.; Lee, P.; Monaco, M.E. Lipid metabolism in prostate cancer. *Am. J. Clin. Exp. Urol.* **2014**, *2*, 111–120. [PubMed]

73. Dang, Q.; Chen, Y.-A.; Hsieh, J.-T. The dysfunctional lipids in prostate cancer. *Am. J. Clin. Exp. Urol.* **2019**, *7*, 273–280. [PubMed]

74. Murai, T. Cholesterol lowering: Role in cancer prevention and treatment. *Biol. Chem.* **2015**, *396*, 1–11. [CrossRef] [PubMed]

75. Yang, J.; Wang, L.; Jia, R. Role of de novo cholesterol synthesis enzymes in cancer. *J. Cancer* **2020**, *11*, 1761–1767. [CrossRef]

76. Kloudova, A.; Guengerich, F.P.; Soucek, P. The Role of Oxysterols in Human Cancer. *Trends Endocrinol. Metab. Tem* **2017**, *28*, 485–496. [CrossRef]

77. Gargiulo, S.; Gamba, P.; Testa, G.; Leonarduzzi, G.; Poli, G. The role of oxysterols in vascular ageing. *J. Physiol.* **2016**, *594*, 2095–2113. [CrossRef]

78. He, H.; Lancina, M.G.; Wang, J.; Korzun, W.J.; Yang, H.; Ghosh, S. Bolstering cholesteryl ester hydrolysis in liver: A hepatocyte-targeting gene delivery strategy for potential alleviation of atherosclerosis. *Biomaterials* **2017**, *130*, 1–13. [CrossRef]

79. Radwan, A.A.; Alanazi, F.K. Targeting cancer using cholesterol conjugates. *Saudi. Pharm. J.* **2014**, *22*, 3–16. [CrossRef]

80. Wang, H.; Feng, Z.; Wu, D.; Fritzsching, K.J.; Rigney, M.; Zhou, J.; Jiang, Y.; Schmidt-Rohr, K.; Xu, B. Enzyme-Regulated Supramolecular Assemblies of Cholesterol Conjugates against Drug-Resistant Ovarian Cancer Cells. *J. Am. Chem. Soc.* **2016**, *138*, 10758–10761. [CrossRef]

81. Dalpiaz, A.; Paganetto, G.; Pavan, B.; Fogagnolo, M.; Medici, A.; Beggiato, S.; Perrone, D. Zidovudine and Ursodeoxycholic Acid Conjugation: Design of a New Prodrug Potentially Able To Bypass the Active Efflux Transport Systems of the Central Nervous System. *Mol. Pharm.* **2012**, *9*, 957–968. [CrossRef] [PubMed]

82. Nieschlag, E. Current topics in testosterone replacement of hypogonadal men. *Best Pract. Res. Clin. Endocrinol. Metab.* **2015**, *29*, 77–90. [CrossRef] [PubMed]

83. Baer, A.; Colon-Moran, W.; Bhattarai, N. Characterization of the effects of immunomodulatory drug fingolimod (FTY720) on human T cell receptor signaling pathways. *Sci. Rep.* **2018**, *8*, 10910. [CrossRef] [PubMed]

84. Stuurman, F.E.; Voest, E.E.; Awada, A.; Witteveen, P.O.; Bergeland, T.; Hals, P.A.; Rasch, W.; Schellens, J.H.; Hendlisz, A. Phase I study of oral CP-4126, a gemcitabine derivative, in patients with advanced solid tumors. *Investig. New Drugs* **2013**, *31*, 959–966. [CrossRef]

85. Rizzieri, D.; Vey, N.; Thomas, X.; Huguet-Rigal, F.; Schlenk, R.F.; Krauter, J.; Kindler, T.; Gjertsen, B.T.; Blau, I.W.; Jacobsen, T.F.; et al. A phase II study of elacytarabine in combination with idarubicin and of human equilibrative nucleoside transporter 1 expression in patients with acute myeloid leukemia and persistent blasts after the first induction course. *Leuk. Lymphoma* **2014**, *55*, 2114–2119. [CrossRef] [PubMed]

86. Sparreboom, A.; Wolff, A.C.; Verweij, J.; Zabelina, Y.; van Zomeren, D.M.; McIntire, G.L.; Swindell, C.S.; Donehower, R.C.; Baker, S.D. Disposition of docosahexaenoic acid-paclitaxel, a novel taxane, in blood: In vitro and clinical pharmacokinetic studies. *Clin. Cancer Res. Off. J. Am. Assoc. Cancer Res.* **2003**, *9*, 151–159.

87. Rochon, L.S.; Devarakonda, K.; Martinez, J.; Williams, K.; Hamilton, E.P.; Shapiro, G.; Sukari, A. Abstract CT230: A phase I first in human dose escalation trial of MNK-010 in subjects with advanced solid tumors. *Cancer Res.* **2015**, *75*, CT230. [CrossRef]

88. Schluep, T.; Lickliter, J.; Hamilton, J.; Lewis, D.L.; Lai, C.L.; Lau, J.Y.; Locarnini, S.A.; Gish, R.G.; Given, B.D. Safety, Tolerability, and Pharmacokinetics of ARC-520 Injection, an RNA Interference-Based Therapeutic for the Treatment of Chronic Hepatitis B Virus Infection, in Healthy Volunteers. *Clin. Pharmacol. Drug Dev.* **2017**, *6*, 350–362. [CrossRef]

89. Cruz, M.P. Aripiprazole Lauroxil (Aristada): An Extended-Release, Long-Acting Injection For the Treatment of Schizophrenia. *Pharm. Ther.* **2016**, *41*, 556–559.

90. Morris, M.T.; Tarpada, S.P. Long-Acting Injectable Paliperidone Palmitate: A Review of Efficacy and Safety. *Psychopharmacol. Bull.* **2017**, *47*, 42–52.

91. Dahan, A.; Markovic, M.; Aponick, A.; Zimmermann, E.M.; Ben-Shabat, S. The prospects of lipidic prodrugs: An old approach with an emerging future. *Future Med. Chem.* **2019**, *11*, 2563–2571. [CrossRef] [PubMed]

92. Andresen, T.L.; Davidsen, J.; Begtrup, M.; Mouritsen, O.G.; Jørgensen, K. Enzymatic Release of Antitumor Ether Lipids by Specific Phospholipase A2 Activation of Liposome-Forming Prodrugs. *J. Med. Chem.* **2004**, *47*, 1694–1703. [CrossRef] [PubMed]

93. Alekseeva, A.A.; Moiseeva, E.V.; Onishchenko, N.R.; Boldyrev, I.A.; Singin, A.S.; Budko, A.P.; Shprakh, Z.S.; Molotkovsky, J.G.; Vodovozova, E.L. Liposomal formulation of a methotrexate lipophilic prodrug: Assessment in tumor cells and mouse T-cell leukemic lymphoma. *Int. J. Nanomed.* **2017**, *12*, 3735–3749. [CrossRef] [PubMed]

MDPI

St. Alban-Anlage 66

4052 Basel

Switzerland

Tel. +41 61 683 77 34

Fax +41 61 302 89 18

www.mdpi.com

International Journal of Molecular Sciences Editorial Office

E-mail: ijms@mdpi.com

www.mdpi.com/journal/ijms